Lecture Notes in Electrical Engineering

Volume 360

About this Series

"Lecture Notes in Electrical Engineering (LNEE)" is a book series which reports the latest research and developments in Electrical Engineering, namely:

- Communication, Networks, and Information Theory
- Computer Engineering
- Signal, Image, Speech and Information Processing
- Circuits and Systems
- Bioengineering

LNEE publishes authored monographs and contributed volumes which present cutting edge research information as well as new perspectives on classical fields, while maintaining Springer's high standards of academic excellence. Also considered for publication are lecture materials, proceedings, and other related materials of exceptionally high quality and interest. The subject matter should be original and timely, reporting the latest research and developments in all areas of electrical engineering.

The audience for the books in LNEE consists of advanced level students, researchers, and industry professionals working at the forefront of their fields. Much like Springer's other Lecture Notes series, LNEE will be distributed through Springer's print and electronic publishing channels.

More information about this series at http://www.springer.com/series/7818

Yingmin Jia · Junping Du · Hongbo Li
Weicun Zhang
Editors

Proceedings of the 2015 Chinese Intelligent Systems Conference

Volume 2

Editors
Yingmin Jia
Beihang University
Beijing
China

Hongbo Li
Tsinghua University
Beijing
China

Junping Du
Beijing University of Posts and Telecommunications
Beijing
China

Weicun Zhang
University of Science and Technology Beijing
Beijing
China

ISSN 1876-1100 ISSN 1876-1119 (electronic)
Lecture Notes in Electrical Engineering
ISBN 978-3-662-56919-1 ISBN 978-3-662-48365-7 (eBook)
DOI 10.1007/978-3-662-48365-7

Springer Heidelberg New York Dordrecht London

Softcover re-print of the Hardcover 1st edition 2016

Printed on acid-free paper

Springer-Verlag GmbH Berlin Heidelberg is part of Springer Science+Business Media
(www.springer.com)

Contents

Chapter 1
Robust Finite-Time Stabilization of Fractional-Order Extended Nonholonomic Chained Form Systems

Hua Chen, Da Yan, Xi Chen, Yan Lei and Yawei Wang

Abstract We discuss the robust, finite-time stabilization of fractional-order extended nonholonomic chained form systems for the first time in this article. By applying sliding mode variable structure theory and stability theorem of finite-time control, the three-step switching control scheme is proposed to deal with the presence of system uncertainties and external disturbance, so that the closed-loop system is finite time stable at the origin equilibrium point within any given settling time. Finally, simulation results show the effectiveness of the presented controller.

Keywords Nonholonomic control systems · Fractional-order · Extended chained-form systems · Finite-time stabilization

1.1 Introduction

Systems with nonholonomic or nonintegrable constraints exist in the motion of various mechanical bodies extensively, such as wheeled mobile robots [1–6], car-like vehicles [7, 8], under-actuated mechanical systems [9, 10] and so on. However, as

H. Chen (✉)
Mathematics and Physics Department, Hohai University, Changzhou Campus, Changzhou 213022, China
e-mail: chenhua112@163.com

H. Chen
Changzhou Key Laboratory of Special Robot and Intelligent Technology, Changzhou 213022, China

D. Yan
College of Mechanical and Electrical Engineering, Hohai University, Changzhou 213022, China

X. Chen
Mathematics Department, Changzhou Institute of Light Industry Technology, Changzhou 213164, China

Y. Lei · Y. Wang
College of Internet of Things Engineering, Hohai University, Changzhou 213022, China

Y. Jia et al. (eds.), *Proceedings of the 2015 Chinese Intelligent Systems Conference*, Lecture Notes in Electrical Engineering 360,
DOI 10.1007/978-3-662-48365-7_1

Brockett [11] has pointed out that any pure smooth state feedback law cannot be used to stabilize such controllable nonholonomic systems. A number of challenging control approaches have been obtained to conquer the Brockett difficulty, which mainly includes time-varying feedback law [12–14] and discontinuous feedback law [15–18]. These research results mentioned above are all based on integral order system model, in fact, to the best knowledge of the authors, there is no literature investigated the nonintegral order issue for nonholonomic systems before.

In engineering practice, some scholars have identified that many systems in interdisciplinary fields, such as viscoelastic materials [19], micro-electromechanical systems [20], signal processing [21], electromagnetism [22], mechanics [23–26], image processing [27], bioengineering [28], automatic control [29, 30], chaotic systems [31, 32], and robotic [33, 34] can be sophisticatedly described using fractional calculus tools. This is because the fractional model provides an excellent approach to describe the properties of the so-called memory and hereditary of various materials and processes, then the advantages of using the fractional-order system model are that we have more degrees of freedom in the model with a "memory".

Additionally, in order to drive a system to the equilibrium point with a fast convergence rate, finite-time stability theory has become a studying focus recently [35–37]. Compared with the traditional asymptotic convergence, the finite-time stabilization makes the closed-loop systems convergent to the origin in a finite settling time, it has a lot of advantages such as fast response, high tracking precision, and perturbation rejection properties.

Motivated by the discussions above, this paper considers the robust finite-time stabilization problem for a class of fractional-order uncertain extended nonholonomic chained form systems for the first time. Our main highlights can be summarized as follows:

(1) It is the first time to discuss the robust stabilization for fractional-order version of uncertain nonholonomic control systems in this article.
(2) Based on the sliding mode variable structure design method and finite-time stability theorem, a discontinuous three-step switching controller is proposed such that all states of the fractional-order extended nonholonomic chained systems can be stabilized to zero in a settling time.

The organization of this paper is as follows: Introduction of the fractional calculus and the statement of the problem considered is given in Sect. 1.2. Section 1.3 provides the three-step switching controller together with the stability analysis. Section 1.4 illustrates some simulation results. And finally, a conclusion remark is drawn in Sect. 1.5.

1.2 Preliminaries and Problem Statement

1.2.1 Basic Definitions of Fractional Calculus and Finite-Time Stabilization

There are many ways to define the fractional integral and derivative, among which Riemann-Liouvill definition and Caputo definition are generally used in recent literatures.

Definition 1 ([38]) The αth-order Riemman-Liouville fractional integration of function $f(t)$ is given by

$$ {}_{t_0}I_t^{\alpha}f(t) = \frac{1}{\Gamma(\alpha)}\int_{t_0}^{t}\frac{f(\tau)}{(t-\tau)^{1-\alpha}}d\tau, $$

where $\Gamma(\alpha)$ is the Gamma function and t_0 is the initial time.

Definition 2 ([38]) Let $n-1<\alpha\leq n,\ n\in N$, the Riemann-Liouville fractional derivative of order α of function $f(t)$ is defined as follows:

$$ {}_{t_0}D_t^{\alpha}f(t) = \frac{d^{\alpha}f(t)}{dt^{\alpha}} = \frac{1}{\Gamma(n-\alpha)}\frac{d^n}{dt^n}\int_{t_0}^{t}\frac{f(\tau)}{(t-\tau)^{\alpha-n+1}}d\tau = \frac{d^n}{dt^n}I^{n-\alpha}. $$

Definition 3 ([38]) The Caputo fractional derivative of order α of a continuous function $f(t)$ is defined as follows:

$$ {}_{t_0}D_t^{\alpha}f(t) = \begin{cases} \frac{1}{\Gamma(n-\alpha)}\int_{t_0}^{t}\frac{f^{(n)}(\tau)}{(t-\tau)^{\alpha-n+1}}d\tau, & n-1<\alpha<n, \\ \frac{d^n f(t)}{dt^n}, & \alpha=n, \end{cases} $$

where n is the smallest integer number, larger than α.

The version of fractional derivative in Definition 3 is adopted in this paper, and in the rest of our work, the notation D^{α} indicates the Caputo fractional derivative with order α.

Definition 4 ([35, 39]) Consider the system

$$ \dot{y}(t) = g(y(t)),\ g(0)=0,\ y(t)\in R^n, \tag{1.1} $$

where $g: D\rightarrow R^n$ is continuous on an open neighborhood $D\subset R^n$. The equilibrium point $y=0$ of system (1.1) is (locally) finite time stable if (i) it is asymptotically stable in D; (ii) it is finite-time convergent in D. Moreover, if $D=R^n$, the origin $y=0$ is globally finite-time stable.

The following lemmas are needed for our controller design later.

Lemma 1 *For the fractional-order chained form systems*

$$\begin{cases} D^\alpha z_1 = z_2, \\ \quad\vdots \\ D^\alpha z_{n-1} = z_n, \\ D^\alpha z_n = \bar{u}, \end{cases} \tag{1.2}$$

where $\alpha \in (0,1)$ is the order of the systems, $Z = [z_1, z_2 \dots, z_n]^T \in R^n$ is the state vector, and $\bar{u} \in R$ is the control input, respectively. If we take the following controller

$$\bar{u} = -D^{\alpha-1}\left\{\left[\sum_{i=1}^{n} \tilde{k}_i(|z_i| + |z_i|^{\tilde{\rho}}) + \sum_{i=1}^{n-1}\left(sgn(z_i)D^{1-\alpha}z_{i+1}\right)\right]sgn(z_n)\right\},$$

where $sgn(\cdot)$ is a sign function, $\tilde{k}_i > 0$, $\tilde{\rho} \in (0,1)$ are design parameters, then system (1.2) can be stabilized to the origin equilibrium point in a finite time.

Proof As in [39], choosing the following positive definite function as a Lyapunov function candidate

$$V_1 = \|Z(t)\|_1 = \sum_{i=1}^{n} |z_i(t)|. \tag{1.3}$$

Take the time derivative of V_1 along system (1.2), we have

$$\begin{aligned} \dot{V}_1 &= \sum_{i=1}^{n} sgn(z_i)\dot{z}_i = \sum_{i=1}^{n} sgn(z_i)\left(D^{1-\alpha}(D^\alpha z_i)\right) \\ &= \sum_{i=1}^{n-1} sgn(z_i)\left(D^{1-\alpha}(D^\alpha z_i)\right) + sgn(z_n)\left(D^{1-\alpha}(\bar{u})\right), \end{aligned}$$

Substituting the controller $\bar{u}$ into the formula above, we have

$$\dot{V}_1 = -\sum_{i=1}^{n} \tilde{k}_i(|z_i| + |z_i|^{\tilde{\rho}}) \le -\tilde{k}\sum_{i=1}^{n}(|z_i| + |z_i|^{\tilde{\rho}}) \le -\tilde{k}\sum_{i=1}^{n}|z_i| = -\tilde{k}V_1, \tag{1.4}$$

where $\tilde{k} = \min\{\tilde{k}_i\}(i = 1, 2, \dots, n)$. Equation (1.4) means that the origin point $Z = 0$ is asymptotically stable.

On the other hand, because

$$\left(\sum_{i=1}^{n}|z_i|\right)^{\tilde{\rho}} \le \sum_{i=1}^{n}|z_i|^{\tilde{\rho}}, \ \textit{i.e.,} \ \|Z(t)\|_1^{\tilde{\rho}} \le \sum_{i=1}^{n}|z_i|^{\tilde{\rho}},$$

then from (1.3) and (1.4), one has

$$\dot{V}_1 \leq -\tilde{k}\sum_{i=1}^{n}(|z_i| + |z_i|^{\tilde{\rho}}) \leq -\tilde{k}\big(\|Z(t)\|_1 + \|Z(t)\|_1^{\tilde{\rho}}\big) = -\tilde{k}\big(V_1 + V_1^{\tilde{\rho}}\big),$$

By simple calculations, we have $\frac{dV_1}{V_1+V_1^{\tilde{\rho}}} \leq -\tilde{k}dt$. So

$$dt \leq -\frac{dV_1}{\tilde{k}\big(V_1 + V_1^{\tilde{\rho}}\big)} = -\frac{1}{\tilde{k}(1-\tilde{\rho})}\frac{dV_1^{1-\tilde{\rho}}}{\big(1 + V_1^{1-\tilde{\rho}}\big)} \tag{1.5}$$

Taking integral of both sides of (1.5) from 0 to t_s and noting that $Z(t_s) = 0$ ($V_1(t_s) = 0$), we have

$$t_s \leq -\frac{1}{\tilde{k}(1-\tilde{\rho})}\ln\big(1 + V_1^{1-\tilde{\rho}}(t)\big)\Big|_0^{t_s} = \frac{1}{\tilde{k}(1-\tilde{\rho})}\ln\big(1 + V_1^{1-\tilde{\rho}}(0)\big) < +\infty.$$

Therefore, all the states of system (1.2) will converge to zero in the finite time $t_s \leq \frac{1}{\tilde{k}(1-\tilde{\rho})}\ln\big(1 + V_1^{1-\tilde{\rho}}(0)\big)$. According to Definition 4, this completes the proof of the lemma.

Lemma 2 ([41]) *Consider system (1.1), suppose there exists a continuous differential positive-definite function $V(y(t)) : D \to R$, real numbers $p > 0, 0 < \eta < 1$, such that*

$$\dot{V}(y(t)) + pV^{\eta}(y(t)) \leq 0, \ \forall y(t) \in D.$$

Then, the origin of system (1.1) is a locally finite-time stable equilibrium, and the settling time, depending on the initial state $y(0) = y_0$, satisfies $T(y_0) \leq \frac{V^{1-\eta}(y_0)}{p(1-\eta)}$. In addition, if $D = R^n$ and $V(y(t))$ is also radially unbounded, then the origin is a globally finite-time stable equilibrium of system (1.1).

1.2.2 Problem Statement

For the first time, we introduce the dynamic fractional-order nonholonomic extended chained form systems as follows,

$$\begin{cases} D^{\alpha}x_1 = u_1, \\ \quad\vdots \\ D^{\alpha}x_{n-1} = x_n u_1, \\ D^{\alpha}x_n = u_2, \\ D^{\alpha}u_1 = \tau_1 + \Delta f_1(x,u,t) + d_1(t), \\ D^{\alpha}u_2 = \tau_2 + \Delta f_2(x,u,t) + d_2(t), \end{cases} \tag{1.6}$$

where $\alpha \in (0, 1)$ is the order of the system, $x = [x_1, x_2, \ldots, x_n]^T \in R^n$ is the state vector, $\Delta f_i(x, u, t) \in R$, $d_i(t) \in R, (i = 1, 2)$ denote unknown mode uncertain and external disturbances of the system, respectively, and $\tau_i \in R, (i = 1, 2)$ are the dynamic (torque) control inputs of u $(= [u_1, u_2]^T \in R^2)$.

Assumption 1 The uncertainty terms $\Delta f_i(x, u, t)$ and the external disturbances $d_i(t)$, $(i = 1, 2)$ are bounded with known constants as their bounds,

$$\left|\Delta f_i(x, u, t)\right| \leq \bar{f}_i, \quad \left|d_i(t)\right| \leq \bar{d}_i, \quad (i = 1, 2).$$

Although the integral version of the dynamic nonholonomic extended chained form systems of (1.6) has been very well researched [42–44], the previous control method is not valid for the fractional case, and some new control strategy should be found to deal with the stabilization problem for system (1.6).

The control task is to design discontinuous switching fractional controllers $\tau_i, (i = 1, 2)$ for system (1.6) under Assumption 1 such that state x of the corresponding closed loop system can be stabilized to zero in a finite time.

1.3 Main Results

First, the switching design idea for system (1.6) is stated in detail. Motivated by the conclusions of Lemma 1, two subsystems of (1.6) are considered, one is

$$\begin{cases} D^\alpha x_1 = u_1, \\ D^\alpha u_1 = \tau_1 + \Delta f_1(x, u, t) + d_1(t), \end{cases} \tag{1.7}$$

and the other is

$$\begin{cases} D^\alpha x_2 = x_3 u_1, \\ \quad \vdots \\ D^\alpha x_{n-1} = x_n u_1, \\ D^\alpha x_n = u_2, \\ D^\alpha u_2 = \tau_2 + \Delta f_2(x, u, t) + d_2(t). \end{cases} \tag{1.8}$$

System (1.7) or (1.8) can be seen as a more generalized version of (1.2) with uncertainties, then how to design robust controllers to stabilize such uncertain systems in a finite time is one of the most significant innovations in this section, based on which, the switching controller for system (1.6) can be proposed.

In fact, observing system (1.7), we may design the power control input τ_1 such that u_1 can be stabilized to a fixed point in a finite time. In the next step, substituting the fixed u_1 into (1.8), and designing the dynamic control input τ_2 such that $(x_2, \ldots, x_n)$ converges to zero in a finite time. Finally, designing the finite-time stabilizer τ_1 for system (1.7) again to make the last state variable x_1 can be stabilizer to zero in a settling time.

Next, we will give the main results.

Theorem 1 *Under* Assumption 1, *consider the uncertain fractional-order extended nonholonomic chained form system (1.6), taking the following switching control law:*
Step 1: *Let* $\tau_2 = 0$,

$$\tau_1 = -\Big\{ (\bar{f}_1 + \bar{d}_1)sgn(s_1) + D^{\alpha-1}\Big[\Big(k_{11}(|x_1| + |x_1|^{\rho_1}) + k_{12}(|u_1 - 1| + |u_1 - 1|^{\rho_1})$$

$$+ sgn(x_1)D^{1-\alpha}(u_1 - 1)\Big)sgn(u_1 - 1)\Big] + k_{13}|s_1|^{\delta_1}sgn(s_1)\Big\},$$

where

$$s_1 = D^{\alpha-1}(u_1 - 1) + D^{\alpha-2}\Big[\Big(k_{11}(|x_1| + |x_1|^{\rho_1}) + k_{12}(|u_1 - 1| + |u_1 - 1|^{\rho_1})$$

$$+ sgn(x_1)D^{1-\alpha}(u_1 - 1)\Big)sgn(u_1 - 1)\Big],$$

where $k_{1i} > 0, (i = 1, 2, 3)$, $\rho_1, \delta_1 \in (0, 1)$ *are design parameters. Until* $u_1 \equiv 1$ *after a finite time* T_1, *then goto* Step 2;
Step 2: *Let*

$$\tau_1 = -\Big\{ (\bar{f}_1 + \bar{d}_1)sgn(s_1) + D^{\alpha-1}\Big[\Big(k_{11}(|x_1| + |x_1|^{\rho_1}) + k_{12}(|u_1 - 1| + |u_1 - 1|^{\rho_1})$$

$$+ sgn(x_1)D^{1-\alpha}(u_1 - 1)\Big)sgn(u_1 - 1)\Big] + k_{13}|s_1|^{\delta_1}sgn(s_1)\Big\},$$

$$\tau_2 = -\Big\{ (\bar{f}_2 + \bar{d}_2)sgn(s_2) + D^{\alpha-1}\Big[\Big(\sum_{i=2}^{n} k_{2i}(|x_i| + |x_i|^{\rho_2}) + k_{2(n+1)}(|u_2| + |u_2|^{\rho_2})$$

$$+ \sum_{i=2}^{n-1} sgn(x_i)D^{1-\alpha}x_{i+1} + sgn(x_n)D^{1-\alpha}u_2\Big)sgn(u_2)\Big] + k_{2(n+2)}|s_2|^{\delta_2}sgn(s_2)\Big\},$$

where

$$s_2 = D^{\alpha-1}(u_2) + D^{\alpha-2}\Big[\Big(\sum_{i=2}^{n} k_{2i}(|x_i| + |x_i|^{\rho_2}) + k_{2(n+1)}(|u_2| + |u_2|^{\rho_2})$$

$$+ \sum_{i=2}^{n-1} sgn(x_i)D^{1-\alpha}x_{i+1} + sgn(x_n)D^{1-\alpha}u_2\Big)sgn(u_2)\Big],$$

where $k_{2i} > 0, (i = 2, \ldots, n+2)$, $\rho_2, \delta_2 \in (0,1)$ *are design parameters. Until* $x_j \equiv 0, (j = 2, \ldots, n)$ *after a finite time* T_2*, then goto* Step 3;

Step 3: *Let* $\tau_2 = 0$,

$$\tau_1 = -\Bigg\{ (\bar{f}_1 + \bar{d}_1) sgn(s_3) + D^{\alpha-1}\Bigg[\Big(k_{31}(|x_1| + |x_1|^{\rho_3}) + k_{32}(|u_1| + |u_1|^{\rho_3}) + sgn(x_1)D^{1-\alpha}u_1\Big) sgn(u_1)\Bigg] + k_{33}|s_3|^{\delta_3} sgn(s_3)\Bigg\},$$

where

$$s_3 = D^{\alpha-1}u_1 + D^{\alpha-2}\Bigg[\Big(k_{31}(|x_1| + |x_1|^{\rho_3}) + k_{32}(|u_1| + |u_1|^{\rho_3}) + sgn(x_1)D^{1-\alpha}u_1\Big) sgn(u_1)\Bigg],$$

where $k_{3i} > 0, (i = 1, 2, 3)$, $\rho_3, \delta_3 \in (0,1)$ *are design parameters. Until* $x_1 = u_1 \equiv 0$ *after a finite time* T_3*, stop.*

Then system (1.6) can be stabilized to the origin equilibrium point in a finite time $(T = T_1 + T_2 + T_3)$ *by the switching controller* Step 1$\sim$ Step 3.

Proof In first step, let $\bar{u}_1 = u_1 - 1$, then we have

$$\begin{cases} D^\alpha x_1 = \bar{u}_1, \\ D^\alpha \bar{u}_1 = \tau_1 + \Delta f_1(x,u,t) + d_1(t). \end{cases} \tag{1.9}$$

Selecting a Lyapunov function for system (1.9),

$$V_1(t) = |s_1(t)|.$$

Take the time derivative of $V_1(t)$ along system (1.9), we have

$$\begin{aligned} \dot{V}_1(t) &= sgn(s_1)\dot{s}_1(t) \\ &= sgn(s_1)\Bigg\{ D^\alpha \bar{u}_1 + D^{\alpha-1}\Bigg[\Big(k_{11}(|x_1| + |x_1|^{\rho_1}) + k_{12}(|\bar{u}_1| + |\bar{u}_1|^{\rho_1}) \\ &\qquad + sgn(x_1)D^{1-\alpha}\bar{u}_1\Big) sgn(\bar{u}_1)\Bigg]\Bigg\} \\ &= sgn(s_1)\Bigg\{ \tau_1 + \Delta f_1(x,u,t) + d_1(t) + D^{\alpha-1}\Bigg[\Big(k_{11}(|x_1| + |x_1|^{\rho_1}) \\ &\qquad + k_{12}(|\bar{u}_1| + |\bar{u}_1|^{\rho_1}) + sgn(x_1)D^{1-\alpha}\bar{u}_1\Big) sgn(\bar{u}_1)\Bigg]\Bigg\} \end{aligned}$$

Substituting τ_1 into the formula above, we have

$$\begin{aligned}\dot{V}_1(t) &= sgn(s_1)\Big\{ -(\bar{f}_1+\bar{d}_1)sgn(s_1) + \Delta f_1(x,u,t) + d_1(t) - k_{13}|s_1|^{\delta_1}sgn(s_1)\Big\}\\ &\le -(\bar{f}_1+\bar{d}_1) + |\Delta f_1(x,u,t)| + |d_1(t)| - k_{13}|s_1|^{\delta_1}\end{aligned}$$

By *Assumption 1*, one can obtain

$$\dot{V}_1(t) + k_{13}|V_1(t)|^{\delta_1} \le 0. \tag{1.10}$$

In fact, the variable $s_1(t)$ can be seen as some sliding mode dynamic surface, then according to variable structure control theory of sliding mode, from (1.10) and Lemma 2, there exists a finite time $T_0 < +\infty$, such that

$$s_1(t) = 0,\ \ \dot{s}_1(t) = 0,\ \ t \ge T_0.$$

As $t \ge T_0$, from $\dot{s}_1(t) = 0$, we have

$$D^\alpha \bar{u}_1 + D^{\alpha-1}\Big[\Big(k_{11}(|x_1| + |x_1|^{\rho_1}) + k_{12}(|\bar{u}_1| + |\bar{u}_1|^{\rho_1}) + sgn(x_1)D^{1-\alpha}\bar{u}_1\Big)sgn(\bar{u}_1)\Big] = 0,$$

from which, we have

$$\begin{aligned}D^\alpha x_1 &= \bar{u}_1,\\ D^\alpha \bar{u}_1 &= -D^{\alpha-1}\Big[\Big(k_{11}(|x_1| + |x_1|^{\rho_1}) + k_{12}(|\bar{u}_1| + |\bar{u}_1|^{\rho_1})\\ &\qquad + sgn(x_1)D^{1-\alpha}\bar{u}_1\Big)sgn(\bar{u}_1)\Big].\end{aligned} \tag{1.11}$$

From Lemma 1, there exists a finite time $T_0' < +\infty$, such that $(x_1, \bar{u}_1)$ of system (1.11) converges to zero as $t \to T_0'$, and $x_1 = \bar{u}_1 \equiv 0, t \ge T_0'$. This implies $u_1 \equiv 1$ as $t \ge T_1 = T_0 + T_0'$, then goes to the second controller designing.

In *Step 2*, system (1.8) has been changed into the following system:

$$\begin{cases} D^\alpha x_2 = x_3,\\ \quad\vdots\\ D^\alpha x_{n-1} = x_n,\\ D^\alpha x_n = u_2,\\ D^\alpha u_2 = \tau_2 + \Delta f_2(x,u,t) + d_2(t). \end{cases} \tag{1.12}$$

Choose a Lyapunov function for system (1.12) $V_2(t) = |s_2(t)|$, its time derivative along (1.12) is

$$
\begin{aligned}
\dot{V}_2(t) &= sgn(s_2)\dot{s}_2(t) \\
&= sgn(s_2)\Big\{ \tau_2 + \Delta f_2(x,u,t) + d_2(t) + D^{\alpha-1}\Big[\Big(\sum_{i=2}^{n} k_{2i}(|x_i| + |x_i|^{\rho_2}) \\
&\quad + k_{2(n+1)}(|u_2| + |u_2|^{\rho_2}) + \sum_{i=2}^{n-1} sgn(x_i)D^{1-\alpha}x_{i+1} + sgn(x_n)D^{1-\alpha}u_2\Big)sgn(u_2)\Big]\Big\},
\end{aligned}
$$

Substituting τ_2 into it, we have

$$
\begin{aligned}
\dot{V}_2(t) &= sgn(s_2)\Big\{ -(\bar{f}_2 + \bar{d}_2)sgn(s_2) + \Delta f_2(x,u,t) + d_2(t) - k_{2(n+2)}|s_2|^{\delta_2}sgn(s_2)\Big\} \\
&\le -(\bar{f}_2 + \bar{d}_2) + |\Delta f_2(x,u,t)| + |d_2(t)| - k_{2(n+2)}|s_2|^{\delta_2},
\end{aligned}
$$

which means $\dot{V}_2(t) + k_{2(n+2)}|V_2(t)|^{\delta_2} \le 0$.

Hence, there exists a finite time $T_{00} < +\infty$, such that

$$
s_2(t) = 0, \;\; \dot{s}_2(t) = 0, \;\; t \ge T_{00}.
$$

As $t \ge T_{00}$, from $\dot{s}_2(t) = 0$, we have

$$
\begin{aligned}
D^{\alpha}u_2 + D^{\alpha-1}\Big[\Big(\sum_{i=2}^{n} k_{2i}(|x_i| + |x_i|^{\rho_2}) + k_{2(n+1)}(|u_2| + |u_2|^{\rho_2}) \\
+ \sum_{i=2}^{n-1} sgn(x_i)D^{1-\alpha}x_{i+1} + sgn(x_n)D^{1-\alpha}u_2\Big)sgn(u_2)\Big] = 0
\end{aligned}
$$

Therefore, when $t \ge T_{00}$, we have

$$
\begin{aligned}
&D^{\alpha}x_2 = x_3, \\
&\quad \vdots \\
&D^{\alpha}x_{n-1} = x_n, \\
&D^{\alpha}x_n = u_2, \\
&D^{\alpha}u_2 = -D^{\alpha-1}\Big[\Big(\textstyle\sum_{i=2}^{n} k_{2i}(|x_i| + |x_i|^{\rho_2}) + k_{2(n+1)}(|u_2| + |u_2|^{\rho_2}) \\
&\qquad\qquad + \textstyle\sum_{i=2}^{n-1} sgn(x_i)D^{1-\alpha}x_{i+1} + sgn(x_n)D^{1-\alpha}u_2\Big)sgn(u_2)\Big],
\end{aligned}
$$

by using the conclusions of Lemma 1, there exists a finite time $T'_{00} < +\infty$ such that

$$
x_i(t), u_2(t) \to 0, \; (i = 2, \ldots, n), \;\; t \ge T'_{00}.
$$

As $t \ge T_2 = T_{00} + T'_{00}$, goes to *Step 3*.

In the last step, consider subsystem (1.7) again, by using the similar analysis, it is not difficult to prove that the state (x_1, u_1) can be stabilized to zero in a finite time T_3 under the controller τ_1 in *Step 3*, and thus completes the proof of this theorem.

1.4 Simulations

The approximate numerical technique for solving the fractional differential equations have been developed by many researchers, see e.g. Diethelm et al. [44], which is the generalized version of the Adams-Bashforth-Moulton algorithm.

Based on the numerical algorithms of fractional-order systems, we consider the simulation for a simple example $n = 3$ of system (1.6), and we take the fractional order is $\alpha = 0.8$, assume that the initial condition is $(-0.6, -1.0, 0.3, 0.5, -1.2)$, the

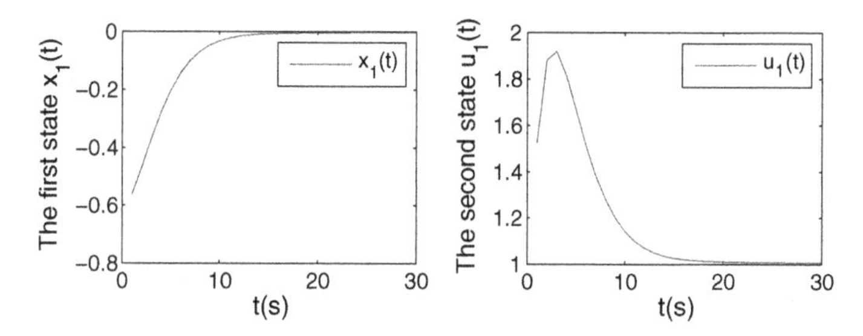

Fig. 1.1 Response of the state (x_1, u_1) with respect to time in first step

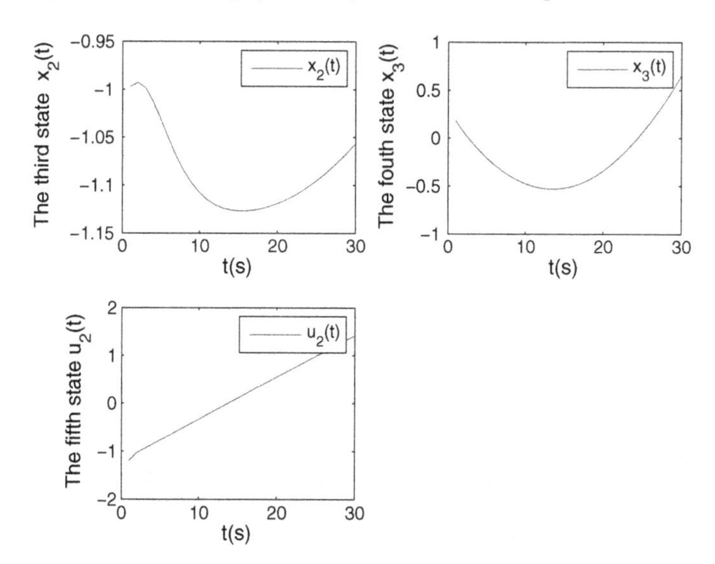

Fig. 1.2 Response of the state (x_2, x_3, u_2) with respect to time in first step

bounds of uncertain terms are $\bar{f}_i = 2.0, \bar{d}_i = 1.0, (i = 1, 2)$. According to Theorem 1, choosing design parameters $k_{1i} = 1.0, k_{2(i+1)} = 1.5, k_{3i} = 1.2, (i = 1, 2, 3)$, $\rho_j = \delta_j = 0.5, (j = 1, 2)$.

By using the numerical algorithm [44], with the sampling interval being $h = 0.002$ s, next we present the simulation result to show the finite time convergence of $(x_1, x_2, x_3, u_1, u_2)$.

From Fig. 1.1, we observe that the state u_1 of subsystem (1.7) is stabilized to the point $u_1 = 1$ as $t \geq 30$ s in *Step 1* by the controller τ_1, the other state x_1 is driven to zero in a finite time $t \leq 20$ s. And Fig. 1.2 shows the response of the state (x_2, x_3, u_2) of subsystem (1.8) with respect to time in first step.

As $t \geq 30$ s, switching to *Step 2*, in Fig. 1.3, it shows that (x_1, u_1) keeps unchanged, which means that subsystem (1.8) is equal to system (1.12) in this step. Figure 1.4 describes the finite time stabilization of system (1.12) under the controller τ_2 proposed in *Step 2*. We can observe that (x_2, x_3, u_2) is stabilized to zero as $30\,\text{s} \leq t \leq 55$ s.

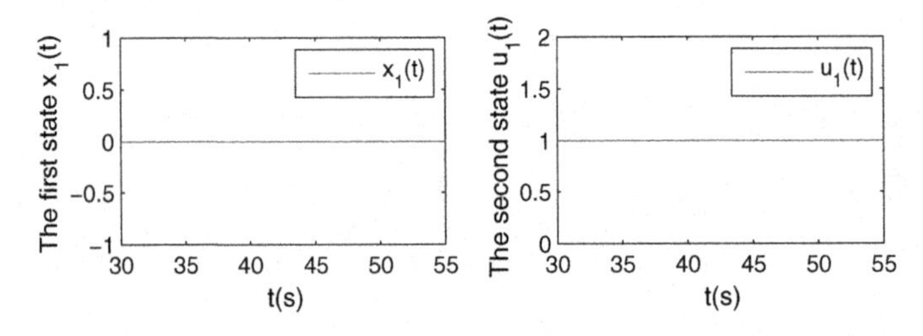

Fig. 1.3 Response of the state (x_1, u_1) with respect to time in second step

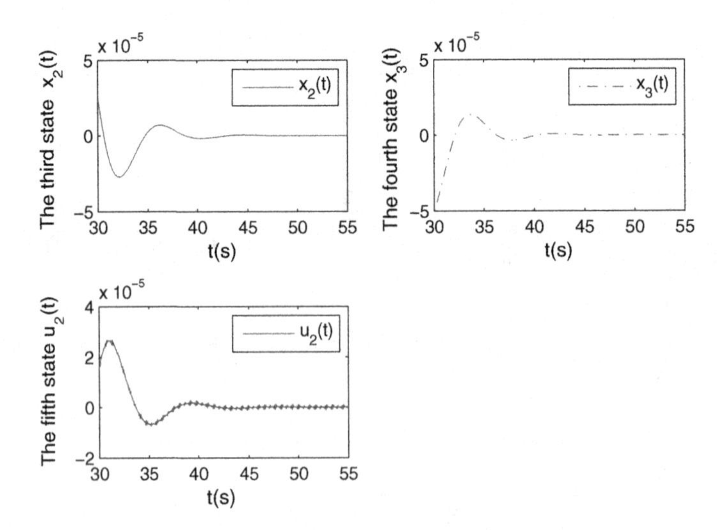

Fig. 1.4 Response of the state (x_2, x_3, u_2) with respect to time in second step

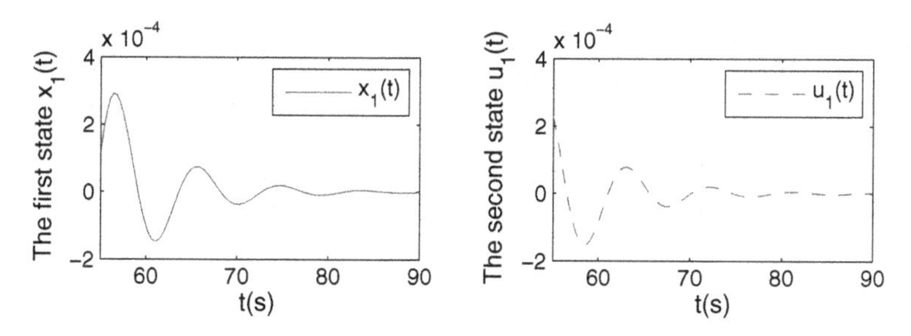

Fig. 1.5 Response of the state (x_1, u_1) with respect to time in the last step

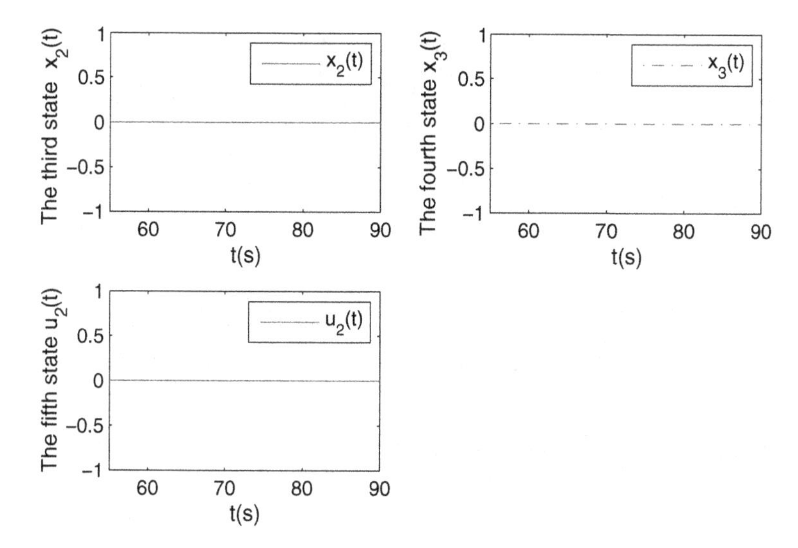

Fig. 1.6 Response of the state (x_2, x_3, u_2) with respect to time in the last step

As $t \geq 55$ s, Figs. 1.5 and 1.6 show that all the states of the original fractional-order extended nonholonomic system (1.6) can be stabilized to zero equilibrium point at a finite time $t \leq 90$ s in the last step.

1.5 Conclusions

A three-step switching controller is proposed in this article in order to address the robust finite-time stabilization problem for the fractional-order extended nonholonomic chained form systems. The stability analysis is given based on the Lyapunov Theorem, stability theory of finite-time stabilization and the theory of variable structure control. Finally, a simple numerical example is adopted to show the effectiveness of our control approach.

Acknowledgments This work was supported by the Natural Science Foundation of China (61304004), the China Postdoctoral Science Foundation funded project (2013M531263), the Jiangsu Planned Projects for Postdoctoral Research Funds (1302140C), the Foundation of Changzhou Key Laboratory of Special Robot and Intelligent Technology (CZSR2014005), and the Foundation of China Scholarship Council (201406715056).

References

1. Fang Y, Liu X, Zhang X (2012) Adaptive active visual servoing of nonholonomic mobile robots. IEEE Trans Ind Electron 59:486–497
2. Liu T, Jiang Z (2013) Distributed formation control of nonholonomic mobile robots without global position measurements. Automatica 49:592–600
3. Mohareri O, Dhaouadi R, Rad AB (2012) Indirect adaptive tracking control of a nonholonomic mobile robot via neural networks. Neurocomputing 88:54–66
4. Chen H, Wang C, Yang L, Zhang D (2012) Semiglobal stabilization for nonholonomic mobile robots based on dynamic feedback with inputs saturation. J Dyn Syst Meas Control 134:1 041006–8–041006
5. Chen H, Wang C, Zhang B, Zhang D (2013) Saturated tracking control for nonholonomic mobile robots with dynamic feedback. Trans Inst Meas Control 35:105–116
6. Hua C, Jinbo Z, Bingyan C, Baojun L (2013) Global practical stabilization for non-holonomic mobile robots with uncalibrated visual parameters by using a switching controller. IMA J Math Control Inf. doi:10.1093/imamci/dns044
7. Yoo K, Chung W (2010) Pushing motion control of n passive off-hooked trailers by a car-like mobile robot. In: 2010 IEEE International Conference on Robotics and Automation (ICRA), pp 4928–4933
8. Demirli K, Khoshnejad M (2009) Autonomous parallel parking of a car-like mobile robot by a neuro-fuzzy sensor-based controller. Fuzzy Sets Syst 160:2876–2891
9. Das S, Sinha M, Kumar KD, Misra A (2010) Reconfigurable magnetic attitude control of earth-pointing satellites. Proc Inst Mech Eng Part G: J Aerosp Eng 224:1309–1326
10. Chwa D (2011) Global tracking control of underactuated ships with input and velocity constraints using dynamic surface control method. IEEE Trans Control Syst Technol 19:1357–1370
11. Brockett RW (1983) Asymptotic stability and feedback stabilization. In: Brockett RW, Millman RS, Sussmann HJ (eds) Differ Geom Control Theory. Birkhauser, Boston, pp 181–208
12. Tian YP, Li S (2002) Exponential stabilization of nonholonomic dynamic systems by smooth time-varying control. Automatica 38:1139–1146
13. Yuan HL, QU Z (2008) Continuous time-varying pure feedback control for chained nonholonomic systems with exponential convergence. In: Proceedings of the 17th IFAC World Congress, pp 15203–15208
14. Shojaei K, Shahri AM (2012) Adaptive robust time-varying control of uncertain nonholonomic robotic systems. IET Control Theory Appl 6:90–102
15. Astolfi A (1996) Discontinuous control of nonholonomic systems. Syst Control Lett 27:37–45
16. Marchand N, Alamir M (2003) Discontinuous exponential stabilization of chained form systems. Automatica 39:343–348
17. Laiou M, Astolfi A (2004) A local transformations to generalized chained forms. In: The 16th International Symposium on Mathematical Theory of Networks and Systems, MTNS
18. Ge SS, Wang Z, Lee TH (2003) Adaptive stabilization of uncertain nonholonomic systems by state and output feedback. Automatica 39:1451–1460
19. Luo Y, Chen YQ, Pi YG (2011) Experimental study of fractional order proportional derivative controller synthesis for fractional order systems. Mechatronics 21:204–214

20. Aghababa MP (2012) Chaos in a fractional-order micro-electromechanical resonator and its suppression. Chin Phys B 21:100505
21. Mandelbrot B, VanNess JW (1968) Fractional Brownian motions fractional noises and applications. SIAM Rev 10:422–437
22. Engheta N (1996) On fractional calculus and fractional multipoles in electromagnetism. IEEE Trans Antennas Propag 44:554–566
23. Baleanu D, Golmankhaneh AK, Nigmatullin R (2010) Fractional newtonian mechanics. Cent Eur J Phys 8:120–125
24. Sun H, Chen W, Wei H, Chen YQ (2011) A comparative study of constant-order and variable-order fractional models in characterizing memory property of systems. Eur Phys J-Spec Top 193:185–193
25. Sun H, Chen Y, Chen W (2011) Random-order fractional differential equation models. Signal Process 91:525–530
26. Chen W, Lin J, Wang F (2011) Regularized meshless method for nonhomogeneous problems. Eng Analwith Bound Elem 35:253–257
27. Oustaloup A (1995) In: Synthese et Application S. La Derivation Non Entiere: Theorie, Editions Hermes, Paris
28. Magin R (2004) Fractional calculus in bioengineering Part 1–3. Crit Rev Bioeng 32
29. Podlubny I (1999) Fractional-order systems and $PI^{\lambda}D^{\mu}$-controllers. IEEE Trans Autom Control 44:208–214
30. Tavazoei MS, Haeri M, Jafari S, Bolouki S, Siami M (2008) Some applications of fractional calculus in suppression of chaotic oscillations. IEEE Trans Ind Electron 55:4098–4101
31. Li CG, Chen GR (2004) Chaos in the fractional order Chen system and its control. Chaos Solitions Fractals 22:549–554
32. Chen H, Chen W, Zhang B, Cao H (2013) Robust synchronization of incommensurate fractional-order chaotic systems via second-order sliding mode technique. J Appl Math 2013:11 Article ID 321253
33. Linares H, Baillot C, Oustaloup A, Ceyral C (1996) Generation of a fractional ground: application in robotics. In International Congress IEE-Smc CESA'96 IMACS Multi-conference, Lille, July 1996
34. Duarte FBM, Macado JAT (2002) Chaotic phenomena and fractional-order dynamics in the trajectory control of redundant manipulators. Nonlinear Dyn 29:315–342
35. Hong Y (2002) Finite-time stabilization and stabilizability of a class of controllable systems. Syst Control Lett 46:231–236
36. Bhat SP, Bernstein DS (1998) Continuous finite-time stabilization of the translational and rotational double integrators. IEEE Trans Autom Control 43:678–682
37. Haimo VT (1986) Finite-time controllers. SIAM J Control Optim 24:760–770
38. Podlubny I (1999) Fractional differential equations. Academic Press, New York
39. Bhat SP, Bernstein DS (2000) Finite-time stability of continuous autonomons systems. SIAM J Control Optim 38:751–766
40. Aghababa MP (2012) Robust finite-time stabilization of fractional-order chaotic systems based on fractional Lyapunov stability theory. J Comput Nonlinear Dyn 7:1-5–021010
41. Jiang ZP (2000) Robust exponential regulation of nonholonomic systems with uncertainties. Automatica 36:189–209
42. Kolmanovsky I, Reyhanoglu M, McClamroch NH (1996) Switched mode feedback control laws for nonholonomic systems in extended power form. Syst Control Lett 27:29–36
43. Wu YQ, Wang B, Zong GD (2005) Finite-time tracking controller design for nonholonomic systems with extended chained form. IEEE Trans Circuits Syst Express Briefs 52:798–802
44. Diethelm K, Ford NJ, Freed AD (2002) A predictor-corrector approach for the numerical solution of fractional differential equations. Nonlinear Dyn 29:3–22

Chapter 2
Performance of Microstrip Patch Antennas Embedded in Electromagnetic Band-Gap Structure

Fangming Cao and Suling Wang

Abstract Electromagnetic band-gap structures are widely used in antenna and microwave element designs due to their unique electromagnetic features which are shown in two aspects: forbidden band gap and in-phase reflection. In this paper, three different antennas are designed and simulated by Ansoft HFSS to know the use of the EBG structure in microstrip antenna engineering. Radiation characteristics such as gain, radiation patterns and s-parameters of these microstrip antennas are performed. The simulated results verify that the gain has been increased noticeably, the radiation pattern has been improved and the sidelobe and backlobe levels have been reduced by using the mushroom-like EBG structures. Especially, the microstrip antenna over the EBG structures not only shows the best electromagnetic characteristics but also maintains its small dimensions.

Keywords EBG structures · Radiation characteristics · Patch antenna · HFSS

2.1 Introduction

Recently, perhaps there is nothing more that has attracted people extremely than Electromagnetic band gap (EBG) structure which is originated from a kind of new artificial electromagnetic material called photonic crystal in the electromagnetic and antenna community. As one special electromagnetic material, electromagnetic band gap structure is composed of cyclical array of dielectric or metal elements [1, 2]. EBG structures have two distinctive electromagnetic properties including the

F. Cao (✉) · S. Wang
School of Electrical Engineering and Automation, Henan Polytechnic University, Jiaozuo 454000, China
e-mail: 957091639@qq.com

Y. Jia et al. (eds.), *Proceedings of the 2015 Chinese Intelligent Systems Conference*, Lecture Notes in Electrical Engineering 360,
DOI 10.1007/978-3-662-48365-7_2

frequency band gap and the phase reflection [3]. The application of EBG structure is focused on these two aspects in microwave antenna designs. First the property of surface-wave suppression enables one to design the microstrip patch antenna with high gain, suppress high order harmonic and depress the mutual coupling of plannar antennas; second some typical applications of the in-phase reflection property are to raise the the antenna gain as a reflection plate, inhibit the rear lobe and design low profile antennas [4, 5]. This paper concentrates on the mushroom-like EBG structure whose compactness property is more attractive compared to other EBG structures.

The strong points of microstrip patch antennas including small size, ease of integration, light weight and low price have led to their numerous favorable applications in wireless communication field [6]. Miniaturization has become one of the essential antenna design requirements with the revolutionary development of the related field. All kinds of methods have been put into effect to make the antenna small in size. Raising the permitivity of the substrate is recognised as the most common measure [7, 8], yet applying the substrate with high permittivity to the microstrip antenna will deteriorate the antenna performance [9]. In this paper, we apply the mushroom-like electromagnetic band gap structures to the microstrip antennas in order to inhibit the surface waves generated by the dielectric substrates with higher permittivity.

2.2 Design Microstrip Patch Antennas

First a traditional patch antenna is designed for comparison. We make the reference antenna work at 1.8 GHz so that the center frequency of the antenna can be covered with the band gap region of the mushroom-like EBG structure. The patch is fabricated on a ground plane with a relative permittivity of 10.2 to remain its compact size. Figure 2.1 presents this conventional antenna. Its resonant frequency is at 1.8 GHz when the area of the patch is $L \times W = 24.8 \times 38.3\,\text{mm}^2$, and the feed location L_1 is 3.34 mm, L_1 is the distance between the feed and the center of the patch [6].

Then we make the radiation patch surrounded by a layer of mushroom-like EBG structure. The mushroom-like EBG structure revealed in Fig. 2.2 is fabricated on the ordinary microwave substrate and the metal patches printed on the surface of substrate are connected to the ground floor through the metal hole [10]. Figure 2.3 depicts the antenna surrounded by the EBG structures which consist of patches and vias array [11]. The EBG structure is devised so that the antenna resonant frequency falls into its surface wave band gap and hence the surface wave excited by this patch antenna is inhibited from propagation. The dimensions of the EBG structures are displayed in Fig. 2.4. The antenna can achieve a good match of 50Ω around 1.8 GHz after continuous optimizations. The L is 24.8 mm, W is 38.3 mm and the L_1 is 3.65 mm. It is worthwhile to point out that the period gap of electromagnetic

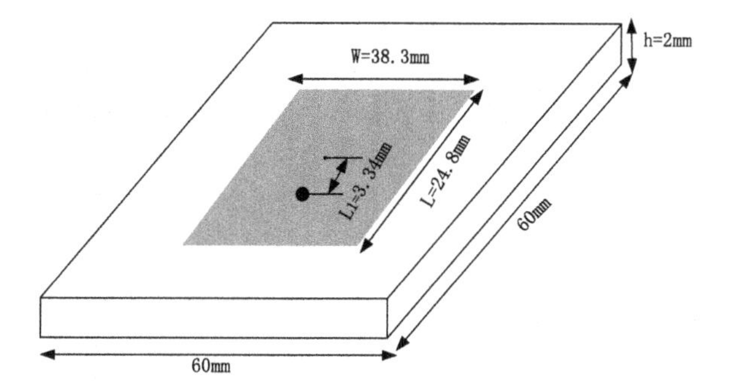

Fig. 2.1 Conventional antenna dimensions

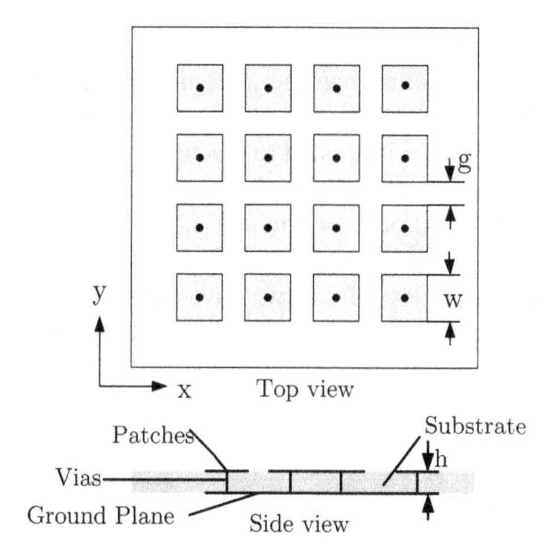

Fig. 2.2 Geometry of mushroom-like EBG structure

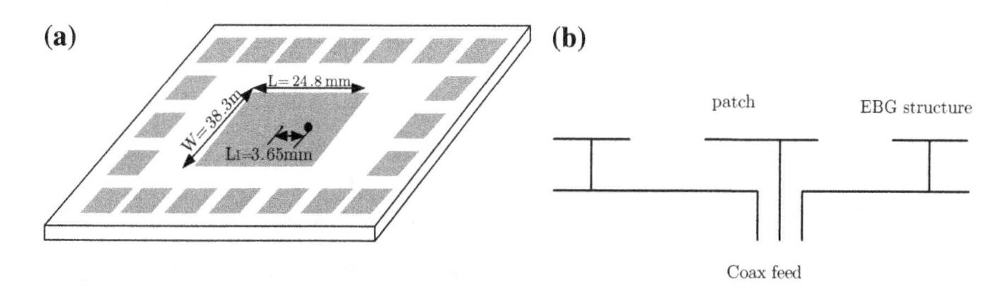

Fig. 2.3 Patch antenna surrounded by mushroom-like EBG structure: **a** geometry and **b** cross section

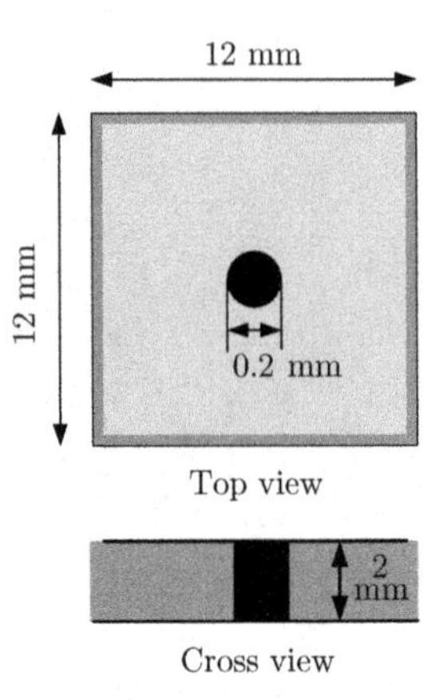

Fig. 2.4 Cell dimensions

band gap cells should be smaller than the interval between the EBG cells and the patch in order to obtain a good radiation characteristic [6].

Generally the EBG structure is applied to placing around the patch antenna in the patch antenna design. But the problem is that the size of the patch will become so large that it can't reach the purpose of miniaturization. The EBG structure attached to the back of the patch as the ground not only inhibits the surface wave but also miniaturizes the antenna. In view of this, we have designed the antenna as shown in Fig. 2.5. When the patch length is 32.1 mm and width is 28 mm, the antenna can work at 1.8 GHz.

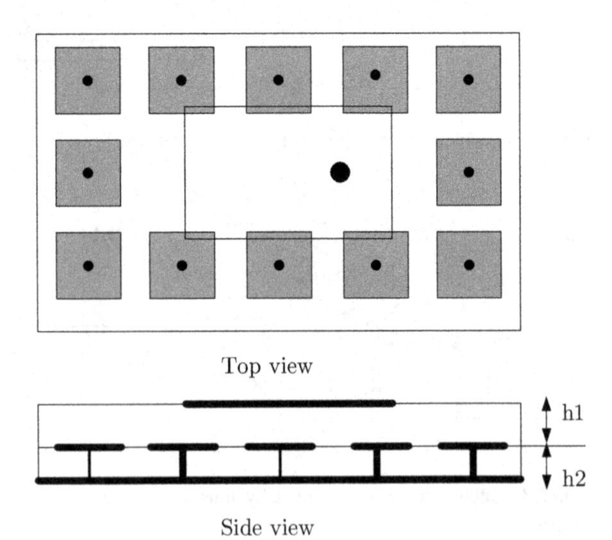

Fig. 2.5 Microstrip antenna over EBG structure

2.3 Simulated Results and Discussion

The simulated s-parameters for these three structures are compared in Fig. 2.6. Curve1 stands for the S_{11} of the traditional antenna, curve2 is on behalf of the S_{11} of the antenna surrounded by the EBG structure, curve3 means the S_{11} of the antenna over the EBG structure. From the S_{11} plot, it can be observed that all the three patches are tuned to resonate at the same frequency 1.8 GHz with better return loss less than-10 dB, especially the microstrip antenna on top of the EBG structure.

Figure 2.7 presents the directional pattern in E-plane. Curve1 stands for the the directional pattern of the conventional antenna, curve2 is on behalf of the the directional pattern of the antenna surrounded by the mushroom-like EBG structure, curve3 means the directional pattern of the antenna over the EBG structure. We can

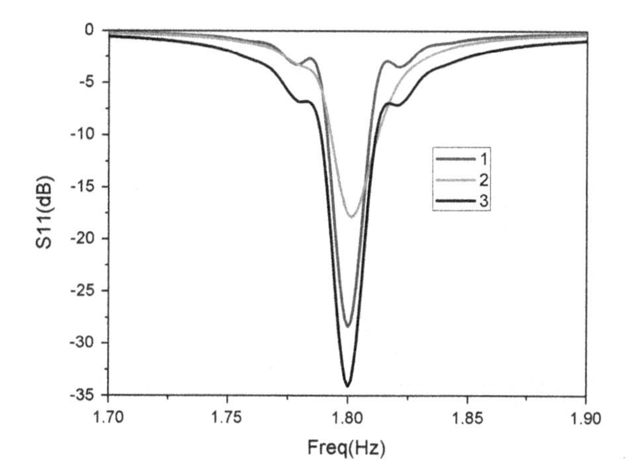

Fig. 2.6 Return loss

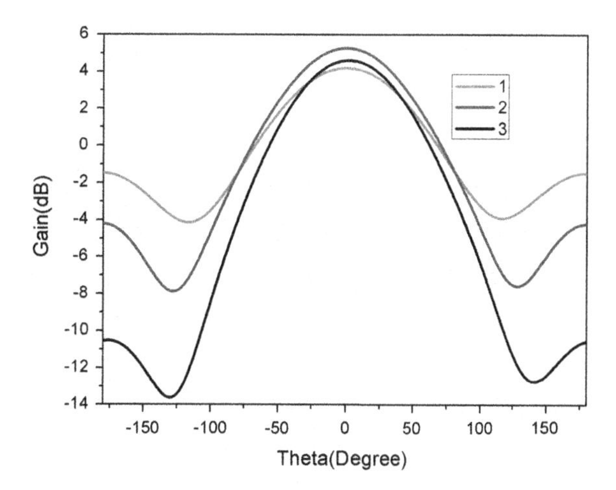

Fig. 2.7 E-plane radiation patterns of patch antennas

conclude that the antenna pattern is influenced by the EBG structure, curve 3 has the lowest back radiation which is more than 9 dB lower than curve 1, though its front radiation only increases 0.41 dB than curve 1. Curve 2 has the highest front radiation which is 5.27 dB. It is apparent that surface-wave suppression improves the back radiation by reducing edge scattering waves.

Some key values are listed in Table 2.1. Figure 2.8 depicts the simulation models of microstrip patch antennas with EBG structure. The area of this microstrip antenna over the EBG structure surface is about half of the area of the microstrip antenna surrounded with the EBG structure by calculating. From above comparisons it can be concluded that the microstrip antenna on top of the EBG has the best radiation performances on basis of miniaturization.

Table 2.1 Antenna characteristics

Structure	s_{11} (%)	Front radiation (dB)	Back radiation (dB)
1	−28.32	4.19	−1.47
2	−17.60	5.27	−4.21
3	−34.08	4.60	−10.55

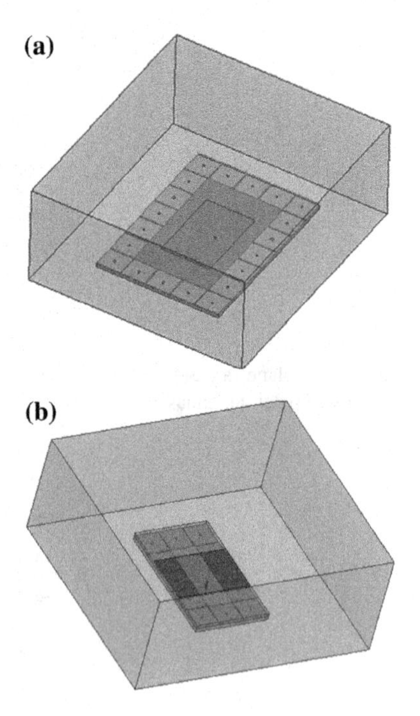

Fig. 2.8 Simulation models of antennas with EBG structures. **a** Simulation model of patch antenna surrounded by mushroom-like EBG structure. **b** Simulation model of patch antenna on top of EBG structure

2.4 Conclusion

In this paper, three kinds of antennas have been designed and simulated by Ansoft HFSS.We conclude that the EBG structures may hinder the propagation of surface wave and thus improve the radiation performance of an antenna. In particular, the patch antenna over the EBG structure has the best performance while keeping its small dimensions. This also specifies that EBG structures can compensate for the shortcomings of the microstrip antenna on the substrate with higher permittivity.

References

1. Yang F, Rahmat-Samii Y (2002) Applications of electromagnetic band-gap (EBG) structures in microwave antenna designs[j]. Microw Millim Wave Technol 2:580–583
2. Liang J, Yang, HYD (2007) Radiation characteristics of a microstrip patch over an electromagnetic bandgap surface[J]. In: IEEE International symposium on antennas and propagation, pp 1691–1697
3. Yang F, Rahmat-Samii Y (2008) Electromagnetic band gap structures in antenna engineering [M]. Cambridge University Press, Cambridge
4. Yang Fan, Rahmat-Samii Yahya (2003) Reflection phase characterizations of the EBG ground plane for low profile wire antenna applications[J]. IEEE Trans Antennas Propag 51 (10):2691–2703
5. Zaman MI, Hamedani FT, Amjadi H (2012) A new EBG structure and its application on microstrip patch antenna[J]. In: IEEE International symposium on ANTEM, pp 1–3
6. Silval F, Pinho P (2011) Performance of microstrip antennas in the presence of EBG in an indoor localization system[J]. In: IEEE International symposium on antennas and propagation, pp 3095–3098
7. Yang L, Feng Z (2011) Advanced methods to improve compactnessin EBG design and utilization[J]. IEEE International symposium on antennas and propagation, 20–25
8. Hanafi NMT, Islam MT, Misran N (2012) Analysis of mushroom-like EBG structure and UC-EBG for SAR reduction[J]. In: IEEE International symposium on antennas and propagation, pp 696–699
9. Qu D, Shafai L (2005) Wide band microstrip patch antenna with EBG substrates[J]. In: IEEE International symposium on antennas and propagation society, pp 594–597
10. Yang F, Rahmat-Samii Y (2001) Mutual coupling reduction of microstrip antennas using electromagnetic band-gap structure[J]. IEEE International symposium on antennas and propagation society, pp 478–481
11. Su Y, Xing L, Cheng ZZ et al (2010) Mutual coupling reduction in microstrip antennas by using dual layer uniplanar compact EBG (UC-EBG) structure[J]. IEEE International conference on microwave and millimeter wave technology, pp 180–183

Chapter 3
Terminal Sliding Mode Control of a Boost Converter

Meimei Xu, Li Ma and Shihong Ding

Abstract In this paper, a terminal sliding mode (TSM) control method for the boost converter is studied. First of all, the nonlinear model of boost converter is established. Then a change of coordinate is employed such that the nonlinear model will be transformed into a linear system. On this basis, a TSM composite controller including the state feedback control and the disturbance feedforward compensation is given. Under the composite controller, the reference voltage can be tracked in a finite time. The effectiveness of the paper is shown with a simulation example.

Keywords Boost converter · Terminal sliding mode control

3.1 Introduction

In recent years, the boost converters have been widely used as telecommunication equipment, DC motor drives, etc. The mathematical model of switch-based boost converter is in essence nonlinear, and thus the control of boost converter is difficult due to its non-minimum phase nature [1]. Consequently, the design of high performance control strategy for boost converter is usually a challenging issue.

Some nonlinear control strategies have been employed to control boost converters in recent years, such as [2, 3]. By assuming the exponential form of the linear

This work was supported in part by the National Natural Science Foundation of China (No. 61203054), Open Foundation of Key Laboratory of Measurement and Control of Complex Systems of Engineering, Ministry of Education (No MCCSE2014A01) and the Priority Academic Program Development of Jiangsu Higher Education Institutions.

M. Xu · L. Ma (✉) · S. Ding
School of Electrical and Information Engineering, Jiangsu University,
Zhenjiang 212013, China
e-mail: mali@mail.ujs.edu.cn

L. Ma
Key Laboratory of Measurement and Control of Complex Systems of Engineering,
Ministry of Education, Southeast University, Nanjing Jiangsu 210096, China

Y. Jia et al. (eds.), *Proceedings of the 2015 Chinese Intelligent Systems Conference*, Lecture Notes in Electrical Engineering 360,
DOI 10.1007/978-3-662-48365-7_3

multiloop controller, a nonlinear control scheme has been proposed in [2], which will provide an additional tuning parameter to modify the output response. Later, an adaptive voltage regulator is proposed in [3] to adjust its parameter and steady-state voltage of the closed loop system under different operation, and a slightly under-damped response can be obtained.

In this paper, a finite-time composite control scheme based on terminal sliding mode (TSMC) and disturbance observer has been developed for the sake of controlling the output voltage of a boost converter to track the reference value in a finite time. The finite-time composite controller includes a TSMC based state feedback control and a disturbance observer based feedforward compensation. As a matter of fact, the TSM feedback controller can be directly used to stabilize the Boost converter system. However, due to the lumped disturbance, the control effort will be large. To this end, the lumped disturbance is estimated with a finite-time disturbance observer, and then the estimated value, as a feedforward term, is used to cancel the lumped disturbance. Simulation results validate the proposed method.

3.2 Modeling and Problem Formulation

A boost converter consists of an input voltage U_i, a switch S, an output diode D, a load resistor R, a filter inductor L and a filter capacitor C, connected as in Fig. 3.1.

When the switch opens and closes, we obtain the Equivalent circuits as shown in Fig. 3.2.

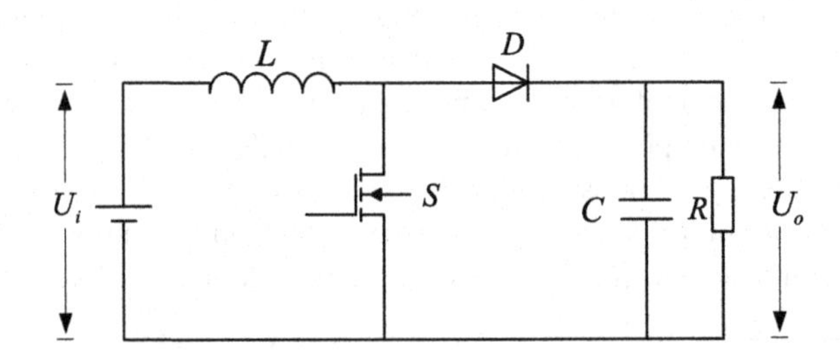

Fig. 3.1 The diagram of boost converter

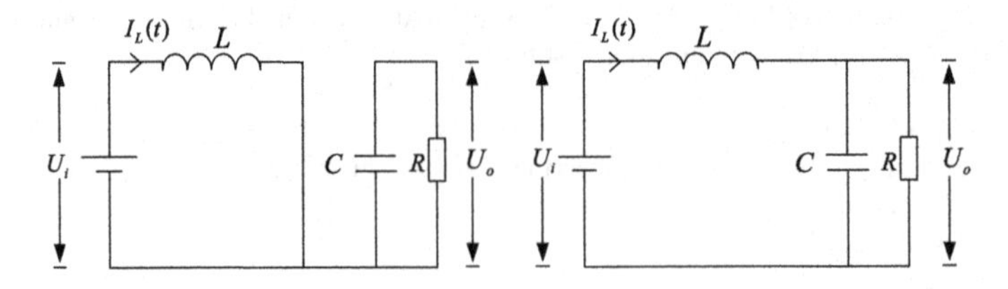

Fig. 3.2 Equivalent circuits when the switch opens and closes

When the switching frequency is high enough, the average mode of the boost converter can be expressed as [4]

$$\dot{x}_1 = -\frac{1}{L}[(1-u)x_2 - U_i] + d_1(t)$$
$$\dot{x}_2 = \frac{1}{C}[(1-u)x_1 - \frac{x_2}{R}] + d_2(t) \tag{3.1}$$

where $x_1 = i_L, x_2 = U_0$ are the inductive current and output voltage respectively, $u \in (0,1)$ is the control input and $d_1(t), d_2(t)$ are the lumped disturbances including system uncertainties and external disturbances. The objective of the boost converter is to control the output voltage U_o finite-time tracking the reference voltage U_{oref} under the following assumption:

Assumption 1 There exist positive constants d_1 and d_2 such that the continuous disturbances $d_1(t)$ and $d_2(t)$ satisfy the following condition almost everywhere

$$|\ddot{d}_1(t)| \leq d_1, |\dot{d}_2(t)| \leq d_2.$$

It is worth noting that when $U_0 = U_{0ref}$, the state x_1 will also be kept in a constant value x_{1ref} which can be calculated from (3.1) as

$$x_{1ref} = \frac{x_{2ref}^2}{U_i R}. \tag{3.2}$$

The control task of the paper can be rewritten as follows: design a controller u such that x_2 will track the reference signal x_{2ref} in finite time.

3.3 Controller Design

3.3.1 Coordinate Transformation

Introduce two new state variables as follows

$$z_1 = \frac{1}{2}Lx_1^2 + \frac{1}{2}Cx_2^2, \quad z_2 = U_i x_1 - \frac{x_2^2}{R}. \tag{3.3}$$

Then, based upon (3.3), a linear system (Brunovsky standard model) can be obtained

$$\dot{z}_1 = z_2 + \bar{d}_1(t), \quad \dot{z}_2 = v_1 + \bar{d}_2(t) \tag{3.4}$$

with

$$v_1 = \frac{U_i^2}{L} + \frac{2x_2^2}{R^2C} - (1-u)(\frac{U_i x_2}{L} + \frac{2x_1x_2}{RC})$$
$$\bar{d}_1(t) = Ld_1(t)x_1 + Cd_2(t)x_2, \;\; \bar{d}_2(t) = d_1(t)U_i + d_2(t)\frac{2x_2}{R}. \tag{3.5}$$

Generally speaking, the voltage and current of a boost converter are bounded. In this case, based on Assumption 1, we know that $\ddot{\bar{d}}_1(t), \dot{\bar{d}}_2(t)$ are also almost bounded everywhere. Thus, we let

$$|\ddot{\bar{d}}_1(t)| \leq \bar{d}_1, \;\; |\dot{\bar{d}}_2(t)| \leq \bar{d}_2 \tag{3.6}$$

with positive constants $\bar{d}_1$ and $\bar{d}_2$. According to (3.3), we define the reference signal as

$$z_{1ref} = \frac{1}{2}Lx_{1ref}^2 + \frac{1}{2}Cx_{2ref}^2, \;\; z_{2ref} = U_i x_{1ref} - \frac{x_{2ref}^2}{R}. \tag{3.7}$$

Substituting Eqs. (3.2) into (3.7) yields

$$z_{1ref} = \frac{1}{2}Lx_{1ref}^2 + \frac{1}{2}Cx_{2ref}^2, \;\; z_{2ref} = 0. \tag{3.8}$$

If we can design the virtual controller v_1 such that z_1 and z_2 will converge to z_{1ref} and z_{2ref}, respectively, then the following holds

$$z_1 = \frac{1}{2}Lx_1^2 + \frac{1}{2}Cx_2^2 = \frac{1}{2}Lx_{1ref}^2 + \frac{1}{2}Cx_{2ref}^2 = z_{1ref}$$
$$z_2 = U_i x_1 - \frac{x_2^2}{R} = U_i x_{1ref} - \frac{x_{2ref}^2}{R} = 0. \tag{3.9}$$

By a simple calculation, we obtain four groups of solutions from (3.9) as ($x_1 = \pm x_{1ref}$, $x_2 = \pm x_{2ref}$). Obviously, only a set of solution ($x_1 = x_{1ref}$, $x_2 = x_{2ref}$) is positive value and desired. The rest of the solution is not in conformity with the actual situation and should be abandoned.

Define $z_{1e} = z_1 - z_{1ref}, z_{2e} = z_2 - z_{2ref}$. By (3.4), the error dynamic system is obtained as

$$\dot{z}_{1e} = z_{2e} + \bar{d}_1(t), \;\; \dot{z}_{2e} = v_1 + \bar{d}_2(t). \tag{3.10}$$

It is clear that if the stabilization of system (3.10) is achieved in a finite time, then the output voltage U_0 can track the desired voltage U_{oref} in a finite time.

3.3.2 Composite Controller Design

In this subsection, based on TSM control technique and disturbance compensation, a composite controller will be constructed to render system (3.1) (i.e. (3.10)) finite-time stable.

A sliding mode surface for system (3.10) is first designed as [5]

$$s = \ddot{z}_{1e} + c_2 \dot{z}_{1e}^{\alpha_2} + c_1 z_{1e}^{\alpha_1} \tag{3.11}$$

where $c_1 > 0$, $c_2 > 0$, $\alpha_2 = \frac{2\alpha_1}{1+\alpha_1}$, $\alpha_1 \in (0, 1)$ is a ratio of positive integers, and $\ddot{z}_{1e}$ is the second time derivative of z_{1e}.

Once $s = 0$ is established, one has

$$\ddot{z}_{1e} = -c_2 \dot{z}_{1e}^{\alpha_2} - c_1 z_{1e}^{\alpha_1}. \tag{3.12}$$

Define $y_1 = z_{1e}$, $y_2 = \dot{z}_{1e}$, we obtain

$$\dot{y}_1 = y_2, \;\; \dot{y}_2 = -c_2 y_2^{\alpha_2} - c_1 y_1^{\alpha_1}. \tag{3.13}$$

According to [6], system (3.13) is globally finite-time stable which implies that the error $(z_{1e}, \dot{z}_{1e})$ will converge to the equilibrium point from any initial condition in a finite time [6].

Based on the sliding mode surface designed above, we have the following main result.

Theorem 1 *Considering system (3.10), if the TSM controller is designed as*

$$v_1 = -c_1 z_{1e}^{\alpha_1} - c_2 (z_{2e} + \varsigma_2)^{\alpha_2} + v_2, \;\; \dot{v}_2 = -k sign(s) \tag{3.14}$$

where k is a positive constant and ς_2 is generated by the following observer

$$\begin{aligned} \dot{\varsigma}_1 &= v_0 + z_{2e}, \; v_0 = -\lambda_1 \Gamma^{\frac{1}{2}} |\varsigma_1 - z_{1e}|^{\frac{1}{2}} sign(\varsigma_1 - z_{1e}) + \varsigma_2 \\ \dot{\varsigma}_2 &= -\lambda_0 \Gamma sign(\varsigma_2 - v_0) \end{aligned} \tag{3.15}$$

with Γ being a large positive constant and $0 < \lambda_0 < \lambda_1$, then the closed-loop system is finite-time stable.

Proof Firstly, we will prove that the disturbance can be estimated with the disturbance observer (3.15) in finite time.

Let $\sigma_0 = \frac{\varsigma_1 - z_{1e}}{\Gamma}$ and $\sigma_1 = \frac{\varsigma_2 - \bar{d}_1(t)}{\Gamma}$. It can be easily verified that

$$\dot{\sigma}_0 = -\lambda_1 |\sigma_0|^{1/2} sign(\sigma_0) + \sigma_1 \tag{3.16}$$

Similarly, we also obtain

$$\dot{\sigma}_1 = \frac{-\lambda_0 \Gamma sign(\varsigma_2 - v_0) - \dot{\bar{d}}_1(t)}{\Gamma} \tag{3.17}$$

Note that $\varsigma_2 - v_0 = \Gamma(\sigma_1 - \dot{\sigma}_0)$. This, together with (3.17), yields

$$\dot{\sigma}_1 = -\lambda_0 sign(\sigma_1 - \dot{\sigma}_0) - \frac{\dot{\bar{d}}_1(t)}{\Gamma} \tag{3.18}$$

By (3.4), we know the lumped disturbance $\bar{d}_1(t)$ has a Lipschitz constant Γ, which implies $|\dot{\bar{d}}_1(t)| \leq \Gamma$. Then, the following differential inclusion holds

$$\dot{\sigma}_1 \in -\lambda_0 sign(\sigma_1 - \dot{\sigma}_0) + [-1, 1] \tag{3.19}$$

which is understood in the Filippov sense [7].

Combining (3.16) and (3.19) together, we have the following differential inclusion

$$\dot{\sigma}_0 = -\lambda_1 |\sigma_0|^{1/2} sign(\sigma_0) + \sigma_1, \quad \dot{\sigma}_1 \in -\lambda_0 sign(\sigma_1 - \dot{\sigma}_0) + [-1, 1] \tag{3.20}$$

Then, by Theorem 6.4 in [8], we conclude that system the observer error σ_0, σ_1 can converge to zero in finite time. That's to say, we have $\varsigma_2 = \bar{d}_1(t)$ after a finite time T.

Design a sliding surface as (3.11), which can also be rewritten as

$$s = v_1 + \bar{d}_2(t) + \dot{\bar{d}}_1(t) + c_2(z_{2e} + \bar{d}_1(t))^{\alpha_2} + c_1 z_{1e}^{\alpha_1} \tag{3.21}$$

Substituting controller (3.14) into (3.21) yields

$$s = v_2 + \bar{d}_2(t) + \dot{\bar{d}}_1(t) + c_2(z_{2e} + \bar{d}_1(t))^{\alpha_2} - c_2(z_{2e} + \varsigma_2)^{\alpha_2} \tag{3.22}$$

Note that for $t \geq T$ we have $\varsigma_2 = \bar{d}_1(t)$. One obtains

$$s = v_2 + \bar{d}_2(t) + \dot{\bar{d}}_1(t). \tag{3.23}$$

From (3.14), we can obtain the time derivative of (3.23) as

$$\dot{s} = \dot{v}_2 + \dot{\bar{d}}_2(t) + \ddot{\bar{d}}_1(t) = -k \cdot sign(s) + \dot{\bar{d}}_2(t) + \ddot{\bar{d}}_1(t) \tag{3.24}$$

If there exists $\eta > 0$ such that

$$s\dot{s} \leq -\eta|s|, \qquad \forall |s| \neq 0 \tag{3.25}$$

then one has $s \equiv 0$ for $t \geq T$. Based on (3.6), we have

$$s\dot{s} = -k \cdot sign(s)s + \dot{\bar{d}}_2(t)s + \ddot{\bar{d}}_1(t)s \leq -k|s| + (\bar{d}_2 + \bar{d}_1)s \tag{3.26}$$

Let $k > \bar{d}_1 + \bar{d}_2$. By (3.26) and let $\eta = k - \bar{d}_1 - \bar{d}_2$, it can be concluded that (3.25) holds. This implies that the ideal sliding surface $s \equiv 0$ can be achieved in finite time T.

Next, we will prove that the observer errors will not drive the states and sliding surface to infinity over the time interval $[0, T]$.

Defining a Lyapunov function $V(z_{1e}, z_{2e}) = \frac{1}{2}z_{1e}^2 + \frac{1}{2}z_{2e}^2$, and taking the time derivative of $V(z_{1e}, z_{2e})$, one gets

$$\begin{aligned}\dot{V}(z_{1e}, z_{2e}) &= z_{1e}(z_{2e} + \bar{d}_1(t)) + z_{2e}(v_1 + \bar{d}_2(t)) \\ &\leq \frac{z_{1e}^2 + z_{2e}^2}{2} + \frac{|\bar{d}_1(t)|^2 + z_{1e}^2}{2} + \frac{z_{2e}^2 + 2(v_1^2 + \bar{d}_2^2(t))}{2}.\end{aligned} \tag{3.27}$$

In addition, we also have

$$\begin{aligned}v_1^2 &\leq (|c_1 z_{1e}^{\alpha_1}| + |c_2(z_{2e} + \varsigma_2)|^{\alpha_2} + |v_2|)^2 \\ &\leq 3c_1^2 z_{1e}^2 + 6c_2^2 z_{2e}^2 + 3c_1^2 + 3c_2^2 + 6c_2^2\varsigma_2^2 + 3v_2^2.\end{aligned} \tag{3.28}$$

Substituting (3.28) into (3.27) yields

$$\dot{V} \leq m_1(z_{1e}^2 + z_{2e}^2) + m_2 = 2m_1 V + m_2 \tag{3.29}$$

where $m_1 = \max\{1 + 3c_1^2, 1 + 6c_2^2\}$, $m_2 = \frac{|\bar{d}_1(t)|^2}{2} + \bar{d}_2^2(t) + 3c_1^2 + 3c_2^2 + 6c_2^2\varsigma_2^2 + 3v_2^2$. It is easy to show that m_1 and m_2 are bounded constants. We can finally conclude that all the signals of the system (3.10) are bounded for $\forall t \in [0, T]$. Hence, the proof of Theorem 1 is completed.

3.4 Simulation

In this section, we will show the effectiveness of the method proposed in this paper with a simulation example. The parameters of Boost converter are chosen as $R = 10\Omega$, $L = 0.004H$, $C = 0.03F$, $U_i = 8\,\text{V}$, $U_{0ref} = 12\,\text{V}$. Suppose the expected output voltage be 12 V and let the control parameters be $c_1 = 2$,$c_2 = 3$, $\alpha_1 = \frac{1}{9}$, $\alpha_2 = \frac{1}{5}$, $\lambda_0 = 2$, $\lambda_1 = 5$, $\Gamma = 1$ and $k = 100$. The disturbance are chosen as $d_1(t) = 2sin(2t)$, $d_2(t) = 2sin(2t)$. The initial state vector is chosen as $(x_1(0), x_2(0), u(0), v_2(0), \varsigma_1(0), \varsigma_2(0)) = (1, 8, 0.5, 1, 1, 0)$.

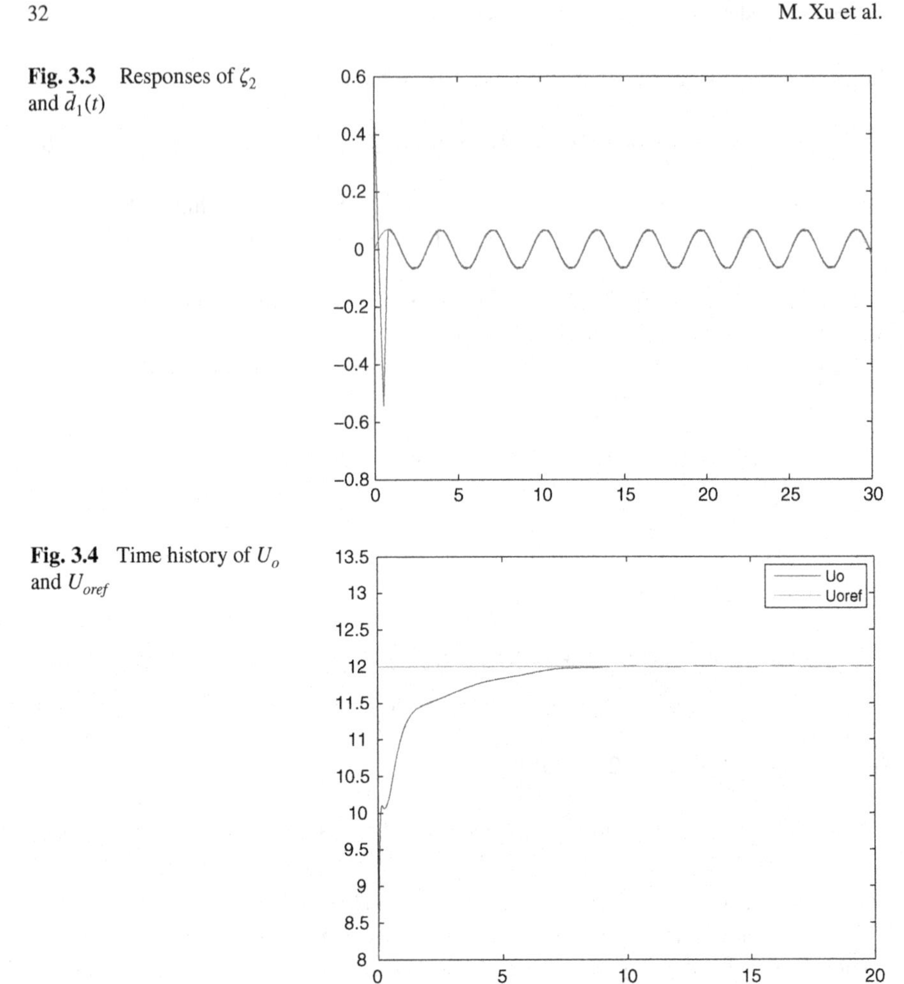

Fig. 3.3 Responses of ζ_2 and $\bar{d}_1(t)$

Fig. 3.4 Time history of U_o and U_{oref}

By Fig. 3.3, we can see clearly that the finite-time disturbance observer work well. In addition, Fig. 3.4 tells us that the desired reference voltage U_{0ref} can be tracked by the output voltage U_0. The simulation results indicate that the proposed TSM control scheme can efficiently control the output voltage of the Boost converter.

3.5 Conclusions

This paper studies the TSM control method for the boost converter. A finite-time disturbance observer is employed to estimate the disturbance. Based upon the disturbance observer, a continuous composite controller is designed to achieved the finite-time tracking of the reference value. A Lyapunov function is selected to verify that all the signals of closed-loop system are bounded for $t \in R^+$. Simulation results for boost converter have shown that superior tracking performance can be achieved by using the proposed controller.

References

1. Song TT, Chung HS (2008) Boundary control of boost converter using state-energy plane. IEEE Trans Power Electron 23(2):551–563
2. Chan CY (2007) A nonlinear control for DC-DC power converters. IEEE Trans Power Electron 22(1):216–222
3. Segaran D, McGrath BP, Holmes DG (2010) Adaptive dynamic control of a bi-directional DC-DC converter. In: IEEE Energy Conversion Congress and Exposition, pp 1442–1449 (2010)
4. Singh P, Purwar S (2012) Sliding mode controller for PWM based buck-boost DC/DC converter as state space averaging method in continuous conduction mode. In: International Conference on Powerr, Control and Embedded Systems, pp 1–5 (2012)
5. Feng Y, Hang F, Yu X (2014) Chattering free full-order sliding-mode control. Automatica 50(4):1310–1314
6. Bhat SP, Bernstein DS (2005) Geometric homogeneity with applications to finite-time stability. Math Control Signals Syst 17(2):101–127
7. Filippov AF (1998) Different Equations with Discontinuous Right-hand Side. Kluwer Academic Publisher, The Netherlands
8. Shtessel Y, Edwards C, Fridman L, Levant A (2013) Sliding Mode Control and Observation. Birkh*ä*user, Boston

Chapter 4
Research on Trajectory Tracking Strategy of Roadheader Cutting Head Using ILC

Fuzhong Wang, Ying Gao and Fukai Zhang

Abstract To realize the precision of cross-section shaping, iterative learning control (ILC) is studied in the trajectory tracking strategy of roadheader cutting head in this paper. The dynamic model of roadheader cutting arm is established, and then an appropriate PD-type iterative learning controller is designed. The simulation results show that the tracking error decreases with the increasing number of iterations, and the tracking curve approaches to the desired target trajectory gradually. The control method is effective for trajectory tracking, which meets the requirement of accuracy in the actual site. Moreover, the method lays a theoretical foundation for automatic cross-section shaping research.

Keywords Dynamic model · Trajectory tracking · PD-type ILC

4.1 Introduction

With the improvement of mechanization and intelligence in tunnels digging underground, the roadheader is applied more and more extensively, especially the boom-type roadheader with flexibility and high cutting power. However, the cutting head deviates from the predefined trajectory in the cutting process, since the presence of transverse and longitudinal cutting resistance and non-constant pressure caused by hydraulic cylinder leakage and so on. The phenomenon of over-excavation or under-excavation is appeared, which influences the quality of

---

F. Wang · Y. Gao (✉)
School of Electrical Engineering and Automation, Henan Polytechnic University, Jiaozuo 454000, China
e-mail: 1187220580@qq.com

F. Zhang
Mengjin Coal Mine, Yima Coal Industry Group Co Ltd., Mengjin 472300, China

Y. Jia et al. (eds.), *Proceedings of the 2015 Chinese Intelligent Systems Conference*, Lecture Notes in Electrical Engineering 360,
DOI 10.1007/978-3-662-48365-7_4

the cross-section shape directly. In order to improve the accuracy of automatic cutting control, intelligent control should be used in trajectory tracking strategy of roadheader cutting head.

There are few literatures about automatic cutting control of roadheader have been reported at present [1, 2]. The dynamic modeling of transverse swing is established for the cutting part in [3], and then the sliding mode control algorithm is used for cutting trajectory tracking. However, section cutting is compounded control combining with transverse and longitudinal motion, and it needs further study to track running trajectory more accurately.

Since the repeatability, nonlinearity, disturbance and uncertainty existing in the cutting arm control system, iterative learning control algorithm is put forward to studying the trajectory tracking strategy of roadheader cutting head [4]. ILC is an approach to achieve accurate control effect by suppressing these uncertainties. It takes the last tracking error as this input, until predetermined trajectory is achieved by the successive iterations [5]. Since the expansion amount of cutting arm is unadjusted, the horizontal and vertical swinging of roadheader cutting arm is studied here. According to the working principle of roadheader, the dynamic model of cutting arm is established using Lagrange equation, and then the PD-type iterative learning controller is designed. After the convergence of algorithm is verified, the validity of the algorithm is proved by simulation analysis.

4.2 Dynamic Model of Cutting Arm

The cutting arm is a fixed length, since it doesn't stretch in the cutting process. The simplified 3D-model of the cutting arm is shown in Fig. 4.1.

In Fig. 4.1, O_1 denotes the rotating joint point of horizontal turntable, O_2 denotes the panel point between lifting arm and turntable, the gravity point of turntable and cutting arm are A and B, the mass of them are m_1 and m_2, and the weight of them are G_1 and G_2, respectively. The turning angle of horizontal joint and vertical joint are θ_1 and θ_2, and θ_1 is the angle between the projection on $X_1O_1Y_1$ plane cast by the cutting arm and the direction of axial X_1, θ_2 is the angle between the cutting arm and the direction of axial X_2. The radius of turntable is r, O_1A equals a, O_2B equals l, and the total length of cutting arm is l_b.

The dynamic model of the cutting arm is given by the following steps, according to the Lagrange method [6].

(1) Determine the joint variables of cutting arm: q_1, q_2.

$$q_1 = \theta_1, \quad q_2 = \theta_2 \tag{4.1}$$

(2) Give the generalized torque F_1, F_2, which corresponds with q_1, q_2.

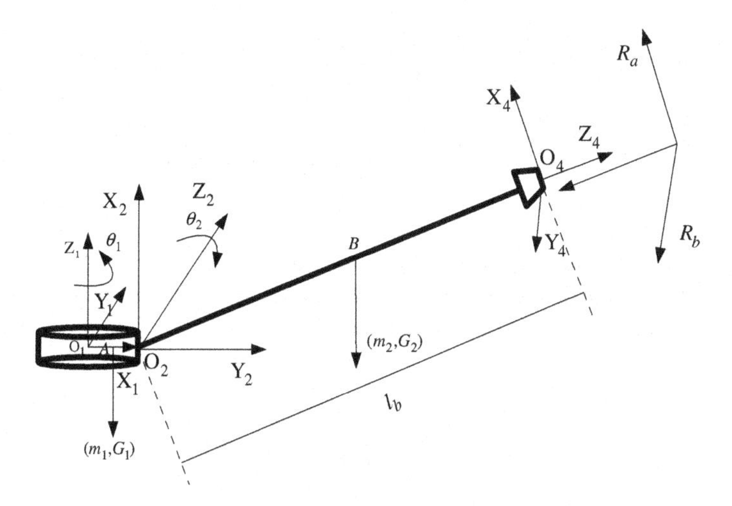

Fig. 4.1 Simplified 3D-model of cutting arm

The generalized torque F_1 of horizontal rotating joint 1 in cutting arm system is given as following:

$$F_1 = M_1 - k_b R_\mathrm{b} \tag{4.2}$$

where, M_1 denotes the driving torque of horizontal swing, R_b denotes the transverse cutting resistance, k_b denotes the resistance coefficient. The generalized torque F_2 of vertical swinging joint 2 is given as follows, when it swings upward:

$$F_2 = M_2 - m_2 g l \sin\theta_2 - k_a R_\mathrm{a} \tag{4.3}$$

where, M_2 denotes the driving torque of vertical swing, R_a denotes the longitudinal cutting resistance, k_a denotes the resistance coefficient.
(3) Obtain the kinetic energy and potential energy of each joint mechanism, and build the Lagrange function of cutting arm.
The kinetic energy of the cutting arm system is expressed as follows:

$$\begin{aligned} K &= 1/2\mathrm{m}_1(\dot{x}_1^2 + \dot{y}_1^2) + 1/2\mathrm{m}_2(\dot{x}_2^2 + \dot{y}_2^2 + \dot{z}_2^2) \\ &= 1/2\mathrm{m}_1 a\dot{\theta}_1^2 + 1/2\mathrm{m}_2[l^2\dot{\theta}_2^2 + r^2\dot{\theta}_1^2 + l^2\dot{\theta}_1^2 \sin\theta_2 {*} (r + l\sin\theta_2)] \end{aligned} \tag{4.4}$$

The horizontal plane of turntable's center is selected as the reference surface, which represents the plane of zero potential. The total potential energy of the cutting arm system can be expressed as:

$$P = \mathrm{m}_2 g l \cos\theta_2 \tag{4.5}$$

The Lagrange operator of system is obtained as follows:

$$\begin{aligned} L &= K - P \\ &= 1/2(\mathrm{m}_1 a\dot{\theta}_1^2 + \mathrm{m}_2 l^2\dot{\theta}_2^2 + \mathrm{m}_2 r^2\dot{\theta}_1^2 + \mathrm{m}_2 lr\dot{\theta}_1^2 \sin\theta_2 + \mathrm{m}_2 l^2\dot{\theta}_1^2 \sin\theta_2 - 2\mathrm{m}_2 gl\cos\theta_2) \end{aligned} \tag{4.6}$$

(4) Obtain the dynamic equation of the cutting arm system.

The relevant variables are substituted into the Lagrange equation. After a series of operations and simplifications, the differential motion equation of the system can be obtained:

$$\begin{cases} M_1 = D_{11}\ddot{\theta}_1 + D_{112}\dot{\theta}_1\dot{\theta}_2 + D_1 \\ M_2 = D_{22}\ddot{\theta}_2 + D_{211}\dot{\theta}_1^2 + D_2 \end{cases} \tag{4.7}$$

The elements in formula (4.7) are written as follows:

$$\begin{cases} D_{11} = \mathrm{m}_1 a + m_2 r^2 + m_2 lr\sin\theta_2 + m_2 l^2 \sin^2\theta_2 \\ D_{112} = m_2 lr\cos\theta_2 + m_2 l^2 \sin(2\theta_2) \\ D_1 = k_b R_b \\ D_{22} = m_2 l^2 \\ D_{211} = m_2 l^2 \sin\theta_2\cos\theta_2 - 1/2 m_2 rl\cos\theta_2 \\ D_2 = 2m_2 gl\sin\theta_2 + k_a R_a \end{cases} \tag{4.8}$$

The upward swing of vertical joint is discussed here, and the downward swing can be obtained by the same way. Therefore, the dynamic model (4.7) of cutting arm can be simplified in the following form:

$$M(\theta)\ddot{\theta} + C(\theta,\dot{\theta})\dot{\theta} + G(\theta) = U(t) \tag{4.9}$$

where, $\mathrm{M}(\theta) \in R^{2\times 2}$ denotes the inertia matrix of roadhead cutting arm, $C(\theta,\dot{\theta}) \in R^{2\times 1}$ expresses the matrix of Coriolis force and centrifugal force, $\mathrm{G}(\theta) \in R^{2\times 1}$ denotes the gravity. $U(t)$ denotes input torque, as input of control system. $\theta = \theta_1, \theta_2$, as the outputs, which represent the angular displacement of horizontal and vertical joint, respectively.

4.3 Design of Iterative Learning Controller

Design an appropriate PD-type iterative learning controller. The convergence of algorithm is proved. Firstly, the dynamic model (4.9) of the cutting arm system is transformed into the state equation.

Suppose

$$\begin{aligned} x_1(t) &= \theta, x_2(t) = \dot{\theta} \\ \mathrm{x}(t) &= [x_1(t) \quad x_2(t)]^T \end{aligned} \tag{4.10}$$

We have:

$$\begin{aligned} \dot{\mathrm{x}}(t) = \begin{bmatrix} \dot{\theta}_1 \\ \dot{\theta}_2 \end{bmatrix} = \begin{bmatrix} x_2(t) \\ -M^{-1}(x_1(t))[C(x_1(t), x_2(t))x_2(t) + G(x_1(t)) \end{bmatrix} + \begin{bmatrix} 0 \\ -M^{-1}(x_1(t)) \end{bmatrix} U(t) \\ \mathrm{y} = C(t)x(t) \end{aligned} \tag{4.11}$$

Suppose

$$\begin{aligned} f(x(t)) &= \begin{bmatrix} x_2(t) \\ -M^{-1}(x_1(t))[C(x_1(t), x_2(t))x_2(t) + G(x_1(t))] \end{bmatrix} \\ B(x(t)) &= \begin{bmatrix} 0 \\ J^{-1}(x_1(t)) \end{bmatrix}, C(t) = I \end{aligned} \tag{4.12}$$

Then formula (4.9) can be transformed into the state equation:

$$\begin{cases} \dot{x}(t) = f(x(t)) + B(x(t))U(t) \\ y = C(t)x(t) \end{cases} \tag{4.13}$$

According to the state Eq. (4.13) of the cutting arm system, the dynamic equations can be obtained when the system runs at the k th iteration:

$$\begin{cases} \dot{x}_k(t) = f(x_k(t)) + B(t)U_k(t) \\ y_k(t) = C(t)x_k(t) \end{cases} \tag{4.14}$$

At the kth iteration, the output error can be expressed as:

$$e_k(t) = y_\mathrm{d}(t) - y_k(t) \tag{4.15}$$

The closed-loop PD-type iterative learning control law is used in this paper, as the formula (4.16).

$$u_{k+1}(t) = u_k(t) + K_p(t)e_{k+1}(t) + K_d(\dot{t})\, e_{k+1}(t) \tag{4.16}$$

where, (x, y, u, f, g, C, I) denotes the vector or matrix with appropriate dimensions. x_d, y_d, u_d denotes the state, control output and control input of desired trajectory, K_p,K_d denotes the learning gain of P-type and D-type, respectively.

Since the dynamic equation of the controlled system is expressed as the formula (4.13), $t \in [0, T]$, which meets the following conditions:

(1) $f(x(t))$ satisfies the Lipschitz condition about x, that is, when $M > 0$, there is $\|f(x_1(t)) - f(t, x_2(t))\| \leq M\|x_1(t) - x_2(t)\|$, $\forall t \in [0, T]$, $\forall x_1, x_2 \in R^n$.
(2) The initial error $\tilde{x}_k(0)$ is a sequence converging to zero at each run.
(3) The existing control input $u_d(t)$ makes the system state and output to be the expected value.
(4) With $\dot{C}(t)$ existing in the system, $B(t), C(t), \dot{C}(t)$ is bounded.
(5) $(I + K_d(t)C(t)B(t))$ is reversible, $t \in [0, T]$.

Theorem 1 *Toward the initial control $u_0(t)$ given arbitrarily and the initial state $x_k(0)$ at each run, a closed loop PD-type iterative learning control law is taken as the learn law. If the sequence $x_k(t)|_{k\geq 0}, y_k(t)|_{k\geq 0}, u_k(t)|_{k\geq 0}$ uniform convergence to $x_d(t), y_d(t), u_d(t)$ about t, the sufficient conditions is that the spectral radius meets the following condition:*

$$\rho[(I + K_d(t)C(t)B(t))^{-1}] < 1, \forall t \in [0, T] \tag{4.17}$$

The necessary conditions:

$$\rho[(I + K_d(t)C(t)B(t))^{-1}]_{t=0} < 1 \tag{4.18}$$

The proof process of convergence can be seen in [7], and it is omitted here. We know the algorithm is convergent, which can be used in the trajectory tracking strategy of roadheader cutting head.

4.4 Simulation

For the cutting arm control system, M files are written to analysis and simulate the control algorithm by Matlab software. Considering the disturbance in system, the dynamic system model (4.9) of roadheader's cutting arm is written as:

$$M(\theta)\dot{\theta} + C(\theta, \dot{\theta})\dot{\theta} + G(\theta) = U(t) - \mathrm{D}(t) \tag{4.19}$$

where, $D(t)$ denotes the disturbance in system.

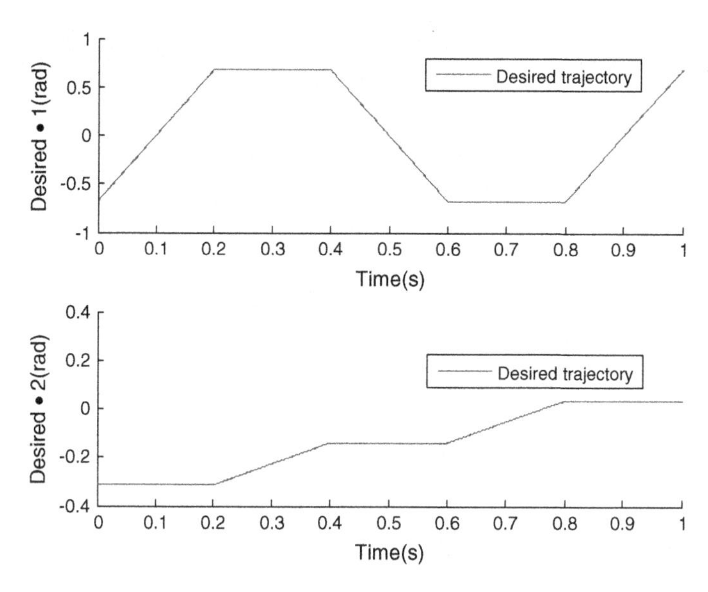

Fig. 4.2 Cutting arm joint angle desired input curve

The cutting arm swings from the bottom gradually, when it swings vertically in simulation process. According to the actual parameters of roadheader, the relevant parameters are set as follows:

$$m_1 = 5, m_2 = 15, r = 0.06, a = 0.05, g = 9.8, R_b = 1$$
$$k_a = l_b = 0.2, R_a = 1, k_b = 0.4 + 0.2\sin(\theta_2)$$

The desired trajectory is shown in Fig. 4.2. The initial state of the system is assumed as $\theta(0) = [\,-0.68 \quad 6.8\,]^T, \dot{\theta}(0) = [\,-0.314 \quad 0\,]^T$. $D(t) = [\,d_1 \quad d_2\,]^T$ denote interference signals, as the sum of random and repetitive disturbance signal. In the condition of meeting the theorem, $K_p = 10E_{2\times 2}, K_d = 10E_{2\times 2}$ is selected by trial and error method in simulation, and the time interval is $t \in [\,0 \quad 1\,]$.

The total number of iterations is set as 30 times, and the execution time is 1 at each experiment. The interference is assumed as the periodic disturbance, that is, $\mathrm{D}(t) = [\,2^* \sin(t) \quad 2^* \cos(t)\,]^T$. The results of simulation are shown as follows.

The desired trajectory of joint angles about horizontal and longitudinal swing is given in Fig. 4.2, respectively. The corresponding tracking curves are shown in Fig. 4.3, and the maximum errors of horizontal and longitudinal joint angle are seen in Fig. 4.4. The simulation results indicate that the PD-type iterative learning controller completes the task of tracking desired trajectory accurately when the

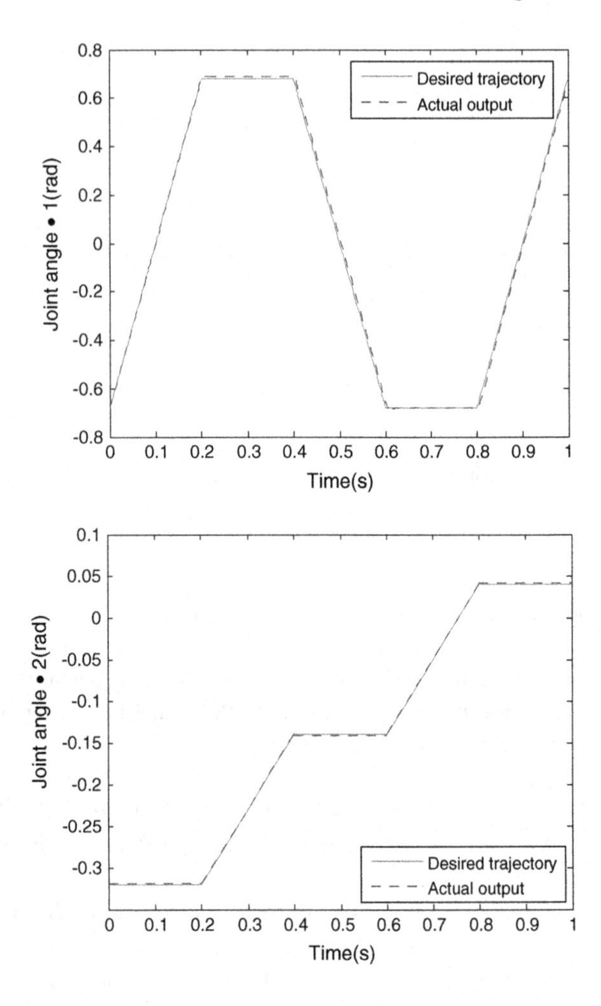

Fig. 4.3 Cutting arm joint angle trajectory tracking curve

repeated interference exists in the system. The tracking error decreases with the increasing number of iterations, when the tracking curves approach to the desired target trajectory gradually. After the stability is achieved, the track error of horizontal joint angle is less than 0.0005 rad (that is 0.3°), and the error of longitudinal joint angle tends to 0 after the 4th iteration. Since the permitted error is about 7.5 % in the manual cutting field, which is about 4.5° converted into the error of joint angle. Therefore, the tracking error of less than 0.3° meets the demand of precision, which means that the cutting arm can track the desired trajectory accurately.

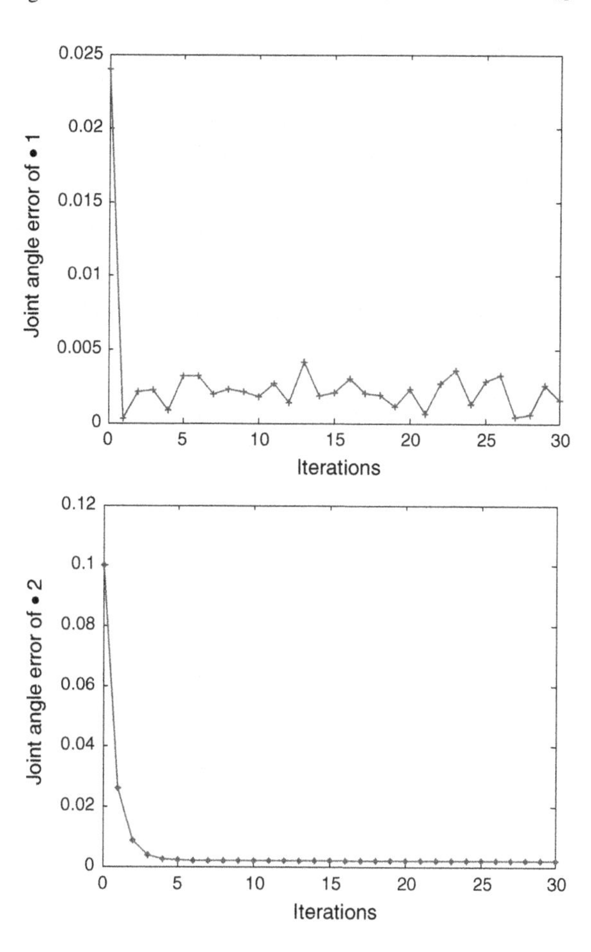

Fig. 4.4 Cutting arm joint angle learning error curve

4.5 Conclusion

By analysing dynamic characteristics of roadhead cutting arm, the dynamics model is established firstly. The PD-type iterative learning controller is studied in the trajectory tracking strategy of roadheader cutting head. According to the simulation results, the PD-type iterative learning control law can track target trajectory perfectly. The tracking error decreases with the increasing number of iterations, when tracking curves approach to the desired target trajectory gradually. The research results indicate that the method of ILC achieves the precise track of cutting trajectory, since the errors meet the requirements of accuracy in practice. It provides an effective method for the automatic cross-section shaping and has an important guiding significance for the practical application. However, there are still tracking errors in the simulation results, the aperiodic and uncertain disturbance factors existed in cutting arm control system is not considered as well. Hence, the further study is necessary.

References

1. Xuecheng Wang, Weiguo Zhang, Yinglin Liu (2010) Initial analysis on present state and recent advance of roadheader [J]. Coal Mine Mach 08:1–3
2. Fukai Zhang, Fuzhong Wang, Qinghua Gao (2014) Cutting arm trajectory control of roadheader based on adaptive iterative learning[J]. J Electron Measure Instrum 12:1355–1362
3. Rui Liang, Yi Chai, Dajie Li (2009) Tunneling cutting robot arm modeling and control [J]. Syst Simul 23:7601–7604
4. Yu G, Weili H (2004) Trajectory tracking of wheeled robot based on iterative learning control [C]. In: Proceedings of the international conference of intelligent mechatronics and automation, pp 228–232
5. Jiufang P, Hai W, Dezhang X (2012) Tracking control of mobile robot based on iterative learning control[J]. Comput Eng Appl 48(9): 222–225
6. Wei Huo (2005) Robot dynamics and control[M]. Higher Education Press, Beijing
7. Xiaofeng Y, Xiaoping F, Shengyue Y (2002) Open-closed loop pd-type iterative learning control for nonlinear system and its application in robot. [J] J Changsha Railway Univ (1):78–84

Chapter 5
Containment Control of Second-Order Multi-Agent Systems with Jointly-Connected Topologies and Varying Delays

Fuyong Wang, Hongyong Yang and Fuyun Han

Abstract The containment problem of second-order multi-agent systems with varying delays is investigated. Supposing system topology is dynamic changed and jointly-connected, the control algorithm of double-integrator systems with multiple leaders is proposed. The stability of the control algorithm is analyzed on Lyapunov-Krasovskii method. Finally, a simulation example is provided to prove the effectiveness of the conclusion.

Keywords Containment control · Second-order multi-agent systems · Multiple leaders · Varying delays · Jointly-connected

5.1 Introduction

Recently, distributed cooperative control of multi agent systems gradually becomes a hot topic of researched in the domains of control theory, applied mathematics, computer science, etc. Containment control has been paid much attention as a kind of distributed cooperative control problem with multiple leaders, which aims to allow followers eventually converge to a target area by designing a control protocol of follower.

In [1], by regarding the collection of leaders as a virtual node, the distributed control of first-order system with multiple leaders can be achieved when network topology is connected. In [2], a containment control algorithm suitable for any finite dimensional state vector is proposed with LaSalle's invariance principle in switching system. In [3], double-integrator systems with multiple stationary leaders and mul-

F. Wang · H. Yang (✉) · F. Han
School of Information and Electrical Engineering, Ludong University,
Yantai 264025, China
e-mail: hyyang@yeah.net

Y. Jia et al. (eds.), *Proceedings of the 2015 Chinese Intelligent Systems Conference*, Lecture Notes in Electrical Engineering 360,
DOI 10.1007/978-3-662-48365-7_5

tiple dynamic leaders are investigated. In [4], the containment problem of multi agent systems in a noisy communication spaces is investigated. When system topology is not connected, it is important significance to investigate the distributed control problem of intelligent systems with multiple leaders and communication delays.

In this paper, a distributed containment problem of double-integrator dynamic systems with varying delays and switching topologies is studied. The containment control algorithm is presented on the supposition that the communication network is jointly-connected, and the stability of the algorithm is analyzed based on the Lyapunov-Krasovskii approach.

5.2 Preliminaries

The communication topology of the system in this paper is modeled by an undirected graph $G=(W,\omega)$, where $W=\{w_1,w_2,\ldots,w_n\}$ is the node's set, $\omega=\{(w_i,w_j): w_i,w_j\in W\}$ is the edge's set. Let $N_i=\{w_j\in W|(w_i,w_j)\in\omega,\ j\neq i\}$ is the set of the neighbors of node w_i. The union of a collection of graphs $G_1,G_2,\ldots,G_m$, each with note set W, is defined as the graph G_{1-m} with the note set W and edge set equaling the union of the edge sets of all the graphs. We say $G_1,G_2,\ldots,G_m$ is jointly-connected if G_{1-m} is a connected graph [5].

The weighted adjacency matrix $A=[a_{ij}]\in\mathbb{R}^{n\times n}$ of undirected graph G satisfying $a_{ij}\geq 0$, and $a_{ij}>0$ if $(w_i,w_j)\in\omega$, $a_{ij}=0$ otherwise. The *Laplacian* matrix $L=[l_{ij}]\in\mathbb{R}^{n\times n}$ of undirected graph G is defined as $l_{ii}=\sum_{j\in N_i} a_{ij}$ and $l_{ij}=-a_{ij}$ for $i\neq j$.

Consider a nonempty sequence of bounded, infinite and contiguous time-intervals $[t_s,t_{s+1})$, $s=1,2,\ldots,$ with $t_1=0$ and $t_{s+1}-t_s\leq T_a$ for some constant $T_a>0$. Suppose that there is a finite sequence of non-overlapping subintervals $[t_{s,r},t_{s,r+1})\in[t_s,t_{s+1})$, $r=1,2,\ldots,m_s$, where $t_{s,1}=t_s$, $t_{s,m_s+1}=t_{s+1}$, $t_{s,r+1}-t_{s,r}\geq T_b$ for some integer $m_s\geq 1$ and some constant $T_b>0$. Suppose that the system topology described by $G_{s,r}$ changes at $t_{s,r}$ and it remains unchanged during each subinterval $[t_{s,r},t_{s,r+1})$. Let $\delta(t)$: $[0,+\infty)\to\Gamma$, $\Gamma=\{1,2,\ldots,N\}$ (N denotes the total number of all possible topologies), be a piecewise constant switching function.

5.3 Main Results

Considering second-order system of m leaders and n following agents, and the followers have the following dynamics

$$\begin{aligned} \dot{q}_i(t)&=p_i(t-\tau(t)),\\ \dot{p}_i(t)&=u_i(t-\tau(t)),\quad i=1,\ldots,n. \end{aligned}\tag{5.1}$$

where $q_i(t) \in \mathbb{R}$ is the position vector, $p_i(t) \in \mathbb{R}$ is the velocity vector, and $u_i(t) \in \mathbb{R}$ is the control input of follower. $\tau(t)$ is the time-varying communication delay. The leaders have the following dynamics

$$\begin{aligned} \dot{q}_{0,l}(t) &= p_{0,l}, \\ \dot{p}_{0,l} &= 0, \quad l = 1, \ldots, m. \end{aligned} \tag{5.2}$$

where $q_{0,l}(t) \in \mathbb{R}$ is the position vector, and $p_{0,l} \in \mathbb{R}$ is the velocity vector of leader. Consider the control protocol of double-integrator systems as follows

$$\begin{aligned} u_i(t) = {} & \alpha\{\sum_{j \in N_i} a_{ij}[q_j(t) - q_i(t)] + \sum_{k=1}^{m} b_{i,k}[q_{0,k}(t) - q_i(t)]\} \\ & + \beta\{\sum_{j \in N_i} a_{ij}[p_j(t) - p_i(t)] + \sum_{k=1}^{m} b_{i,k}[p_{0,k} - p_i(t)]\}, \quad i = 1, 2, \ldots, n. \end{aligned} \tag{5.3}$$

where $\alpha > 0$, $\beta > 0$, and $b_{i,k} > 0$ if follower i is linked to leader $k (k = 1, \cdots, m)$, $b_{i,k} = 0$, otherwise.

Assumption 1 The time-varying communication delay $\tau(t)$ in the multi-agent system (5.1) is bounded, i.e., there exists $h > 0$ satisfying: $0 \le \tau(t) < h$, $t \ge 0$.

Assumption 2 The communication topologies generated by n followers and m leaders, in each interval $[t_s, t_{s+1})$, $s = 1, 2, \ldots,$ are jointly connected.

Definition 1 *([6])*. Let $\chi = \{x_1, x_2, \ldots, x_m\} \subseteq \mathbb{R}$ be a real vector set, and let $CO(\chi)$ is a convex hull of the set χ, where $CO(\chi) = \{\sum_{i=1}^{m} \beta_i x_i | x_i \in \chi, \beta_i \ge 0, \sum_{i=1}^{m} \beta_i = 1\}$.

From (5.1) to (5.3), the system can be written as

$$\begin{aligned} \dot{q}(t) &= p(t - \tau(t)) \\ \dot{p}(t) &= \alpha[-Hq(t - \tau(t)) + B(I_m \otimes 1_n) q_0(t - \tau(t))] \\ &\quad + \beta[-Hp(t - \tau(t)) + B(I_m \otimes 1_n) p_0] \end{aligned} \tag{5.4}$$

where $q(t) = [q_1(t), \ldots, q_n(t)]^{\mathrm{T}}$, $q_0(t) = [q_{0,1}(t), \ldots, q_{0,m}(t)]^{\mathrm{T}}$, $p_0 = [p_{0,1}, \ldots, p_{0,m}]^{\mathrm{T}}$, $p(t) = [p_1(t), \ldots, p_n(t)]^{\mathrm{T}}$, and $H = L + B(1_m \otimes I_n)$, with $\otimes$ being the Kronecker product. L is the *Laplacian* matrix consisted of n following agents, $B = [B_1, \ldots, B_m] \in \mathbb{R}^{n \times nm}$, $B_k = diag\{b_{1,k}, b_{2,k}, \ldots, b_{n,k}\} \in \mathbb{R}^{n \times n}$, where $b_{i,k} \ge 0$, $i = 1, 2, \ldots, n$, $k = 1, 2, \ldots, m$.

Lemma 1 *Let* $q_0(t) = [q_{0,1}(t), q_{0,2}(t), \ldots, q_{0,m}(t)]^{\mathrm{T}}$, $q(t) = [q_1(t), q_2(t), \ldots, q_n(t)]^{\mathrm{T}}$, *if* $q(t) \to H^{-1}B(I_m \otimes 1_n) q_0(t)$, *then the distributed control of systems with multiple leaders can be achieved.*

Proof From [1] and Definition 5.1, we can see $H^{-1}B(I_m \otimes 1_n)$ is a stochastic matrix, and $H^{-1}B(I_m \otimes 1_n) q_0(t)$ is a convex hull formed by those dynamic leaders.

Thus, if $q(t) \to H^{-1}B(I_m \otimes 1_n)q_0(t)$, then the conclusion of this paper can be achieved. □

Let $\tilde{q}(t) = q(t) - H^{-1}B(I_m \otimes 1_n)q_0(t)$, $\tilde{p}(t) = p(t) - H^{-1}B(I_m \otimes 1_n)p_0$, Eq. (5.4) can be transformed into the following dynamics

$$\begin{aligned} \dot{\tilde{q}}(t) &= \tilde{p}(t - \tau(t)) \\ \dot{\tilde{p}}(t) &= -\alpha H\tilde{q}(t - \tau(t)) - \beta H\tilde{p}(t - \tau(t)) \end{aligned} \tag{5.5}$$

From (5.5), the system can be transformed into the following dynamics

$$\dot{x}(t) = -Fx(t - \tau(t)) \tag{5.6}$$

where $x(t) = (\tilde{q}^{\mathrm{T}}(t), \tilde{p}^{\mathrm{T}}(t))^{\mathrm{T}}$, $F = \begin{bmatrix} 0 & -I_n \\ \alpha H & \beta H \end{bmatrix}$.

Suppose the communication topology $\bar{G}_\delta$ with n following agents and m leaders on subinterval $[t_{s,r}, t_{s,r+1})$ has $n_\delta \geq 1$ connected subgraphs $\bar{G}^i_\delta$, $i = 1, 2, \cdots, n_\delta$, and each connected subgraph $\bar{G}^i_\delta$ has $d^i_\delta = d^i_{\delta F} + d^i_{\delta L}$ nodes, where $d^i_{\delta F} \geq 1$ represents the number of followers, $d^i_{\delta L} \geq 1$ represents the number of leaders. The subgraph G^i_δ is composed of $d^i_{\delta F}$ followers, and its *Laplacian* matrix is denoted by L^i_δ. Then there exists a permutation matrix $E_\delta \in \mathbb{R}^{n \times n}$ such that

$$E^{\mathrm{T}}_\delta L_\delta E_\delta = diag\{L^1_\delta, L^2_\delta, \ldots, L^{n_\delta}_\delta\} \tag{5.7}$$

$$E^{\mathrm{T}}_\delta B_\delta E_\delta = diag\{B^1_\delta, B^2_\delta, \ldots, B^{n_\delta}_\delta\} \tag{5.8}$$

$$E^{\mathrm{T}}_\delta I_n E_\delta = diag\{I_{d^1_\delta}, I_{d^2_\delta}, \ldots, I_{d^{n_\delta}_\delta}\} \tag{5.9}$$

$$\tilde{q}^{\mathrm{T}}(t)E_\delta = [\tilde{q}^{1\mathrm{T}}_\delta(t), \tilde{q}^{2\mathrm{T}}_\delta(t), \ldots, \tilde{q}^{n_\delta \mathrm{T}}_\delta(t)] \tag{5.10}$$

$$\tilde{p}^{\mathrm{T}}(t)E_\delta = [\tilde{p}^{1\mathrm{T}}_\delta(t), \tilde{p}^{2\mathrm{T}}_\delta(t), \ldots, \tilde{p}^{n_\delta \mathrm{T}}_\delta(t)] \tag{5.11}$$

$$q^{\mathrm{T}}_0 E_\delta = [q^{1^{\mathrm{T}}}_0, q^{2^{\mathrm{T}}}_0, \ldots, q^{n^{\mathrm{T}}_\delta}_0] \tag{5.12}$$

$$p^{\mathrm{T}}_0 E_\delta = [p^{1^{\mathrm{T}}}_0, p^{2^{\mathrm{T}}}_0, \ldots, p^{n^{\mathrm{T}}_\delta}_0] \tag{5.13}$$

$$x^{\mathrm{T}}(t)E_\delta = [x^{1\mathrm{T}}_\delta(t), x^{2\mathrm{T}}_\delta(t), \cdots, x^{n_\delta \mathrm{T}}_\delta(t)] \tag{5.14}$$

Assumption 3 The graph $\bar{G}^i_\delta$ of system is connected, it means that at one agent in each component of G^i_δ is connected to the virtual node formed by leaders.

According to the above described, under the dynamic switching topology (5.6) can be transformed into the following dynamics

$$\dot{x}(t) = -F_{\delta}x(t-\tau(t)) \tag{5.15}$$

The system dynamics (5.15) can be written as the following subsystems dynamic in each subinterval $[t_{s,r}, t_{s,r+1})$:

$$\dot{x}_{\delta}^{i}(t) = -F_{\delta}^{i}x_{\delta}^{i}(t-\tau(t)), \quad i = 1, 2, \ldots, n_{\delta} \tag{5.16}$$

where $x_{\delta}^{i}(t) = (\tilde{q}_{\delta}^{iT}(t), \tilde{p}_{\delta}^{iT}(t))^{T}$, $\tilde{q}_{\delta}^{i}(t) = q_{\delta}^{i}(t) - H_{\delta}^{i-1}B_{\delta}^{i}(1_{d_{\delta L}^{i}} \otimes I_{d_{\delta F}^{i}})q_{0,\delta}^{i}(t)$, $\tilde{p}_{\delta}^{i}(t) = p_{\delta}^{i}(t) - H_{\delta}^{i-1}B_{\delta}^{i}(1_{d_{\delta L}^{i}} \otimes I_{d_{\delta F}^{i}})p_{0}^{i}$, $q_{\delta}^{i}(t) = [q_{\delta 1}^{i}(t), q_{\delta 2}^{i}(t), \ldots, q_{\delta d_{\delta F}^{i}}^{i}(t)]^{T}$, $p_{\delta}^{i}(t) = [p_{\delta 1}^{i}(t), p_{\delta 2}^{i}(t), \ldots, p_{\delta d_{\delta F}^{i}}^{i}(t)]^{T}$, $q_{0,\delta}^{i}(t) = [q_{0,\delta 1}^{i}(t), q_{0,\delta 2}^{i}(t), \ldots, q_{0,\delta d_{\sigma L}^{i}}^{i}(t)]^{T}$, $p_{0}^{i} = [p_{0,1}^{i}, p_{0,2}^{i}, \ldots, p_{0,d_{\delta L}^{i}}^{i}]^{T}$, $F_{\delta}^{i} = \begin{bmatrix} 0 & -I_{d_{\delta F}^{i}} \\ \alpha H_{\delta}^{i} & \beta H_{\delta}^{i} \end{bmatrix}$, $H_{\delta}^{i} = L_{\delta}^{i} + B_{\delta}^{i}(1_{d_{\delta L}^{i}} \otimes I_{d_{\delta F}^{i}})$, $B_{\delta}^{i} = [B_{\delta 1}^{i}, \ldots, B_{\delta d_{\delta L}^{i}}^{i}] \in \mathbb{R}^{d_{\delta F}^{i} \times d_{\delta F}^{i} d_{\delta L}^{i}}$, $B_{\delta q}^{i} = diag\{b_{1,q}^{i}, b_{2,q}^{i}, \ldots, b_{d_{\delta F}^{i},q}^{i}\} \in \mathbb{R}^{d_{\delta F}^{i} \times d_{\delta F}^{i}}$, $b_{p,q}^{i} \geq 0$, $p = 1, 2, \ldots, d_{\delta F}^{i}$, $q = 1, 2, \ldots, d_{\delta L}^{i}$, $i = 1, 2, \ldots, n_{\delta}$.

Theorem 1 *Consider second-order dynamic systems of n following agents with dynamics (1) and m dynamic leaders with dynamics (2) under switching topologies. Suppose Assumption 1, Assumption 2 and Assumption 3 holds, the containment control of double-integrator systems with time-varying delays can be solved by using control protocol (3), if for any time* $t \in (0, +\infty)$,

$$h < \frac{2}{\lambda_{n}(H_{\delta}) \times (\beta + \sqrt{\beta^{2} - 4\alpha^{2}})}, \beta > 2\alpha \tag{5.17}$$

where $\lambda_{n}(H_{\delta})$ is the maximum eigenvalue of H_{δ}.

Proof Construct a Lyapunov-Krasovskii function for dynamics (5.15) as follows

$$V(t) = x^{T}(t)x(t) + \int_{t-h}^{t}(s-t+h)\dot{x}^{T}(s)\dot{x}(s)ds \tag{5.18}$$

From (5.14), the function $V(t)$ can be transformed into

$$V(t) = \sum_{i=1}^{n_{\delta}}\{x_{\delta}^{iT}(t)x_{\delta}^{i}(t) + \int_{t-h}^{t}(s-t+h)\dot{x}_{\delta}^{iT}(s)\dot{x}_{\delta}^{i}(s)ds\} \tag{5.19}$$

Calculating the differentiation of $V(t)$ along the trajectory of the solution of system (5.16)

$$\begin{aligned} \dot{V}(t) = \sum_{i=1}^{n_{\delta}}\{ & -x_{\delta}^{iT}(t-\tau(t))F_{\delta}^{iT}x_{\delta}^{i}(t) - x_{\delta}^{iT}(t)F_{\delta}^{i}x_{\delta}^{i}(t-\tau(t)) \\ & + hx_{\delta}^{iT}(t-\tau(t))F_{\delta}^{iT}F_{\delta}^{i}x_{\delta}^{i}(t-\tau(t)) - \int_{t-h}^{t}\dot{x}_{\delta}^{iT}(s)\dot{x}_{\delta}^{i}(s)ds\} \end{aligned} \tag{5.20}$$

According to [8] and Assumption 1, we can get

$$-\int_{t-\tau(t)}^{t} \dot{x}_{\delta}^{iT}(s)\dot{x}_{\delta}^{i}(s)ds \le -\frac{1}{h}[x_{\delta}^{i}(t)-x_{\delta}^{i}(t-\tau(t))]^{T}[x_{\delta}^{i}(t)-x_{\delta}^{i}(t-\tau(t))] \quad (5.21)$$

From (5.21), we have

$$\begin{aligned}\dot{V}(t) \le & \sum_{i=1}^{n_\delta}\{-x_{\delta}^{iT}(t-\tau(t))F_{\delta}^{iT}x_{\delta}^{i}(t)-x_{\delta}^{iT}(t)F_{\delta}^{i}x_{\delta}^{i}(t-\tau(t))+hx_{\delta}^{iT}(t-\tau(t))\times \\ & F_{\delta}^{iT}F_{\delta}^{i}x_{\delta}^{i}(t-\tau(t))-\frac{1}{h}[x_{\delta}^{i}(t)-x_{\delta}^{i}(t-\tau(t))]^{T}[x_{\delta}^{i}(t)-x_{\delta}^{i}(t-\tau(t))]\} \\ = & -\sum_{i=1}^{n_\delta} y_{\delta}^{iT}\Xi_{\delta}^{i}y_{\delta}^{i}\end{aligned}$$

where $y_{\delta}^{i}=[x_{\delta}^{iT}(t),x_{\delta}^{iT}(t-\tau(t))]^{T}$, $\Xi_{\delta}^{i}=\begin{bmatrix}\frac{1}{h}I_{d_{\delta F}^{i}} & F_{\delta}^{i}-\frac{1}{h}I_{d_{\delta F}^{i}} \\ F_{\delta}^{iT}-\frac{1}{h}I_{d_{\delta F}^{i}} & \frac{1}{h}I_{d_{\delta F}^{i}}-hF_{\delta}^{iT}F_{\delta}^{i}\end{bmatrix}$.

Next, we will prove Ξ_{δ}^{i} is positive definite. According to [9], since $\frac{1}{h}>0$, we will just prove $\Omega_{\delta}^{i}=(\frac{1}{h}I_{d_{\delta F}^{i}}-hF_{\delta}^{iT}F_{\delta}^{i})-(F_{\delta}^{iT}-\frac{1}{h}I_{d_{\delta F}^{i}})\times hI_{d_{\delta F}^{i}}\times(F_{\delta}^{i}-\frac{1}{h}I_{d_{\delta F}^{i}})= \; -2hF_{\delta}^{iT}F_{\delta}^{i}+(F_{\delta}^{i}+F_{\delta}^{iT})$ is positive definite, where $F_{\delta}^{i}=\begin{bmatrix}0 & -I_{d_{\delta F}^{i}} \\ \alpha H_{\delta}^{i} & \beta H_{\delta}^{i}\end{bmatrix}$, $\lambda_{g}(F_{\delta}^{i})=\frac{(\beta\pm\sqrt{\beta^{2}-4\alpha^{2}})}{2}\times\lambda_{g}(H_{\delta}^{i})$. Since Ω_{δ}^{i} is symmetric, these eigenvalues of Ω_{δ}^{i} are real, If $-2h\lambda_{g}^{2}(F_{\delta}^{i})+2\lambda_{g}(F_{\delta}^{i})>0$, i.e., $\lambda_{g}(H_{\delta}^{i})<\frac{2}{(\beta+\sqrt{\beta^{2}-4\alpha^{2}})h}$ and $\beta>2\alpha$, then Ω_{δ}^{i} is positive definite. Therefore, if Eq. (5.17) holds, Ξ_{δ}^{i} is positive definite, i.e., $\dot{V}(t)<0$. The dynamics (5.15) is gradually stable.

Then by [10], we have $\lim_{t\to+\infty}(-\sum_{i=1}^{n_\delta}y_{\delta}^{iT}\Xi_{\delta}^{i}y_{\delta}^{i})=0$. Since Ξ_{δ}^{i} is positive definite, we have $y_{\delta}^{i}=0$ as $t\to+\infty$, then $q_{\delta}^{i}(t)=q_{\delta}^{i}(t-\tau(t))=H_{\delta}^{i^{-1}}B_{\delta}^{i}(I_{d_{\delta L}^{i}}\otimes 1_{d_{\delta F}^{i}})q_{0}^{i}(t)$, $p_{\delta}^{i}(t)=p_{\delta}^{i}(t-\tau(t))=H_{\delta}^{i^{-1}}B_{\delta}^{i}(I_{d_{\delta L}^{i}}\otimes 1_{d_{\delta F}^{i}})p_{0}^{i}$.

Furthermore, in any subinterval $[t_{s,r},t_{s,r+1})\subset[t_{s},t_{s+1})$, $r=1,2,\ldots,m_{s}$, $\lim_{t\to+\infty}q_{\delta}^{i}(t)=\lim_{t\to+\infty}q_{\delta}^{i}(t-\tau(t))=H_{\delta}^{i^{-1}}B_{\delta}^{i}(I_{d_{\delta L}^{i}}\otimes 1_{d_{\delta F}^{i}})q_{0}^{i}(t)$, $i=1,2,\cdots,n_{\delta}$. Therefore, in the connected portion of intelligent systems on the subinterval $[t_{s,r},t_{s,r+1})$ and $[t_{s,r+1},t_{s,r+2})$, i.e., in the subinterval $[t_{s,r},t_{s,r+2})$, $\lim_{t\to+\infty}q_{\delta}^{i'}(t)=\lim_{t\to+\infty}q_{\delta}^{i'}(t-\tau(t))=H_{\delta}^{i^{-1'}}B_{\delta}^{i'}(I_{d_{\delta L}^{i}}\otimes 1_{d_{\delta F}^{i}})q_{0}^{i'}(t)$ still holds, $i=1,2,\ldots,n_{\delta}^{'}$, $t\to\infty$. Then by induction, According to Assumption 2, $\lim_{t\to+\infty}q(t)=\lim_{t\to+\infty}q(t-\tau(t))=H^{-1}B(I_{m}\otimes 1_{n})q_{0}(t)$. □

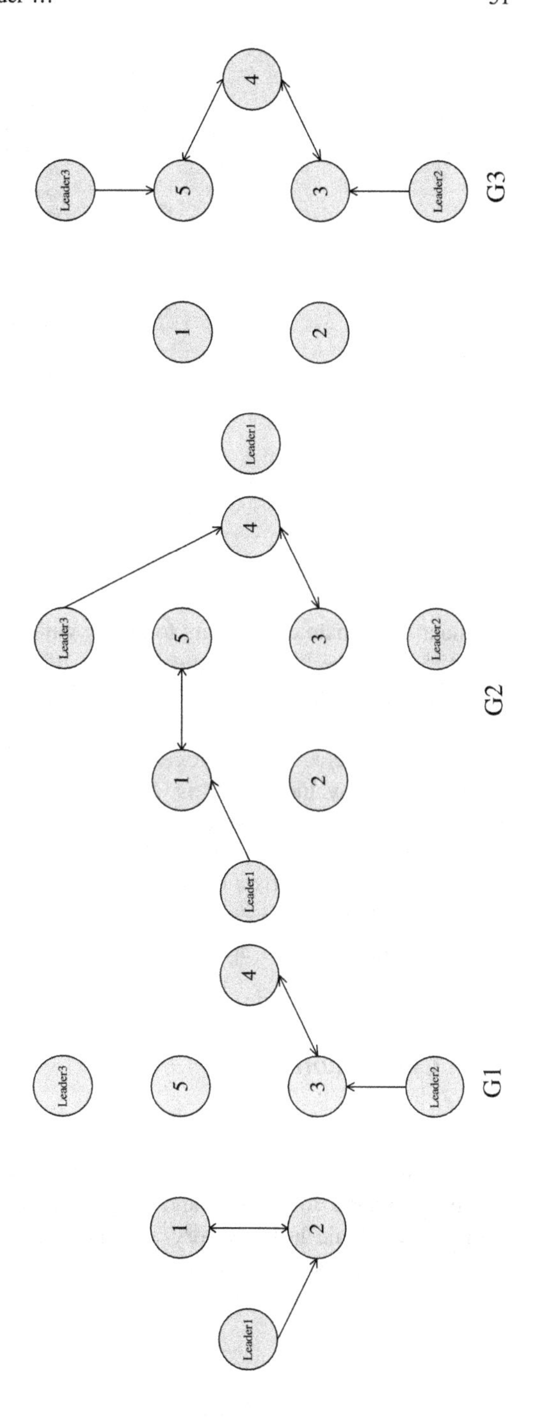

Fig. 5.1 Possible interaction topologies for three leaders and five followers

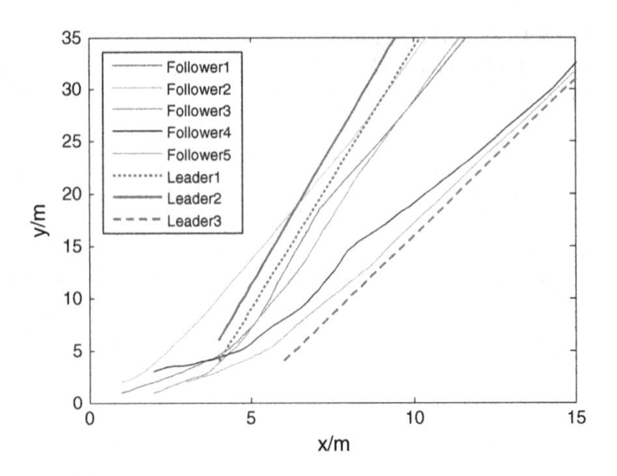

Fig. 5.2 State trajectories of all agents

5.4 Simulations

Consider the dynamic switching topology with 5 followers and 3 leaders are shown in Fig. 5.1, where the connection weight of each edge is 1. Supposing the communication network of intelligent systems with leaders randomly changes in $G1$ to $G3$ at $t = kT$, $k = 0, 1, \ldots$, where $T = 0.5s$.

From the interaction topologies, the system matrix can be obtained, and the maximum eigenvalue of H_σ is 4.618 by calculating. Assume that the control parameter of system is taken $\alpha = 1.0$ and $\beta = 2.1$. According to the constraints of Theorem in this paper, the allowed upper bound of the delays is 0.158. In experiments, we let $\tau(t) = 0.07 + 0.07 \sin t$ is the time-varying delays. The initial velocity vector and initial position vector of all agents are taken ramdomly. The trajectories of leaders and followers are shown in Fig. 5.2.

5.5 Conclusion

The distributed containment problem of intelligent systems with dynamic leaders is studied, and a containment control algorithm of double-integrator systems with jointly-connected topologies and varying delays is proposed in this paper. By applying algebraic graph theory and modern control theory, the convergence of intelligent systems with the topology of communication delays is analyzed on Lyapunov-Krasovskii method.

References

1. Hu J, Yuan H (2009) Collective coordination of multi-agent systems guided by multiple leaders [J]. Chin Phys B 18(09):3777–3782
2. Notarstefano G, Egerstedt M, Haque M (2011) Containment in leader-follower networks with switching communication topologies [J]. Automatica 47(5):1035–1040
3. Cao Y, Stuart D, Ren W, Meng Z (2011) Distributed containment control for multiple autonomous vehicles with double-integrator dynamics: algorithms and experiments [J]. IEEE Trans Control Syst Technol 19(4):929–938
4. Wang Y, Cheng L, Hou Z, Tan M, Wang M (2014) Containment control of multi-agent systems in a noisy communication environment [J]. Automatica 50:1922–1928
5. Lin P, Jia Y (2010) Consensus of a class of second-order multi-agent systems with time-delay and jointly-connected topologies[J]. IEEE Trans Autom Control 55(3):778–784
6. Rockafellar RT (1972) Convex Analysis. Princeton University Press, New Jersey
7. Li Z, Ren W, Liu X, Fu M (2013) Distributed containment control of multi-agent systems with general linear dynamics in the presence of multiple leaders. Int J Robust Nonlinear Control 23:534–547
8. Sun Y, Wang L, Xie G (2008) Average consensus in networks of dynamic agents with switching topologies and multiple time-varying delays[J]. Syst Control Lett 57(2):175–183
9. Boyd B, Ghaoui L, Feron E et al (1994) Linear matrix inequalities in system and control theory [M]. SIAM, Philadelphia, pp 28–29
10. Lin P, Qin K, Zhao H, Sun M (2012) A new approach to average consensus problems with multiple time-delays and jointly-connected topologies [J]. J Franklin Inst 349:293–304

Chapter 6
Robust Visual Tracking Via Part-Based Template Matching with Low-Rank Regulation

Fei Teng, Qing Liu, Langqi Mei and Pingping Lu

Abstract This paper presents a simple yet effective visual tracking method to attack the challenge when the target object undergoes partial or even full occlusion. First, a fixed number of image patches are sampled as the template set around current object location. In the detection stage, candidate image patches are sampled as the candidate set around the object location in the previous frame. Second, both the template set and candidate set patches are divided into sub-regions and features can be efficiently extracted via random projections. The confidence score for a specific candidate patch is computed through compressive features' low-rank regulation with the template set patches. The lowest confidence score in the current frame indicates the new object location. The encouraging experimental results show that our proposed method outperforms several state-of-the-art algorithms, especially when the target object suffers partial or even full occlusion.

Keywords Visual tracking · Part-based model · Random projections · Low-rank regulation

6.1 Introduction

Visual tracking remains an extensively popular research topic in the computer vision community in the past decades. It has many potential applications such as visual surveillance and event detection, scene analysis and recognition and so forth. Despite numerous algorithms have been proposed over past decades, tracking arbitrary object in diverse real-word scenes still remains challenging due to various

F. Teng (✉) · Q. Liu
School of Energy and Power Engineering, Wuhan University of Technology, Wuhan 430070, China
e-mail: changeerhao_love@126.com

L. Mei · P. Lu
School of Automation, Wuhan University of Technology, Wuhan 430070, China

Y. Jia et al. (eds.), *Proceedings of the 2015 Chinese Intelligent Systems Conference*, Lecture Notes in Electrical Engineering 360,
DOI 10.1007/978-3-662-48365-7_6

factors. Specifically, scenarios that contain abrupt illumination change, heavy partial or full occlusion or extreme scale variation are particularly difficult to keep tracking of the target object.

Recently, random projections have been widely applied in visual tracking domain with promising results. Zhang et al. [1] proposed a novel feature extraction technique based on random projections, and further came up with real-time compressive tracking (CT) framework. They first employed a very sparse non-adaptive random measurement matrix to extract the low-dimensional features. Then the compressive features were classified via a naive Bayes classifier with online update. Reference [2] proposed fast compressive tracking (FCT) by extending CT with a two-step search strategy (i.e., coarse search and fine search) to speed up the time-consuming detection procedure. It demonstrated that FCT achieved favorably high efficiency and accuracy. Wu et al. [3] developed a MSCT tracker that utilized a 2-order transition model to estimate the current scale status through the target velocity. In [4], a MSRP method that developed their appearance model via efficient feature extraction in the compressive domain, mostly in order to track in scenes with light and texture changing. Teng and Liu [5] proposed a multiple compressive features fusion (MCFF) method that employed two kinds of random measurement matrices to extract two complementary good features to track the target object. The extensive evaluations on numerous challenging videos demonstrated that MCFF obtained favorable performances, but not stable enough when the target object suffered occlusion.

Motivated by the above-mentioned discussions, in this paper, we set down to track the target object under partial or even full occlusion. We organize the rest of this paper as follows: Sect. 6.2.1 describes our proposed method in details. Section 6.2.2 conducts extensive experiments on several benchmark video sequences to illustrate the encouraging performances of the proposed algorithm. Finally, Sect. 6.2.3 presents a brief summary.

6.2 Proposed Algorithm

The proposed tracker is summarized in Table 6.1. Following Table 6.1, we will describe the main steps in Sects. 6.2.1–6.2.3 sequentially.

6.2.1 Sample the Template and Candidate Set and Divide Those Patches into Sub-regions

In step 1, we assume N represents the total number of sub-regions, $W(w)$ and $H(h)$ denote the width and height of the target object and sub-region respectively. It is straightforward that higher w or h cause the anti-occlusion effectiveness to

Table 6.1 Basic flow of our tracking algorithm

Initialization (in the first frame):

Step 1: Manually choose the target object, sample the template set and divide those patches into sub-regions according to Fig. 1

Step 2: For each sub-region, extract the compressive features via random projections (see Eq. 2)

For frame $t = 2, 3\cdots$

  Step 3: Extract the candidate set patches according to Fig. 2

  For candidate set patch $i = 1,2,\ldots,M$

    Step 4: Divide the candidate patch into sub-regions

    For sub-region $j = 1,2,\ldots,N$

      Step 5: Extract the compressive features via random projections (see Eq. 2)

      Step 7: Perform the low-rank regulation (see Eq. 4)

    End For

    Get the confidence score for each candidate patch

  End For

  Get the tracking result by minimizing the confidence scores

  Sample patches to update the template set

End For

deteriorate, and otherwise the extracted compressive features from sub-regions tend to be not discriminative and stable. Moreover, larger N is more likely to obtain good tracking result, but certainly to increase the computational burden. For the sake of effectiveness and efficiency, we employ a simple yet effective manner to uniformly divide the target object patch into 4 parts, the appearances of which are illustrated in Fig. 6.1. Similarly, in the detection stage, candidate image patches are sampled as the candidate set within a search radius around the object location in the previous frame as illustrated in Fig. 6.2.

6.2.2 *Random Projections*

Compressive trackers [1–5] usually represent the sample $z \in \mathbb{R}^{w*h}$ by convolving z with a pre-defined rectangle filter bank defined as:

$$h_{p,q}(x,y) = \begin{cases} 1, & x_i \leq x \leq x_i + p, y_i \leq y \leq y_i + q \\ 0, & otherwise \end{cases} \tag{6.1}$$

where (x_i, y_i) represents the upper left corner of the rectangle filter, $p \in [1, w]$, $q \in [1, h]$ denote the width and height of the rectangle filter bank. Furthermore, a sparse random measurement matrix is employed to extract the low-dimensional features based on compressive sensing theory:

Fig. 6.1 Sample the template set and divide those patches into sub-regions

Fig. 6.2 Sample the candidate set

$$\mathbf{y} = R\mathbf{x} \tag{6.2}$$

where $\mathbf{x} \in \mathbb{R}^{\mathrm{m}}$ corresponds to a high-dimensional feature space, $\mathbf{y} \in \mathbb{R}^{\mathrm{n}}$ corresponds to a low-dimensional feature space and $R \in \mathbb{R}^{n*m}$ represents the random measurement matrix with entries defined as:

$$r_{i,j} = \sqrt{s} * \begin{cases} 1 & \textit{with probability} \quad \frac{1}{2s} \\ 0 & \textit{with probability} \quad 1 - \frac{1}{s} \\ -1 & \textit{with probability} \quad \frac{1}{2s} \end{cases} \tag{6.3}$$

Under the condition that R satisfies the Johnson-linden Strauss (JL) lemma, $\mathbf{x}$ has potentially high possibility to be reconstructed with minimum error from $\mathbf{y}$ if $\mathbf{x}$ is image or audio signal [6]. A variant condition of JL lemma in compressive tracking is restricted isometry property (RIP), which aims to retain the distances between signals when projecting those high-dimensional signals onto low-dimensional space [6].

6.2.3 Low-Rank Regulation

To simplify the notation, we define a matrix $F^i = (f_1^i, f_1^i, \ldots f_N^i) \in \mathbb{R}^{d*N}, \forall i \in \{1, 2, \ldots, M\}$, where i is the index of candidate patches and M represent the total number of candidate patches, d denotes the length of the compressive features and N denotes the total number of sub-regions. Let $T_j = (f_j^1, f_j^2, \ldots, f_j^M), \forall j \in \{1, 2, \ldots, N\}$, then T_j contains the corresponding parts from each target template and the current candidate patch. Low-rank regulation preserves similarity by using the rank minimization on T_j and vice versa, the feature vector in T_j can be enforced to be similar. Therefore, the problem of optimizing the prediction can be formulated as following:

$$min_j \, rank(T_j) \text{ where} j \in \{1, 2, \ldots, N\} \tag{6.4}$$

As rank($\cdot$) is a non-convex and discrete function, we can use the nuclear norm $||\cdot||_*$ (sum of the singular values) to further optimize it. Thus the objective of minimization will then be reformulated as:

$$min_j \, ||T_j||_* \text{ where} j \in \{1, 2, \ldots, N\} \tag{6.5}$$

6.3 Experiment Results

In order to demonstrate the effectiveness of our proposed method, our tracker is evaluated with other 5 state-of-the-art trackers on several benchmark sequences which are publicly available. The 5 evaluated trackers are FCT tracker [2], MCFF tracker [5], MSRP tracker [4], MSCT tracker [3] and TLD tracker [7]. The 6 challenging sequences are taken from [8], including coke, david_1, david_2, freeman, jumping and woman. In the experiment, the number of template patches is 15, the number of sub-regions is 4, and the length of the compressive features for each sub-region is 100. Our tracker runs about 15 frames per second (FPS) in MATLAB on an Intel I7 4 Core 2.5 GHz, CPU with 8 GB RAM machine.

We use two popular metrics to make quantitative comparisons. The first one is the overlap score defined as score $= \frac{area(ROI_T \cap ROI_G)}{area(ROI_T \cup ROI_G) - area(ROI_T \cap ROI_G)}$, where ROI_T represents the tracking result and ROI_G denotes the manually labeled ground truth bounding box. The plot that shows the ratio of successful frames over different overlap thresholds is termed as success plot. The other metric is Euclidean distance between the tracking result and the ground truth, which is termed as center location error (CLE). The corresponding plot over different error thresholds is termed as precision plot. Figure 6.3 shows the averaging performances in terms of both success and precision rate. Motivated by [8], three different kinds of initialization,

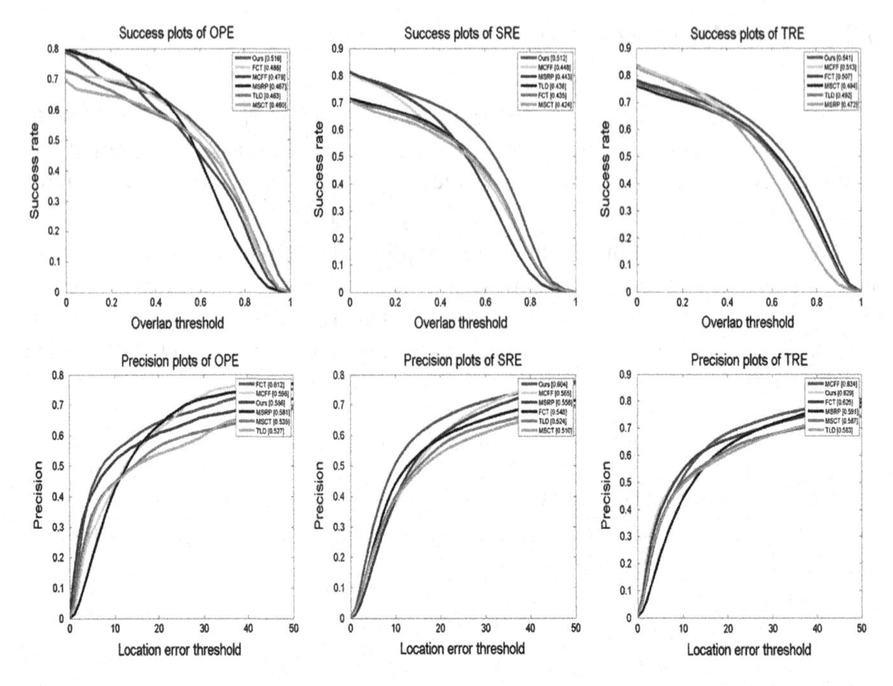

Fig. 6.3 Success and precision plots for OPE, TRE and SRE

one-pass evaluation (OPE), temporal robustness evaluation (TRE) and spatial robustness evaluation (SRE) are further considered to test the tracker's robustness to initialization.

From Fig. 6.3, it is clear that our proposed tracker achieves the best or second best performance in almost all success and precision plots for OPE, TRE and SRE. Its success rate is 0.516 in OPE, 0.541 in TRE, 0.512 in SRE, significantly outperforms the second best performance 0.488 in OPE, 0.513 in TRE and 0.448 in SRE respectively. Meanwhile, in the precision plot, the proposed tracker ranks 3rd, 1th and 2nd in OPE, TRE, and SRE respectively, but with only a narrow margin to higher-ranked trackers.

Figure 6.4 presents some screenshots from some of the sampled tracking results. For sequences David_1, David_2, Freeman and Jumping where the target object suffers illumination variation and cluttered background, FCT and MCFF perform better than MSRP, MSCT and TLD. Our tracker performs well on these sequences due to three aspects: first, random projection has strong ability to model the target appearance based on the compressing theory. Furthermore, discriminative models with local features have been demonstrated intrinsic advantage to handle pose variation well. The last but not the least, the objective of tracking is to locate the object of interest from frame to frame and the objective of classification is to predict the possibility of the instance label as positive or negative. Therefore, for trackers [2–5], these two objectives are not consistent during tracking, which may lead to

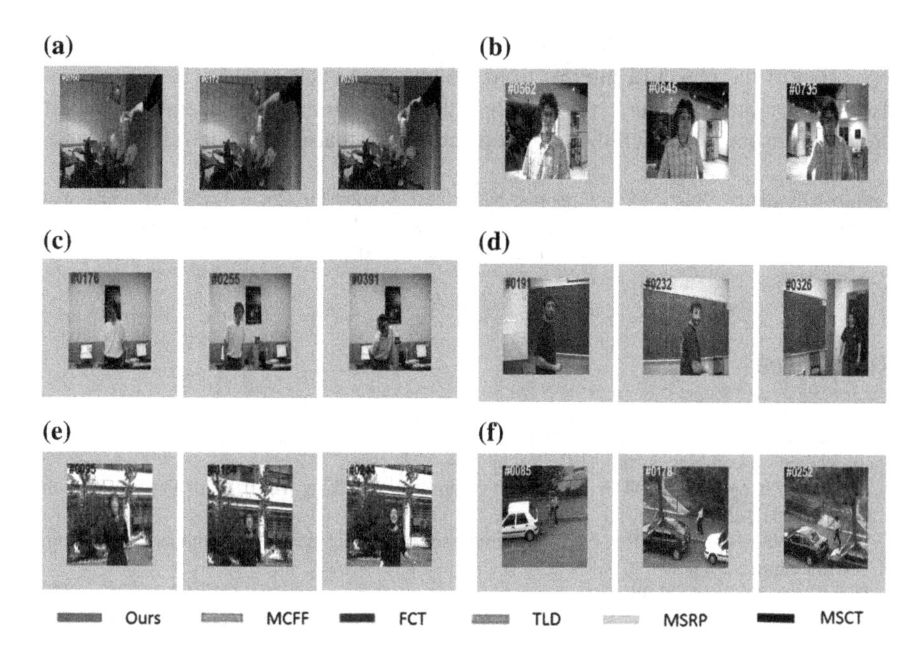

Fig. 6.4 Screenshots from some of the sampled tracking results. **a** Coke. **b** David_1. **c** David_2. **d** Freeman. **e** Jumping. **f** Woman

inaccurate estimation by maximizing the classifier response. For sequences coke and woman where the target object undergoes partial or even full occlusion, our tracker significantly outperforms other trackers. On one hand, our appearance model is part-based which has demonstrated robustness to partial or even full occlusion. On the other hand, minimizing nuclear norm is a good surrogate for minimizing the rank, which further facilitates the separation of the target object from the cluttered background.

6.4 Conclusion

This paper presents a simple yet effective visual tracking method based on compressive sensing theory. If the random measurement matrix satisfies RIP, the distances between any pairs of sparse signals are approximately preserved when projecting those high-dimensional signals onto low-dimensional space. Under this very strong theoretic support from compressive sensing, we make the straightforward assumption that similar image patches are more likely to formulate low-rank regulation. Part-based strategy is further employed to enhance its robustness against occlusion. Then we can simply perform template matching for each and every candidate patch and the candidate patch with the lowest confidence score denotes the object location in the current frame. The experiment shows minimizing nuclear norm is a good surrogate for minimizing the rank, which further facilitates the separation of the object of interest from the cluttered background. The experiment also demonstrates that our proposed algorithm outperforms several state-of-the-art algorithms especially when the target object sufferes partial or even full occlusion.

Acknowledgments This work is supported by Chinese Scholarship Council (CSC), the National Natural Science Foundation of China (NSFC 51279152, 61104158) and the Seed Foundation of Wuhan University of Technology (No. 145211005, 155211005).

References

1. Zhang K, Zhang L, Yang M-H (2012) Real-time compressive tracking. In: 12th European conference on computer vision (ECCV), pp 864–877
2. Zhang K, Zhang L, Yang M-H (2014) Fast compressive tracking. IEEE Trans Patt Anal Mach Intell 99:2012–2015
3. Wu YX, Ni J, Sun JP (2014) Real-time multi-scale tracking based on compressive sensing. Visual Comput 25(5):1–11
4. Teng F, Liu Q (2014) Multi-scale ship tracking via random projections. SIViP 8(6):1069–1076
5. Teng F, Liu Q (2015) Robust multi-scale ship tracking via multiple compressed features fusion. Sig Process Image Commun 31:76–85
6. Achlioptas D (2003) Database-friendly random projections: Johnson-Lindenstrauss with binary coins. J Comput Syst Sci 66(4):671–687
7. Kalal Z, Mikolajczyk K, Matas J (2011) Tracking-learning-detection. IEEE Trans Pattern Anal Mach Intell 34(7):1409–1422
8. Yi W, Jongwoo L, Ming-Hsuan Y (2013) Online object tracking: a benchmark. In: IEEE conference on computer vision and pattern recognition (CVPR), pp 2411–2418

Chapter 7
Modeling and Optimization of Coal Moisture Control System Based on BFO

Xiaobin Li, Haiyan Sun and Yang Yu

Abstract Coal moisture control process is a critical process in energy saving for pollution reduction and improving production efficiency and the quality of coke. The RBF artificial neural network approach for modeling is used to achieve precise control of coal moisture control system and against their strong coupling nonlinear systems with time-delay characteristics. The bionic BFO (Bacterial Foraging Optimization) is used to the fitness to optimize the RBF Neural network parameters. In order to achieve better results the RBF Neural network performance is optimized by these bionic BFO. This method provides a theoretical basis for accurate control of coal moisture process. The reduction of energy and pollution with improving the quality of coke is established.

Keywords Coal moisture control · BFO · Modeling · Optimization

7.1 Introduction

CMC (coal moisture control) is a key procedure in coking to save energy, reduce pollution and improve productivity and quality. It is through direct or indirect heating that coal moisture is controlled within about 6 % before putting into furnace and remained stable to ensure production with high efficiency and low consumption.

Coal moisture control system is a multi-parameter, strong coupling and non-linear system with large time delay, of which the model can hardly be estab-

X. Li (✉) · Y. Yu
School of Electrical and Electronic Engineering, Shanghai Institute of Technology, Shanghai 201418, China
e-mail: lixiaobinauto@163.com

H. Sun
School of Ecological Technology and Engineering, Shanghai Institute of Technology, Shanghai 201418, China

Y. Jia et al. (eds.), *Proceedings of the 2015 Chinese Intelligent Systems Conference*, Lecture Notes in Electrical Engineering 360,
DOI 10.1007/978-3-662-48365-7_7

lished through mechanism. Online moisture monitoring on the dryer gate rarely provides reliable data for the disturbance of moisture and dust. In production, there is about 15 h delay in detecting the moisture of coal out of the dryer and provides little reference to real production. Such large time delay can be handled when the moisture is stable, but will influence the effect of moisture control system when the coal moisture out of the dryer is quite sensitive to the weather in open-air coal ground in rainy season in southern China. Therefore, it is of great importance to find relationships between such parameters as steam consumption and flue gas temperature, moisture of coal in and out of the dryer under different conditions by analyzing and thus establishing control system to realize automatic online control of coal moisture.

For the modeling of coal moisture control, Ergun [1] adopts linear partial differential equation to describe the mathematical model of coal drying in prompt dropping phase assuming that evaporation only occurs in the surface of solid, which is not satisfactory to coal with complicated construction. For the strong coupling of coal moisture control, Gao Jianjun [2] provides a mathematical model based on equilibrium between moisture and coal and thermal balance of control volume with moisture control reactor as the control volume, to calculate the flue gas volume, temperature of in and out waste gas and effect of coal moisture before treating on the humidified coal powder and waste gas moisture. The study adopts Bacterial Foraging Optimization (BFO) algorithm to establish and optimize the RBF Neural Network Model based on thermal equilibrium theory and relevant analysis so as to solve the problems occurs in other modeling.

7.2 Process and Modeling

7.2.1 Process

Coal moisture controlling system in Bao Steel is studied and indirect type steam tube rotary dryer, dust catcher, fluid reservoir, flash tank, draught fan, pipeline, steam tube and conveyer belt and etc., are involved in the system. Figure 7.1 shows its process.

In the process, there are over 30 factors that influence directly or indirectly the coal moisture after drying, including output, stream pressure, pressure of condensate tank, steam temperature, real steam flow, steam pipeline flow, main motor current, current of feed screw and discharge screw, current of draught fan, output of draught fan, CO strength in flue gas, steam flow heated by recycle gas, temperature of recycle gas, outlet gas flow rate, recycle gas flow rate, blast volume, pressure of dryer inlet, temperature of blast, pressure differences of dust catcher, gas temperature of dryer outlet, gas temperature of inset and outlet, gas pressure of inlet and outlet, bag temperature of catcher, dust hopper temperature, temperature and pressure of reverse nitrogen, oxygen content of desiccant and inlet coal moisture.

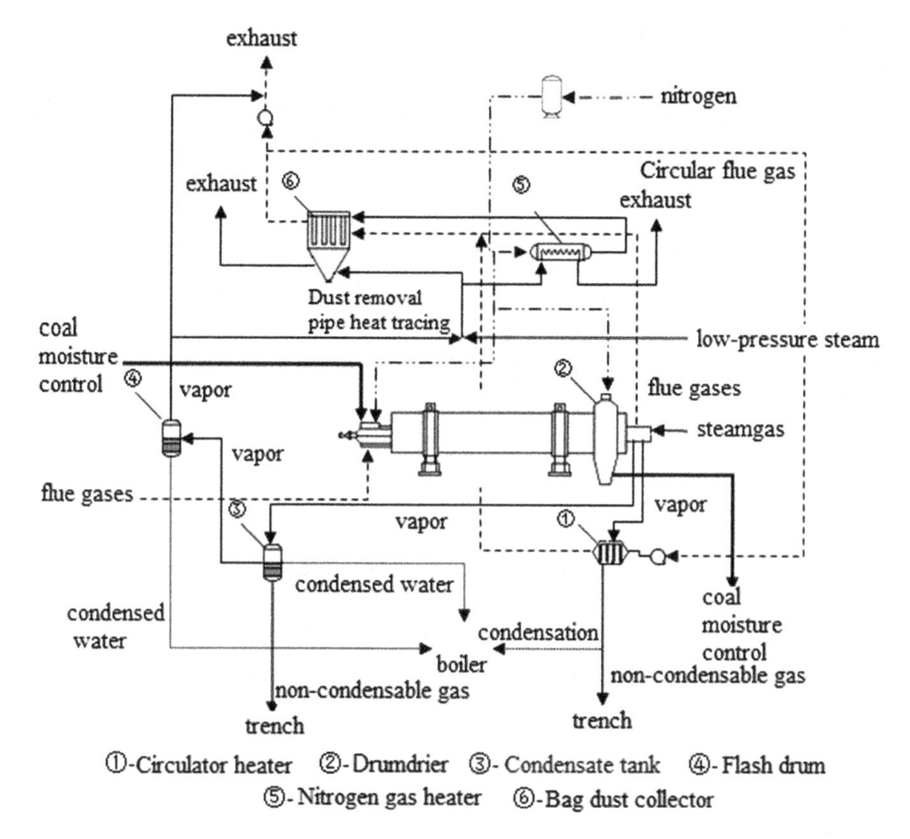

Fig. 7.1 Coal moisture control process

By analyzing the data of coal moisture controlling based on thermal equilibrium mechanism and mathematical statistics theory and relationships between variables with the aim of saving energy and reducing consumption with high control accuracy, such independent variables most relevant to steam volume are identified to be as input and output of the control system, including coal to be treated, gas temperature of dryer outlet, inlet coal moisture, outlet coal moisture.

7.2.2 Modeling

For the characteristics of non-linear, multi-input and strong coupling and low accuracy of online detecting of outlet coal moisture, RBF artificial neural network is adopted for modeling. Figure 7.2 shows the structure, of which x_1, x_2 … are inputs, c_1, c_2, c_k are center vectors, $\Phi(\|x - c_k\|, s_k)$ is Gauss function, w_{ki} is weight.

Fig. 7.2 RBF artificial neural network

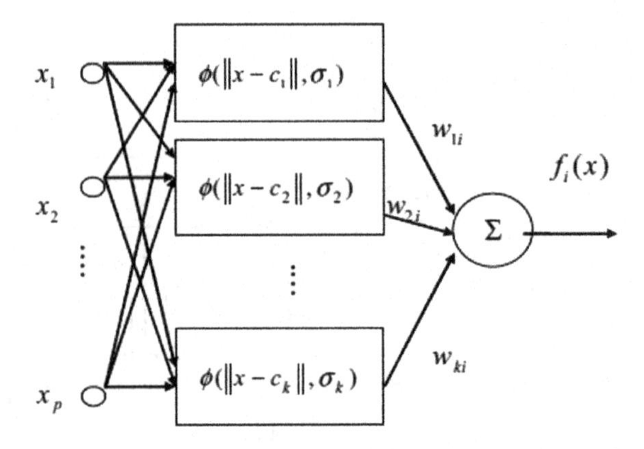

Number of units in input layer: as the input layer is the cache of outer data and depends on the dimensions of input vectors, the number of units in input layer is set as 3.

Number of units in output layer: to simplify the system and save the cost of calculation, the number of units in output layer is set as 1.

Number of hidden units: is set as 8 after training 100 set data.

7.3 Data Acquisition

The parameters of environment, operation and working of coal moisture controlling system and the period of coal in the rotary kiln are recorded in the field and sample coal is taken back to lab. Figure 7.3 shows the monitoring process.

720 set of data is recoded at the same time lap by workers of 4 round working time in Bao Steel work shop, of which 310 set is chosen as training samples. Seen as in Table 7.1.

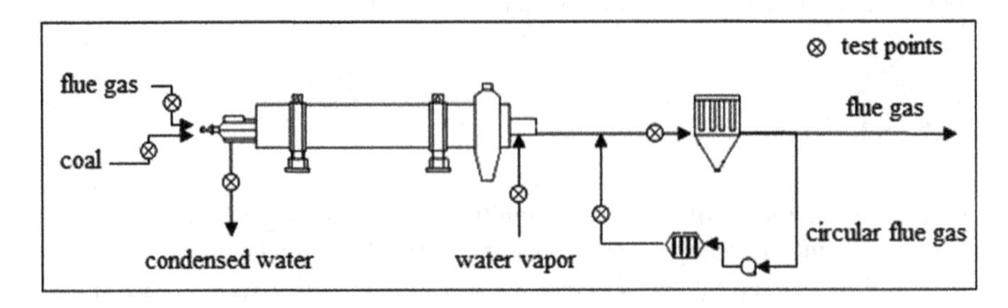

Fig. 7.3 Testing points of the system

Table 7.1 Training samples

Output (ton/h)	Outlet temperature of dryer (centigrade)	Inlet coal moisture (%RH)	Steam flow rate (kg/h)
247.9291	74.9978	12.1207	11088
251.2817	70.5444	12.0825	10939
247.2381	81.2406	12.1175	11043
249.4915	76.2792	12.0942	10948
243.3044	81.7253	12.2356	11510
252.8835	77.3085	12.0127	10829
247.4501	81.5356	12.1405	11146
247.0261	81.6410	12.1640	11234
249.7035	81.5356	12.1405	11768
248.5415	81.5778	12.1399	11142
245.2909	81.5988	12.2353	11502
244.1760	79.8083	12.2839	11888
248.0547	81.0720	12.1752	11262
…	…	…	…

7.4 Parameters Identification and Optimization [3–5]

7.4.1 Adaptive BFO Algorism

In the initial period of searching optimum, large step helps to avoid local optima and find the area where optima is while in the half period, small step helps to improve accuracy.

Evaluate fitness to the step in chemotaxis to adjust step which each bacteria takes.

Step1: Evaluate fitness

$$V = \frac{J_i}{J_{\max}}(X_{\max} - X_{\min})*rand \tag{7.1}$$

where V is fitness, $X_{\max}$, $X_{\min}$ are the boundaries of variables and J is fitness.

Step2: Tumble: generate a random direction unit $\Delta(i)$ to optimize steps according to the following function:

$$\theta^{i}(j+1,k,l) = \theta^{i}(j,k,l) + C(i)\frac{\Delta(i)}{\sqrt{\Delta^{T}(i)\Delta(i)}} \tag{7.2}$$

to update the location and its corresponding fitness of each bacteria.

Step3: Swim: optimized by the following function:

$$C(i) = C(i)*V \tag{7.3}$$

Step4: Linear gradient fitness is represented by the following fuction:

$$V = \frac{step_{\max} - step_{\min}}{step_{\max}} * V \tag{7.4}$$

7.4.2 *Identification and Optimization*

To improve the prediction effect of the model, BFO with step of 0.1 and 0.5, adaptive BFO algorism are used to optimize RBF network and the following are the procedures:

Step1: Train weight of each unit by RBF algorism;
Step 2: Calculate J = MSE of each network as its fitness;
Step 3: Search optima of fitness J by 5 algorisms to find the minimum value of which the corresponding weight is the optima of the network. As seen in Fig. 7.4.

As shown in Fig. 7.4, when optimized by BFO, the convergence rate of parameters in RBF is 0.5 BFO, ABFO, 0.1 BFO, of which 0.5 BFO is faster.

For the parameters with high dimensions (the RBF network has 40 dimensions), 0.1 BFO leads to local optima (Table 7.2).

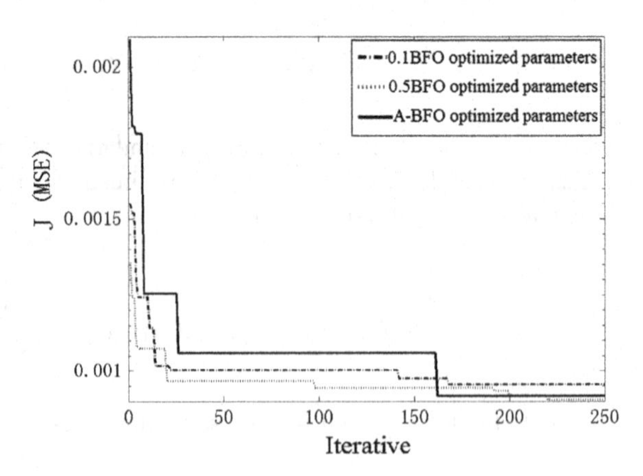

Fig. 7.4 BFO optimized RBF parameters

Table 7.2 Samples checking

Output (ton/h)	Outlet temperature of dryer (centigrade)	Inlet coal moisture (%RH)	Steam flow rate (kg/h)
249.7035	77.0774	12.0613	10900
250.6064	71.4266	12.0960	10977
299.3966	82.9053	10.9383	9702
248.1332	81.3881	12.0948	10963
253.0326	81.7253	11.9765	10777
250.9363	73.5483	12.0954	10969
257.5866	81.3881	11.8725	10729
245.5814	81.3038	12.1985	11360
251.5879	74.0945	12.0960	10977
250.4573	75.3969	12.1302	11122
246.6650	81.1774	12.1991	11367
250.8892	81.6831	12.0831	10940
248.5415	81.1353	12.1558	10612.52
…	…	…	…

7.4.3 Model Checking

From 720 sets data, 310 are chosen in the prediction model.

As Figs. 7.5, 7.6, 7.7, 7.8 show that the difference of RBF network without optimization reaches as high as 18 % with small difference in only limited areas, which leads to large difference in overall control. The accuracy of 0.5 BFO is high than that of 0.1 BFO (0.1 BFO leads to local optima), of which the difference remains within 10 and 13 % respectively. The difference of ABFO remains within 11 %, falling in between of above.

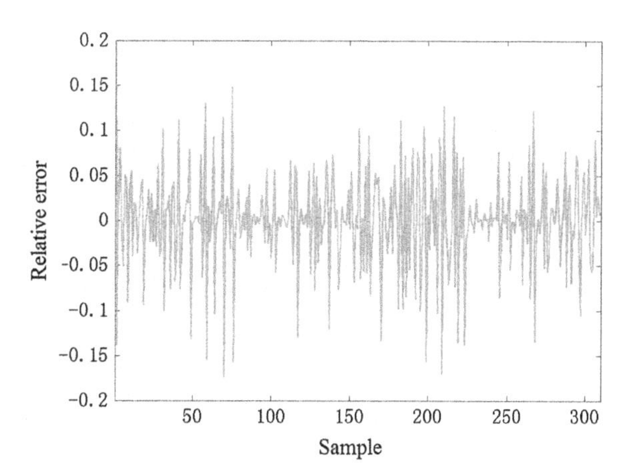

Fig. 7.5 RBF system prediction error

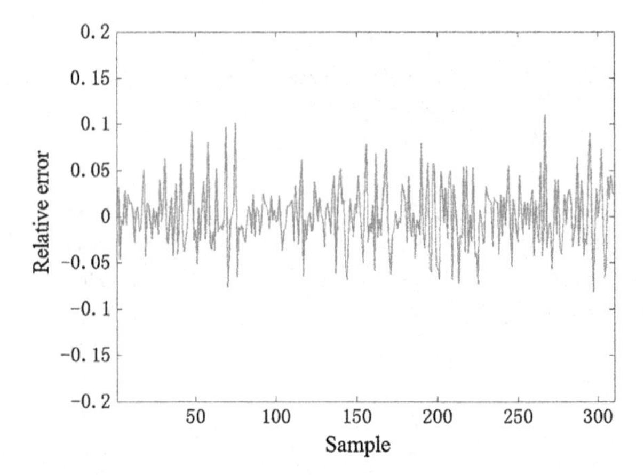

Fig. 7.6 Step 0.1 BFO optimized RBF error

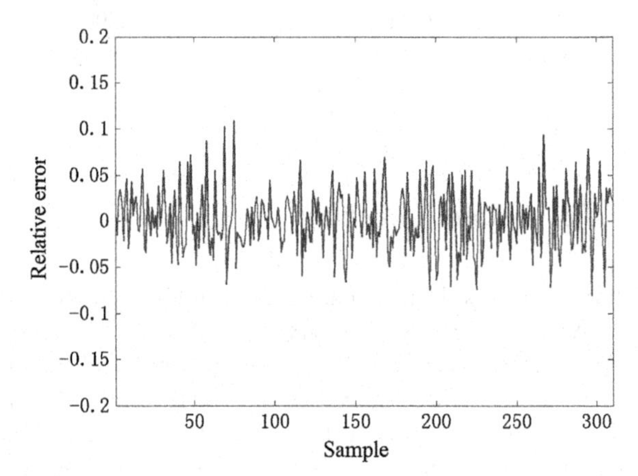

Fig. 7.7 ABFO optimized RBF error

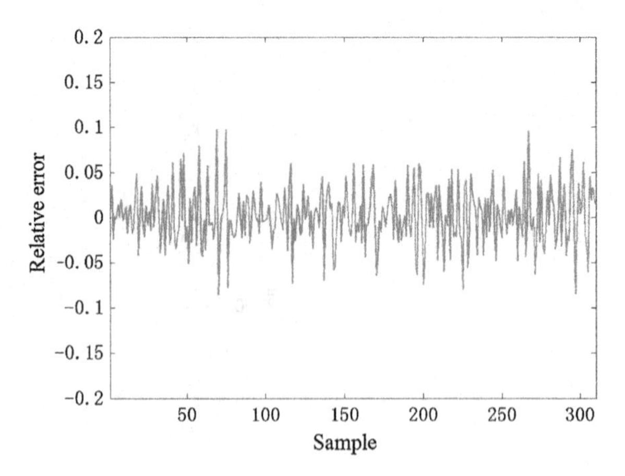

Fig. 7.8 Step 0.5 BFO optimized RBF error

Figures 7.9, 7.10, 7.11, 7.12 show the difference between real steam flow rate and prediction output of RBF network based on 310 sets data, of which the real steam flow rate is within the range of from 8000 kg/h to 20000 kg/h.

Compared with real steam flow rate, the predicted output of RBF network without optimization fluctuates at as much as 4000 kg/h and 0.1 BFO optimized RBF fluctuates at 2300 kg/h, adaptive BFO at 2250 kg/h and 0.5 BFO one at 2200 kg/h.

From above, the RBF network optimized by BFO with 0.5 step yields the best result, adaptive BFO falls in between 0.1 BFO and 0.5 BFO.

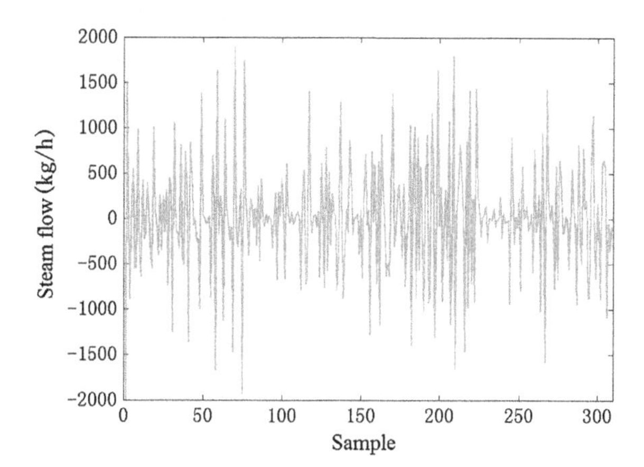

Fig. 7.9 Real and prediction output error

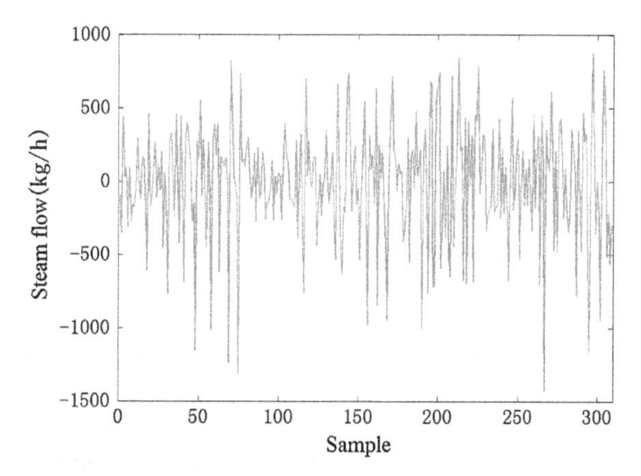

Fig. 7.10 Real and step 0.1 BFO * RBF output error

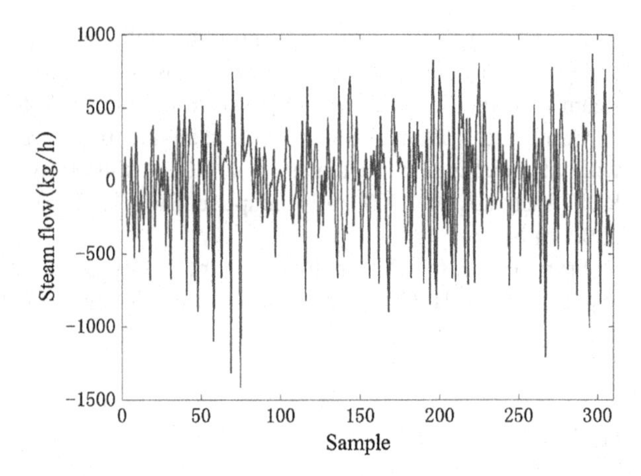

Fig. 7.11 Real and ABFO* RBF output error

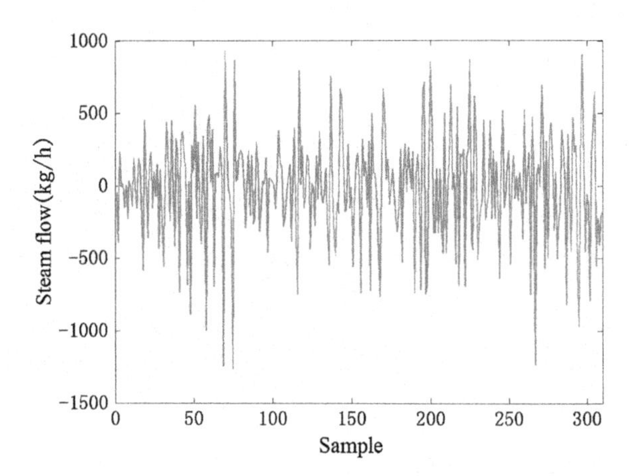

Fig. 7.12 Real and step 0.5 BFO * RBF output error

7.5 Conclusion

(1) A nature-inspired BFO modeling and optimization of coal moisture controlling system is studied and provided by analyzing its process and existed models.
(2) When optimizing RBF network by BFO, the convergence rate of BFO with 0.5 step is fastest and followed by ABFO and BFO with 0.1 step.
(3) When identifying RBF network by BFO, the accuracy is 0.5 BFO, ABFO and 0.1 BFO in sequence, with 0.5 BFO being the highest.
(4) BFO usually leads to local optima to RBF network with high dimensions (the RBF network has 40 dimensions). In the study, RBF network optimized by 0.1 BFO falls in local optima, of which the accuracy is lower than that of 0.5 step BFO.

Acknowledgment This work is partially supported by Shanghai key scientific research project No.11510502700, and science and technology innovation focus of SHMEC No.12ZZ189.

References

1. Ergun G, Somer TG, Kisakurek B (1984) Analysis Of Drying Rate Mechanism In Complex Solids[J]. Soc. Chem. Eng. 4:127–132
2. Jianjun G, Yuhua G, Hemin Z, Haifeng W (2012) Modeling and Calculation of parameters of coal moisture[J]. Fuels Chem. 43:1–7
3. Passino KM (2002) Biomimicry of bacterial foraging for distributed optimization and control [J]. IEEE Control Syst. Mag. 22:52–67
4. Dasgupta S, Biswa A, Abraham A et al (2009) Adaptive computational chemotaxis in bacterial foraging optimization: an analysis[J]. IEEE Trans. Evol. Comput. 13:919–941
5. Jie H (2012) Study on improvement and application of BFO algorism. [D] Wuhan University of Science and Technology, Wuhan

Chapter 8
Image Super-Resolution Based on MCA and Dictionary Learning

Kun Zhang, Hongpeng Yin and Yi Chai

Abstract Image super-resolution focuses on achieving the high-resolution version of single or multiple low-resolution images. In this paper, a novel super-resolution approach based on morphological component analysis (MCA) and dictionary learning is proposed in this paper. The approach can recover each hierarchical structure well for the reconstructed image. It is integrated mainly by the dictionary learning step and high-resolution image reconstruction step. In the first step, the high-resolution and low-resolution dictionary pairs are trained based on MCA and sparse representation. In the second step, the high-resolution image is reconstructed by the fusion between the high-resolution cartoon part and texture part. The cartoon is acquired by MCA from the interpolated source image. The texture is recovered by the dictionary pairs. Experiments show that the desired super-resolution results can be achieved by the approach based on MCA and dictionary learning.

Keywords Image super-resolution · Dictionary learning · Morphological component analysis · Cartoon · Texture

8.1 Introduction

Image super-resolution (SR) can produce the high-resolution (HR) enhancement version of single or multiple observed low-resolution (LR) images without changing the low-cost image acquisition sensors [1]. Therefore, image SR attracts

K. Zhang · H. Yin (✉) · Y. Chai
College of Automation, Chongqing University, Chongqing 400044, China
e-mail: yinhongpeng@gmail.com

H. Yin
Key Laboratory of Dependable Service Computing in Cyber Physical Society, Ministry of Education, Chongqing 400030, China

Y. Chai
Key Laboratory of Power Transmission Equipment and System Security, Chongqing 400044, China

Y. Jia et al. (eds.), *Proceedings of the 2015 Chinese Intelligent Systems Conference*, Lecture Notes in Electrical Engineering 360,
DOI 10.1007/978-3-662-48365-7_8

much attention and always plays an important role in various applications, such as the remote sensing, the satellite image application, the medical imaging diagnosis system, the digital television system. As a result, various SR methods have been proposed, mainly including the interpolation-based methods [2], the regularized reconstruction-based methods [1–3] and the learning-based methods [4–12].

The learning-based SR methods have been comprehensively researched since the example-based approach in [4] was proposed. In these methods, the prior relationship between the HR and LR training set can usually be learned in terms of the image structure and content. The relationship is very beneficial to recovering high-resolution version of the LR test image. Consequently, various learning-based SR approaches are put forward [5–12]. Referring to manifold learning strategy, Chang et al. [5] utilize the principle of local linear embedding (LLE) in SR. Their approach is helpful in decreasing the scale of training set, but the fitting problem makes the approach imperfect in image sharpness [7]. In [6], the contourlet transform is introduced into the single-image SR to capture the smoothness contours by directional decompositions. The dictionary-based learning approach via sparse representation is further introduced into image SR by Yang et al. [7, 8]. After the work of Yang, some improved approaches based on sparse representation are also proposed. In [9], a texture constrained sparse representation is proposed. However, the approach is sensitive to noise. In [10], the dictionary is trained by the difference between the HR images and LR images. The approach performs well for image denoising and SR reconstruction. In [11], a hierarchical clustering algorithm is applied to optimize the parameter for the dictionary learning way in [7].

However, the aforementioned SR approaches focus to the omitted high-frequency texture component and should pay more attention to the other component. Inspired by the morphological component analysis (MCA) [12], a novel image SR approach based on MCA and dictionary learning is applied. In our approach, MCA is utilized to decompose the interpolated LR test image into the cartoon and texture component. The texture component is discarded. The cartoon component is reserved as the HR cartoon part of the expected HR image. The expected HR image's texture is estimated by means of the HR\LR dictionary pairs established from training image set. The reconstructed image result can be obtained by combining the HR cartoon and the HR texture. Thus, each component of the reconstructed HR image can be both focused on well. Experimental results show the desired super-resolution image quality can be achieved by the approach based on MCA and dictionary learning.

In the outline of this paper, Sect. 8.2 describes our approach based on MCA and dictionary learning in detail, Sect. 8.3 conducts some experiments and analyzes the results, and Sect. 8.4 shows the conclusion for this paper briefly.

8.2 Our Approach for Image Supper-Resolution

In the proposed approach, two steps are necessary: the dictionary learning step and the SR reconstruction step. The former can provide a pair of dictionaries for the later to obtain a high-resolution enhancement version of the LR test image. In each step, the MCA theory is utilized to decompose images.

8.2.1 *Image Decomposition Using MCA*

In the view of MCA, an original image X is the overlap of M different morphology layers $\{X_i\}_{i=1}^{M}$, as shown in Eq. (8.1).

$$X = \{\mathrm{X_i}\}_{\mathrm{i}=1}^{M} = \mathrm{X_1} + \mathrm{X_2} + \cdots + \mathrm{X_M} \tag{8.1}$$

$$X_t = T_t \alpha_t \,, \quad X_s = T_s \alpha_s \tag{8.2}$$

In the SR application, two morphology layers $\{X_t, X_s\}$ should be separated from X. X_t denotes the texture and X_s is for the needed cartoon. Correspondingly, T_t is the dictionary for X_t and T_s is for X_s. Then, the layers should be optimally represented by sparse representation $\{\alpha_t, \alpha_s\}$ in Eq. (8.2), where, $X_t, X_s \in R^N, T_t, T_s \in R^{N \times L} (L \gg N)$. To realize sparse representation, the noise-constraint solution via the basis pursuit (BP) is often used in Eq. (8.3) [12]. Here, ε stands for the noise level as the prior information in the decomposed image.

$$\{\alpha_t^{opt}, \alpha_s^{opt}\} = Arg \min_{\{\alpha_t, \alpha_s\}} ||\alpha_t||_1 + \min||\alpha_s||_1 \quad s.t.||X - T_t\alpha_t - T_s\alpha_s|| \le \varepsilon \tag{8.3}$$

$$X_s = X - T_t \alpha_t^{opt} \tag{8.4}$$

In the dictionary learning step for SR, the approach in Eq. (8.4) is adopted to further leave all the noise in the cartoon component after MCA image decomposition by Eq. (8.3). The solution is robust to generate the texture part with little noise from the HR training image in the super-resolution.

$$\{\alpha_t^{opt}, \alpha_s^{opt}\} = Arg \min_{\{\alpha_t, \alpha_s\}} ||\alpha_t||_1 + \min||\alpha_s||_1 + \lambda||X - T_t\alpha_t - T_s\alpha_s||_2^2 + \gamma TV\{T_s\alpha_s\} \tag{8.5}$$

However, Eq. (8.3) is not suitable to extract smooth cartoon in SR reconstruction step. So Eq. (8.5) is adopted in this step [12]. In this solution, the noise-constraint condition is replaced by the unconstrained penalized item $\lambda||X - T_t\alpha_t - T_s\alpha_s||_2^2$. Besides, a total variation (TV) penalty is adopted to obtain the cartoon structure with pronounced edge. Here, $TV\{T_s\alpha_s\}$ is the l_1-norm of the gradient. Figure 8.1

Fig. 8.1 Image decomposition by MCA. *Left* original image, *Middle* cartoon, *Right* texture

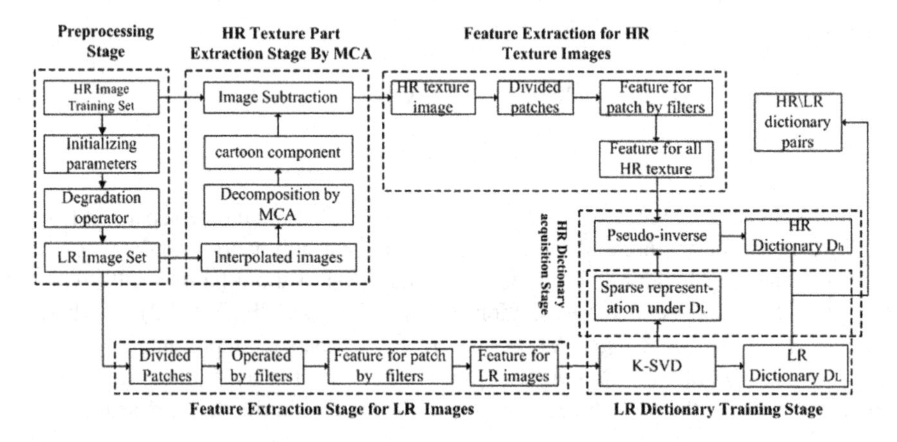

Fig. 8.2 The flowchart for dictionary learning step

shows image decomposition result by MCA with the TV penalty and the original image can be found in http://pan.baidu.com/s/1c0Ix6Zu. The cartoon component is completed and piecewise smooth. Thus, it can help the SR reconstruction step recover the cartoon part of the expected HR image efficiently.

8.2.2 *The Training of the HR/LR Dictionary Pairs*

In the dictionary training, MCA and dictionary learning method via sparse representation are combined to describe the feature of HR/LR images. The flowchart for this step is shown in Fig. 8.2.

In the feature extraction for LR dictionary, each LR image is interpolated and divided into patches with $N \times N$ pixels and Eq. (8.6) are used as the derivative feature of image patches [5, 7]. Thus, the feature vector obtained for each LR patch by concatenating the four vectors has a length of $4N^2$. To reduce the complexity, the principal components analysis (PCA) is employed to reduce feature space. The reduced feature y_i can reflect the corresponding low-resolution patch. Eventually, LR set can be described as shown in Eq. (8.7).

K-SVD [13] is used for training LR dictionary. The algorithm aims to iteratively improve the initial dictionary and achieve optimal sparse representations A for the feature set Y_l in Eq. (8.8) [13]. Here, α_i is the ith columm vector of sparse matrix A and denotes the sparse code for y_i, T_0 is the target sparsity constraint. Eventually, the optimal LR dictionary and the homologous sparse representations can be obtained by the algorithm.

In the HR texture extraction, Bicubic interpolation [2] are employed to magnify LR images into the HR image's size. The MCA with the noise-constraint in Eq. (8.3) are applied to separate cartoon component from the interpolated image. Thus, the HR texture is extracted from the HR image by image subtraction.

$$\begin{aligned} f_1 &= [-1, 0, 1], \qquad & f_2 &= f_1^T \\ f_3 &= [1, 0, -2, 0, 1], \qquad & f_4 &= f_3^T \end{aligned} \tag{8.6}$$

$$Y_l = \{y_i\}_{i=1}^{n} = \{y_1, y_2, \ldots, y_n\} \tag{8.7}$$

$$\min_{D_L, A}\{\|Y_l - D_L A\|_F^2\}, \quad s.t. \forall i \quad \|\alpha_i\|_0 \le T_0 \tag{8.8}$$

$$X_h = \{x_i\}_{i=1}^{n} = \{x_1, x_2, \ldots, x_n\} \tag{8.9}$$

In the HR feature extraction, each texture image is divided into patches and the feature is extracted by the same method in the feature extraction for LR images. Eventually, the HR texture image set can be described in Eq. (8.9).

The HR dictionary is acquired by assuming that the patches can be represented by the same sparse matrix A under the HR/LR dictionary pairs. The HR dictionary

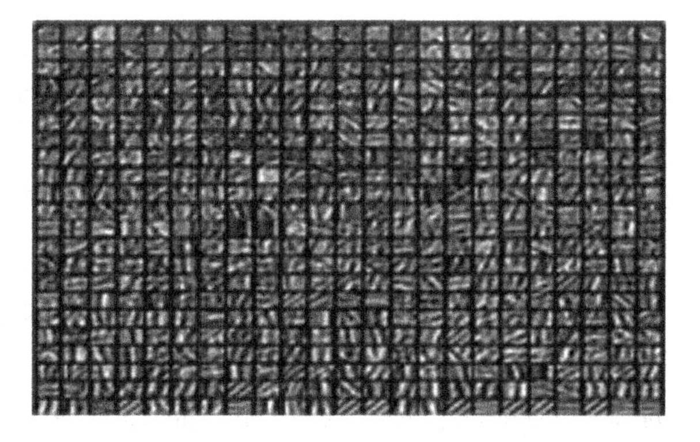

Fig. 8.3 The visual version of some HR dictionary's atoms from our experiment

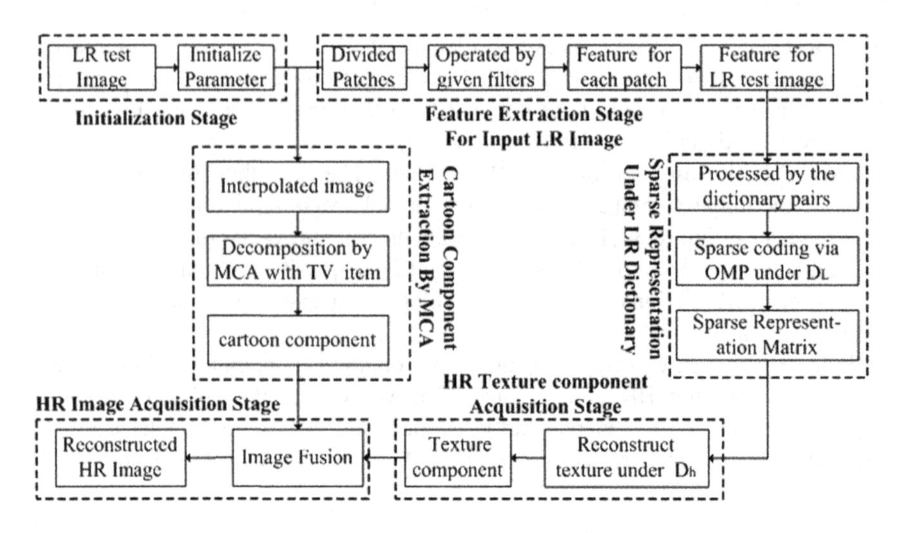

Fig. 8.4 The flowchart for SR reconstruction step

in Eq. (8.10) can be considered to solve the pseudo-inverse problem in Eq. (8.11). Figure 8.3 shows some HR dictionary's atoms obtained by the approach.

$$D_{\mathrm{h}} = \arg\min \|X_h - D_{\mathrm{h}} A\|_F^2 \tag{8.10}$$

$$D_{\mathrm{h}} = X_h A^+ = X_h A^T \left(AA^T\right)^{-1} \tag{8.11}$$

8.2.3 *The Reconstruction for Super-Resolution Image*

In this section, an approach based on MCA and sparse representation is presented to obtain HR image using inputted LR image. The flowchart for the SR image reconstruction step is shown in Fig. 8.4.

In the step, the magnified LR image is decomposed into the cartoon and texture component by the MCA method with TV penalty item in Eq. (8.5). The decomposed cartoon component is retained as the HR cartoon part of the expected HR image. The same methods from the dictionary learning step are applied to extract the feature vector z_i of each LR patch. The input LR image can be described as shown in Eq. (8.12).

$$Z_{test} = \{z_i\}_{i=1}^{n_1} = \{z_1, z_2, \ldots, z_{n_1}\} \tag{8.12}$$

In the sparse representation, the OMP algorithm is applied in sparse coding for the LR feature by iteratively approximating the solution of the problem in

Eq. (8.13) [13].Thus, the sparse representation matrix for the input LR image can be obtained under the LR dictionary.

$$\gamma_i = \arg\min_{\gamma_i} \|z_i - D_L \gamma_i\|_2^2 \quad s.t. \ \|\gamma_i\|_0 \le T_0 \tag{8.13}$$

To estimate HR texture, each texture patch p_i can be calculated using the relevant sparse code and the HR dictionary in Eq. (8.14). Thus, the texture of the expected HR can be obtained using all the texture image patches.

$$p_i = D_h \ \gamma_i \tag{8.14}$$

Finally, the cartoon part from MCA image decomposition and the texture part from the sparse representation under the dictionary pairs are fused into the expected HR image.

8.3 Experiment and Result Analysis

60 training images (*t1.bmp* ~ *t60.bmp* in the folder '*CVPR08-SR/Data/Training*') are download from http://www.ifp.illinois.edu/~jyang29/ScSR.htm in the experiment. Each training image is divided into patches as the same way in [7] to train the HR\LR dictionary pairs. Four LR test images (*Lena, Peppers, Comic and Butterfly*) are used for SR reconstruction, which are the down-sampled images from http://pan.baidu.com/s/1c0Ix6Zu. For the color images, only the luminance channel is processed by various SR methods and the others are processed using Bicubic interpolation [2].

8.3.1 *The Quality Evaluation for SR Reconstruction*

In this section, some experiments are implemented for image super-resolution by different SR methods, including NN [2], Bicubic interpolation [2], Yang's in [8] and our approach. For each method, the magnification is always set as 3 and three evaluation indicators are employed to evaluate the results: PSNR, SSIM [14] and RMSE. The smaller RMSE and the bigger PSNR or SSIM mean the better image quality. The reconstructed HR images are shown in Fig. 8.5.

As shown in Fig. 8.5, the edges are usually jagged and some speckle rags stay in the SR results from NN and Bicubic interpolation. Yang's result is a little smooth and the local texture isn't enough prominent. In the result by our approach, the quality of the edges and texture is improved, such as the hat edges in *Lena*, the complex fold structures in the drawn rectangular region or the gangly pepper in *Peppers*. Besides, our approach shows the least freckles and jagged blocks among

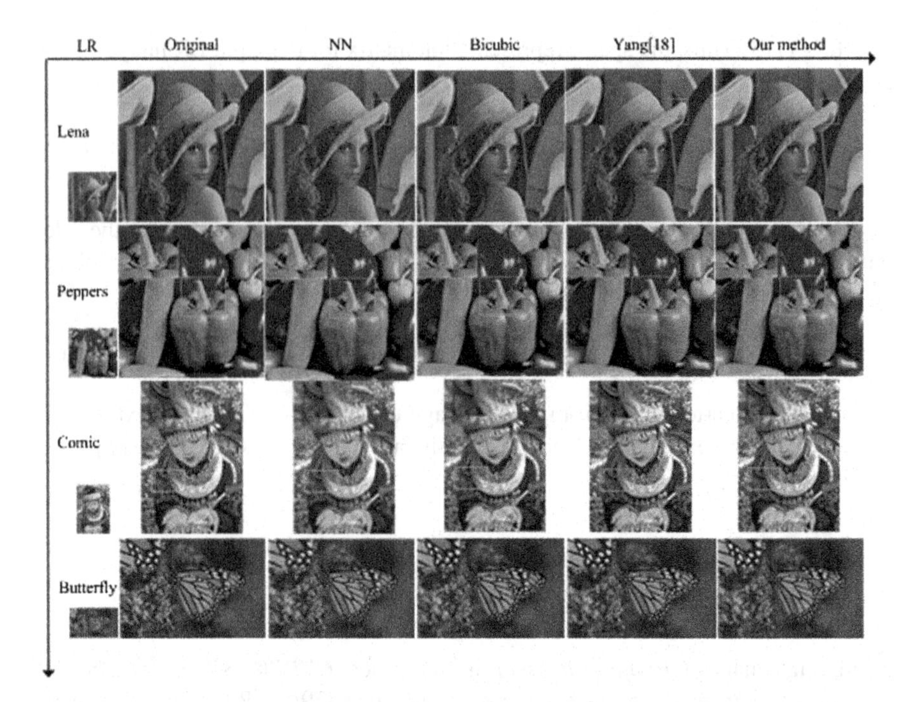

Fig. 8.5 The SR results using different methods. From *left* to *right* in each row: LR image, original HR image, the reconstructed results by NN, Bicubic, Yang's and our approach

Table 8.1 The evaluation values for the reconstruction results in different approaches

LR Image	Evaluation	NN	Bicubic	Yang [8]	Our
Lena (170*170 pixel)	PSNR	29.4889	31.6688	32.6294	32.9133
	SSIM	0.9430	0.9528	0.9563	0.9656
	EMSE	8.5525	6.6543	5.9576	5.7660
Peppers (170*170 pixel)	PSNR	29.8185	32.3616	33.3092	33.9546
	SSIM	0.9589	0.9681	0.9638	0.9770
	EMSE	8.2341	6.1441	5.5091	5.1146
Comic (83*120 pixel)	PSNR	22.0139	23.1121	23.9000	23.9522
	SSIM	0.6452	0.6987	0.7555	0.7603
	RMSE	20.2230	17.8211	16.2757	16.1781
Butterfly (256*170 pixel)	PSNR	27.0356	29.4222	30.7031	31.0197
	SSIM	0.9599	0.9700	0.9762	0.9803
	RMSE	11.3438	8.6185	7.4368	7.1706

Table 8.2 PSNR for reconstructed images with additive Gaussian noise

Standard deviation	NN	Bicubic	Yang [8]	Our
$\sigma = 1$	22.0019	23.1025	23.8752	23.9276
$\sigma = 2$	21.9696	23.0708	23.8201	23.8538
$\sigma = 3$	21.9185	23.0290	23.6907	23.7302
$\sigma = 4$	21.8447	22.9644	23.5461	23.5848
$\sigma = 5$	21.7559	22.8870	23.3651	23.3751
$\sigma = 6$	21.6438	22.7923	23.1353	23.1587
$\sigma = 7$	21.5235	22.6853	22.8707	22.8872

these methods, such as the texture fold, the curved boundaries of the garment and the detailed decoration around the neck in *Comic*, and the marking of the butterfly's wings in *Butterfly*.

To avoid the visual error, PSNR, RMSE and SSIM are compared between result and original HR image in Table 8.1. PSNR in our approach are bigger and RMSE are smaller obviously. It certifies that the recovered HR images by our approach are more approximated to the original HR image. Besides, the bigger SSIM means that our results are more similar to the original HR image in terms of image structures [14].

8.3.2 The Reconstruction for Images with Additive Noises

In this section, several experiments are conducted to demonstrate that the proposed approach is more robust to the image with additive noises. The *Comic* image is selected in this section, because more complex textures and hierarchical structures are included in this image.

In these experiments, the LR *Comic* image with different additive noises are magnified into HR image with 249*360 pixels from 83*120 pixels. In each experiment, various Gaussian white noises with constant mean 0 and different standard deviations σ are added into the inputted original test image. The PSNR are shown in Table 8.2. In Table 8.2, the PSNR from our approach is biggest obviously. It proves that the proposed method is more robust to noise than these conventional methods.

8.4 Conclusions

In this paper, a novel approach is proposed for SR image based on MCA and dictionary learning. The MCA theory is used to decompose images well both in the dictionary learning step and HR image reconstruction step. The LR dictionary is trained by the KSVD from the extracted feature of LR image patches. The HR dictionary is calculated directly via the sparse representation. It's helpful to reduce the complexity of the dictionary learning step. In the reconstruction step, the cartoon part can be obtained well by MCA with TV penalty item. The dictionary pairs make a contribution to recovering the texture part for the expected HR image. A series of experiments and results verify the validity and robustness of our approach.

Acknowledgement We'd like to thank all the researchers in the references for their related work and meaningful comments on image super-resolution. Further more, the work in this paper is funded by Chongqing University Postgraduates' Innovation Project, Project Number: CYS15026.

References

1. Park SC, Park MK, Kang MG (2003) Super-resolution image reconstruction: a technical overview. IEEE Sig Process Mag 20(3):21–36
2. Hadhoud MM, Abd EI-samie FE, EI-Khamy SE (2004) New trends in high resolution image processing. In: The Workshop on Photonics and Its Application, pp 2–23
3. Li YR, Dai DQ, Shen LX (2010) Multiframe super-resolution reconstruction using sparse directional regularization. IEEE Trans Circ Syst Video Technol 20(7):945–956
4. Freeman WT, Jones TR, Pasztor EC (2002) Example-based super-resolution. Comput Graph Appl 22(2):56–65
5. Chang H, Yeung DY, Xiong YM (2004) Super-resolution through neighbor embedding. In: Proceedings of the IEEE Computer Society Conference on Computer Vision and Pattern Recognition 1:1275–1282
6. Jiji CV, Chaudhuri S (2006) Single-frame image super-resolution through contourlet learning. EURASIP J Appl Sig Process
7. Yang JC, Wright J, Huang T, Ma Y (2008) Image super-resolution as sparse representation of raw image patches. In: Computer Vision and Pattern Recognition, CVPR 2008
8. Yang JC, Wright J, Huang TS, Ma Y (2010) Image super-resolution via sparse representation. IEEE Trans Image Process 19(11):2861–2873
9. Yin HT, Li ST, Hu JW (2011) Single image super-resolution via texture constrained sparse representation. In: Proceedings of International Conference on Image Processing, ICIP, pp 1161–1164
10. Zheng ZH, Wang B, Sun K (2011) Single remote sensing image super-resolution and denoising via sparse representation. In: 2011 International Workshop on Multi-Platform/ Multi- Sensor Remote Sensing and Mapping
11. Hu WG, Hu TB, Wu T, Zhang B, Liu QX (2011) Sea-surface image super-resolution based on sparse representation. In: Proceedings of 2011 International Conference on Image Analysis and Signal Processing, p 102–107
12. Elad M, Starck JL, Querre P, Donoho DL (2005) Simultaneous cartoon and texture image inpainting using morphological component analysis (MCA). Appl Comput Harmonic Anal 19 (3):340–358
13. Aharon M, Elad M, Bruckstein A (2006) K-SVD: an algorithm for designing overcomplete dictionaries for sparse representation. IEEE Trans Signal Process 54(11):4311–4322
14. Zhou W, Alan CB, Hamid RS, Eero PS (2004) Image quality assessment: from error visibility to structural similarity. IEEE Trans Image Process 13(4):600–612

Chapter 9
Scene Classification Based on Regularized Auto-Encoder and SVM

Yi Li, Nan Li, Hongpeng Yin, Yi Chai and Xuguo Jiao

Abstract Scene classification aims at grouping images into semantic categories. In this article, a new scene classification method is proposed. It consists of regularized auto-encoder-based feature learning step and SVM-based classification step. In the first step, the regularized auto-encoder, imposed with the maximum scatter difference (MSD) criterion and sparse constraint, is trained to extract features of the source images. In the second step, a multi-class SVM classifier is employed to classify those features. To evaluate the proposed approach, experiments based on 8-category sport events (LF data set) are conducted. Results prove that the introduced approach significantly improves the performance of the current popular scene classification methods.

Keywords Scene classification · Feature learning · Regularized auto-encoder · MSD · SVM

9.1 Introduction

In the last decades, scene classification has been an active and important research topic in image understanding [1]. It manages to automatically label an image among several categories. Scene classification can be applied to a wide spread application,

Y. Li · N. Li · H. Yin (✉) · Y. Chai · X. Jiao
School of Automation, Chongqing University, Chongqing 400044, China
e-mail: yinhongpeng@gmail.com

H. Yin
Key Laboratory of Dependable Service Computing in Cyber Physical Society, Ministry of Education, Chongqing 400030, China

Y. Chai
Key Laboratory of Power Transmission Equipment and System Security, Chongqing 400044, China

Y. Jia et al. (eds.), *Proceedings of the 2015 Chinese Intelligent Systems Conference*, Lecture Notes in Electrical Engineering 360,
DOI 10.1007/978-3-662-48365-7_9

such as image indexing, object recognition, and intelligent robot navigation. Many approaches have been successfully adopted in scene classification [2]. However, in view of the variability between different classes and the similarity within one class, scene classification still remains a challenging issue.

In this paper, a novel scene classification method based on regularized auto-encoder and SVM is presented. It is simple but effective. The regularized auto-encoder is imposed with the Maximum Scatter Difference (MSD) criterion [3] and the sparsity constraint [4]. Scene classification based on this novel method mainly contains two stages, feature learning and classifier designing. The feature learning stage is based on the regularized auto-encoder. Features learned in this step are sufficient and appropriate to represent the source images. The classifier designing is based on a multi-class SVM, which is an acknowledged classifier that can achieve good performance in classification tasks. Experiments based on the LF data set demonstrate that the introduced method outperfoms traditional methods in scene classification. There are two main contributions within our proposed scene classification method.

(1) The auto-encoder with the sparsity constraint automatically extracts features from the source images. The learned features are sufficient and appropriate to describe the scene images. It is independent of prior knowledge, which can largely reduce the calculation cost.
(2) The MSD criterion considers the similarity between different scenes and the dissimilarity within a scene. In the MSD work, images belonging to different categories can be easily classified by finding the best projection direction.

The next sections of the paper are organized as follows. The details of the novel scene classification approach are elaborated in Sect. 9.2. Experiment based on the LF data set is performed and result analysis is illustrated in Sect. 9.3. Conclusions and further research are summarized in Sect. 9.4.

9.2 The Details of the Proposed Scene Classification Approach

In this part, the novel scene classification approach is described in detail. This new proposed method contains steps of feature learning and classification. In the feature learning step, sufficient and appropriate features of source images are learned by the regularized auto-encoder imposed with the sparsity constraint and the MSD criterion. In the classification step, a multi-class SVM is adopted. Particularly, several categories of scene images are classified with the trained SVM classifiers. SVMs are trained by the 1-vs-1 strategy: constructing one SVM for each pair of the classes.

9.2.1 Feature Learning

Training images are disposed into small image patches. These patches suffer from measures of normalization and whitening. Then a regularized auto-encoder is trained by these pre-processed patches. Considering the big similarity between different categories and the small dissimilarity within a category, the proposed model is applied to extract sufficient and appropriate feature from the scene images.

The regularized auto-encoder model described in this paper is a 3-layer neural network, with an input layer and output layer of equal dimension, and a single hidden layer with k nodes. In particular, in response to the input patches set $X=\{x_1, \ldots, x_m\}, x_i \in R^N$, the feature mapping $a(x)$ of the hidden layer with nodes, i.e. the encoding function is defined by Eq. (9.1).

$$a(x)=g(W_1+b_1) \tag{9.1}$$

where $g(z)=1/(1+\exp(-z))$ is the non-linear sigmoid activation function applied component-wise to the vector z. And $a(x)\in R^K$, $W_1 \in R^{K\times N}$, $b_1 \in R^K$ are the output values, weights and bias of the hidden layer, respectively. W_1 is the bases learned from input patches. In order to make the input patches set less redundant, pre-processing is applied to X. In particular, after X is normalized by substracting the mean and dividing by the standard deviation of its elements, the entire patches set may be whitened [5]. After pre-processing, the linear activation function of the output layer, i.e. the decoding function is

$$\tilde{x}=W_2 a(x)+b_2 \tag{9.2}$$

where $\tilde{x} \in R^N$ is the output value of the output layer. $W_2 \in R^{N\times K}$ and $b_2 \in R^N$ are the weight matrix and bias vector of the third layer, respectively. Thus the training of the regular auto-encoder model turns out to be the following optimization problem.

$$J_{re}=0.5\sum_{i=1}^{m}\|x_i-\tilde{x}_i\|^2 \tag{9.3}$$

Minimizing the above squared reconstruction error function with the back propagation algorithm, the weight matrices W_1, W_2 and bias b_1, b_2 are adapted. In order to overcome the over-fitting proble, a weight decay term, Eq. (9.4) is added to the cost function.

$$J_{wd}=0.5\lambda(\|W_1\|_F^2+\|W_2\|_F^2) \tag{9.4}$$

where $\|\bullet\|_F$ is the F-norm of W_1 and W_2. Besides, λ represents the weight decay parameter.

The regular auto-encoder talked above relied on the number of hidden units being small. But there is a larger number of hidden units than the input pixels, it fails to discover interesting structure of the input. Particularly, when applied in multi-classification tasks, the similarity between different scenes and the dissimilarity within a scene largely reduce the classification accuracy. To handle the presented problems, we impose the sparsity constraint and the MSD criterion respectively.

(1) The sparsity constraint

The regular auto-encoder can easily obtain sufficient features of the input data. However, when the number of hidden units is large, the performance of regular auto-encoder is affected by the large computational expense. To discover interesting structure of the input data in the situation of lots of hidden units, a sparsity constraint is imposed on the hidden units. This novel auto-encoder model is known as sparse auto-encoder (SAE) [4]. Within the SAE framework, a neuron is treated as being "active" sparsity is imposed by restricting the average activation of the hidden units to a desired constant ρ. Specifically, this is achieved by adding a penalty term with the form $\sum_{i=1}^{k} KL(\rho||\hat{\rho}_i)$ to the cost function, where KL is the KL-divergence between ρ and $\hat{\rho}_i$. And is the sparsity parameter, $\hat{\rho}_i = (1/m)\sum_{i=1}^{m} a_j(x_i)$ is the average activation of hidden node j (averaged over the training set), and β controls the weights of the sparsity penalty term.

Thus the training of the regularized auto-encoder model turns out to be the following optimization problem:

$$J_{SAE} = \min(0.5\sum_{i=1}^{m}||x_i - \tilde{x}_i||^2 + 0.5\lambda(||W_1||_F^2 + ||W_2||_F^2) + \beta\sum_{j=1}^{K} KL(\rho||\hat{\rho}_j)) \quad (9.5)$$

(2) The MSD criterion

The Maximum Scatter Difference (MSD) norm [6] is a normalization of fisher discriminant criterion. It tries to seek a best projection direction, which can easily divide the categories of samples.

Assuming there are N pattern classes leave to be recognized, $l_1, l_2, \ldots l_N$, the intra-class scatter matrix S_b and inter-class scatter matrix S_w are defined as:

$$S_b = \frac{1}{C}\sum_{i=1}^{N} C_i(t_i - t_{mean})(t_i - t_{mean})^T \quad (9.6)$$

$$S_w = \frac{1}{C}\sum_{i=1}^{N}\sum_{j=1}^{C_i}(x_i^j - t_i)(x_i^j - m_i)^T \quad (9.7)$$

where C represents the training samples number, similarly C_i is the training samples number in class i, in which the jth training sample is denoted by x_i^j. Furthermore, t_{mean} and t_i are the mean vector of all training samples and training samples in class i respectively. Different categories are divided remotely with larger S_b value. Meanwhile, images from the same category get closer with smaller S_w value.

Reviewing the classic Fisher discriminant analysis [7], samples can be easily separated when the ratio of S_b and S_w, or their difference value gets maximal value. Combining with the above contents, we adopt the following format of MSD criterion:

$$J_{MSD} = w^T S_b w - \zeta \cdot w^T S_w w = w^T (S_b - \zeta \cdot S_w) \tag{9.8}$$

The two items $w^T S_w w$ and $w^T S_b w$ are balanced by the nonnegative constant ζ.

Based on the concept of Rayleigh quotient and its extreme property, the optimal solution of quotion (9.6), referring to the eigenvectors $w_1, w_2, \ldots, w_k$, can be acquired by the characteristic equation $(S_b - \zeta \cdot S_w) w_j = \lambda_j w_j$, in which the first largest eigenvalues meet the requirements $\lambda_1 \geq \lambda_2 \geq \ldots \lambda_k$.

Comparing with Fisher discriminant analysis method, the calculation of $S_w^{-1} S_b$ is replaced with $S_b - \zeta \cdot S_w$ within the framework of MSD criterion. Thus it will be fine weather S_w is a singular matrix or not. This makes computing becomes more effective.

Given the basic idea of auto-encoder, imposed with the sparsity constraint, the MSD criterion is imposed by taking the weight W_1, which contacts the input layer with the hidden layer, as the projection matrix. Thus the proposed model turns to be the following optimization problem

$$J(w) = J_{SAE} + J_{MSD} \tag{9.9}$$

where J_{SAE} and J_{MSD} are described there-in-before. By a number of iteration steps, a balanced value of the weight and projection direction is obtained. With the proposed feature learning model, sufficient and appropriate features from training and testing images can be extracted.

9.2.2 *Classifier Designing*

In this part, a multi-class support vector machine (SVM) classifier is designed for scene classification. It is a kind of supervised machine learning. SVM is often used to learn high-level concepts from low-level image features. In this section, the one-against-one strategy is applied to train $l(l-1)/2$ non-linear SVMs, where l is the number of the scene categories.

Given the training data $v_i \in R^n$, $i = 1, \ldots, s$ in two classes, with labels $y_i \in \{-1, 1\}$. Each SVM solves the following constraint convex optimization problem

$$\min_{\psi, b, \xi} (0.5\|\psi\|^2 + C \sum_{i=1}^{S} \xi_i), s.t. y_i(\psi^T \phi(v_i) + b) \geq 1 - \xi_i, \xi_i > 0 \tag{9.10}$$

where $C \sum_{i=1}^{S} \xi_i$ is the regulation term for the non-linearly separable data set, $(\psi^T \phi(v) + b)$ is the hyper-plane. There are two main parameters that play an important role in SVM classification, C and γ. Parameter C represents the cost of penalty, which has great influence on the classification outcome. The selection of γ can affect the partitioning outcome in the feature space. Parameters C and γ make SVM achieve significant performance in classification tasks.

9.3 Experiment and Results Analysis

In this section, experiment on LF data set is conducted to evaluate the consequence of the introduced method. Experimental results show that better performance over current approaches is achieved in scene classification task (Fig. 9.1).

Experiment on LF data set

All these RGB images are firstly transferred into gray ones. Then feature learning is conducted with the proposed regularized auto-encoder model, convolution and mean pooling method. The obtained features are fed to multi-class SVM classifier to classify features. Figure 9.2 displays the confusion matrix of the sport data set.

The average performance obtained in 10 independent experiments is 91.56 %.

As shown in columns of Fig. 9.2, wrong classifications often occur in "badminton" and "bocce", which demonstrates that in the feature space of the proposed

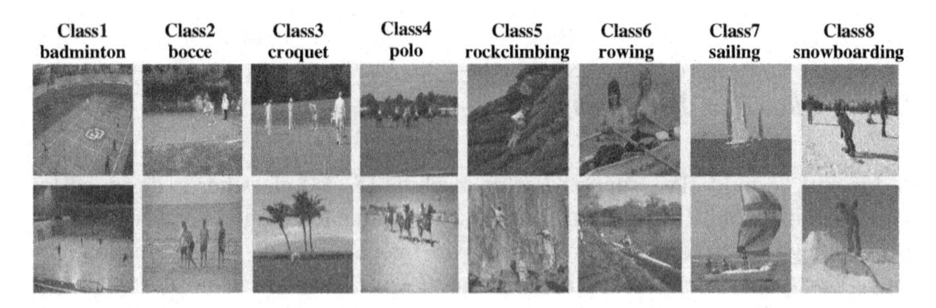

Fig. 9.1 There are 1579 RGB sport images included in LF data set: badminton (200), bocce (137), croquet (236), polo (182), rockclimbing (194), rowing (250), sailing (190), snow boarding (190). Each image has the size of 256 × 256 pixels

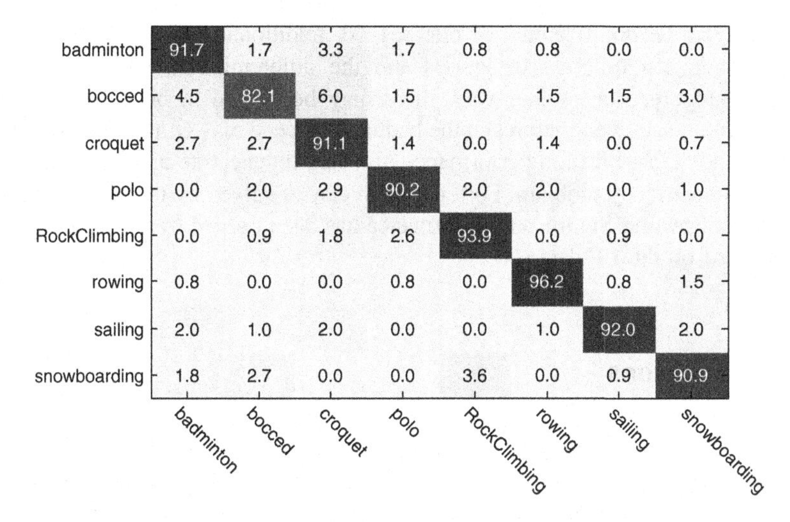

Fig. 9.2 The classification accuracy on LF data set

method, there are some similarities between these two categories with the other six categories.

Effect contrast with other current methods

In this section, experimental results of our method are compared with the previous studies on LF data set. The relevant researches and outcomes are listed in Table 9.1.

Seen from Table 9.1, our proposed method achieves better performance than the best performance Zhou et al. can achieve by far. Within the work of Zhou et al., a multi-resolution bag-of-features model is utilized. And the corresponding performance is an 85.1 % correct. Also can be seen from the outcome table, Li et al. get 73.4 % correct. This is realized by integrating scenes and object categorizations. It is mentionable that before long, Li et al. improve this performance by 4.3 % percentage points. In this method, objects of scene images are regarded as attributes, which makes recognition easier in scene classification.

Table 9.1 Effect contrast with previous studies

No	Methods	Accuracy (%)
1	Li et al. [8]	73.4
2	Wu et al. [9]	84.2
3	Li et al. [10]	77.9
4	Nakayama et al. [11]	84.4
5	Li et al. [12]	76.3
6	Zhou et al. [13]	85.1
7	Gao et al. [14]	84.9
8	**Our approach**	**91.56**

To analyze the improvement of our method, traditional approaches ignore the similarity between different categories and the differentiation between pictures within a category. Put another way, there may be people in bocced scene and bocced scene. Moreover, pictures in the badminton scene may contain people while other may not. Classification performance may be influenced to a certain extent by the object consistency problem. Fortunately, it can be solved by the MSD criterion adopted in our work. Significant performance has been proved by the experimental results based on the LF data set.

9.4 Conclusions

In this paper, a simple but effective scene classification method is proposed. It is based on a regularized auto-encoder model. Comparing with previous approaches, the proposed method achieves better performance due to three major improvements. First, the regular auto-encoder is imposed with the sparsity constraint, which is known as SAE model criterion. It automatically learns image features without relying on prior knowledge. Second, the MSD criterion is added into the SAE model, considering the similarity between different scenes and the dissimilarity within a scene. It offers more discriminative information of the input images. Thus, sufficient and appropriate features with more discriminative information are automatically extracted without prior information. Third, a multi-class SVM classifier is designed with the usually used optimization algorithm. It improves the classification accuracy efficiently. Results on LF data set indicate that this scene classification method gains better classification accuracy than other current methods.

References

1. Serrano-Talamantes JF, Aviles-Cruz C, Villegas-Cortez J, Sossa-Azuela JH (2013) Self-organizing natural scene image retrieval. Expert Syst Appl 40(7):2398–2409
2. Qin J, Yung NHC (2010) Scene categorization via contextual visual words. Pattern Recogn 43:1874–1888
3. Song F, Zhang D, Chen Q, W Jizhong (2007) Face recognition based a novel linear discriminant criterion. Pattern Anal Applic 10:165–174
4. Deng J, Zhang Z, Marchi E, Schuller B (2013) Sparse auto-encoder-based feature transfer learning for speech emotion recognition. In: Affective computing and intelligent interaction, human association conference, pp 511–516
5. Krizhevsky A, Hinton GE (2010) Factored 3-way restricted Boltzmann achiness for modelling natural images. In: International conference on artificial intelligence and statistics, pp 621–628
6. Chen Y, Xu W, Wu J, Zhang G (2012) Fuzzy maximum scatter discriminant analysis in image segmentation. J Convergence Inf Technol 7(5)
7. Yang J, Yang JY (2003) Why can LDA be performed in PCA transformed space. Pattern Recognit 2:563–566

8. Li LJ, Su H, Fei-Fei L (2007) What, where and who? Classification events by scene and object recognition. In: Eleventh IEEE international conference on computer vision, pp 1–8
9. Wu J, Rehg JM (2009) Beyond the Euclidean distance: creating effective visual codebooks using the histogram intersection kernel. In: Twelfth IEEE international conference on computer vision, pp 630–637
10. Li LJ, Su H, Lim YW, Fei-Fei L (2010) Objects as attributes for scene classification. In: European conference computer vision, pp 1–13
11. Nakayama H, Harada T, Kuniyoshi Y (2010) Global Gaussian approach for scene categorization using information geometry. In: IEEE conference on computer vision and pattern recognition, pp 2336–2343
12. Li LJ, Su H, Xing EP, Fei-Fei L (2010) Object bank: a high-level image representation for scene classification & semantic feature sparsification. In: Proceedings of the neural information processing systems
13. Zhou L, Zhou Z, Hu D (2013) Scene classification using a multi-resolution bag-of-features model. Pattern Recogn 46:424–433
14. Gao SH, Tsang IWH, Chia LT (2010) Kernel sparse representation for image classification and face recognition. In: European conference computer vision, pp 1–14
15. Huang GB, Lee H, Learned-Miller E (2012) Learning hierarchical representations for face verification with convolutional deep belief networks. In: IEEE Conference on computer vision and pattern recognition, pp 2518–2525
16. Lin SW, Ying KC, Chen SC, Lee ZJ (2008) Particle swarm optimization for parameter determination and feature selection of support vector machines. Expert Syst Appl 35 (4):1817–1824

Chapter 10
An Adaptive Optimization Algorithm Based on FOA

Qisi Zhu, Zhangli Cai and Wei Dai

Abstract To solve the problem that it is difficult to determine the initial location of the fruit fly in Fruit Fly Optimization Algorithm (FOA), an improved FOA, Adaptive Fruit Fly Optimization Algorithm (AFOA), is proposed in this paper. According to the ranges of variables to be optimized, AFOA can set the initial location of the fruit fly automatically and adjust the step value adaptively during iteration. Finally, the proposed algorithm is applied to Himmelblau's non-linear optimization problem and time series prediction using Echo State Network (ESN). The experimental results imply that AFOA is effective and also show better ability in adaptation and optimization than traditional FOA, Particle Swarm Optimization (PSO) and Genetic Algorithm (GA).

Keywords AFOA · Initial location · Step value

10.1 Introduction

For most systems, their performance will be influenced directly or indirectly by the selected parameters. If the parameters don't fit the system, its performance will reduce. However, in most cases, the optimization algorithms can pick out the fittest parameters for the systems according to the optimization goal. So the parameters optimization is important to improve the systems' performance. Among all of the optimization algorithms, heuristic optimization algorithms, such as Ant Colony Optimization (ACO), Genetic Algorithm (GA), and Particle Swarm Optimization (PSO) and so on, attract peoples' attention by their specific advantages for the non-linear optimization problem. And they are researched and applied widely [1–5]. Recently, a global optimization algorithm, Fruit Fly Optimization Algorithm (FOA) [6], which developed based on the behavior of finding food of the fruit fly,

Q. Zhu · Z. Cai (✉) · W. Dai
School of Automation, ChongQing University, Chongqing 10611, China
e-mail: czl@cqu.edu.cn

Y. Jia et al. (eds.), *Proceedings of the 2015 Chinese Intelligent Systems Conference*, Lecture Notes in Electrical Engineering 360,
DOI 10.1007/978-3-662-48365-7_10

also gets some researches' attention. And it has been applied to many problems, such as seek function extremum [7], financial Z-SCORE model [6], parameters optimization of the generalized regression neural network (GRNN) [8] and support vector machine (SVM) [9] and so on. Compared with ACO, GA and PSO, FOA has the advantages of simple construction to understand and less parameters to design. However, some shortages are also discovered. It easily fails into the local optimum and it's sensitive to the initial location of the fruit fly. When to find the fittest parameters for the system using FOA, if the initial location isn't selected properly, it will give a local optimal result or lead to failure, which is more obviously to the constrained optimization problems.

To overcome the shortages of FOA mentioned above, an improved FOA, Adaptive Fruit Fly Optimization Algorithm (AFOA), is proposed in this paper. AFOA can set the initial location of the fruit fly automatically and avoid the failure that optimized variable(s) is/are out of its/their range(s) because of the improper initial location of the fruit fly. Reference to the literature [10], this paper introduces three-dimensional search space to instead the two-dimensional search space in traditional FOA, and takes the correction coefficient into consideration when to calculate the smell concentration judgment value, which can overcome the shortage that traditional FOA is only valid to the situation presented in paper [7] that the ranges of the optimized variables are greater than zero. When every fruit fly searches food using osphresis, a new parameter, scale factor, is introduced to realize the adaptive adjustment of the step value during iteration.

10.2 Fruit Fly Optimization Algorithm

The Fruit Fly Optimization Algorithm (FOA) is inspired by foraging behavior of the fruit flies. At first, the fruit flies fly towards the high scent direction randomly from their initial flocking location. After a random distance, they use their sharp vision to find other's aggregative location where is closer to the food location and fly to there. Then, they fly towards the higher scent direction randomly from the new flocking location, and repeat the process above until the food is found finally. In a word, the process that the fruit flies find food is a journey from the position of weak scent to the position of high scent. The traditional FOA was designed through the above process and its major steps are as follow [6]:

(1) Set the fruit fly swarm scale (sizepop) and the max iterations (maxgen), and then set the initial location (X_{axis}, Y_{axis}) of the fruit fly randomly.

(2) Get the coordinate of every fruit fly after flying a random distance along the random direction.

$$X_i = X_{axis} + RandomValue \tag{2.1}$$

$$Y_i = Y_{axis} + RandomValue \tag{2.2}$$

(3) Calculate the distance ($Dist_i$) between the location of each individual and the origin, then we can get the smell concentration judgment value (S_i) which is the reciprocal of $Dist_i$.

$$Dist_i = \sqrt{X_i^2 + Y_i^2} \tag{2.3}$$

$$S_i = 1/Dist_i \tag{2.4}$$

(4) According to S_i and Fitness function that can be constructed with the given optimization goal, the smell concentration ($Smell_i$) of the every fruit fly can be calculated.

$$Smell_i = FitnessFun(S_i) \tag{2.5}$$

(5) Pick out the best smell concentration value (*smellbest*) among $Smell_i$ and find out its location (*indexbest*).

(6) Let the all of the fruit flies fly towards the location (*indexbest*) where find out at step (5), and then update the initial location and save *smellbest*.

$$X_{axis} = \mathrm{X}(indexbest) \tag{2.6}$$

$$Y_{axis} = Y(indexbest) \tag{2.7}$$

$$BestSmell = smellbest \tag{2.8}$$

(7) Repeat step (2)–(5) to enter the iteration, and judge whether *smellbest* is better (bigger or smaller is up to your optimization problem) or not, if yes, the step (6) is executed.

10.3 Adaptive Fruit Fly Optimization Algorithm

Since the fruit fly searches food in three-dimension space in real world, so paper [10] extended the FOA's search space to three-dimension from two-dimension, and the correction coefficient is taken into consideration when to compute the smell concentration judgment value, which reduced the possibility to trap into the local optimum and made the algorithm closer to the reality. The paper also points out that the optimal result has to do with the selection of initial location of the fruit fly and the random direction and distance for searching food using osphresis. When we verified the traditional FOA with some experiments, we also found that if the selection of some parameters mentioned above doesn't fit the traditional FOA, the optimal solution will often be missed, which even makes the optimized variables beyond their ranges and gets a wrong result for the constrained optimal problems. Now an improve algorithm, Adaptive Fruit Fly Optimization Algorithm (AFOA), is proposed to try to overcome the shortages above. To eliminate the effect of initial location of the fruit fly on optimization result and the trouble of determining the initial location by trial and error, AFOA sets the initial location with the reciprocal of the ranges of the optimized variables. To reduce the influence of the random direction and distance for searching food on optimization result, AFOA introduces the scale factor to adjust the direction and distance. The main steps of AFOA are as follow:

(1) Set the fruit fly swarm scale (sizepop) and the max iterations (maxgen), and then set the initial location ($X_{axis}, Y_{axis}, Z_{axis}$) randomly according to the ranges ($[V_{min}, V_{max}]$) of the optimized variables(V).

$$X_{axis} = [1/V_{max} + (1/V_{min} - 1/V_{max}) \times rand]/\sqrt{3} \tag{3.1}$$

$$Y_{axis} = [1/V_{max} + (1/V_{min} - 1/V_{max}) \times rand]/\sqrt{3} \tag{3.2}$$

$$Z_{axis} = [1/V_{max} + (1/V_{min} - 1/V_{max}) \times rand]/\sqrt{3} \tag{3.3}$$

where *rand* is the function to generate the random number whose range is [0, 1].

(2) Get the coordinate of every fruit fly after flying a random distance along the random direction.

$$X_i = X_{axis} + \alpha \times X_{axis} \times rands \tag{3.4}$$

$$Y_i = Y_{axis} + \alpha \times Y_{axis} \times rands \tag{3.5}$$

$$Z_i = Z_{axis} + \alpha \times Z_{axis} \times rands \tag{3.6}$$

where $\alpha (0 < \alpha \leq 1)$ is the scale factor to adjust the random direction and distance and *rands* is the function to generate the random number whose range is [−1, 1].

(3) Calculate the distance ($Dist_i$) between the location of each individual and the origin, then we can get the smell concentration judgment value (S_i) that is the sum of the reciprocal of $Dist_i$ and the correction coefficient.

$$Dist_i = \sqrt{X_i^2 + Y_i^2 + Z_i^2} \tag{3.7}$$

$$S_i = 1/Dist_i + B_i \tag{3.8}$$

where $B_i = 1/Dist_i \times (0.5 - rand)$ is the correction coefficient.

(4) If the smell concentration judgment value $S \notin [V_{min}, V_{max}]$ (S is an column vector composed by S_i), a new counter designed to count the times that the optimized variable(s) is/are beyond its range(s) is started and its value will add one. And if the counter's value is greater than 5, it implies that the optimized variables cannot locate in their ranges by step (2), and the correction coefficient will be

replaced by $B_i = Dist_i \times (0.5 - rand)$ and the counter will be reset at the same moment, then the step (2) will be executed to adjust the direction and distance of the fruit fly. If the counter's value is less than 5, the correction coefficient will be unchanged. If the smell concentration judgment value $S \in [V_{min}, V_{max}]$, step (5) will be executed.

(5) According to S_i and Fitness function that can be constructed with the given optimization goal, the smell concentration ($Smell_i$) of the individual can be calculated.

$$Smell_i = FitnessFun(S_i) \tag{3.9}$$

(6) Pick out the best smell concentration value (*smellbest*) among $Smell_i$ and find out its location (*indexbest*).

(7) Let the all of the fruit flies fly towards the location (*indexbest*) where find out at step (6), and then update the initial location and save *smellbest*.

$$X_{axis} = X(indexbest) \tag{3.10}$$

$$Y_{axis} = Y(indexbest) \tag{3.11}$$

$$Z_{axis} = z(indexbest) \tag{3.12}$$

$$BestSmell = smellbest \tag{3.13}$$

(8) Repeat step (2)–(6) to enter the iteration, and judge whether *smellbest* is better (bigger or smaller is up to your optimization problem) or not, if yes, the step (7) is executed.

10.4 Simulation Results and Analysis

To verify AFOA proposed in part 3, Himmelblau's non-linear optimization problem and time series prediction using ESN will be introduced to find the fittest parameters using AFOA for them. By the way, many experiments had been done to try to find the fittest parameters using traditional FOA for those two cases, but they were failed, which applies that, in a sense, traditional FOA isn't fit for multivariable optimization problems, especially for the constrained multivariable optimal problems.

10.4.1 Himmelblau Non-Linear Optimization Problem

This is a multivariable optimization problem, which proposed by Himmelblau [11], and solved using the method of generalized reduced gradient (GRG). In this problem, there are five variables $(x_1, x_2, x_3, x_4, x_5)$ to be optimized. The problem can be stated as follows:

$$min\, f(\mathbf{X}) = 5.3578547x_3^2 + 0.8356891x_1x_5 + 37.29329x_1 - 40792.141 \quad (4.1)$$

$$g_1(\mathbf{X}) = 85.334407 + 0.0056858x_2x_5 + 0.00026x_1x_4 - 0.0022053x_3x_5 \quad (4.2)$$

$$g_2(\mathbf{X}) = 80.51249 + 0.0071317x_2x_5 + 0.0029955x_1x_2 + 0.0021813x_3^2 \quad (4.3)$$

$$g_3(\mathbf{X}) = 9.300961 + 0.0047026x_3x_5 + 0.0012547x_1x_3 + 0.0019085x_3x_4 \quad (4.4)$$

This problem's boundary conditions list as follows:

$$0 \le g_1(\mathbf{X}) \le 92, 90 \le g_2(\mathbf{X}) \le 110, 20 \le g_3(\mathbf{X}) \le 25 \quad (4.5)$$

$$78 \le x_1 \le 102, 33 \le x_2 \le 45, 27 \le x_3 \le 45, 27 \le x_4 \le 45, 27 \le x_5 \le 45 \quad (4.6)$$

where $\mathbf{X} = (x_1, x_2, x_3, x_4, x_5)^T$.

In this paper, the Himmelblau problem will be solved using AFOA with the fruit fly swarm scale (sizepop = 30), the max iterations (maxgen = 1000) and the scale factor ($\alpha = 0.0045$). And then the optimal result will be compared with other results listed in some papers, which show in Table 10.1.

Unlike traditional FOA, it's easy for AFOA to get the final optimal result without trial and error, which also proves the advantage of AFOA in multivariable optimization problem. Among the four optimization algorithms listed in Table 10.1, the AFOA gives the minimum optimal value $f(\mathbf{X}) = -31023.925$, which shows its superior ability. On the other hand, compared with those optimization methods list

Table 10.1 The optimal result for Himmelblau problem using four different methods

Designed variables	AFOA	Carlos A [12]	Abdollah [13]	GRG [11]
x_1	78.0043	78.0495	78.00	78.6200
x_2	33.0068	33.0070	33.00	33.4400
x_3	27.0785	27.0810	29.995	31.0700
x_4	44.9925	45.0000	45.00	44.1800
x_5	44.9559	44.9400	36.776	35.2200
$g_1(X)$	91.9992	91.997635	90.715	90.520761
$g_2(X)$	100.4067	100.407857	98.840	98.892933
$g_3(X)$	20.0010	20.001911	20.00	20.131578
$f(X)$	−31023.925	−31020.859	−30665.5	−30373.949

in Table 10.1, AFOA takes the advantages of simple calculation, less designed parameters and easy implementation.

10.4.2 Time Series Prediction Using ESN

Echo State Network (ESN) is a new style of neural network, and it is an improvement and revolution to the traditional Recurrent Neural Network (RNN). ESN was originally proposed by Herbert Jaeger [14] in 2001, and it overcomes the RNN's shortages of hard to determine the structure and complex training algorithm.

ESN consists of three layers: input layer, reservoir and readout network. The connection weight matrix W_{in} between the input layer and reservoir, W inside the reservoir are generated randomly and unchanged after that. Only the connection weight matrix W_{out} between the reservoir and readout network is trained, which allows ESN to have less amount of calculation and high speed of network training than RNN.

In this paper, ESN's four parameters (the number of reservoir's node (N), spectral radius (SR), input scale factor (IS) and sparsity of reservoir (SD)) will be optimized using AFOA, and then we will test optimized ESN on the Lorenz time series prediction problem. The four parameters of ESN have the same ranges with paper [15], i.e. $n/10 \leq N \leq n/2$, $0.1 \leq SR \leq 0.99$, $0.01 \leq IS \leq 1$, $0.01 \leq SD \leq 0.5$, where n is the number of training samples. The Lorenz system can be described by the three functions as follow:

$$\begin{cases} \dot{x} = \delta(y - x) \\ \dot{y} = rx - y - xz \\ \dot{z} = -bz + xy \end{cases} \tag{4.7}$$

In this experiment, we let $\delta = 10$, $r = 28$, $b = 8/3$ and the initial conditions $x(0) = 10$, $y(0) = 1$, $z(0) = 0$. We will take 3000 groups of data generated by Lorenz system to finish the five steps forward prediction task. The first 2000 groups will be training data and the rest will be testing data. We repeat the experiment 10 times using AFOA with fly swarm scale (sizepop = 20), the max iterations (maxgen = 200) and the scale factor ($\alpha = 1$). Finally, the optimal result will be compared with GA's and PSO's, which shows in Table 10.2.

If the four main parameters of ESN are selected improperly, the precision of the time series prediction will be affected. Although when the best result appears among 10 times, the error of prediction using AFOA is slightly bad with respect to GA's, it is much lower than PSO's and when we take the mean of error (mean (error)) and the variance of error (var(error)) into consideration, we will find the mean of error using AFOA is lowest and the variance of error is lower than GA's which means that AFOA's stability is better than GA's in time series prediction. On the other hand, according to Table 10.2, we can also find that the number of node in

Table 10.2 The optimal result of the Lorenz time series using ESN with three different methods

Designed variables	AFOA		GA		PSO	
	Best result	Worst result	Best result	Worst result	Best result	Worst result
N	779	598	921	690	985	995
SR	0.99	0.99	0.97	0.90	0.10	0.10
IS	0.93	0.98	0.92	0.92	1.00	1.00
SD	0.16	0.25	0.31	0.39	0.31	0.37
error	6.35e-04	1.20e-03	5.83e-04	1.53e-03	8.63e-03	8.77e-03
mean(error)	8.78e-04		1.14e-03		8.68e-03	
var(error)	3.92e-08		7.32e-08		1.74e-09	

reservoir that is obtained by AFOA is least among the three methods, which means less amount of calculation and high speed.

10.5 Conclusion

To overcome the shortages of traditional FOA that it's difficult to determine the initial location of the fruit fly and select the step value during iteration, an improved FOA, AFOA, was proposed in this paper. This algorithm can set the initial location automatically and the step value adaptively, and only the fruit fly swarm scale, the max iterations and a new parameter, scale factor, need the users to determine, which can reduce the uncertainty and improve the adaptation of this algorithm. Finally, the efficiency of AFOA has been proved by two cases in part 4.

Acknowledgement Project Supported by the Fundamental Research Funds for the Central Universities, China (No. CDJZR12170006).

References

1. Dobric Goran, Stojanovic Zoran, Stojkovic Zlatan (2015) The application of genetic algorithm in diagnostics of metal-oxide surge arrester [J]. Electr Power Syst Res 119:76–82
2. Mandloi Manish, Bhatia Vimal (2015) Congestion control based ant colony optimization algorithm for large MIMO detection [J]. Expert Syst Appl 42(7):3662–3669
3. Lee CS, Ayala HVH, dos Santos Coelho L (2015) Capacitor placement of distribution systems using particle swarm optimization approaches [J]. Electr. Power Energy Syst 64: 839–851
4. Deepak Saini;Neeraj Saini (2015) A study of load flow analysis using particle swarm optimization [J]. Int J Eng Res Appl 5(1):125–131
5. Li Y, Wang R, Xu M (2014) Rescheduling of observing spacecraft using fuzzy neural network and ant colony algorithm [J]. Chin J Aeronaut 27(3): 678–687
6. Pan Wen-Tsao (2012) A new fruit fly optimization algorithm: taking the financial distress model as an example [J]. Knowl-Based Syst 26:69–74

7. Dai Hongde, Zhao Guorong, Jianhua Lu, Dai Shaowu (2014) Comment and improvement on "a new fruit fly optimization algorithm: taking the financial distress model as an example" [J]. Knowl-Based Syst 59:159–160
8. Aiqin Huang, Yong Wang (2014) Pressure model of control valve based on LS-SVM with the fruit fly algorithm [J]. Algorithms 7(3):363–375
9. Hong WP, Liao MJ (2013) Application of fruit fly optimization algorithm-least square support vector machine in fault diagnosis of fans [J]. Adv Mater Res 860–863:1510–1516
10. Pan WT (2013) Using modified fruit fly optimization algorithm to perform the function test and case studies [J]. Connection Sci 25(2–3):151–160
11. Himmelblau DM (1972) Applied Nonlinear Programming [M]. McGraw-Hill, New York
12. Carlos A, Coello C (2000) Use of a self-adaptive penalty approach for engineering optimization problems [J]. Comput Ind 41(2):113–127
13. Homaifar Abdollah, Qi Charlene X, Lai Steven H (1994) Constrained optimization via genetic algorithms [J]. Simulation 62(4):242–253
14. Jaeger H (2001) The "Echo State" Approach to analyzing and training recurrent neural network [R]. GMD Reprot 148, GMD-German National Research Institute for Computer Science, Bremen
15. Zhao LuSha (2012) Research on nonlinear time series prediction based on Echo State Networks [D]. Harbin Institute of Technology, HeiLongJiang

Chapter 11
Mirror Image-Based Robust Minimum Squared Error Algorithm for Face Recognition

Yu Wang, Caikou Chen, Ya Gu and Rong Wang

Abstract To address the problems that the minimum squared error (MSE) algorithm is short of sufficient robustness and there are often not enough training samples for face recognition (FR), a mirror image-based robust MSE algorithm which uses mirror face as new training samples is proposed. The solution vector of MSE classification model needs to transform both original training samples and its mirror images into its class labels. Owing to mirror image reflect the change of pose and expression of original face images, the solution of classification model has some robustness. In addition, two classifier schemes are proposed and the second classifier scheme outperforms the first classifier scheme computationally. The experimental results on FERET, Extended YaleB and ORL databases indicate that the proposed approach achieves better robustness on images whose pose and expression are changed than the traditional MSE algorithm.

Keywords Squared error · Face recognition · Mirror image

11.1 Introduction

Minimum squared error (MSE) algorithm has been widely used for pattern recognition [1]. MSE algorithm use both samples and their class labels as input and output of the model [2], and attempt to achieve the mapping that can best transform the input into the corresponding output [3].

Duda [4] has shown that when the number of training samples is infinite, MSE algorithm is completely equivalent to linear discriminant analysis (LDA) for two class classification problems [5]. LDA is a supervised algorithm which uses classification

Y. Wang (✉) · C. Chen · Y. Gu · R. Wang
College of Information Engineering, Yangzhou University,
Yangzhou 225127, China
e-mail: wangdayu19890928@163.com

Y. Jia et al. (eds.), *Proceedings of the 2015 Chinese Intelligent Systems Conference*, Lecture Notes in Electrical Engineering 360,
DOI 10.1007/978-3-662-48365-7_11

information of samples effectively [6]. A kernel minimum squared error (KMSE) algorithm which can solve nonlinear problem was presented by Xu [5].

Recently, in order to overcome lack of training samples in face recognition, mirror image was presented by Yong Xu, so the number of training samples can be expanded [7].

In this paper, robust minimum squared error (RMSE) algorithm is presented. We use every training image to construct the corresponding mirror image so as to expand training set. The solution of classification model has more robustness due to mirror image reflect the change of pose and expression of original face images. In addition, this paper presents two classifier schemes for the final recognition. Finally, the experimental results on FERET, Extended YaleB and ORL face databases verify the effectiveness of RMSE.

11.2 Related Works

Suppose that there are c classes and a collection of N training samples $\{\mathbf{x}_i\}_{i=1}^{n}$. Each training sample $\mathbf{x}_i$ belongs to one of c classes. We use a c-dimensional vector to represent the class label. The row vector $\mathbf{g}_1 = [1, 0, \ldots, 0]$ and $\mathbf{g}_c = [0, 0, \ldots, 1]$ are set as the class label of the first class and the class label of the last class, respectively. That is, suppose the class label of the kth class is $\mathbf{g}_k$ which kth element is one and the other elements are all zeroes. This class label is also called the class label of the kth class.

11.2.1 Minimum Squared Error (MSE) Algorithm

Suppose that matrix $\mathbf{T}$ can approximately transform each training sample into its class label. The equation can be described as follows:

$$\mathbf{XT} = \mathbf{G} \tag{11.1}$$

where $\mathbf{X} = [\mathbf{x}_1^{\mathrm{T}}, \mathbf{x}_2^{\mathrm{T}}, \ldots, \mathbf{x}_N^{\mathrm{T}}]^{\mathrm{T}}, \mathbf{G} = [\mathbf{g}_1^{\mathrm{T}}, \mathbf{g}_2^{\mathrm{T}}, \ldots, \mathbf{g}_N^{\mathrm{T}}]^{\mathrm{T}}$.

Due to Eq. (11.1) cannot be directly solved; we convert it into the following equation:

$$\mathbf{X}^{\mathrm{T}}\mathbf{XT} = \mathbf{X}^{\mathrm{T}}\mathbf{G} \tag{11.2}$$

We can use $\mathbf{T}^* = (\mathbf{X}^{\mathrm{T}}\mathbf{X} + \alpha \mathrm{I})^{-1}\mathbf{X}^{\mathrm{T}}\mathbf{G}$ as the optimization solution, where $\mathbf{I}$ and α denote the identity matrix and a small positive constant respectively.

MSE identifies the class label of a test sample $\mathbf{z}$ as follows: firstly, $\mathbf{g}_{\mathbf{z}} = \mathbf{z}\,\mathbf{T}^*$ is used to calculate predicted label of the test sample. Secondly, we can calculate the distance between $\mathbf{g_z}$ and the class label of all the c classes are calculated. Thirdly, if the class label of the kth class is the closest to $\mathbf{g_z}$, test sample $\mathbf{z}$ will be identified into class k.

11.2.2 Mirror Image

Suppose $\mathbf{X}_i^{\mathrm{o}}$ is a piece of original image which dimension is $p \times q$ and $\mathbf{X}_i$ is the corresponding one dimension row vector, where $i = 1, 2\ldots N$. $\mathbf{Y}_i^{\mathrm{o}}$ is the mirror image of $\mathbf{X}_i^{\mathrm{o}}$ and y_i is a one dimension row vector of $\mathbf{Y}_i^{\mathrm{o}}$, where $i = 1, 2\ldots N$. We can use the following equation to generate $\mathbf{Y}_i^{\mathrm{o}}$:

$$\mathbf{Y}_i^{\mathrm{o}}(i,j) = \mathbf{X}_i^{\mathrm{o}}(i, q - j + 1) \tag{11.3}$$

where $i = 1, 2\ldots p$, $j = 1, 2\ldots q$.

We can obviously see from the construction process of the mirror images that the mirrored images obtained are different from the corresponding original face images and also properly represent some natural variation of the face image under pose and illumination. Therefore the mirrored images can be used to improve the performance of existing face recognition methods.

11.3 Robust Minimum Squared Error (RMSE) Algorithm

11.3.1 Description of the Proposed Method

Suppose $\mathbf{x}_i$ is a p-dimension row vector and $\mathbf{y}_i$ is the corresponding mirror image ($i = 1, 2\ldots N$). RMSE has the following model:

$$\min(||\mathbf{XT} - \mathbf{G}||_2^2 + \gamma||\mathbf{YT} - \mathbf{G}||_2^2) \tag{11.4}$$

where

$$\mathbf{X} = [\mathbf{x}_1^{\mathrm{T}}, \mathbf{x}_2^{\mathrm{T}}, \cdots, \mathbf{x}_N^{\mathrm{T}}]^{\mathrm{T}}, \mathbf{Y} = [\mathbf{y}_1^{\mathrm{T}}, \mathbf{y}_2^{\mathrm{T}}, \cdots, \mathbf{y}_N^{\mathrm{T}}]^{\mathrm{T}}, \mathbf{G} = [\mathbf{g}_1^{\mathrm{T}}, \mathbf{g}_2^{\mathrm{T}}, \cdots, \mathbf{g}_N^{\mathrm{T}}]^{\mathrm{T}}. \tag{11.5}$$

In Eq. (11.4), γ is a small positive constant. It is clear that the dimension of $\mathbf{X}$ and $\mathbf{Y}$ is $N \times p$, $\mathbf{G}$ is an $N \times c$ matrix, and $\mathbf{T}$ is a $p \times c$ matrix and $\mathbf{g}_i$ is the class label of ith training sample ($i = 1, 2\ldots N$). We can obtain $\mathbf{T}$ using

$$\mathbf{T}^* = (\mathbf{X}^{\mathrm{T}}\mathbf{X} + \gamma\mathbf{Y}^{\mathrm{T}}\mathbf{Y} + \beta\mathbf{I})^{-1}(\mathbf{X}^{\mathrm{T}} + \gamma\mathbf{Y}^{\mathrm{T}})\mathbf{G} \tag{11.6}$$

where β is a little positive constant and $\mathbf{I}$ is the identity matrix.

Now, we can use Eq. (11.7) to get the c-dimension predict class label of test sample:

$$\mathbf{g_z} = \mathbf{z}\mathbf{T}^* \tag{11.7}$$

11.3.2 Scheme1 for Classification Stage

Let $\mathbf{g}_i$ is the actual category of ith class label (i = 1, 2... c), thus the distance between $\mathbf{g_z}$ and $\mathbf{g}_i$ can be described as following equation [8]:

$$\mathrm{r}_i = ||\mathbf{g_z} - \mathbf{g}_i||_2 \tag{11.8}$$

Then we can employ Eq. (11.9) to classify the test sample [9]:

$$k = \arg\min_i(\mathrm{r}_i) \tag{11.9}$$

11.3.3 Scheme2 for Classification Stage

Another way to identify the class label of a test sample is that one finds the maximum element of the class label vector $\mathbf{g_z}$ and the class responding to the maximal element is assigned to the test sample. Specifically, suppose that $\mathbf{g_z}$ has c elements, which represents 1, 2... c classes and let the position of maximum element being k, we can decide the test sample belongs to class k. The classifier scheme can be defined as:

$$\mathbf{g_z}(k) = \max_{1\le l\le c}\{\mathbf{g_z}(1), \mathbf{g_z}(2), \ldots, \mathbf{g_z}(l), \mathbf{g_z}(l+1), \ldots, \mathbf{g_z}(c)\} \tag{11.10}$$

where $\mathbf{g_z}(l)$ is the lth element of $\mathbf{g_z}$.

11.3.4 Steps of our Method

The algorithm of RMSE includes the following steps:

Step 1. Use Eq. (11.3) to construct each face image's mirror image.

Step 2. The training sample matrix **X** and mirror image matrix **Y** can be constructed by all the original training samples and its mirror images through Eq. (11.5). Use the same way to generate class label matrix **G**.

Step 3. Use Eq. (11.6) to solve Eq. (11.4), get the optimization transform matrix $\mathbf{T}^*$.

Step 4. Use Eq. (11.7) to calculate the predict class label of test sample.
Step 5. Apply classifier scheme1 or scheme2 to classify the test sample.

11.4 Experimental Results

11.4.1 Experimental Results on the FERET Face Database

The FERET face database has 1400 photos of 200 people and each subject has 7 images. We used down-sampling method to normalize each image to 40 × 40. We first used Eq. (11.3) to generate the mirror image of each training sample. Figure 11.1 shows the 7 sample images of this face database and Fig. 11.2 shows the corresponding mirror images. We use the top *m* images of each individual for training and the other 7-*m* images are employed for testing, where $m = 3, 4$. Thus, we respectively take 200 × *m* original images for training and 200 × (7-*m*) images as test samples. In the final recognition stage, two classifier scheme, called scheme1 and scheme2 are used to identify the class label each test sample belongs to. Tables 11.1 and 11.2 show the experimental results on the FERET face database. The numbers in parentheses is the running time of the proposed algorithm and parameters α, β of MSE and RMSE algorithm are set as $\alpha = \beta = 0.001$.

11.4.2 Experimental Results on the Extended YaleB Face Database

The Extended YaleB face database contains 38 subjects and each subject has 64 images. The dimension of each picture is 192 × 168. We use the 7th, 8th, 9th, 37th,

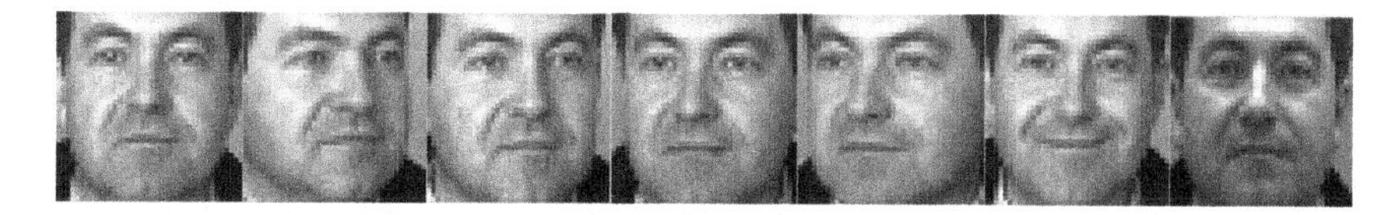

Fig. 11.1 The first 7 images of FERET face database

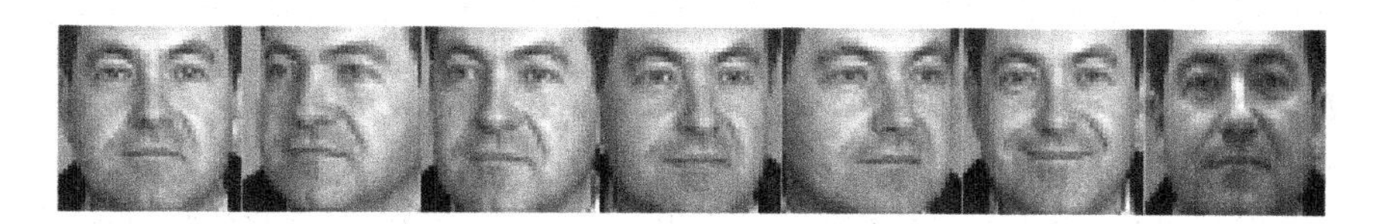

Fig. 11.2 The mirror images of the first 7 images of FERET face database

Table 11.1 Recognition rates (%) of MSE algorithm and RMSE algorithm using the classifier scheme1 on the FERET face database

The value of m	3	4
MSE	37.62	44.67
RMSE ($\gamma = 0.1$)	44.62	49.67
RMSE ($\gamma = 0.5$)	43.87	45.67
RMSE ($\gamma = 0.7$)	44	45.17

Table 11.2 Recognition rates (%) and running time (s) of RMSE using scheme1 and scheme2 with varying values of γ on the FERET face database

The value of m	3	4
RMSE (scheme1, $\gamma = 0.1$)	44.62(0.9069)	49.67(0.7309)
RMSE (scheme2, $\gamma = 0.1$)	44.62(0.2276)	49.67(0.1522)
RMSE (scheme1, $\gamma = 0.5$)	43.87(0.916)	45.67(0.6986)
RMSE (scheme2, $\gamma = 0.5$)	43.87(0.2219)	45.67(0.1442)
RMSE (scheme1, $\gamma = 0.7$)	44(0.9139)	45.17(0.6894)
RMSE (scheme2, $\gamma = 0.7$)	44(0.2118)	45.17(0.1677)

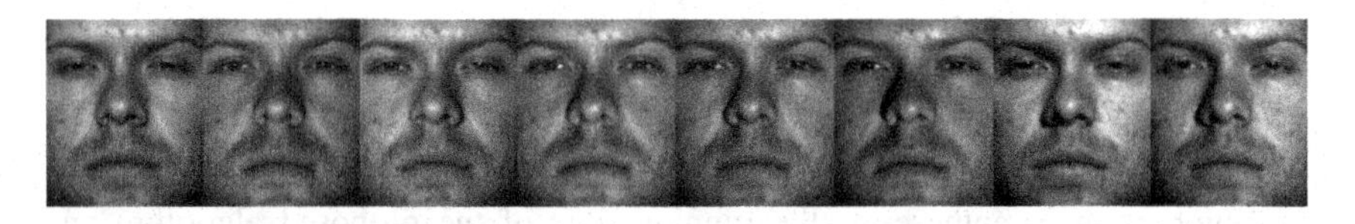

Fig. 11.3 The training samples from the first individual

Fig. 11.4 Test images from the first subject

38th, 40th, 41st and 42nd images of each subject as training set. Figure 11.3 shows some sample images from the first individual. The parameters of this experiment are set as $\alpha = \beta = 0.001$. We choose images of the 18th, 21st, 22nd, 23rd, 47th, 50th, 51st and 52nd from the same subject as the test set. Figure 11.4 shows some test samples. The total number of original training samples and test samples are both 304. The classification rates and running time of MSE and RMSE method is shown in Tables 11.3 and 11.4.

11.4.3 Experimental Results on the ORL Face Database

This database includes 400 pictures of 40 individual, each subject having 10 images. The dimension of each image was resized to 56×46. The top t images of

Table 11.3 Classification rates of MSE and RMSE on Extended YaleB face database. The dimension of each image was resized to 8 × 8

The value of γ	MSE	RMSE (scheme1)	RMSE (scheme2)
0.1	18.09	26.64(0.0206)	26.64(0.0024)
0.5	18.09	30.92(0.0208)	30.92(0.0024)
0.7	18.09	30.26(0.0233)	30.26(0.0025)

Table 11.4 Classification rates of MSE and RMSE on Extended YaleB face database. The dimension of each image was resized to 32 × 32

The value of γ	MSE	RMSE (scheme1)	RMSE (scheme2)
0.1	63.16	65.13(0.0317)	65.13(0.0209)
0.5	63.16	63.49(0.0393)	63.49(0.0141)
0.7	63.16	63.49(0.0598)	63.49(0.0177)

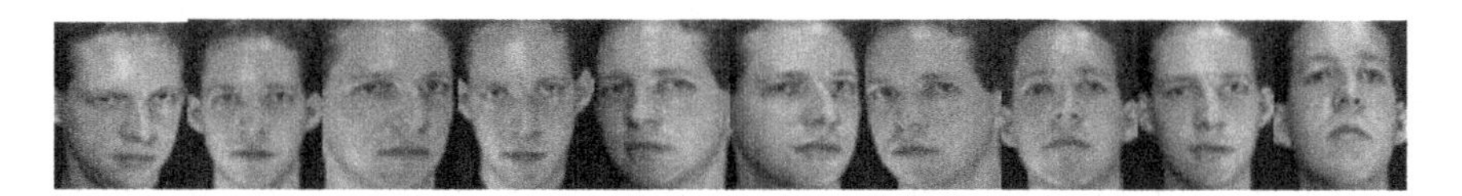

Fig. 11.5 Images from the first subject of ORL face database

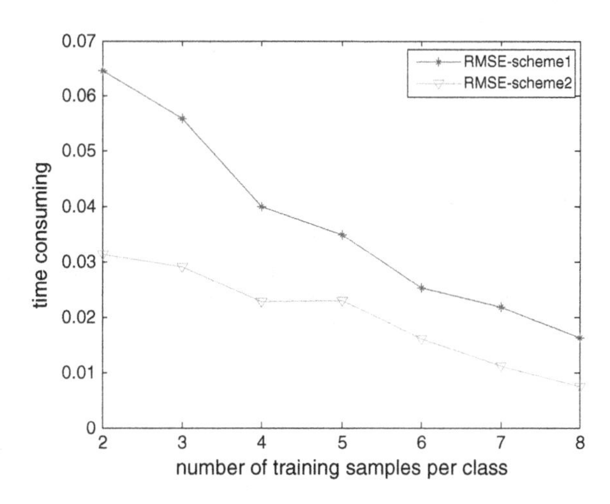

Fig. 11.6 Time consuming of scheme1 and scheme2 on ORL face database

each individual are chosen for training and the rest are used for testing, where $t = 2$, 3, …, 8. Equation (11.3) is used to generate the corresponding mirror images. Figure 11.5 shows some face samples from the first individual. This experiment is performed to just evaluate time consuming for the classifier scheme1 and scheme2. The experimental result is shown in Fig. 11.6.

11.5 Analysis of Experimental Results

From the above experimental results, we can conclude that the proposed RMSE algorithm, which uses mirror images as new training samples, is higher than the classical MSE algorithm in recognition rate. In addition, the proposed new classifier scheme is much more efficient than the traditional MSE classifier. When the classifier scheme2 is used for the final classification, we can get the same recognition rate as scheme1. Finally, through all the experiments we can see that the choice of parameters, especially γ, is one of the important factors affecting the recognition rate.

11.6 Conclusions

A mirror image-based robust minimum squared error algorithm is proposed in this paper. In order to enhance the robustness of traditional MSE method, we construct mirror images to enlarge the set of the original training samples, which can enhance the robustness of the traditional MSE algorithm. In addition, a new classifier scheme is presented and scheme2 is the improvement of scheme1. The proposed classifier scheme is more efficiency than the traditional MSE-based classifier. Extensive Experimental results on FERET, Extended YaleB and ORL face databases show that RMSE is superior to traditional MSE approach.

References

1. Zhang L, Wu X (2005) Color demosaicking via directional linear minimum mean square-error estimation. IEEE Trans (12):2167–2178
2. Dalton LA, Dougherty ER (2012) Exact sample conditioned MSE performance of the Bayesian MMSE estimator for classification error-Part I: representation. IEEE Trans 1:2575–2587
3. Yong X, Xiaozhao F, Qi Z, Yan C, Jane Y, Hong L (2014) Modified minimum squared error algorithm for robust classification and face recognition experiments. Neurocomputing 2:253–261
4. Duda RO, Hart PE, Stork DG (2004) Pattern Classification, 2nd edn. Wiley, New York
5. Jianhua X, Xuegong Z, Yanda L (2001) Kernel MSEC algorithm: a unified framework for KFD, LS-SVM and KRR. Neural Netw 2:1486–1491
6. Tao D, Li X, Wu X, Maybank SJ (2007) General tensor discriminant analysis and Gabor features for gait recognition. IEEE Trans 10:1700–1715
7. Yong X, Xuelong L, Jian Y, David Z (2014) Integrate the original face image and its mirror image for face recognition. Neurocomputing 11:191–199
8. Wright J, Yang AY, Ganesh A, Sastry SS, Ma Y (2009) Robust face recognition via sparse representation. IEEE Trans 2:210–227
9. Zhang L, Yang M, Feng X (2011) Sparse representation or collaborative representation: which helps face recognition? In: ICCV, no 6, pp 471–478

Chapter 12
Flocking of Distributed Multi-Agent Systems with Prediction Mechanism

Zhengguang Ma, Zhongxin Liu, Zengqiang Chen and Mingwei Sun

Abstract Flocking problem is one of the most critical problems of multi-agent systems. In this paper, a modified artificial potential function is proposed, which guarantees the stability of their inter-agent distances and ensures the smooth collision avoidance among neighboring agents. The prediction control algorithm is designed that can accelerate the convergence of speed for the second-order multi-agent systems. Mathematical analysis and simulation examples show the effectiveness of the theory.

Keywords Connectivity preserving · Multi-agent systems (MAS) · Flocking problem · Potential function · Prediction mechanism

12.1 Introduction

When a group of birds are foraging or in flight [1], and there are parallels with the schooling behavior of fish [2], and herd behavior of animals [3], We call that flocking behavior. During recent decades, flocking problem as one of the multi-agent systems' problems has received lots of attention. Scientists from different fields including physics, social sciences, and computer science have participated in the study for the flocking behavior in groups of agents with local interactions [4–7].

In many applications, the connectivity of multi-agent system can be ensured by some conditions. For example, in [9], the connection between agents are guaranteed by adding appropriate weights to the edges in the graph and some performance objective is achieved. To ensure the connectivity, Wang et al. [10] provide a set of decentralized flocking control protocols. Also, Wen et al. investigated the flocking

Z. Ma · Z. Liu (✉) · Z. Chen · M. Sun
Tianjin Key Laboratory of Intelligent Robotics, College of Computer and Control Engineering, Nankai University, Tianjin 300071, China
e-mail: lzhx@nankai.edu.cn

Z. Chen
College of Science, Civil Aviation University of China, Tianjin 300300, China

Y. Jia et al. (eds.), *Proceedings of the 2015 Chinese Intelligent Systems Conference*, Lecture Notes in Electrical Engineering 360,
DOI 10.1007/978-3-662-48365-7_12

problem of multi-agent systems without assuming that the communication topology can remain its connectivity frequently enough in [11]. In [12], the authors analysed the problem of preserving connectivity with bounded control inputs for multi-agent systems as well.

Based on the history information sequence, Zhang et al. [13] studied the collective behavior coordination with prediction mechanism. The prediction mechanism expands the range of feasible sampling periods and saves costly long range communications. In [14], Zhan and Li studied the model predictive control flocking of a networked multi-agent systems based on position measurements only. They proposed the centralized and distributed impulsive MPC flocking algorithms which lead to flocking. However, since the prediction is based on the historical state sequences and used to predict the future states, the decentralized predictive protocol is valid when the topology remains constant within a given changing rate for the topology.

In this paper, a modified control algorithm is proposed to solve the flocking problem of second-order multi-agent systems. With the cost of one time communication, a state predictor is designed, which enlarge the minimum nonzero eigenvalue, to accelerate the convergence. Considering collision avoidance, consensus algorithms containing modified potential function are presented, which guarantees collision avoidance as well as connectivity preserving. Corresponding results are verified with simulation.

The rest of this paper is organized as follows. In Sect. 12.2, some matrix theory and algebraic graph theory used in this paper are formulated and the flocking problem is described. In Sect. 12.3, main results are presented, in which a modified artificial potential function is proposed and the prediction mechanism is added into the control algorithm. The simulation results are presented in Sect. 12.4. Finally, conclusion is given in Sect. 12.5.

12.2 Preliminaries

In this section, some basic concepts in matrix theory and algebraic graph theory are presented. The flocking problem is described as well. For more details, the readers are referred to [15–17].

12.2.1 Algebraic Graph Theory and Matrix Theory

For a distributed control system, we represent the network of interacting agents by a dynamic graph $\mathcal{G}(t) = (\mathcal{V}, \mathcal{E}(t), \mathcal{A}(t))$, where $\mathcal{V}$ is the set of nodes representing the agents, $\mathcal{E}(t) \in \mathcal{V} \times \mathcal{V}$ is the set of edges at time t, and $\mathcal{A}(t)$ is an $N \times N$ adjacency matrix with $a_{ij}(t) \geq 0$ denoting the edge weight from node i to node j at time t. Assuming there is no self-cycle, then $a_{ii}(t) \equiv 0$ for all $i \in \mathcal{V}$. For an undirected graph

$\mathcal{G}(t)$, the adjacency matrix $\mathcal{A}(t)$ is symmetric. The *Laplacian matrix* are defined as $L = \{l_{ij}\}, i, j = 1, \dots, n$, with

$$l_{ij} = \begin{cases} \sum_{j=1, j\neq i}^{N} a_{ij}, \\ -a_{ij}, \end{cases}$$

Once the graph is connected, the smallest eigenvalue of L is always zero [15].We express $d_i = \sum_{j=1, j\neq i}^{N} a_{ij}$ as the degree of vertex i.

In the rest paper, I_m denotes the $m \times m$ unitary matrix for positive integer m. We use $diag(\cdot)$ to denote a diagonal matrix. The norm of a vector $v \in R^m$ is defined as $||v|| = \sqrt{v^T v}$. $||u||_\sigma = \sqrt{1 + ||u||^2} - 1$ denotes a nonnegative σ-norm. And a very useful tool is the *Kronecker product*, which is defined between two matrices $P = [p_{ij}]$ and Q as $P \otimes Q = [p_{ij} Q]$.

12.2.2 Flocking Problem

Consider a multi-agent system with N agents, of which agent i has the dynamics as

$$\begin{cases} \dot{x}_i(t) = v_i(t), \\ \dot{v}_i(t) = u_i(t), \end{cases} \quad (12.1)$$

where $x_i(t)$, $v_i(t)$, $u_i(t) \in \mathbb{R}^m$ denote the position, velocity and control input of the i-th agent respectively.

The distributed control law $u_i(t)$ is often designed by utilizing the state information acquired from agent i and its neighbor agents, which is given by

$$u_i = -\sum_{j \in N_i(t)} a_{ij}(v_i - v_j) - \sum_{j \in N_i(t)} \nabla_{x_i} V_{ij}, \quad (12.2)$$

where $N_i(t)$ represents the set of neighbors of agent i at time t and $N_i(t) = \{j \in \mathcal{V} | a_{ij}(t) \neq 0\} = \{j \in \mathcal{V} | (i, j) \in \mathcal{E}\}$, and $\nabla_{x_i} V_{ij}$ corresponds to a vector in the direction of the reduced gradient of an artificial potential function which guarantees the stability of their inter-agent distance and ensures the collision avoidence among neighboring agents.

Lemma 1 ([17]). *A group of mobile agents is said to (asymptotically) flock, when all agents attain the same velocity, distance between the agents are stabilized and no collision among them occurs.*

Before the representation of the main result, we have the following definition.

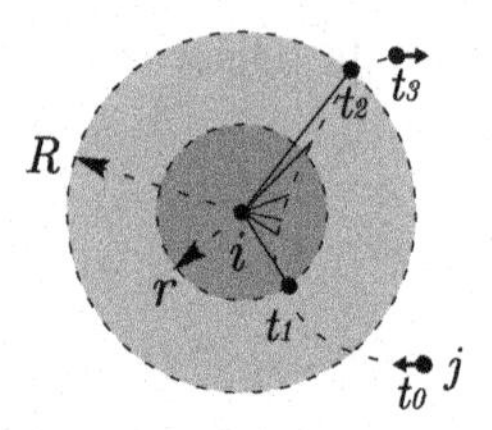

Fig. 12.1 Link creation with hysteresis

Definition 1 A hysteresis in the connectivity creation is shown in Fig. 12.1. For any $0 < r < R$, if agent j comes up close to agent i, and $\| x_i - x_j \| \in (0, r]$, then $(i,j) \in \mathcal{E}(t)$; if agent j deviates from agent i, and $\| x_i - x_j \| \in [R, \infty)$, then $(i,j) \notin \mathcal{E}$. Here R is the radius of the perception, r is the hysteresis radius.

The definition means that there exists hysteresis in the link creation. The effect of the hysteresis will be explained in Sect. 12.3.2.

12.3 Main Results

In this section, several main results are presented.

12.3.1 The New Control Algorithm with Prediction Mechanism

In order to predict the status trends of the system, a state predictor $\hat{v} = -\tilde{L}v$ is proposed, where $\hat{v} = (\hat{v}_1^T, \hat{v}_2^T, \ldots, \hat{v}_N^T)^T$ and $\tilde{L} = L \otimes I_m$. After calculating the predictive state, agent i will transfer $\hat{v}_i$ to its neighbors $j \in N_i$ at the cost of one communication. Then, we add the predictive control item $\hat{v}$ as compensation of velocity to improve the convergence speed of the system.

With the predictive control mechanism above, we define the control law as follows:

$$u_i = -\sum_{j \in N_i(t)} a_{ij}(v_i - v_j) + \gamma \sum_{j \in N_i(t)} a_{ij}(\hat{v}_i - \hat{v}_j) = -\sum_{j \in N_i(t)} a_{ij}(v_i - v_j)$$
$$-\gamma\Bigg(\sum_{j \in N_i(t)} \sum_{p \in N_i(t)} a_{ij}a_{ip}(v_i - v_p) - \sum_{j \in N_i(t)} \sum_{q \in N_j(t)} a_{ij}a_{jq}(v_j - v_q)\Bigg), \tag{12.3}$$

or in compact form,

$$u = -\tilde{L}v - \gamma\tilde{L}^2 v, \tag{12.4}$$

where $\gamma > 0$ is the parameter of influence. $|N_i(t)|$ means number of components of the neighbour set. $\nabla_x V = \left(\nabla_{x_1} V_1^T, \nabla_{x_2} V_2^T, \ldots, \nabla_{x_N} V_N^T\right)^T$, and $V_i = \sum_{j \in N_i(t)} V_{ij}$.

Lemma 2 ([19]). *The second smallest eigenvalue of* Laplacian matrix *is called the algebraic connectivity, which quantifies the speed of convergence of consensus algorithms.*

Theorem 1 *Consider the system* (12.1) *with the protocol* (12.4)*, Assume graph* $\mathcal{G}$ *is connected, then the state predictor can accelerate the convergence by enlarging the algebraic connectivity.*

Proof For the connected graph $\mathcal{G}$, we assume that $0 = \lambda_1 < \lambda_2 \leq \ldots \leq \lambda_N$ denote the eigenvalues of the matrix L. Since L is symmetric, there exists an non-singular matrix P, such that $P^{-1}LP = diag(\lambda_N, \ldots, \lambda_2, 0)$. Then, $P^{-1}(L + \gamma L^2)P = diag(\lambda_N + \gamma\lambda_N^2, \ldots, \lambda_2 + \gamma\lambda_2^2, 0)$.

Knowing that the eigenvalue of L and $L + \gamma L^2$ equal to the eigenvalue of $\tilde{L}$ and $\tilde{L} + \gamma\tilde{L}^2$ respectively under the same communication topology, and $\lambda_2^* = \lambda_2 + \gamma\lambda_2^2 > \lambda_2$, the algebraic connectivity of the system with a predictive control item is greater than the system without that. By *Lemma 2*, the system with predictive control item converges much faster than system without it. ■

Remark 1 From the statement above, we know that, to establish the control algorithm, the neighbors' velocity and two-hop neighbors' velocity information of agent i are needed. Which means that the control algorithm is formed at the cost of one time communication.

Remark 2 As the impact factor of the prediction mechanism, the parameter γ affects the convergence speed. γ represents the proportion of the compensation of the control. With a bigger γ, the control algorithm become more sensitive to the predictive states, might drive the system diverse at last. a suitable gamma can reduce the amplitude fluctuation effectively. Otherwise, the current states are emphasized. Especially when $\gamma = 0$, making the control algorithm equal to the routine one as (12.2).

12.3.2 Connectivity Preserving and Collision Avoidance

The flocking problem has been considered by various potential function techniques, which guarantees the avoidance of collision and the preservation of all existing edges. One example of such potential functions is the following,

$$V_{ij} = \frac{1}{\| x_{ij} \|_\sigma} + \frac{\| x_{ij} \|_\sigma}{\| R \|_\sigma - \| x_{ij} \|_\sigma}, \tag{12.5}$$

where $\| x_{ij} \|_\sigma$ is differentiable everywhere. This property of σ-norms is used for the construction of smooth artificial potential function for the system. According to the hysteresis of Definition 1, $||R||_\sigma - ||x_{ij}||_\sigma \neq 0$. Clearly, V_{ij} grows unbounded when $\| x_{ij} \| \to R^-$. Therefore, the condition in Definition 1 ensures that the potential function is bounded when new connection between agent i and j is created. Note that $0 < ||x_i - x_j||_\sigma$ means that there is no collision between adjacent agents i and j in $\mathcal{G}$.

We now proposed our potential function as follows:

$$\Psi_{ij} = V_{ij} + \Phi(\theta_{ij})V_{ij}, \tag{12.6}$$

where V_{ij} is defined in (12.5), and

$$\Phi(\theta) = \begin{bmatrix} \cos(\theta) & -\sin(\theta) \\ \sin(\theta) & \cos(\theta) \end{bmatrix},$$

is the coordinate rotating matrix, in which the angle θ is defined as follows:

$$\theta_{ij} = \begin{cases} \pi/2, \angle(v_i, v_j) \leq \pi/2; \\ -\pi/2, \angle(v_i, v_j) > \pi/2, \end{cases} \tag{12.7}$$

Then, a distributed protocol with prediction mechanism is proposed as:

$$u_i = -\sum_{j \in N_i(t)} a_{ij}(v_i - v_j) + \gamma \sum_{j \in N_i(t)} a_{ij}(\hat{v}_i - \hat{v}_j) - \sum_{j \in N_i(t)} \frac{1}{|N_i(t)|} \nabla_{x_i} \Psi_{ij}, \tag{12.8}$$

or in compact form,

$$u = -\tilde{L}v - \gamma \tilde{L}^2 v - D(t) \cdot \nabla_x \Psi, \tag{12.9}$$

where $\gamma > 0$ is the parameter of influence. $D(t) = \mathtt{diag}\{\frac{1}{|N_1(t)|}, \frac{1}{|N_2(t)|}, \ldots, \frac{1}{|N_N(t)|}\} \otimes I_m$, $|N_i(t)|$ means number of components of the neighbour set. $\nabla_x \Psi = \left(\nabla_{x_1} \Psi_1^T, \left(\nabla_{x_2} \Psi_2^T, \ldots, \nabla_{x_N} \Psi_N^T\right)^T\right.$, and $\Psi_i = \sum_{j \in N_i(t)} \Psi_{ij}$.

Theorem 2 *Consider the system* (12.1) *with the protocol* (12.9). *if* $\mathcal{G}(0)$ *is connected, then the function* Ψ_{ij} *defined in* (12.6) *is bounded.*

Proof According to the equation (12.7), we know that Ψ_{ij} is a piecewise function. So we discuss the boundedness of the potential function respectively. With lose of generality, we assume that $\angle(v_i, v_j) \leq \pi/2$, which means that $\theta_i = \pi/2$, $\forall j \in N_i$ for agent i.

We define a function $\Psi_{\mathcal{G}} : D_{\mathcal{G}} \times R^{mn}$, such that,

$$\Psi_{\mathcal{G}} = \frac{1}{2}(\| v \|_2^2 + \frac{1}{N}\sum_{i=1}^{N}\Psi_i), \tag{12.10}$$

where $\Psi_i = \sum_{j\in N_i(t)} \Psi_{ij}$, N means the number of all agents and $D_{\mathcal{G}} = \{x \in \mathbb{R}^{mn} | \| x_{ij} \|_2 \in (0, R), \forall (i,j) \in \mathcal{E}\}$. Clearly, $\Psi_{\mathcal{G}}$ is positive semi-definite function of v and x_{ij}. For any $c > 0$, let $\Omega_{\mathcal{G}} = \{(x, v) \in D_{\mathcal{G}} \times \mathbb{R}^{Nm} | \Psi_{\mathcal{G}} \le c\}$ denote the level sets of $V_{\mathcal{G}}$. Then the time derivative of the $\Psi_{\mathcal{G}}(t)$ satisfies

$$\dot{\Psi}_{\mathcal{G}} = \frac{1}{2N}\sum_{i=1}^{N}\dot{\Psi}_i - \sum_{i=1}^{N} v_i^T(\sum_{j\in N_i(t)} a_{ij}(v_i - v_j) - \gamma \sum_{j\in N_i(t)} a_{ij}(\hat{v}_i - \hat{v}_j) + \frac{1}{|N_i|}\nabla_{x_i}\Psi_i),$$

moreover,

$$\begin{aligned}\sum_{i=1}^{N}\dot{\Psi}_i &= \sum_{i=1}^{N}\sum_{j\in N_i(t)} \dot{x}_{ij}^T \nabla_{x_{ij}}\Psi_{ij} = \sum_{i=1}^{N}\sum_{j\in N_i(t)} (\dot{x}_i^T \nabla_{x_{ij}}\Psi_{ij} - \dot{x}_j^T \nabla_{x_{ij}}\Psi_{ij}) \\ &= \sum_{i=1}^{N}\sum_{j\in N_i(t)} (\dot{x}_i^T \nabla_{x_i}\Psi_{ij} + \dot{x}_j^T \nabla_{x_j}\Psi_{ij}) = 2\sum_{i=1}^{N} v_i^T \nabla_{x_i}\Psi_i,\end{aligned} \tag{12.11}$$

with the function Ψ_{ij} is symmetric. Hence,

$$\begin{aligned}\dot{\Psi}_{\mathcal{G}} &= \frac{1}{N}\sum_{i=1}^{N} v_i^T \nabla_{x_i}\Psi_i - \sum_{i=1}^{N} v_i^T(\sum_{j\in N_i} a_{ij}(v_i - v_j) - \gamma \sum_{j\in N_i} a_{ij}(\hat{v}_i - \hat{v}_j) + \frac{1}{|N_i|}\nabla_{x_i}\Psi_i) \\ &\le \frac{1}{N}\sum_{i=1}^{N} v_i^T \nabla_{x_i}\Psi_i - \sum_{i=1}^{N} v_i^T(\sum_{j\in N_i} a_{ij}(v_i - v_j) - \gamma \sum_{j\in N_i} a_{ij}(\hat{v}_i - \hat{v}_j) + \frac{1}{N}\nabla_{x_i}\Psi_i) \\ &\le -\sum_{i=1}^{N} v_i^T \sum_{j\in N_i} a_{ij}\left((v_i - v_j) - \gamma(\hat{v}_i - \hat{v}_j)\right) \le -v^T(\tilde{L} + \gamma\tilde{L}^2)v.\end{aligned} \tag{12.12}$$

By the positive semi-definiteness of the Laplacian matrix $L_{\mathcal{G}} = \tilde{L} + \gamma\tilde{L}^2$, we know $\dot{\Psi}_{\mathcal{G}} \le 0$, which implies that $\Psi_{\mathcal{G}}(t) \le \Psi_{\mathcal{G}}(0) \le \infty$. Therefore, the level set $\Omega_{\mathcal{G}}$ is positively invariant, which means that Ψ_{ij} remains bounded.

The proof for the boundedness of Ψ_{ij} with the condition that $\angle(v_i, v_j) > \pi/2$ can be obtained similarly. ■

Theorem 3 *Consider dynamic system* (12.1) *with the protocol* (12.9). *Assuming $\mathcal{G}(0)$ is connected, then the velocities of all agents become asymptotically the same, meanwhile, connectivity preserving and collision avoidance can be guaranteed. The convergence speed is accelerated.*

Proof From the proof of Theorem 2, we know that Ψ, defined in (12.6), is bounded. If $||x_{ij}|| \to R$ for some $(i,j) \in \epsilon$, $\Psi_{ij} \to \infty$, therefore, $||x_{ij}|| < R$ for all $(i,j) \in \mathcal{E}$. In another way, all links of graph $\mathcal{G}$ are maintained, where t_p for $p = 1, 2, \ldots$ denotes the switching time. Thus, $\mathcal{E}(t_p) \subset \mathcal{E}(t_q)$ with $t_p < t_q$. In a similar way, $x_{ij} \to 0$ can be used for the analysis for collision avoidance.

According to (12.12), we know

$$\dot{\Psi}_{\mathcal{G}} \leq -v^T(\tilde{L} + \gamma\tilde{L}^2)v = -\sum_{j=1}^{m} v_{yj}^T(L + \gamma L^2)v_{yj} \leq 0, \tag{12.13}$$

where $v_{yj} \in \mathbb{R}^N$ is the y components of the agents' velocities. With Lasalle's invariance principle, the solution starting in $\Omega_{\mathcal{G}}$ ultimately converges to the largest invariant set $\{(x, v)|\dot{\Psi}_{\mathcal{G}} = 0\}$, in other way, $\{v|L_{\mathcal{G}}v_{yj} = 0, \forall j = 1, \ldots, m\}$ with equation (13). Based on the properties of the laplacian matrix, one can obtain that $v_{yj} \in span\{1\}$ for all $j = 1, \ldots, m$, which means that the velocities of all agents asymptotically become the same. With the convergence speed has been proved in *Theorem 1*, the proof is thus completed. ■

12.4 Simulation

In this section, several simulation examples for the system with the proposed distributed flocking algorithm are provided. The system has a population size of $N = 10$. we choose the initial position of each agent on a line with distances between adjacent agents equalling to 7. Certainly, the initial position can be chosen randomly with the condition that the topology is connected. The velocities of all agents are chosen randomly in a unit circle, while $R = 10$ and $r = 5$.

In Fig. 12.2, we present simulation results for the flocking problem with connectivity preserving and prediction mechanism for the multi-agent systems. The system consists of the dynamics (12.1) and the control algorithm (12.9) with the potential function (12.6). Obviously, the connectivity of the system can be guaranteed, flocking of the group is achieved and no links are deleted. And the trajectories of all agents in Fig. 12.2 are shown in Fig. 12.3, which present the global trajectories of all the agents when they achieve the flocking objective.

To illustrate the superior of predictive mechanism of our method, we compare the control algorithm of this paper and that in [8]. The corresponding simulation results are presented in Fig. 12.4a, b. Obviously, Fig. 12.4a shows that the velocity of each agent in the system with prediction mechanism converge much faster. And the compensation brought by prediction mechanism reduce the amplitude fluctuation effectively just as *Remark 2* analysed. Similar results can be deduced from Fig. 12.4b.

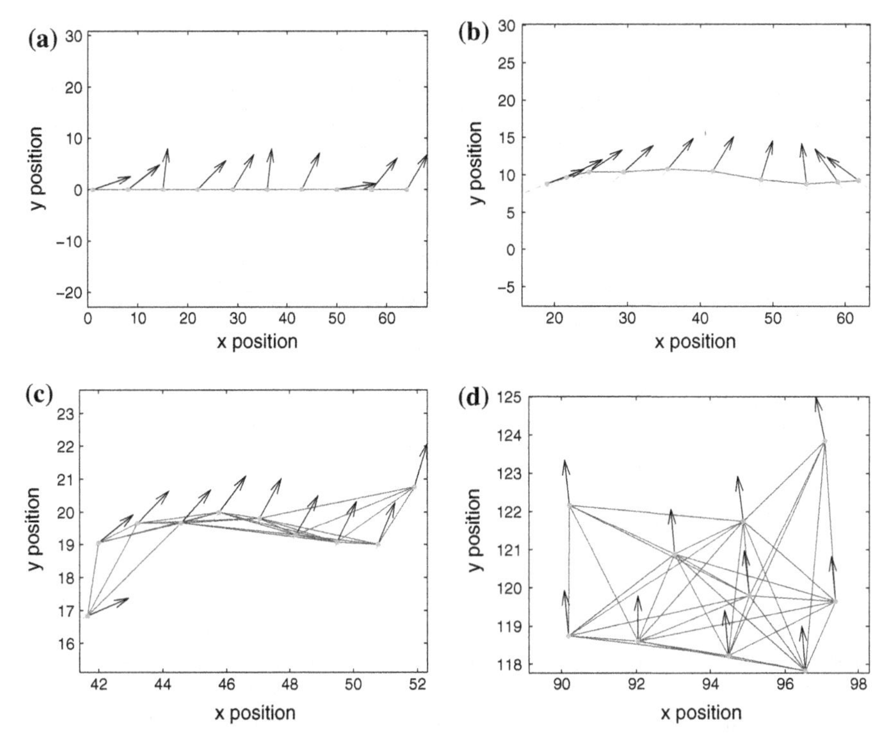

Fig. 12.2 Flocking with prediction mechanism and connectivity preserving, where *red dots* represent the agents, and links between the agents are denoted with *blue solid lines*, the *dash lines* are the motion trajectories of the agents, and *black arrows* correspond to the agents velocity vectors. **a** t = 0. **b** t = 5. **c** t = 10. **d** t = 40

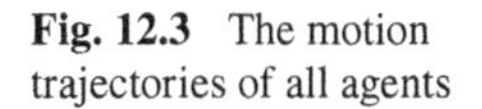

Fig. 12.3 The motion trajectories of all agents

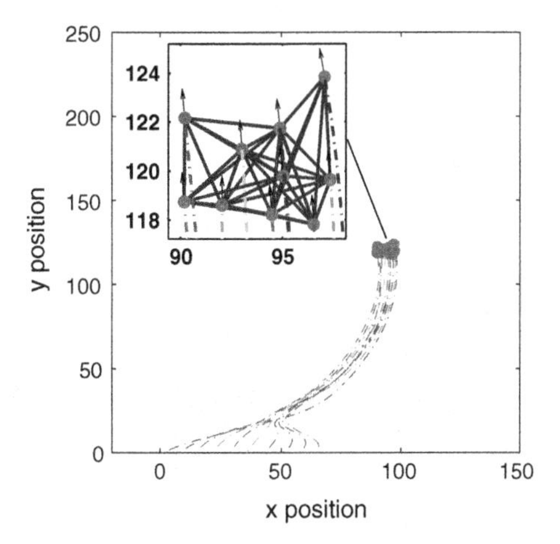

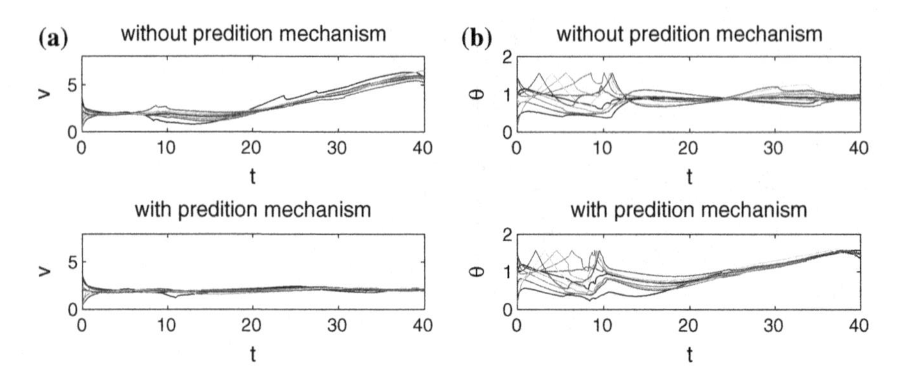

Fig. 12.4 Magnitude and direction comparison of velocity for all agents with and without prediction mechanism. **a** Magnitude comparison of velocity. **b** Direction comparison of velocity

12.5 Conclusions

In this paper, we consider a distributed control framework to the multi-agent flocking problem with prediction mechanism which can simultaneously align their velocity vectors, stabilize their inter-agent distances and accelerate the velocity convergence. Two-hop neighbors' velocity information act as the compensation to stabilize the system more efficiently. Smooth collision-avoiding motion is guaranteed using modified potential function with the coordinate rotating matrix. The simulation results show the superiority of the proposed algorithm.

Acknowledgments This research is supported by the National Natural Science Foundation of China (Grant No.61174094,61273138), and the Tianjin Natural Science Foundation of China (Grant No.14JCYBJC18700, 14JCZDJC39300)

References

1. Reynolds CW (1987) Flocks, herds and schools: a distributed behavioral model. ACM SIGGRAPH Comput Graph 21(4):25–34
2. Shaw E (1975) Fish in schools. Nat Hist 84(8):40–45
3. Okubo A (1986) Dynamical aspects of animal grouping: swarms, schools, flocks, and herds. Adv Biophys 22:1–94
4. Li S, Liu X, Tang W, Zhang J (2014) Flocking of multi-agents following a leader with adaptive protocol in a noisy environment. Asian J Control 16(6):1771–1778
5. Ling G, Nian X, Pan H (2014) Leader-following consensus of multi-agent system with diffusion. Asian J Control 16(1):188–197
6. Liu Z, Chen Z (2012) Discarded consensus of network of agents with state constraint. IEEE Trans Automat Control 57(11):2869–2874
7. Liu J, Chen Z (2012) Robust H_∞ consensus control concerning second-order multi-agent systems with time-varying delays on dynamic topologies. Int J Comput Math 88(12):2485–2501

8. Tanner HG, Jadbabaie A, Pappas GJ (2003) Stable flocking of mobile agents part II: dynamic topology. In: Proceedings of 42nd IEEE Conference on Decision and Control, vol 2, pp 2016–2021
9. Ji M, Egerstedt MB (2007) Distributed coordination control of multiagent systems while preserving connectedness. IEEE Trans Robot 23(4):693–703
10. Wang Q, Fang H, Chen J, Mao Y, Dou L (2012) Flocking with obstacle avoidance and connectivity maintenance in multi-agent systems. In: 2012 IEEE 51st Annual Conference on Decision and Control, pp 4009–4014
11. Wen G, Duan Z, Su HS, Chen G, Yu W (2012) A connectivity-preserving flocking algorithm for multi-agent dynamical systems with bounded potential function. IET Control Theory Appl 6(6):813–821
12. Fan Y, Feng G, Gao Q (2012) Bounded control for preserving connectivity of multi-agent systems using the constraint function approach. IET Control Theory Appl 6(11):1752–1757
13. Zhang HT, Chen MZ, Stan GB, Zhou T, Maciejowski JM (2008) Collective behavior coordination with predictive mechanisms. IEEE Circuits Syst Mag 8(3):67–85
14. Zhan J, Li X (2013) Flocking of multi-agent systems via model predictive control based on position-only measurements. IEEE Trans Ind Inf 9(1):377–385
15. Horn RA, Johnson CR (2012) Matrix Analysis. Cambridge University Press, Cambridge
16. Biggs N (1993) Algebraic Graph Theory. Cambridge University Press, Cambridge
17. Zavlanos MM, Jadbabaie A, Pappas GJ (2007) Flocking while preserving network connectivity. In: 46th IEEE Conference on Decision and Control 2007, pp 2919–2924
18. Olfati-Saber R (2006) Flocking for multi-agent dynamic systems: algorithms and theory. IEEE Trans Automat Control 51(3):401–420
19. Olfati-Saber R, Fax JA, Murray RM (2007) Consensus and cooperation in networked multi-agent systems. Proc IEEE 95(1):215–233
20. Tanner HG, Jadbabaie A, Pappas GJ (2007) Flocking in fixed and switching networks. IEEE Trans Automat Control 52(5):863–868
21. Olfati-Saber R, Murray RM (2004) Consensus problems in networks of agents with switching topology and time-delays. IEEE Trans Automat Control 49(9):1520–1533
22. HouSheng S, Wang X, Chen G (2009) A connectivity-preserving flocking algorithm for multi-agent systems based only on position measurements. Int J Control 82(7):1334–1343

Chapter 13
Dual-Command Operation Generation in Bi-Directional Flow-Rack Automated Storage and Retrieval Systems with Random Storage Policy

Zhuxi Chen and Yun Li

Abstract In the bi-directional flow-rack (BFR) automated storage and retrieval systems (AS/RS), bins slope to opposite directions to make unit-loads be retrieved from half bins and be stored to the other half on the same working face. For random storage policy, a batching-greedy heuristic (BGH) has been proposed to generate dual-command (DC) operations in BFR AS/RS. In this paper, a novel DC operation generation rule specially designed for the BFR AS/RS is introduced to BGH, of which the effectiveness and efficiency are evaluated by simulation experiments.

Keywords Bi-directional flow-rack · AS/RS · Dual-command operation

13.1 Introduction

Automated storage and retrieval systems (AS/RSs) have been widely arranged in unit-load warehouses [1, 2] to obtain high storage accuracy, high throughput capacity and reasonable labor cost. For improving the floor space utilization, double-deep AS/RS [3], flow-rack AS/RS [4], 3D compact AS/RS [5], mobile rack AS/RS [6] and puzzle-based AS/RS [7] have been designed since 2005.

The bi-directional flow-rack (BFR) AS/RS was modified from the flow-rack AS/RS [8]. In a BFR, bins slope to opposite directions to ensure that unit-loads can be retrieved from and stored to both working faces. Therefore, dual-command (DC) operation, which stores a unit-load and retrieves another one within a single cycle to reduce the travel time of handling machine [9], can be generated in BFR AS/RSs. A batching-greedy heuristic (BGH) has been proposed for DC operation generation in BFR AS/RS with random storage policy [8]. However, the generation rule, NN (Nearest Neighbor) [10], SL (Shortest Leg) [10], and SDC (Shortest DC time) [11], applied by BGH for DC operations are designed for single-deep AS/RSs. Therefore,

Z. Chen (✉) · Y. Li
College of Information Engineering, Yangzhou University, Yangzhou, China
e-mail: zxchen@yzu.edu.cn

Y. Jia et al. (eds.), *Proceedings of the 2015 Chinese Intelligent Systems Conference*, Lecture Notes in Electrical Engineering 360,
DOI 10.1007/978-3-662-48365-7_13

a novel rule considered the structure of BFR AS/RS is introduced to BGH and its effectiveness and efficiency are evaluated by simulation experiments in this paper.

The rest part is as follows. In Sect. 13.2, BGH and a novel selection rule are introduced. The effectiveness and efficiency of the selection rule are evaluated in Sect. 13.3, followed by some conclusions in Sect. 13.4.

13.2 BGH and the Novel Generation Rule

An $L \times H \times M$ BFR is illustrate in Fig. 13.1. The BFR contains L columns (L is even) and H rows of bins. Each bin consists of M segments to store at most M unit-load. Bins slope from F_B to F_A in Column $1, 3, \ldots, L-1$ and from F_A to F_B in Column $2, 4, \ldots, L$. Correspondingly, unit-loads slide from F_B to F_A in odd columns and from F_A to F_B in even columns driven by gravity. In each bin, unit-loads follow FIFO (first-in-first-out) rule, which implies that outgoing unit-loads are blocked by blocking unit-loads with certain probability. Blocking unit-loads must be removed before retrieving requested unit-loads. Removed blocking unit-loads are stored to bins on the same working face.

Two handling machines are arranged to F_A and F_B, respectively. t_h is the travel time between two horizontal adjacent bins and t_v is the travel time between two vertical adjacent bins. Let (x_i, y_i) represent bin i locates at Column x_i and Row y_i and (x_j, y_j) represent bin j locates at Column x_j and Row y_j. Let $c_i = \max\{t_h x_i, t_v(y_i - 1)\}$ ($c_j = \max\{t_h x_j, t_v(y_j - 1)\}$) be the travel time between i (j) and the P/D station. Similarly, $c_{ij} = \max\{|x_i - x_j|t_h, |y_i - y_j|t_v\}$ is the travel time from i (j) to j (i). The normalization and the shape factors of an $L \times H \times M$ BFR are $T = \max\{t_h L, t_v H\}$ and $b = \min\{\frac{t_h L}{T}, \frac{t_v H}{T}\}$, respectively.

In the random storage policy, any stored unit-load can be requested and an incoming unit-load can be stored to any available bin (a bin contains at least one idle

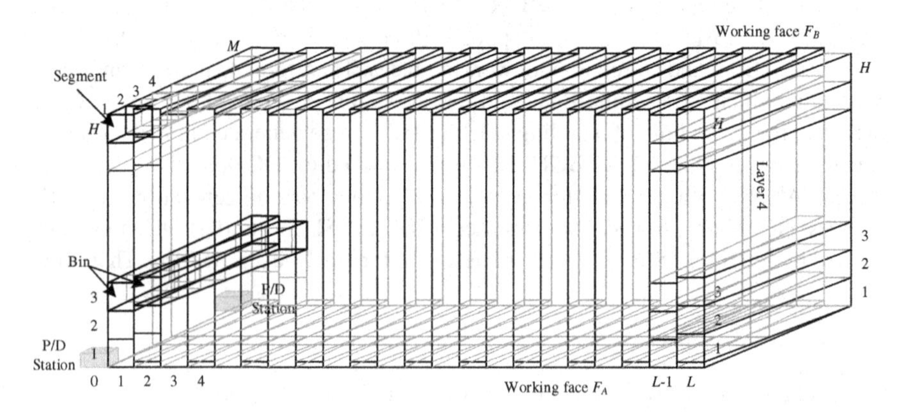

Fig. 13.1 An $L \times H \times M$ BFR

segment). The sequencing of unit-loads is denoted as $S = (s_1, s_2, \ldots, s_N)$, in which incoming unit-loads must be handled one by one. As well, the set of outgoing unit-loads is represented as $R = \{r_1, r_2, \ldots, r_N\}$, in which outgoing unit-loads can be retrieved by any order.

The position of an idle segment is an opening and the position of an outgoing unit-load is a retrieval. I is the set of openings. $J = \{j_1, j_2, \ldots, j_N\}$ is the set of retrievals, in which r_k locates at j_k where $k \in \{1, \ldots, N\}$. Let I_j be the set of openings located on the same working face of j. C_j is the travel time of removing and re-storing blocking unit-loads for retrieval j.

In a BFR AS/RS, a unit-load retrieved or removed from F_A (F_B) releases an occupied segment on F_B (F_A). Therefore, some openings on F_A (F_B) become available after retrieving or removing unit-loads from F_B (F_A). If such an opening is selected, the process time may be increased because the handling machine may stop to wait for the releasing.

BGH separates incoming unit-loads into batches, for each of which a sequence of DC operations are generated. The released openings in a batch are employed in the following batches to ensure the continuous working of handling machines. Let $P = \lceil \frac{N}{n} \rceil$ be the number of batches. In the pth batch, a total of n or $N - (P - 1)n$ incoming unit-loads and the same number of outgoing unit-loads are conducted by DC operations. BGH is described in Algorithm 1.

Algorithm 1 Batching-greedy Heuristic

Input: S, R, I, J
Output: sequence of DC operations
1: **for** ($p = 1$ to P) **do**
2: Build S_p;
3: $S \leftarrow S - S_p$, $I_p \leftarrow \emptyset$.
4: **while** ($S_p \neq \emptyset$) **do**
5: $s \leftarrow Head(S_p)$, $S_p \leftarrow S_p - \{s\}$.
6: Select $j \in J$ and $i \in I_j$ with minimal D_{ij}.
7: $I \leftarrow I - \{i\}$.
8: **while** (blocking unit-loads exist) **do**
9: Select $i \in I_j$ with minimal c_{ij}.
10: $I \leftarrow I - \{i\}, I_p \leftarrow I_p \cup \{j\}$.
11: $R \leftarrow R - \{r\}, J \leftarrow J - \{j\}, I_p \leftarrow I_p \cup \{j\}$.
12: $I \leftarrow I \cup I_p$.
13: **return**

In Step 6 of BGH, an opening and a retrieval are selected based on the value of D_{ij}. In NN, SL and SDC, $D_{ij} = c_{ij}$, $D_{ij} = c_i + c_{ij}$ and $D_{ij} = c_i + c_{ij} + c_j$, respectively. The time complexity of Step 6 is $O(LHN)$ for NN, SL, and SDC. Let $\rho = \frac{N_u}{LHM}$ be the load rate, in which N_u is the number of stored unit-loads. For a bin, there are a number of K stored unit-loads. Because of the random storage policy, the expected value of K is $E(K) = \rho M$. Therefore, the expected number of blocking unit-loads for a retrieval is $\sum_{k=0}^{K-1} \frac{1}{K} = \frac{\rho M-1}{2}$. The time complexity of choosing openings for blocking unit-loads of a retrieval is $O(LH\frac{\rho M-1}{2})$. In normal, $\frac{\rho M-1}{2} < N$, which makes the time complexity of generating a DC operation in BGH with NN, SL, and SDC is $O(LHN)$. Because there are a total of N DC operations to be generated, the time complexity of BGH with NN, SL, and SDC is $O(LHN^2)$.

Because blocking unit-loads are re-stored in DC operations, SDCB (Shortest DC time with Blocking unit-loads) is designed, in which $D_{ij} = c_i + c_{ij} + C_j + c_j$. The time complexity of Step 6 of BGH with SDCB is $O(\frac{\rho M+1}{2}LHN)$. Correspondingly, the time complexity of BGH with SDCB is $O(\frac{\rho M+1}{2}LHN^2)$.

13.3 Performance Evaluation

In this section, the effectiveness and efficiency of SDCB are evaluated by simulation experiments. Because SL is better than NN and SDC [8], BGH with SL is executed in simulation experiments for comparing. Simulation experiments are coded in C++. A PC with 2.5 GHz CPU and 16 GB RAM is applied for running simulation experiments.

Let $t_h = 1.0$ and $t_v = 2.0$. In each experiment, a total of 100 instances are conducted. For each instance, let $N = 100$, i.e., 100 DC operations are generated to store 100 incoming unit-loads and retrieve 100 outgoing unit-loads. The batch size takes value of $\{2, 4, 10, 20\}$. The DC operation travel times of instances are recorded, of which the average value is $\mathcal{Z}_{DC}$. As well, the process times of instances are saved, of which the average value is $\mathcal{Z}_{PT}$. In order to narrow the tables, the values of $\frac{\mathcal{Z}_{DC}}{N}$ and $\frac{\mathcal{Z}_{PT}}{N}$ are listed.

Firstly, the impacts of b are analyzed, in which $L \times H \approx 200$, $M = 5$, and $\rho = 0.8$. The values of $\frac{\mathcal{Z}_{DC}}{N}$ and $\frac{\mathcal{Z}_{PT}}{N}$ obtained by BGH with SL and SDCB with different values of b are illustrated in Table 13.1.

Table 13.1 Values of $\frac{\mathcal{Z}_{DC}}{N}$ and $\frac{\mathcal{Z}_{PT}}{N}$ obtained by SL and SDCB with different values of b ($L \times H \approx 200$, $M = 5$ and $\rho = 0.8$)

b		SL				SDCB				CPU time				
		2	4	10	20	2	4	10	20		2	4	10	20
1.00	$\frac{\mathcal{Z}_{DC}}{N}$	30.33	30.43	30.77	31.25	30.71	30.82	30.97	31.20	SL	0.01	0.01	0.01	0.01
	$\frac{\mathcal{Z}_{PT}}{N}$	26.02	23.45	19.79	18.51	24.50	22.14	19.53	18.41	SDCB	2.14	2.12	2.10	1.87
0.67	$\frac{\mathcal{Z}_{DC}}{N}$	31.72	31.82	32.22	32.72	32.05	32.12	32.32	32.48	SL	0.01	0.01	0.01	0.01
	$\frac{\mathcal{Z}_{PT}}{N}$	27.71	24.70	20.71	19.49	25.71	23.19	20.40	19.15	SDCB	1.85	1.77	1.75	1.66
0.38	$\frac{\mathcal{Z}_{DC}}{N}$	38.09	38.24	38.80	39.35	38.27	38.34	38.55	38.89	SL	0.01	0.01	0.01	0.01
	$\frac{\mathcal{Z}_{PT}}{N}$	32.64	28.50	24.54	23.42	30.30	27.12	24.34	22.89	SDCB	1.81	1.78	1.64	1.58
0.25	$\frac{\mathcal{Z}_{DC}}{N}$	45.53	45.75	46.31	46.96	45.85	45.92	46.16	46.53	SL	0.01	0.01	0.01	0.01
	$\frac{\mathcal{Z}_{PT}}{N}$	38.22	33.09	29.08	27.70	36.03	32.22	28.81	27.51	SDCB	1.97	1.95	1.85	1.82

Table 13.1 indicates that the values of $\frac{z_{DC}}{N}$ increase and the values of $\frac{z_{PT}}{N}$ decrease with the increase of batch size. The values of $\frac{z_{DC}}{N}$ obtained by SDCB become smaller than those gained by SL in large size of batch. The values of $\frac{z_{PT}}{N}$ obtained by SDCB are smaller than those gained by SL. SDCB demands much longer CPU time than that spent by SL, which follows the time complexity. With the increase of n, the CPU time required by SDCB decreases.

Then, the impacts of different values of layers are examined and the experimental results are illustrated in Tables 13.2 and 13.3, respectively. In Table 13.2, let $L \times H \times M \approx 1{,}000$, $M \in \{10, 8, 6, 5\}$, $b = 1.0$, and $\rho = 0.8$. In Table 13.3, let $L = 20$, $H = 10$, $\rho = 0.8$, and $M \in \{5, 6, 7, 8\}$.

In Table 13.2, the values of n, $\frac{z_{DC}}{N}$ and $\frac{z_{PT}}{N}$ have the similar changes as those in Table 13.1. SL returns shorter travel time than that gained by SDCB with small batch size and obtains higher travel time than that gained by SDCB with large batch size.

Table 13.2 Values of $\frac{z_{DC}}{N}$ and $\frac{z_{PT}}{N}$ obtained by SL and SDCB with different values of M ($L \times H \times M \approx 1{,}000$, $b = 1.0$, and $\rho = 0.8$)

M		SL				SDCB				CPU time				
		2	4	10	20	2	4	10	20		2	4	10	20
10	$\frac{z_{DC}}{N}$	25.78	25.91	26.53	27.89	26.59	26.85	27.49	28.69	SL	0.01	0.01	0.01	0.01
	$\frac{z_{PT}}{N}$	22.96	20.92	18.08	16.77	21.65	19.65	17.87	17.50	SDCB	1.36	1.24	1.20	1.17
8	$\frac{z_{DC}}{N}$	27.49	27.63	28.16	29.22	28.19	28.41	28.86	29.64	SL	0.01	0.01	0.01	0.01
	$\frac{z_{PT}}{N}$	24.12	21.93	19.00	17.52	22.79	20.72	18.46	17.86	SDCB	1.72	1.60	1.59	1.51
6	$\frac{z_{DC}}{N}$	28.39	28.50	28.86	29.46	28.90	29.01	29.27	29.65	SL	0.01	0.01	0.01	0.01
	$\frac{z_{PT}}{N}$	24.55	22.23	19.01	17.57	23.12	20.98	18.68	17.52	SDCB	1.85	1.76	1.75	1.53
5	$\frac{z_{DC}}{N}$	30.33	30.43	30.77	31.25	30.71	30.82	30.97	31.20	SL	0.01	0.01	0.01	0.01
	$\frac{z_{PT}}{N}$	26.02	23.45	19.79	18.51	24.50	22.14	19.53	18.41	SDCB	2.14	2.12	2.10	1.87

Table 13.3 Values of $\frac{z_{DC}}{N}$ and $\frac{z_{PT}}{N}$ obtained by SL and SDCB with different values of M ($L = 20$, $H = 10$, and $\rho = 0.8$)

M		SL				SDCB				Run time				
		2	4	10	20	2	4	10	20		2	4	10	20
5	$\frac{z_{DC}}{N}$	30.33	30.43	30.77	31.25	30.71	30.82	30.97	31.20	SL	0.01	0.01	0.01	0.01
	$\frac{z_{PT}}{N}$	26.02	23.45	19.79	18.51	24.50	22.14	19.53	18.41	SDCB	2.14	2.12	2.10	1.87
6	$\frac{z_{DC}}{N}$	31.29	31.43	31.81	32.35	31.69	31.79	31.96	32.31	SL	0.02	0.02	0.02	0.02
	$\frac{z_{PT}}{N}$	26.89	24.06	20.51	19.14	25.04	22.68	20.20	18.99	SDCB	3.00	2.90	2.84	2.72
7	$\frac{z_{DC}}{N}$	32.23	32.38	32.84	33.37	32.80	32.89	33.15	33.40	SL	0.03	0.03	0.03	0.03
	$\frac{z_{PT}}{N}$	27.56	24.96	21.08	20.07	26.11	23.56	20.98	19.66	SDCB	3.71	3.65	3.56	3.40
8	$\frac{z_{DC}}{N}$	33.41	33.56	34.11	34.82	33.87	34.03	34.29	34.68	SL	0.04	0.04	0.04	0.04
	$\frac{z_{PT}}{N}$	28.89	26.00	22.30	21.05	27.02	24.35	21.87	20.76	SDCB	4.77	4.70	4.62	4.49

Table 13.4 Values of $\frac{z_{DC}}{N}$ and $\frac{z_{PT}}{N}$ obtained by SL and SDCB with different values of ρ ($L = 20$, $H = 10$, and $M = 10$)

ρ		SL				SDCB				CPU time				
		2	4	10	20	2	4	10	20		2	4	10	20
0.75	$\frac{z_{DC}}{N}$	33.88	33.99	34.43	34.90	34.48	34.58	34.81	35.03	SL	0.03	0.03	0.03	0.03
	$\frac{z_{PT}}{N}$	29.55	26.34	22.45	21.04	27.61	24.97	22.33	20.98	SDCB	5.65	5.58	5.46	5.44
0.80	$\frac{z_{DC}}{N}$	35.10	35.27	35.94	36.74	35.71	35.83	36.16	36.69	SL	0.03	0.03	0.03	0.03
	$\frac{z_{PT}}{N}$	30.40	27.36	23.30	22.06	28.42	25.57	22.81	21.81	SDCB	3.29	3.28	3.22	3.15
0.85	$\frac{z_{DC}}{N}$	37.51	37.76	38.83	40.44	38.04	38.37	39.07	40.42	SL	0.02	0.02	0.02	0.02
	$\frac{z_{PT}}{N}$	32.59	29.33	25.02	24.07	30.24	27.65	25.00	24.04	SDCB	5.22	5.16	5.01	4.77
0.90	$\frac{z_{DC}}{N}$	41.46	41.97	43.77	46.48	42.03	42.66	44.79	47.89	SL	0.01	0.01	0.01	0.01
	$\frac{z_{PT}}{N}$	36.33	32.58	28.82	28.04	33.59	30.63	28.86	29.12	SDCB	3.26	3.20	3.05	2.79

SDCB can obtain shorter process time than that gained by SL. As well, SL runs faster than SDCB. The CPU time demanded by SDCB decreases with the increase of n.

Table 13.3 demonstrates that the travel time increases and the process time decreases with the increase of n. SL obtains shorter travel time than that gained by SDCB. SDCB gains shorter process time that obtained by SL. SDCB requires much longer CPU time that demanded by SL. More layers in the BFR implies more blocking unit-loads, which causes the increase of operational cost and the increase of process time.

At last, the impacts of different values of ρ are evaluated by simulation experiments, in which let $L = 20$, $H = 10$, $M = 5$, and $\rho \in \{0.75, 0.80, 0.85, 0.90\}$. The experimental results are listed in Table 13.4.

Table 13.4 shows that the operational cost increases and the process time decreases with the increase of n. SL obtains lower operational cost than that gained by SDCB. SDCB gains shorter process time that obtained by SL. SDCB requires much longer CPU time that demanded by SL. Larger value of ρ implies more blocking unit-loads. Correspondingly, the values of $\frac{z_{DC}}{N}$ and $\frac{z_{PT}}{N}$ obtained by SL and SDCB increase with the increase of ρ.

In summary, the size of batch must be carefully evaluated to balance the operational cost and the process time of DC operations. SL obtains lower travel time and SDCB gains shorter process time. The process time may be more important than the travel time because the process time determines the response time, which is the key parameter of the quality of service (QoS) of a BFR AS/RS.

13.4 Conclusions

In this paper, the SDCB is introduced to the BGH for DC operation generation in BFR AS/RS with random storage policy. Experimental results illustrate that SDCB obtains shorter process time that gained by SL and SL gains shorter total travel time than that obtained by SDCB. The size of batch should be carefully considered to balance the operational cost and the process time. BGH with SDCB is more suitable than BGH with SL in applications where to improve the response time is more important than to reduce the operational cost.

In the future, dual-shuttle machine, which carries two unit-loads simultaneously and has been deployed to flow-rack AS/RSs for improving the retrieval performance [12], can be applied in BFR AS/RSs to enhance the throughput capacity.

Acknowledgments This research was supported by the Chinese National Natural Science Foundation (61402396), Natural Science Foundation of Jiangsu Higher Education Institutions (13KJB520027, 13KJB510039), The High-level Talent Project of "Six Talent Peaks" of Jiangsu Province (2012-WLW-024), Joint Innovation Fund Project of Industry, Education and Research of Jiangsu Province (BY2013063-10) and The Talent Project of Green Yangzhou & Golden Phoenix (2013-50).

References

1. Van Der Berg JP (1999) A literature survey on planning and control of warehousing systems. IIE Trans 31(8):751–762
2. Roodbergen KJ, Vis IFA (2009) A survey of literature on automated storage and retrieval systems. Eur J Oper Res 194(2):343–362
3. Lerher T, Sraml M, Potrc I, Tollazzi T (2010) Travel time models for double-deep automated storage and retreival systems. Int J Prod Res 48(11):3151–3172
4. Sari Z, Saygin C, Ghouali N (2005) Travel-time models for flow-rack aotumated storage and retrieval systems. Int J Adv Manuf Technol 25(9):979–987
5. De Koster MBM, Le-Duc T, Yu Y (2008) Optimal storage rack design for a 3-dimensional compact AS/RS. Int J Prod Res 46(6):1495–1514
6. Chang TH, Fu HP, Hu KY (2006) Innovative application of an intergrated multi-level conveying device to a mobile storage system. Int J Adv Manuf Technol 29(9–10):962–968
7. Uludag O (2014) GridPick: a high density puzzle based prder picking system with decentralized control. Ph.D. Thesis, Auburn University
8. Chen Z, Li X, JND Gupta (2015) A bi-directional flow-rack automated storage and retrieval system for unit-load warehouses. Int J Prod Res 53(14):4176–4188
9. Gagliardi JP, Renaud J, Ruiz A (2014) On sequencing policies for unit-load automated storage and retrieval systems. Int J Prod Res 52(4):1090–1099
10. Han MH, McGinnis LF, Shieh JS, White JA (1987) On sequencing retrievals in an automated storage/retrieval system. IIE Trans 19(1):56–66
11. Lee HF, Samaniha SK (1996) Retrieval sequencing for unit-load automated storage and retrieval systems with multiple openings. Int J Prod Res 34(10):2943–2962
12. Chen Z, Xiaoping L (2015) Sequencing method for dual-shuttle flow-rackautomated storage and retrieval systems. J Southeast Univ (English Edition) 31:31–37

Chapter 14
Influenza Immune Model Based on Agent

Shengrong Zou, Yujiao Zhu, Zijun Du and Xiaomin Jin

Abstract All along, the immune system has been a hotspot and difficulty in the field of biological research. Traditional experimental immunology can observe the overall reaction of the immune system, but would be difficulty on the some details research, such as the recognition principle between antigen and antibody. In this paper, we use the binary string to express the gene of antigen and antibody, and use binary string matching to express the immune recognition process. Then, we would use the Agent computer model and the computer simulation method to study the microscopic properties of the immune system and some important details. The results of our study provide powerful basis to establish accurate perfect model of the immune system. Using the simulation model of immune responses to influenza virus, include the interactions between cells, some basic rules of the immune system are obtained.

Keywords Immune system · Computer simulation · Agent model · Influenza virus

14.1 Introduction

The immune system is a complex system. It attacks foreign invaders which will cause diseases and damage body systems. A network of cells work together to against infectious organisms and invaders. This series of steps are called the immune response. The immune response consists of primary response and secondary response. The lymphocytes monitor viruses and bacteria that might cause problems for the body. When a new kind of antigens is detected for the first time, lymphocytes can recognize them and proliferate, differentiate, produce antibodies to keeping people healthy and preventing infections. And the whole process called

---

S. Zou (✉) · Y. Zhu · Z. Du · X. Jin
College of Information Engineering, Yangzhou University, Yangzhou 225127, China
e-mail: srzou@qq.com; 460046777@qq.com

Y. Jia et al. (eds.), *Proceedings of the 2015 Chinese Intelligent Systems Conference*, Lecture Notes in Electrical Engineering 360,
DOI 10.1007/978-3-662-48365-7_14

primary response. In addition, in the process of primary immune, the lymphocytes will produce some memory cells. Memory cells can exist in a person's body for a long time, so when the immune system encounter same antigen again, the memory cells can produce antibodies and overcome antigens more rapidly. This process called secondary response. This can explain why people will not get some kind of disease again, like smallpox.

In the past, many methods have been used to research different details of the immune system [1], but researchers still can not describe all of its characteristics perfectly. In order to measure their own characteristics and interaction between a large number of immune cells more accurately and conveniently, method of computer simulation is imperative.

When the immune system was attacked by one kind of virus, its details and features will be fully displayed. In this paper, we chose the data of influenza A to assess whether our model is realistic enough or not.

In the following section, a brief description of the nature of influenza and related works has been given. After that, we has been presented an immune system model, and the model is based on the agent parameter and behavior of influenza. Finally, we expand on the biological meaning of the model.

14.2 About Influenza

Influenza is an infectious disease. It is caused by influenza virus. Humans and birds all have the chance to be infected. There are three types of influenza virus: Type A, B and C. The mutation rate and aggressive symptoms of type A is highest. All known Influenza pandemics is caused by type A. As most viruses, Influenza can grow and reproduce by taking over the replicative machinery of host cells, with the help of cunning strategies, they can evade immune defenses. Lymphocytes T cells (Tc) are crucial in the elimination of viral infection. It can recognize and remember viruses and produce specific antibody to destroy them [2–5].

14.3 Related Work

There are many common implements to understand the natural world. Such as modeling and simulating biological phenomena. Many methods have been used to research the immune system. The behavior of virus can be shown by the simulation of the immune system.

Different immune system simulations would utilize various methods [6–8]. With the help of mathematical models, much work has been done to capture the dynamics of influenza A. These models can help us to make wise public health decisions: when, whom and how to quarantine, vaccinate, treat with antivirals, and what and how much to stockpile [9–16]. Cellular Automata for simulation was used by ImmSim [17], C-ImmSim and ParImm [18], ImmunoGrid and C-ImmSim [19], SIS and SIS-I [20] and CyCells [21]. Other used different methods, such as AbAIS [22] used agent and Multi Agent System like.

Recently, the interest in using agent-based models to model the immune system has been renewed. Compared with a specific model, it is more of a language for describing models and a modeling technique. In a discrete simulation space, according to predefined rules, all kinds of discrete entities can interact with others. Now this method is more and more popular, and has been applied to various fields. In this way, we can simulate the complexity of biological entities. For example, we can better simulate the immune system's response to viral invasion. In order to guarantee the space and time are discrete. The particles are live in a mesh, meanwhile, their states are updated at discrete time-steps.

14.4 The Agent Model

People tend to choose mathematical modeling methods to understand complex systems. It can gain the most likely outcome. But on the one hand, the immune system has certain regularity at the macro level, on the other hand, the cell interaction is random at the micro level. So the mathematical model is not suitable. The model based on the Agent can solve this problem. Randomness is an important feature of the immune response. With the randomness we can study all kinds of possible results. Compared with the most likely outcome of the mathematical model, it is more effective in the study of the immune response. The application of randomness in the modeling also can better simulate the nonlinear dynamic relationship between the components. Secondly, the model based on the Agent can be implemented on computer efficiently. In this paper, we use the Agent model to research the immune response process of influenza A virus.

The key element of the Agent model is Agent. In our model, that is cells entity. According to the real system description, different types of entities are defined. At the same time, each Agent can move freely in the simulated space, interact with others, and update themselves. Related regulations also be set.

(A) Entities Representation

Our model considers four entities. Ag(antigen):Ag may be any substance that causes immune system to produce antibodies against it. In this paper, we choose Influenza A as antigen; APC(epithelial cells):Epithelial cells can be collided and infected by antigen. After that it will have the function of antigen presenting cells. And it can present antigen to T lymphocyte; In(infected cells):The epithelial cells which have been infected by antigen; T cell(lymphocytes):T cell belong to a group of white blood cells known as lymphocytes. It allow the immune system to recognize previous invaders and help the body destroy them by attacking virus-infected cells, foreign cells and cancer cells etc.

All kinds of entities are live in a limited area and have an independent position. We use coordinates (x-axis, y-axis) to express the position of each cell. They can move from one position to another position freely. Of course, collision and interaction is inevitable. But those interacted entities are implemented under the system rules.

(B) Entities Attributes:

As biological factors, all kind of entities have the basic biological properties. Such as birth, aging, illness and death. In this paper, we use different parameters to represent different properties.

When a new entity appeared, it needs a unique position. Position is a crucial parameter because it defines whether they can interact. At the same time, it also needs a parameter (called lifecycle) to determine how long it can exist, a parameter (called HP) to reflect its biological activity. In the process of growth, it will mature, split and generate daughter cells. The parameter (called splitCellNum) decided the number of daughter cells that one cell can produce.

Different cell types not only have the common properties, but also each entity has their unique properties. For example, because Ag cannot complete by own replication, so the splitCellNum of Ag is zero. With the help of In, Ag can get daughter cells. One of the most important attributes is genetic properties. And only T cell and Ag has this attribute. As is known to all, the primary function of the immune system is recognition function. With the help of genetic properties, T cell can identify the Ag, and remember the gene of Ag in order to cope with the secondary invasion. In this paper, we use binary string to express the gene character.

(C) Regulation of Interaction:

Interaction is another important function of the immune system. As mentioned above, entity can move from one position to another position. Collision and interaction is inevitable. When collision occurs, the processing rules of collision and interaction become crucial.

When there is no antigen intrusion, there are only APC and T cell in the immune system, and the immune response does not occur. When one cell collides with other cells, only some simple physical damage will happen. In other word, both of them will be hurt. We use the attack to indicate the extent of the hurt. But they also have certain abilities of self-protection. We use the def to express the extent of the self-protection. Hurt and self-protection are associated with biological activity (HP). So the value of the biological activity (HP) will change. The formula of the HP can be expressed as: new hp = hp − attack + def. If the value of HP downs to zero, this entity will die.

When there is antigen intrusion, immune response will play its function. So except the general rules mentioned above, when a collision occurs in certain types of cells, there are some specific rules.

(a) APC:

When APC collides with Ag, if the biological activity (HP) of Ag is strong enough, the APC will has some chance of being bundled and become In. No matter

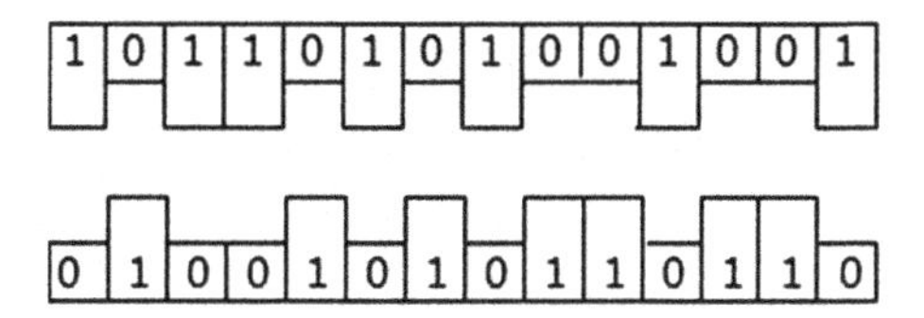

Fig. 14.1 Binary string matching

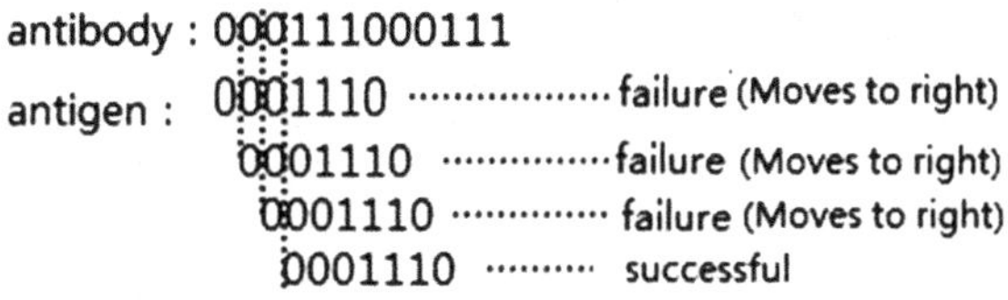

antibody : 000111000111
antigen : 0001110 .... failure (Moves to right)
0001110 .... failure (Moves to right)
0001110 .... failure (Moves to right)
0001110 .... successful

Fig. 14.2 Matching principle

whether been bundled or not, this APC will have the function of transinformation. When it collides with T cell or T memory cells, it can transmit the information of Ag invasion to them, and stimulate them to proliferate, differentiate and produce antibodies rapidly.

(b) T Cell:

When T cell collides with In, it will identify it and promote the apoptosis of In. So the value of the biological activity (HP) of In will drop rapidly. Meanwhile, the value of def of In will down to zero.

The most important rules are the processing rules between T cell and Ag. In the immune response, T cells use genetic matching to determine whether it is the first invasion of this kind of Ag or not. The rules of gene matching as shown in Fig. 14.1. If the 1 and 0 mapping from each other, the matching is successful.

It is worth noting, in biology, the virus only have a short period of single RNA, but T cell have a full set of double-stranded DNA. So the length of T cell genes is much longer than that of Ag. In the process of matching, Ag genes should be matched with any pieces of T cell genes. Until we find successful matching pieces (Fig. 14.2). Once the matching successful, Ag will be swallowed up by the T cell, it's biological activity (HP) will reduce greatly, and the value of attack and def will become zero. The value of T cell' splitCellNum will increase. If there is no piece of T cell genes can match with Ag, T cell will optimize itself through remembering the complementary genes of Ag genes. When the invasion occurs again, it can match, identify and overcome it rapidly.

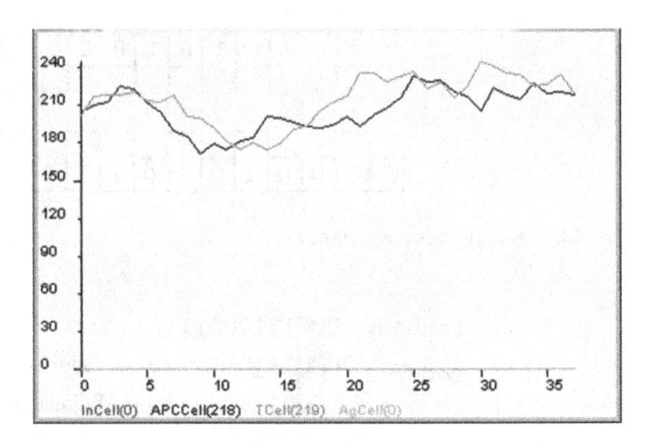

Fig. 14.3 Cell number curve without virus invasion

14.5 Results and Discussion

The key step in the immune simulation is to simulate the immune process. Through injecting an influenza virus at two subsequent times, we can observe system operation. According to the results of operation, we can find some rules.

When the program running, the system can generate a function curve. The horizontal axis represents time scales, and each time point represents the reality of 4.5 h. The vertical axis represents the number of cells.

Figure 14.3 shows the cell number curve without virus invasion. Because the immune response does not occur, so T cell and APC are in a state of dynamic stability.

In primary immune response (Fig. 14.4), because the virus which invaded the body needs some time to infect APC, so at the beginning, the number of In is zero. With the pass of time, the virus began to invade APC. In began to appear. And some viruses which failed to invade are killed by T cell. So the number of virus fells to a lower end. Meanwhile, some T cells have identified the virus invasion, and the number of T cell began a modest growth. A moment later, virus complete self-replicating in In. And the In begin to release the virus. The number of free virus in the system begin to rise. As a result, there have been more and more T cell identify virus invasion. The number of T cell began to grow rapidly. At 35 points

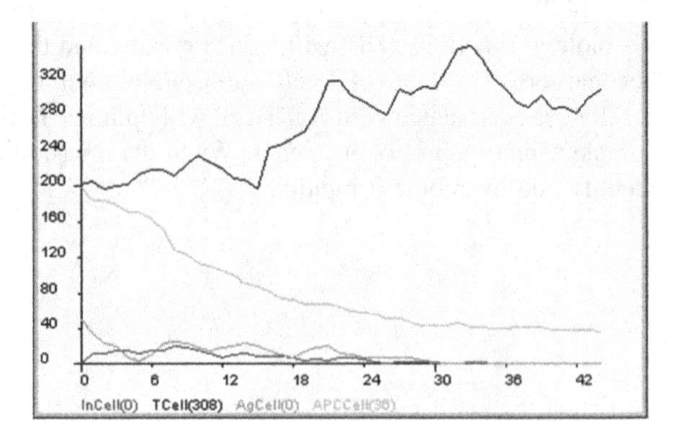

Fig. 14.4 Cell number curve of primary immune response

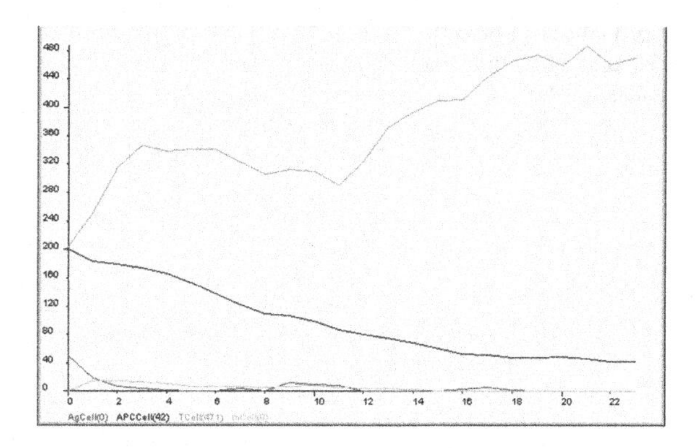

Fig. 14.5 Cell number curve of secondary immune response

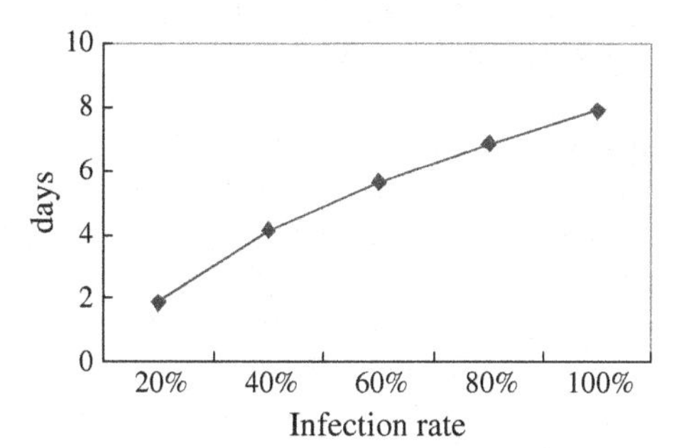

Fig. 14.6 Time of virus clearance under different infection rate

(Approximately 6.56 days), the number of virus down to zero, which means the virus are overcome.

In secondary response (Fig. 14.5), with the help of T memory cells, the time of virus clearance down to only 18.5 time-scales (Approximately 3.47 days). Because once detected virus invasion, these memory cells can produce secondary response fast and strong. T memory cells will split and multiply immediately. And the matching rate between T memory cells and viruses is more higher.

As you can see from Figs. 14.4 and 14.5, the simulation results is consistent with the disease phenomenon in medical observations. This shows that our simulation model is effective.

Immune system is a complex system that composed of many actors. Some changes of details might have great influence on the result. So we use the method of undetermined coefficients to analyze some related parameters.

Firstly, after the virus invade into the body, some of them are killed by T cell directly, others can infect APC and become In. We name the ratio of successful Ag as infection rate. Figure 14.6 provides the curve of the immune response cycle under different infection rate. The X axis represents different infection of Ag, and Y-axis indicates time of virus clearance. We can find that with the increase of infection rate, the immune cycle showing an increasing trend. Therefore, it is difficult for the immune system to deal with some viruses that have more toxic and higher infection rate.

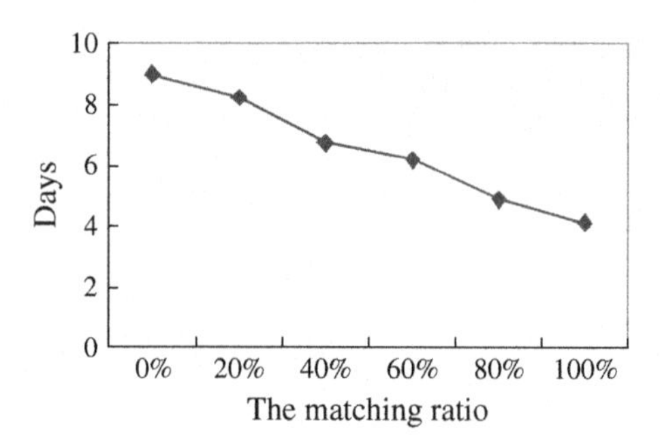

Fig. 14.7 Time of virus clearance under different matching ratio

The matching ratio between T cell and viruses also have a great effect on virus removal. Figure 14.7 shows the curve of the immune response cycle under different matching ratio. The X axis represents different value of matching ratio, and Y-axis indicates time of virus clearance. We can find that, when the value of matching ratio increase, the time required to remove virus will be less. In other words, the immune cycle become shorter. This can explain why the immune second response cycle is shorter than primary immune response cycle. The T memory cells which produced in primary immune response still retain the properties and characteristics of the virus, when meet the same virus again, the matching ratio between them is really high, so the second response is faster, and less time required. This also reminds us that it is more challenging and dangerous for our immune system to fight with new virus.

14.6 Conclusions

Up to now, there are still many unanswered questions related to the dynamics of the immune system. The diversity form of cell and factor makes it hard to find an unified specification to describe the immune system and establish the corresponding theoretical model. Computer simulation provides a new approach to this issue. This article adopts the method of computer simulation to simulate the interaction of all participants in the immune system. Through the analysis of simulation results we can prove that the dynamic behavior of model is similar to the basic laws of biology, and the simulation model is effective. With the help of simulation model, we can understand the immune system and explain the detail problems more easily.

The simulation describes the action of T cell, APC, and Ag in immune response, and we find some parameters which play a major role in viral immune process, and how can they affect the immune system.

Modeling and simulation of the immune system is a process of constantly improve. Many new methods and new ideas can be used to describe the immune system completely and accurately. With the help of computer simulation, some problems which cannot be solved in existing biological experiments will have the chance to be overcome, and biological theory research will get better.

References

1. Beauchemina C, Samuelb J, Tuszynskia J (2004) A simple cellular automaton model for influenza A viral infections, 12 Feb 2004
2. Fachada N, Lopes VV, Rosa A (2008) Simulations of Antigenic Variability in Influenza A, 14 Sep 2008
3. World Health Organization.http://www.who.int/mediacentre/factsheets/fs211/en/. Accessed March 2014
4. Hampson A, Mackenzie J (2006) Med J Australia 185(10 Suppl):S39–S43
5. Ryan K, Ray C (2004) Sherris medical microbiology: an introduction to infectious diseases. McGraw-Hill Medical, London (2004)
6. Miller JH, Page SE (2007) Complex adaptive systems: an introduction to computational models of social life. Princeton University Press, Princeton (2007)
7. Funk G, Barbour A, Hengartner H, Kalinke U (1998) Mathematical model of a virus neutralizing immunglobulin response. J Theor Biol 195(1):41–52
8. Forrest S, Beauchemin C (2007) Computer immunology. Immunol Rev 216(1):176–197
9. Ferguson NM, Cummings DAT, Fraser C, Cajka JC, Cooley PC, Burke DS (2006) Strategies for mitigating an influenza pandemic. Nature 442(7101):448–452
10. Germann TC, Kadau K, Longini IM, Macken CA (2006) Mitigation strategies for pandemic influenza in the United States. Proc Natl Acad Sci USA 103(15):5935–5940
11. Halloran ME, Ferguson NM, Eubank S, Longini IM, Cummings DAT, Lewis B, Xu S, Fraser C, Vullikanti A, Germann TC, Wagener D, Beckman R, Kadau K, Barrett C, Macken CA, Burke DS, Cooley P (2008) Modeling targeted layered containment of an influenza pandemic in the United States. Proc Natl Acad Sci USA 105(12):4639–4644
12. McVernon J, McCaw CT, Mathews JD (2007) Model answers or trivial pursuits The role of mathematical models in influenza pandemic preparedness planning. Influenza Other Respir Viruses 1(2):43–54
13. Hatchett RJ, Mecher CE, Lipsitch M (2007) Public health interventions and epidemic intensity during the 1918 influenza pandemic. Proc Natl Acad Sci USA 104(18):7582–7587
14. Kim SY, Goldie SJ (2008) Cost-effectiveness analyses of vaccination programmes: a focused review of modelling approaches. Pharmacoeconomics 26(3):191–215
15. Wu JT, Leung GM, Lipsitch M, Cooper BS, Riley S (2009) Hedging against antiviral resistance during the next influenza pandemic using small stockpiles of an alternative chemotherapy. PLoS Med 6(5):e1000085
16. Coburn BJ, Wagner BG, Blower S (2009) Modeling influenza epidemics and pandemics: insights into the future of swine flu (H1N1). BMC Med 7:30
17. Fachada N (2005) SimulIm: an application for the modelling and simulation of complex systems, using the immune system as an example. Graduation project report, Higher Technical Institute, Technical University of Lisbon
18. Bernaschi M, Castiglione F (2001) Design and implementation of an immune system simulator. Comput Biol Med 31(5):303–313
19. Emerson A, Rossi E (2007) ImmunoGrid—the virtual human immune system project. Stud Health Technol Inform 126:87–92
20. Mata J, Cohn M (2007) Cellular automata-based modeling program: synthetic immune system. Immunol Rev 216(1):198–212
21. Warrender C (2004) Modeling intercellular interactions in the peripheral immune system. Ph. D. thesis, The University of New Mexico, Dec 2004
22. Grilo A, Caetano A, Rosa A (2001) Agent based Artificial Immune System. In: Proceedings of GECCO-01, vol LBP, pp 145–151

Chapter 15
The Influenza Virus Immune Model on the Android Platform

Shengrong Zou, Xiaomin Jin, Na Zhong, Jundong Yan and Lin Yu

Abstract In biological experiments, it has been impossible that we just use experimental apparatus to deal with the complex problems in immune cells. And the traditional mathematics and the physics model have some limitations, like lacking of microcosmic performance description of unit cells [1]. In this article, we do detail design after analyzing the requirements of the immune system. Then, combining with the related data of influenza virus, we use the Android platform application development to simulate the system. Android platform's simple style of page, the application of interactive interface and the easy management can bring us different experiences. With the help of the computer program simulation, the experimental result is consistent with the model of immune response in the immune system.

Keywords The immune system · Simulation · The influenza virus · The android platform

15.1 Introduction

The immune system is a kind of the body's important immune response and immune function system. This system mainly consists of Immune organs, organizations, and other components of the immune cells. The system is the most effective weapon [2]. It can find and remove foreign bodies and foreign invasion of pathogenic microorganism [3] etc.

Most early immune system modeling is developed by the subject of a single way of research and development. At this stage, the immune system modeling is more and more various in form, also more and more novelty in the methods, but no significant breakthroughs in theory [4]. In this paper, with the help of the concepts

S. Zou (✉) · X. Jin · N. Zhong · J. Yan · L. Yu
School of Information Engineering, Yangzhou University, Yangzhou 225127, China
e-mail: 460046777@qq.com

Y. Jia et al. (eds.), *Proceedings of the 2015 Chinese Intelligent Systems Conference*, Lecture Notes in Electrical Engineering 360,
DOI 10.1007/978-3-662-48365-7_15

of bioinformatics, the immune system is regarded as an information system for research. We try to use the method of the computer to simulate the interaction between cells. But there are few domestic relevant papers in this field.

As the years went by, smartphones have been fully into our lives, because the function is very powerful, and it's more close to the computer. However, in smartphones, the development trend of the proportion of the fastest-growing is undoubtedly the Android phone. Based on the Android platform, the immune system model arises at the historic moment. Concise and profile page style of Android, interactive interface and easy application management during simulation can bring us different experiences [5].

15.2 The Description of the Immune System

Because the immune system is composed of immune cells, immune cells are the keys to design. The process of immune cells removing foreign bodies is known as the cellular immunity and the involved cells are called immune effector cells or sensitization [6]. Immune cells are mediated by T cells. When T cells suffer antigens stimulation, they will proliferate and differentiate into T effector cells. When the same antigens enter the body's cells again, T effector cells will kill them more strongly and release cytokines that have coordinated role [7, 8]. Figure 15.1 shows the immune system.

An initial T cell without differentiation can survive for many years. These initial cells move in cycles in the blood and lymph organs, and supervise physical changes. It is generally accepted that normal immune response to extrinsic antigens depends on coordinating interaction between T cells and antigen-presenting cells (APC) [9]. When a particular APC stops moving, the initial T cells can recognize

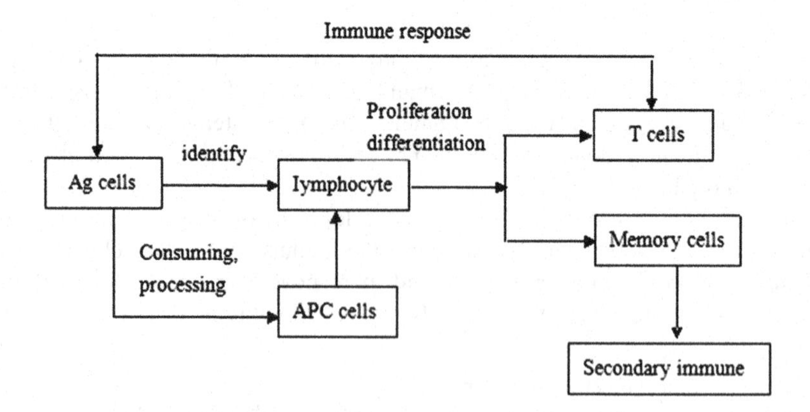

Fig. 15.1 The immune system

the specificity of the surface antigen and activate, proliferate and differentiate into effector T cells with the function of guarding. They will eventually clear antigens [10, 11].

15.3 The Requirements Analysis

A part of the immune system cell cases have the following different kinds:

Invasion; Antigen-presenting; Processing; Identification (recognition); Activation; Proliferation; Differentiation; Remove antigen (kill); Bind.

For the promotion of T cells or inhibition (promote, restrain) [12–15]:

Use cases and participants have the following relations: when the body suffered antigenicity stimulation such as the influenza virus. Firstly, APC will process and present antigens. Secondly, specific lymphocytes identify the antigens, which will cause the corresponding lymphocytes activate, proliferate and differentiate. Thirdly, through a series of immunological effects, antigens will be cleared.

The main object is the Cell, which includes the general properties of the cells. Ag cells, T cells and APC cells inherit the Cell. As Fig. 15.2 shown.

The related parameter settings:

Number of cell division [16], attack force [17] and defensive power [18]. (Once they meet different cells, the loss will be produced. The process is like when cells in collision, they can make the other cells reduce a certain life values. We use attack force to explain. At the same time, cells also have certain abilities of self-protection for the acting force, and their life values will increase. We use defensive power to explain. We use attack force and defensive power to represent the interactive force between cells. During this time, there will be cell aging and death in the mutual collision. And new cells are produced.) And so on.

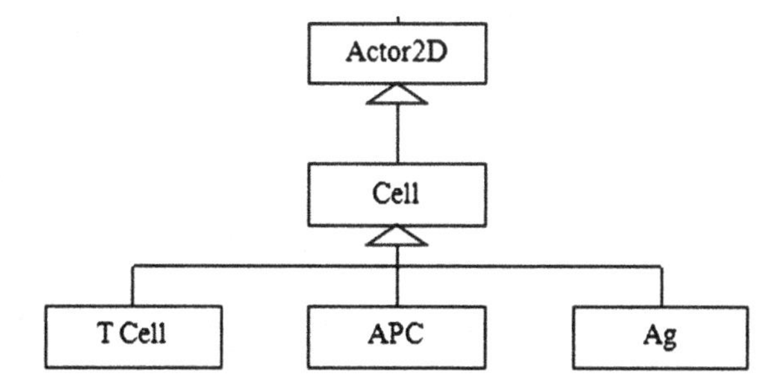

Fig. 15.2 The class diagram

15.4 Simulation on Android Platform

The important thoughts of the illustration on Android platform:

(1) The location of the new generation cells:

The size of the cells in the human body is different, and so is the form [18]. Every day, there are many cells that die in our bodies. At the same time, many new cells generate and exist. It can't happen that cells exist in the same place [19], so we have to determine whether there is a cell somewhere or not. If so, new cells can't be generated.

As show below in Fig. 15.3, black spots represent cells. The right triangle of the hypotenuse made of WIDTH and HEIGHT represents the center distance between the two cells. So if the distance is greater than the sum of two cell radius, which called overlapping cells, new cells can't generate here. By the same token, in order to avoid the new and old cells overlap in the great circle, there cannot generate new cells.

(2) Cells in the collision:

As illustrated in Fig. 15.4, the top left corner in the display acts like the origin of coordinates. The width is X axis and the length is Y axis. The 'this' and 'other' represent two motorial cells. The arrow is behalf of the direction of motion respectively. We determine whether there is a collision on the basis of that if borders overlapping happened. The collision is divided into four ways to discuss: the X same direction, the X opposite direction, the Y the same direction, and the Y opposite direction. In different situations, the rate and direction of motorial cells do the corresponding changes.

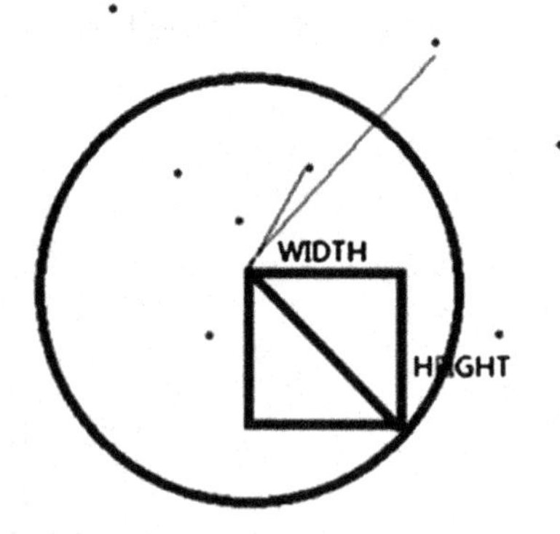

Fig. 15.3 The location of the new generation cells

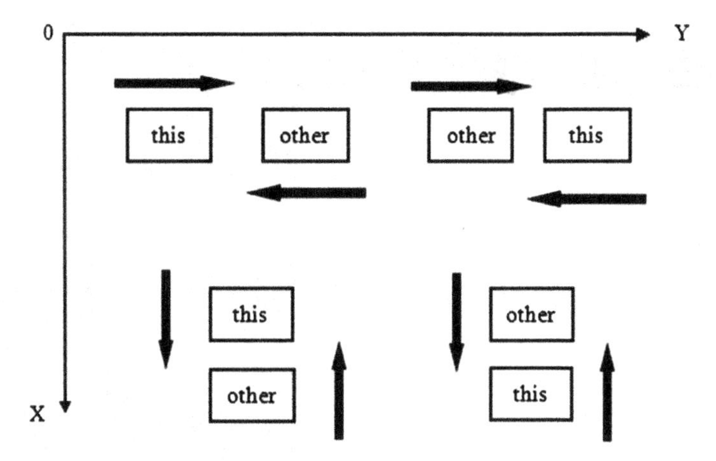

Fig. 15.4 Cells in the collision

15.5 System Simulation and the Result Analysis

Figure 15.5 is a description about the process of immune response. In the figure, the X axis represents time scale and each point represents 4 h in reality. The Y axis represents the number of cells. It can be seen from the diagram that when Ag virus in 42 points (7 days), its number turned 0, which means the virus are cleared. The result is consistent with the discovery that influenza is a short-term infection lasting about 7 days in most cases.

There are many factors that can affect the cellular immune cycle. The age is a major factor. Young children and the elderly are the groups of low immunity commonly. Because a baby, just a half year old, can take active immune substances from his mother, so his immune function is generally strong. It denotes more APCs existing. But after half a year, the maternal immune active material consume gradually and his own immune system is still in the stage of growth and development. The function is not perfect, which can cause low immunity. After aging, the function of human's organs decline and immune organs gradually atrophy [20, 21]. It means less APCs. As shown in Fig. 15.6, the horizontal axis shows virus Ag cell number, the vertical axis represents the APC number.

Figure 15.7 shows that different T cell life cycles also can affect immune cycle time. The horizontal axis shows T different life cycle value and the longitudinal axis is reactive time. The figure shows that in a certain range, the value of the T cell life cycle and reactive time is a positive correlation. In clinical practice, people who are strong can often recover faster than the people in poor health. Because people who are strong have the fast metabolism, which means T cells have short life cycle [22].

Figure 15.8 shows the time changing under different attack force. The horizontal axis shows the attack force of T cells and the vertical axis represents reactive time. It can be seen that with the increased damage, immune cycle shows a trend of

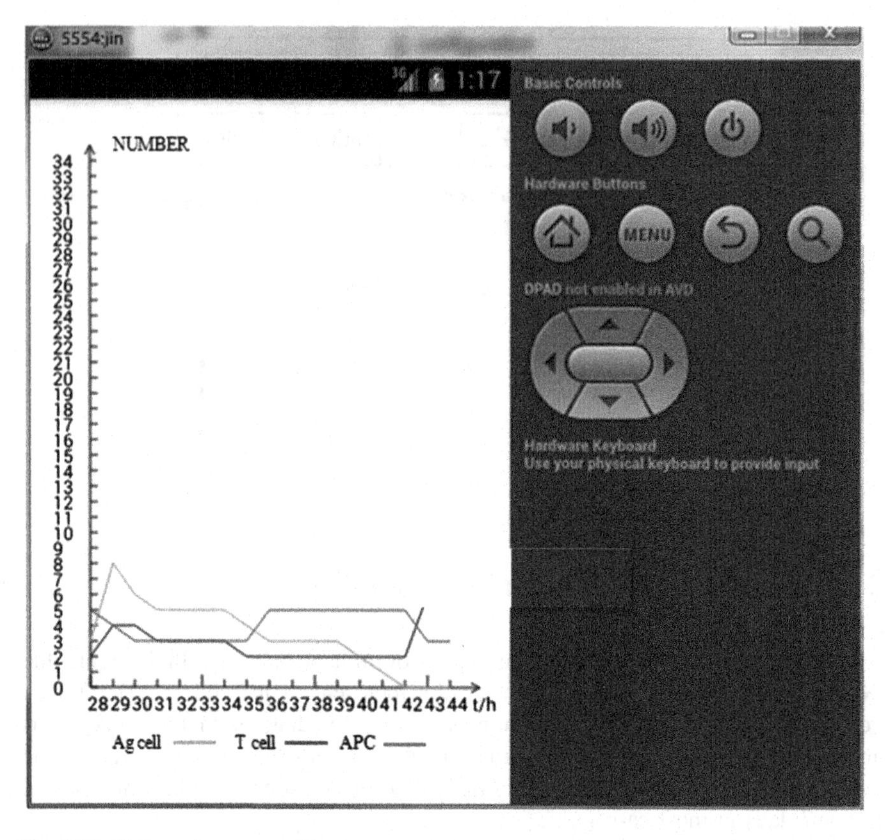

Fig. 15.5 Cell number curve of immune response

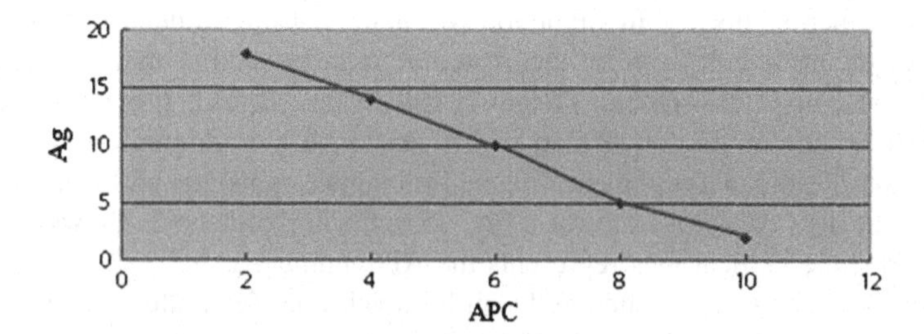

Fig. 15.6 The influence of APC cells on Ag cell reproduction

decline. It also means that within a certain threshold, the greater the damage, the shorter response time of virus clearance [22]. It fits the medical observation that different age populations with different immune capacity and have different influenza recovery period.

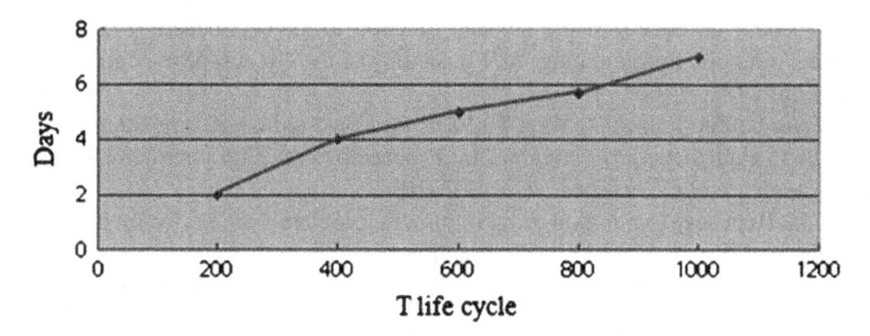

Fig. 15.7 Time of virus clearance under different T life cycle

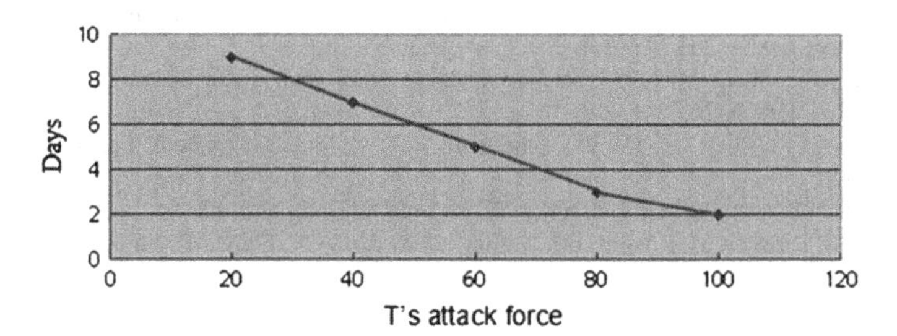

Fig. 15.8 Time of virus clearance under different attack force of T

15.6 Conclusion

In this paper, according to the process of software engineering, we analyze and design the immune system. For the immune model, we do a detailed design. Then we use the JAVA language on the Android platform to complete the code. And we set the previous defined class attributes with the relevant parameters of the influenza virus. Finally, we simulate the system. This article makes some attempts and explorations in theory. But there are still several issues to be further analysis and research, such as in system implementation part. Because we just selected a few cells that play a leading role, there are still many other cells. So our next step is adding more different cells to the system implementation, which can improve the accuracy of the system model.

References

1. Ferreira C (2001) Gene expression programming: a new adaptive algorithm for solving problems. Complex Syst 13(2):87–129
2. Ferreira C (2006) Gene Expression Programming: Mathematical Modeling by an Artificial Intelligence. Springer-Verlag, Berlin

3. Ferreira C (2002) Discovery of the boolean function to the best density—classification rules using gene expression programming. In: Proceeding of the 4th european conference on genetic programming, 2278:51–60
4. Yuan C, Tang C, Zuo J, chen A, Wen Y (2006) Attribute reduction function mining algorithm based on gene expression programming. In: Proceedings of 2006 international conference on machine learning and cybernetics, 2(7):1007–1012
5. Ferreira C (2001) Gene Expression programming in problem solving. In: Invited tutorial of the 6 the online world conference on soft computing in industrial applications
6. Keith MJ, Martin MC (1994) Genetic Programming in C ++:Implementation Issue. MIT press, Cambridge, p 97
7. Solomonoff R, Rapoport A (1951) Connectivity of random networks. Bull Math Biophys 13:107–117
8. Bollobas B (2001) RandomGraphs, 2nd edn. AcademicPress, NewYork
9. Newman MEJ, Watts DJ (1999) Renormalization group analysis of the small-world network model. Phys Lett A 263:341–346
10. Newman MEJ, Watts DJ (1999) Renormalization group analysis of the small-world network model. Phys Lett A 263:341–346
11. Barabasi AL, Albert R (1999) Emergence of scaling in random networks. Science 286:509–512
12. Kerkhove MDV, Asikainen T, Becker NG, Bjorge S, Desenclos JC, dos Santos T, Fraser C, Leung GM, Lipsitch M, Longini IM, McBryde ES, Roth CE, Shay DK, Smith DJ, Wallinga J, White PJ, Ferguson NM, Riley S (2010) W H O informal network for mathematical modelling for pandemic influenza H1N1 2009 (Working Group on Data Needs): studies needed to address public health challenges of the 2009 H1N1 influenza pandemic: insights from modeling. PLoS Med 7(6):e1000275
13. Miller JH, Page SE (2007) Complex Adaptive Systems: An Introduction to Computational Models of Social Life. Princeton University Press, Princeton
14. Funk G, Barbour A, Hengartner H, Kalinke U (1998) Mathematical model of a virus neutralizing immunglobulin response. J Theor Biol 195(1):41–52
15. Forrest S, Beauchemin C (2007) Computer immunology. Immunol Rev 216(1):176–197
16. Fachada N (2005) SimulIm: An application for the modelling and simulation of complex systems, using the immune system as an example. Graduation project report, Higher Technical Institute, Technical University of Lisbon
17. Bernaschi M, Castiglione F (2001) Design and implementation of an immune system simulator. Comput Biol Med 31(5):303–313
18. Emerson A, Rossi E (2007) ImmunoGrid—the virtual human immune system project. Stud Health Technol Inf 126:87–92
19. Mata J, Cohn M (2007) Cellular automata-based modeling program: synthetic immune system. Immunol Rev 216(1):198–212
20. Warrender C (2004) Modeling intercellular interactions in the peripheral immune system. Ph. D. thesis, The University of New Mexico
21. Ballet P, Tisseau J, Harrouet F (1997) A multiagent system to model a human humoral response. IEEE International conference on computational, cybernetics and simulation 10:357–362
22. Daigle J (2006) Human immune system simulation: a survey of current approaches. ACM J

Chapter 16
An Extend Set-Membership Filter for State Estimation in Power Systems

Yi Chai, Ping Deng, Nan Li and Shanbi Wei

Abstract In order to improve the accuracy and reliability of nonlinear state estimation problems with unknown but bounded noises in power system, an extend set-membership filter for state estimation in power systems is present in this article. The method is based on three sampling sine wave relational model. It overcomes the poor robustness, divergence and weak traceability of kalman filter, avoids the complex calculation process of traditional extend set-membership filter. Compared with kalman filter, the simulation results show that extend set-membership filter algorithm can track signals faster and more accurately.

Keywords State estimation · Set-membership filter · Power system

16.1 Introduction

To ensure the power system is running in desired states, state estimation has always been a fundamental function for Energy Management Systems (EMS) of electric power networks [1]. The knowledge of the state vector at each bus, i.e., voltage magnitude and phase angle, enables the EMS to perform various crucial tasks, such as bad data detection, optimizing power flows, maintaining system stability and reliability, etc. [2].

Several techniques can be exploited to solve the aforementioned estimation problems. A classic approach is to use a Kalman Filter (KF), with zero mean, Gaussian noise assumptions, to estimate a state, and to use a linear quadratic regulator for control based on the state estimate [3, 4]. However, the approaches

Y. Chai
State Key Laboratory of Power Transmission Equipment and System Security and New Technology, Chongqing 4000430, China

Y. Chai · P. Deng · N. Li · S. Wei (✉)
College of Automation, Chongqing University, Chongqing 400030, China
e-mail: weishanbi@cqu.edu.cn

Y. Jia et al. (eds.), *Proceedings of the 2015 Chinese Intelligent Systems Conference*, Lecture Notes in Electrical Engineering 360,
DOI 10.1007/978-3-662-48365-7_16

above always require all the uncertainties (disturbances, model errors) in a stochastic framework deliver the highest probability state estimation [5]. The noise must have a Gaussian distribution. The probabilistic nature of the estimates requires the use of mean and variance to describe the state distributions. Unfortunately, in many practical situations it is not realistic to assume a probability statistical properties for disturbances include correlated, biased noises. Actually, a bound on model errors is often the only available information [6].

To solve the mentioned problems above, the Set-membership Filter (SMF) theory is applied. The SMF is under the assumption conditions that noises are unknown but bounded, rather than requiring the state distributions described by means and variances. The goal of the SMF is to find a feasible set that guarantees the true state of the system is within the computed set and is compatible with all the available information (the measured data, modelling uncertainty and the known noise bounds) in the state space [7]. Extension to nonlinear problems is the extended set-membership (ESMF). Several works in [6, 8, 9] have addressed this issue.

Ellipsoid set is the most common approach to describe the state estimation for the nonlinear discrete-time system. Based on the works on the [10, 11], a more valid and pertinence state estimation method for power system is proposed in this paper. A three sampling sine wave relational model is introduced to avoid the complex calculation` process of the usual ESMF algorithm, to improve the instantaneity of online calculation.

16.2 Signal Model

Consider a discrete-time signal of voltage or current in power system after sampling as described by [12]:

$$y_k = A\cos(k\omega T_s + \phi) + v_k \tag{16.1}$$

where y_k represent instantaneous signal value; A is signal amplitude; k represents the variables at time step k; T_s is sampling period; ω denotes fundamental frequency; ϕ is the phase position; v_k represents the added noise, unknown but bounded.

Rewriting Eq. (16.1) as follows:

$$y_k = \hat{y}_k + v_k \tag{16.2}$$

where $\hat{y}_k$ is the signal to be estimated.

Three consecutive samples of the above single-phase sinusoidal signals satisfy the following relationship:

$$\hat{y}_k - 2\cos\omega T_s \hat{y}_{k-1} + \hat{y}_{k-2} = 0. \tag{16.3}$$

Frequency can be accurately estimated by the formula (16.3), with selection of the state reasonably to eliminate the influence of noise effectively. State estimation vector is as follows:

$$x_k = [2\cos\omega T_s \quad \hat{y}_{k-1} \quad \hat{y}_{k-2}]^T \tag{16.4}$$

$$x_{k+1} = \begin{bmatrix} 1 & 0 & 0 \\ 0 & 2\cos\omega T_s & -1 \\ 0 & 1 & 0 \end{bmatrix} x_k \tag{16.5}$$

The corresponding measurement equations with added noise:

$$\hat{y}_k = [0 \quad 2\cos\omega T_s \quad -1] x_k + v_k. \tag{16.6}$$

where $x_k \in R^n$ represents the state variables, and vector $y_k \in R^m$ represents available measurements.

Rewriting the discrete-time Eqs. (16.5) and (16.6) in a uniform structure for a power system state estimation:

$$x_{k+1} = f(x_k) \tag{16.7}$$

$$\hat{y}_k = g(x_k) + v_k \tag{16.8}$$

where:

$$f(x_k) = [2\cos\omega T_s \quad 2\cos\omega T_s \cdot \hat{y}_{k-1} - \hat{y}_{k-2} \quad \hat{y}_{k-1}]^T \tag{16.9}$$

$$g(x_k) = 2\cos\omega T_s \hat{y}_{k-1} - \hat{y}_{k-2}. \tag{16.10}$$

where $v_k \in R^m$ represents the measurement noise, which is assumed to be unknown but bounded and confined to the box:

$$V_k = \{v_k : |v_k^i| \le \varepsilon_k^i\}, i = 1, \ldots, m \tag{16.11}$$

where v_k^i is the ith component of v_k, and $\varepsilon_k^i > 0$.

16.3 Extended Set-Membership Filter

The ellipsoid E_{k+1} containing the state x_{k+1} at the time instant $k+1$ is

$$E_{k+1} = \left\{x_{k+1}: (x_{k+1} - \hat{x}_{k+1})^T P_{k+1}^{-1} (x_{k+1} - \hat{x}_{k+1}) \leq 1\right\} \tag{16.12}$$

where $\hat{x}_{k+1}$ is the state estimate of x_{k+1} and P_{k+1} defines the shape. In other words, we look for P_{k+1} and $\hat{x}_{k+1}$ such that

$$(x_{k+1} - \hat{x}_{k+1})^T P_{k+1}^{-1} (x_{k+1} - \hat{x}_{k+1}) \leq 1 \tag{16.13}$$

subject to $v_k \in V_k$ assuming that the ellipsoid E_k is obtained.

Considering the Taylor approximation of $f(x_k)$ around $\hat{x}_{k|k}$, Eq. (16.7) equals to:

$$x_{k+1} = \hat{x}_{k+1/k} + F_k \Delta x_k + R_2(\Delta x_k, \bar{X}_k) \tag{16.14}$$

where F_k is the gradient operator:

$$F_k = \left.\frac{\partial f(x)}{\partial x}\right|_{x = \hat{x}_{k|k}} \tag{16.15}$$

where $\Delta x_k = x_k - \hat{x}_{k|k}$, $\bar{X}_k = [\hat{x}_{k|k}^i - \sqrt{P_{k|k}^{i,i}}, \hat{x}_{k|k}^i + \sqrt{P_{k|k}^{i,i}}]$. The Lagrange remainder $R_2(*)$ is processed as a bounded noises by using interval arithmetic, and the detailed derivation are proposed in [13]. The feasible set of $R_2(\Delta x_k, \bar{X}_k)$ is calculated by a box Γ_k [14]:

$$\Gamma_k = \{\bar{w}_k: |w_k^i| \leq b_i, i = 1, .., n\} \tag{16.16}$$

$$b_i = rad(R_2^i(\Delta x_k, \bar{X}_k)) \tag{16.17}$$

where $R_2^i(\Delta x_k, \bar{X}_k) = \frac{1}{2} \Delta x_k^T \frac{\partial^2 f^i(\bar{X}_k)}{\partial x^2} \Delta x_k, i = 1, \ldots, n$. $rad(R_2^i(\Delta x_k, \bar{X}_k))$ denotes the radius of interval variable $\bar{X}_k$.

Therefore, the time-update ellipsoid $E_{k+1|k}$ is the minimum volume ellipsoid containing the ellipsoid $E_{k|k}$ and box Γ_k. A computation cheaper algorithm was proposed in [11, 14], the ellipsoid $E_{k+1|k}$ is obtained using the following algorithm.

Time update:

1. The ellipsoid center $\hat{x}_{k+1/k}$ is calculated by

$$\hat{x}_{k+1|k} = f(\hat{x}_{k|k}) \tag{16.18}$$

2. The matrix of the time-updated ellipsoid is initialized as:

$$P^{0}_{k+1|k} = F_k P_k F_k \tag{16.19}$$

3. Recursion calculation: for $i = 0, \ldots, n-1$,

$$P^{i+1}_{k+1|k} = (1 + p_i) P^{i}_{k+1|k} + (1 + p^{i}_{k|k+1}) b_i^2 I_i I_i^T \tag{16.20}$$

Parameter p_i is the positive root of

$$n p_i^2 + (n-1) a_i p_i - a_i = 0 \tag{16.21}$$

where $a_i = tr(b_i^2 (P^{i}_{k+1|k})^{-1} I_i I_i^T) = b_i^2 I_i^T (P^{i}_{k+1|k})^{-1} I_i$.

4. Finally, $P_{k+1|k} = P^{n}_{k+1|k}$.

Measurement update:
The gradient operator:

$$G_{k+1} = \left. \frac{\partial g(x)}{\partial x} \right|_{x = \hat{x}_{k+1|k}} \tag{16.22}$$

The process of the Lagrange remainder in the Taylor approximation of the measurement equation y_{k+1} is just the same as that of the state equation in (16.14)–(16.17).

As proposed in [14], the measurement-update ellipsoid E_{k+1} should be the minimal-volume ellipsoid which contains ellipsoid-strip intersection of the time-update ellipsoid $E_{k+1|k}$ and strip set R_{k+1} at each iteration. It means $E_{k+1} \supseteq E_{k+1|k} \bigcap R_{k+1}$. The center $\hat{x}_{k+1}$ and the shape matrix P_{k+1} of the measurement-update ellipsoid E_{k+1} are calculated by the following algorithm.

1. Initialization:

$$\hat{x}^{0}_{k+1} = \hat{x}_{k+1|k} \tag{16.23}$$

$$P^{0}_{k+1} = P_{k+1|k} \tag{16.24}$$

2. Recursion calculation: for $i = 1, \ldots, m$,

$$h_i = G^{T}_{k+1,i} P^{i-1}_{k+1} G_{k+1,i} \tag{16.25}$$

$$\phi_i^{+} = (z^{i}_{k+1} - G^{T}_{k+1,i} x^{i-1}_{k+1} + \delta^{i}_{k+1}) / \sqrt{h_i} \tag{16.26}$$

$$\phi_i^{-} = (z^{i}_{k+1} - G^{T}_{k+1,i} x^{i-1}_{k+1} - \delta^{i}_{k+1}) / \sqrt{h_i} \tag{16.27}$$

Where $z_{k+1}^{i} = y_{k+1}^{i} - g_i(\hat{x}_{k+1|k}) + G_{k+1,i}^{T}\hat{x}_{k+1|k}$, $\delta_{k+1}^{i} = rad(R_2^i(\Delta\tilde{x}_{k+1}, \tilde{X}_{k+1})) + \varepsilon_{k+1}^{i}$.

$$R_2^i(\Delta\tilde{x}_{k+1}, \tilde{X}_{k+1}) = \frac{1}{2}\Delta\tilde{x}_{k+1}^{T}\frac{\partial^2 g^i(\tilde{X}_{k+1})}{\partial x^2}\Delta\tilde{x}_{k+1} \tag{16.28}$$

and $\Delta\tilde{x}_{k+1} = x_{k+1} - \tilde{x}_{k+1|k}$, $\tilde{X}_{k+1} = [\hat{x}_{k+1|k}^{i} - \sqrt{P_{k+1|k}^{i,i}}, \hat{x}_{k+1|k}^{i} + \sqrt{P_{k+1|k}^{i,i}}]$.

3. If $\phi_i^+ > 1$, then set $\phi_i^+ = 1$; if $\phi_i^- < -1$, then set $\phi_i^- = -1$; and if $\phi_i^+\phi_i^- \le -1/n$, then $P_{k+1}^{i} = P_{k+1}^{i-1}$, and $\hat{x}_{k+1}^{i} = \hat{x}_{k+1}^{i-1}$. Otherwise for $i = 1, \ldots, m$,

$$\hat{x}_{k+1}^{i} = \hat{x}_{k+1}^{i-1} + q_i\frac{D_{k+1}^{i}G_{k+1,i}e_i}{d_i^2} \tag{16.29}$$

$$P_{k+1}^{i} = (1 + q_i - \frac{q_ie_i^2}{d_i^2 + q_ih_i})D_{k+1}^{i} \tag{16.30}$$

Where

$$D_{k+1}^{i} = P_{k+1}^{i-1} - \frac{q_i}{d_i^2 + q_ih_i}P_{k+1}^{i-1}G_{k+1,i}G_{k+1,i}^{T}P_{k+1}^{i-1} \tag{16.31}$$

And $e_i = \sqrt{h_i}\frac{\phi_i^+ + \phi_i^-}{2}$, $d_i = \sqrt{h_i}\frac{\phi_i^+ - \phi_i^-}{2}$.

The parameter q_i is the positive root of

$$(n-1)h_i^2q_i^2 + [(2n-1)d_i^2 - h_i + e_i^2]h_iq_i + [n(d_i^2 - e_i^2) - h_i]d_i^2 = 0 \tag{16.32}$$

4. Finally, $P_{k+1} = P_{k+1}^{m}$, $\hat{x}_{k+1}^{i} = \hat{x}_{k+1}^{m}$.

16.4 Numerical Experiments

According to the model in state Eqs. (16.7) and (16.9), we derive:

$$x_{k+1} = \begin{bmatrix} x_k^1 & x_k^1x_k^2 - x_k^3 & x_k^2 \end{bmatrix}^T \tag{16.33}$$

$$F_k = [1 \quad 0 \quad 0;\ \hat{x}_{k|k}^2 \quad \hat{x}_{k|k}^1 \quad -1;\ 0 \quad 1 \quad 0] \tag{16.34}$$

The remaining terms are: $R_2^1(\Delta x_k, \bar{X}_k) = 0$, $R_2^3(\Delta x_k, \bar{X}_k) = 0$,

$$R_2^2(\Delta x_k, \bar{X}_k) = \frac{1}{2}[\Delta x_k^1\ \Delta x_k^2\ \Delta x_k^3]\begin{bmatrix} 0 & 1 & 0 \\ 1 & 0 & 0 \\ 0 & 0 & 0 \end{bmatrix}[\Delta x_k^1\ \Delta x_k^2\ \Delta x_k^3]^T \tag{16.35}$$

According to the measurement Eqs. (16.8) and (16.10), we derive:

$$y_k = x_k^1 x_k^2 - x_k^3 + v_k^1 \tag{16.36}$$

$$G_k = \begin{bmatrix} \hat{x}^2_{k|k-1} & \hat{x}^1_{k|k-1} & -1 \end{bmatrix} \tag{16.37}$$

$$R_2^1(\Delta\tilde{x}_{k+1}, \tilde{X}_{k+1}) = = \frac{1}{2}[\Delta x^1_{k+1}\ \Delta x^2_{k+1}\ \Delta x^3_{k+1}] \begin{bmatrix} 0 & 1 & 0 \\ 1 & 0 & 0 \\ 0 & 0 & 0 \end{bmatrix} [\Delta x^1_{k+1}\ \Delta x^2_{k+1}\ \Delta x^3_{k+1}]^T \tag{16.38}$$

The initial feasible-state ellipsoidal set has $P_0 = 10^{-3} I_3$, and initial centre is $\hat{x}_0 = x_0 + [0.02\ 0.02\ 0.02]^T$, which is chosen after simulation experiments. For the three sampling sine wave relational model, signal amplitude $A = 1$, phase position $\phi = \pi/2$, fundamental frequency $\omega_s = 2\pi \times 50Hz$, sampling frequency is 2 kHz. The noise is assumed to be unknown but bounded, and the bounds of process and measurement noise are set as $|v_k^1| \leq 0.04$. The noises are uniformly distributed.

The results of our method are from step 3 to 140. The step is from $k = 3$ for choice of the state vector in (16.4). In Fig. 16.1, the solid red, the solid blue, dashed purple and dashed green lines represent the true values, the estimated values, the hard upper bounds and the hard lower bounds respectively. As illustrated in the figure, the upper and the lower bounds contain the true values all the time as expected. The decreasing bounds indicate that estimate ellipsoid is convergent.

To display the estimation of the feasible solution set better, the transformation of the estimation ellipsoid is illustrated in Fig. 16.2. The three-dimensional ellipsoids are projected onto two-dimensional plane to form ellipses. The estimation values

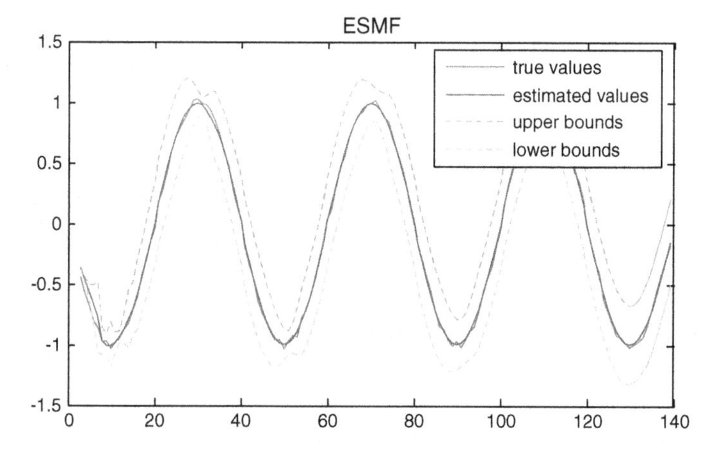

Fig. 16.1 Estimation result using ESMF

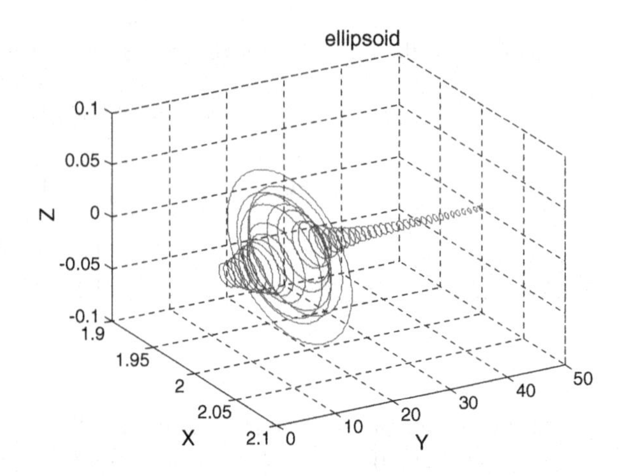

Fig. 16.2 The transformation of the estimation ellipsoid

are located within the ellipses. The ellipse areas are decreasing, which indicate the estimation ranges are reduced at each step.

For comparison, the estimation results using EKF are presented below. The bound of the EKF is calculated by introducing a noise fluctuation 3σ to the estimated values $\hat{x}$ according to [10]. The initial values of states are set equal to that of the ESMF. The covariance matrix $P_0 = 10^{-3} I_3$. The measurement noise is Gaussian white noise with zero mean, and variances are σ, respectively. The covariance matrix is set to $R_k = \sigma^2 = v_k$.

Aimed to evaluate the effectiveness of the algorithms, the performance will be measured by mean-square errors (Mse), which is computed as:

$$\bar{e}_i = \frac{1}{r} \sum_{s=1}^{r} \frac{1}{140} \sum_{k=10}^{140} (y_k^s - \hat{y}_k^s)^2$$

Where r is the times of the simulation run; y_k^s is the true value at time k in run s and $\hat{y}_k^s$ is the center of the corresponding estimated ellipsoid. r is set as 10, and the values before time $k = 10$ are excluded to avoid the influence of the initial conditions.

As shown in Figs. 16.3 and 16.4, the state estimation bounds using EKF fail to contain the estimated values within it, which is less acceptable compared with the hard bounds provided by ESMF. The Mse results are shown in Table 16.1, which implies the error of the ESMF is a bit larger than that of the EKF. Besides, ESMF algorithm provides a hard estimation upper and lower bounds containing the true values all the time what EKF do not have. The results implicate that ESMF are superior to EKF when 100 % confidence is required in state estimation especially for the security and reliability of power systems.

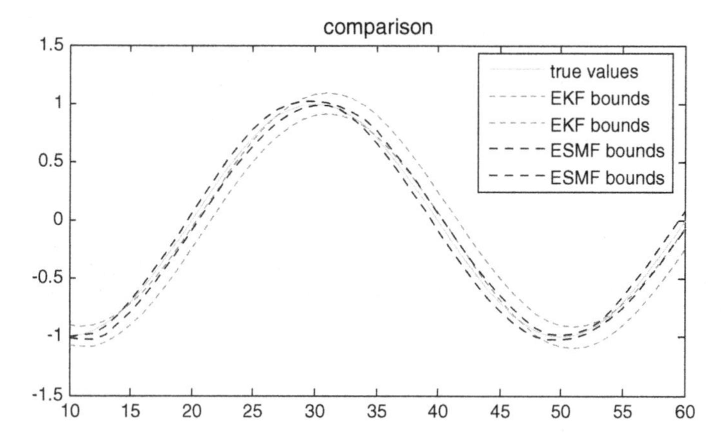

Fig. 16.3 Bounds comparison using ESMF and EKF

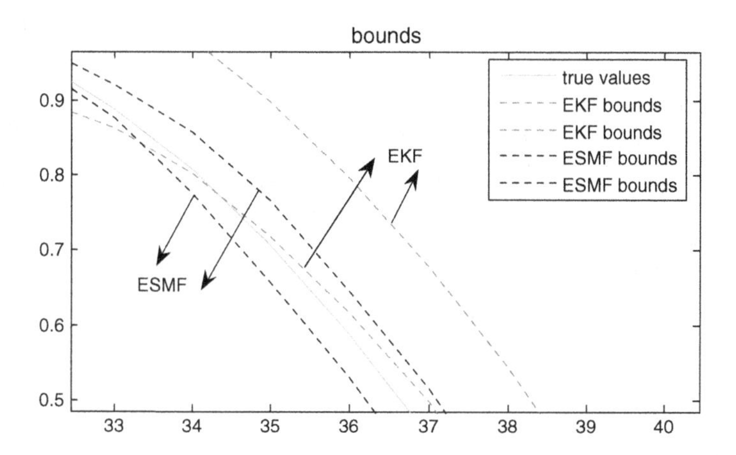

Fig. 16.4 Partial enlarged view of bounds comparison using ESMF and EKF

Table 16.1 Mean-square estimation errors comparison

Method	ESMF	EKF
Method	ESMF	EKF
Mse	0.000978	0.000631

16.5 Conclusions

For the sampled voltage or current discrete-time signal, an extend set-membership filter for state estimation in power systems is present in our paper. For this we built three sampling sine wave relational model. Compared with other extend set-membership filters, our method is simple and easy to implement in amount of computations. Theoretical analysis and simulation results indicate that ESMF can quickly track the states under disturbance of harmonics and noise. Besides, estimation ellipsoid of ESMF can provide accurate real-time estimation range containing the true values compared with EKF. Therefore this method is an effective and accurate method for state estimation of distorted voltage (current) signal in power system.

Acknowledgement We would like to thank the supports by Chongqing city science and technology project (cstc2012ggyyjs00007), National Natural Science Foundation of China (61374135), Specialized Fund for the Basic Research Operating expenses Program of Central College (0215005207007).

References

1. Huang YF, Werner S, Huang J, Kashyap N, Gupta V (2012) State estimation in electric power grids: Meeting new challenges presented by the requirements of the future grid. IEEE Trans Sig Process 29(5):33–43
2. Schweppe FC, Handschin EJ (1974) Static state estimation in electric power systems. Proc IEEE 7(62):972–982
3. Nishiyama Kiyoshi (1997) Nonlinear filter for estimating a sinusoidal signal and parameters in white noise: on the case of a single sinusoid. IEEE Trans Sig Process 45(2):970–981
4. Dash KP, Pradhan KA, Panda G (1999) Frequency estimation of distorted power system signals using extended Kalman filter. IEEE Trans Power Deliv 14(3):761–766
5. Scholte E, Campbell ME (2002) On-line nonlinear guaranteed estimation with application to a high performance aircraft. In: Proceedings of the American Control Conference, pp 184–190
6. Kieffer M, Jaulin L, Walter E (2002) Guaranteed recursive non-linear state bounding using interval analysis. Int J Adapt Control Sig Process 16(3):193–218
7. Maksarov D, Norton J (1996) State bounding with ellipsoidal set description of the uncertainty. Int J Control 65(5):847–866
8. Schole E, Cambell ME (2003) A nonlinear set-membership filter for on-line applications. Int J Robust Nonlinear Control 13(15):1337–1358
9. Bertsekas DP, Rhodes IB (1971) Recursive state estimation for a set-membership description of uncertainty. IEEE Trans Autom Control 16(2):117–128
10. Qing Xiangyun, Yang Fuwen, Wang Xingyu (2013) Extended set-membership filter for power system dynamic state estimation. Electr Power Syst Res 99:56–63
11. Maksarov DG, Norton JP (2002) Computational efficient algorithms for state estimation with ellipsoidal approximations. Int J Adapt Control Sig Process 16(5):411–434
12. Routray A, Pradhan KA, Pao PK (2002) A novel Kalman filter for frequency estimation of distorted signal in power systems. IEEE Trans Instrum Meas 51(3):469–479
13. Jaulin L, Walter E (1993) Set interval via interval analysis for nonlinear bounded-error estimation. Automatica 29(4):1053–1064
14. Chai W, Sun X (2007) Nonlinear set membership filtering using ellipsoids and its application in fault diagnosis. Acta Aeronaut et Astronaut Sin 28(4):948–952

Chapter 17
Anti-disturbance Control for T-S Fuzzy Models Using Fuzzy Disturbance Modeling

Xiangxiang Fan, Zhongde Sun, Yang Yi and Tianping Zhang

Abstract In this paper, an anti-disturbance tracking control scheme is proposed for T-S fuzzy models subject to parametric uncertainties and unknown disturbances. Different with those previous results, exogenous disturbances are also described by T-S disturbance models. Under this framework, a composite observer is constructed to estimate the system state and the disturbances simultaneously. Meanwhile, by integrating the PI-type control algorithm with the estimates of the state and the disturbance, a feedback control input is designed to ensure the system stability and the convergence of the tracking error to zero as well as satisfactory disturbance estimation and attenuation performance.

Keywords T-S fuzzy models · Anti-disturbance control · Tracking control · Disturbance observer · T-S disturbance modeling

17.1 Introduction

It is well known that disturbances exist in all practical processes [1–5]. In recent years, disturbance observer based control (DOBC) strategies have been successfully used in various systems, such as robot manipulators [6–8], high speed direct-drive positioning tables [9], permanent magnet synchronous motors [10] and magnetic hard drive servo systems [11] etc. However, the exogenous disturbances in most DOBC results [12, 13] are supposed to be generated by linear exogenous system, while there are always irregular and nonlinear disturbances in practical systems, which will no longer be effective by using the present model-based disturbance observer design methods.

X. Fan · Y. Yi (✉) · T. Zhang
College of Information Engineering, Yangzhou University, Yangzhou 225127, China
e-mail: yiyangcontrol@163.com

Z. Sun
Yangzhou Meteorologic Bureau, Yangzhou 225009, China

Y. Jia et al. (eds.), *Proceedings of the 2015 Chinese Intelligent Systems Conference*, Lecture Notes in Electrical Engineering 360,
DOI 10.1007/978-3-662-48365-7_17

On the other hand, Takagi-Sugeno (T-S) fuzzy model becomes very popular since it is a powerful tool for approximating a wide class of nonlinear systems, such as descriptor systems [14], networked control systems [15, 17], stochastic systems [16] and time-delay systems [15]. Furthermore, some typical control problems, including dynamic tracking control [18], fault estimation and detection [19], sliding-mode control [20] and filter design [21] have also been considered through T-S fuzzy modeling.

This paper discusses the anti-disturbance tracking control for the T-S fuzzy models with parametric uncertainties and irregular disturbances. Following the T-S fuzzy modeling for unknown irregular disturbances, the composite anti-disturbance controller are designed by combining PI control structure and disturbance observer design method. It is shown that the stability and the favorable tracking performance of augmented systems can achieved by using convex optimization algorithm and Lyapunov analysis method. Finally, simulation results in flight control system are given to show the efficiency of the proposed approach.

17.2 Model Description with Fuzzy Disturbance Modeling

Considering the following T-S fuzzy model with parametric uncertainties and exogenous disturbances

Plant Rule a: If ϑ_1 is M_1^a, and $\cdots$ and ϑ_q is M_q^a, then

$$\begin{cases} \dot{x}(t) = (A_{0a} + \Delta A_{0a})x(t) + (B_{0a} + \Delta B_{0a})[u(t) + d_1(t)] + d_2(t) \\ y(t) = C_{0a}x(t) \end{cases} \tag{17.1}$$

where $\vartheta = [\vartheta_1, \ldots, \vartheta_q]$ and $M_g^a(a = 1, 2, \ldots, p)$ are the premise variables and the fuzzy sets, respectively. p, q are the numbers of If-Then rules and premise variables, respectively. $x(t) \in R^n, u(t) \in R^m, d_1(t) \in R^m, d_2(t) \in R^n$ and $y(t) \in R^{p_1}$ are the control input, modeled disturbance, unmodeled disturbance with bounded peaks and the measurement output, respectively. A_{0a}, B_{0a}, C_{0a} are the coefficient matrices with appropriate dimensions. ΔA_{0a} and ΔB_{0a} represent parametric uncertainties.

The overall fuzzy model can be inferred as follows

$$\begin{cases} \dot{x}(t) = \sum_{a=1}^{p} h_a(\vartheta) \left\{ (A_{0a} + \Delta A_{0a})x(t) + (B_{0a} + \Delta B_{0a})[u(t) + d_1(t)] \right\} + d_2(t) \\ y(t) = \sum_{a=1}^{p} h_a(\vartheta) C_{0a}x(t) \end{cases} \tag{17.2}$$

where $\sigma_a(\vartheta) = \prod_{g=1}^{q} M_g^a(\vartheta_g), h_a(\vartheta) = \sigma_a(\vartheta)/\sum_{a=1}^{p}\sigma_a(\vartheta)$, in which $M_g^a(\vartheta_g)$ is the grade of membership of ϑ_g in M_g^a and $h_a(\vartheta) \geq 0, \sum_{a=1}^{p} h_a(\vartheta) = 1$.

Moreover, the nonlinear disturbances $d_1(t)$ can be generated by the following T-S fuzzy model with r plant rules.

Plant Rule j: If θ_1 is A_1^j, and $\cdots$ and θ_n is A_n^j, then

$$\begin{cases} \dot{w}(t) = W_j w(t) \\ d_1(t) = V_j w(t) \end{cases} \tag{17.3}$$

where W_j and V_j are known coefficient matrices. $\theta_i (i = 1, \ldots, n)$ and $A_i^j (j = 1, 2, \ldots, r)$ are the premise variables and the fuzzy sets, respectively. r is the number of If-Then rules, while n is the number of the premise variables.

By fuzzy blending, the overall fuzzy model can be defined as follows

$$\begin{cases} \dot{w}(t) = \sum_{j=1}^{r} h_j(\theta) W_j w(t) \\ d_1(t) = \sum_{j=1}^{r} h_j(\theta) V_j w(t) \end{cases} \tag{17.4}$$

where $\omega_j(\theta) = \prod_{i=1}^{n} A_i^j(\theta_i), h_j(\theta) = \omega_j(\theta)/\sum_{j=1}^{r}\omega_j(\theta), \theta = [\theta_1, \ldots, \theta_n], j = 1, \ldots, r$ is the membership function of the system with respect to plant rule j, and $h_j(\theta) = \omega_j(\theta)/\sum_{j=1}^{r}\omega_j(\theta)$.

The uncertainties in T-S fuzzy model (1) are assumed to be of the form

$$\Delta A_{0a} = HF(t)E_{1a}, \; \Delta B_{0a} = HF(t)E_{2a} \tag{17.5}$$

where H, E_{1a} and E_{2a} are constant matrices with corresponding dimensions. $F(t)$ is an unknown, real and possibly time-varying matrix with Lebesgue measurable elements satisfying

$$F^T(t)F(t) \leq I, \forall t. \tag{17.6}$$

Lemma 1 *Assume that X and Y are vectors or matrices with appropriate dimension. The following inequality*

$$X^T Y + Y^T X \leq \alpha X^T X + \alpha^{-1} Y^T Y \tag{17.7}$$

holds for any constant $\alpha > 0$.

17.3 Design of DOB PI Composite Controller

In this section, we construct a composite full-state observer to estimate the state $x(t)$ and the disturbance $d_1(t)$ simultaneously.

Based on the above-mentioned T-S fuzzy model (2), we introduce a new state variable

$$\bar{x}(t) := \left[x^T(t), \int_0^t e^T(\tau)d\tau\right]^T \tag{17.8}$$

where the tracking error $e(t)$ is defined as $e(t) = y(t) - y_d$, y_d is the reference output. Then the augmented system can be established as

$$\dot{\bar{x}}(t) = \sum_{a=1}^{p} h_a(\vartheta)\{A_a\bar{x}(t) + B_a[u(t) + d_1(t)]\} + Cy_d + Jd_2(t) \tag{17.9}$$

where $A_a = \begin{bmatrix} A_{0a} + HFE_{1a} & 0 \\ C_{0a} & 0 \end{bmatrix}, B_a = \begin{bmatrix} B_{0a} + HFE_{2a} \\ 0 \end{bmatrix}, C = \begin{bmatrix} 0 \\ -I \end{bmatrix}, J = \begin{bmatrix} I \\ 0 \end{bmatrix}$.

By combining the exogenous disturbance model (4) with T-S fuzzy model (2), the augmented system can be further constructed by

$$\begin{cases} \dot{z}(t) = \sum_{a=1}^{p}\sum_{j=1}^{r} h_a(\vartheta)h_j(\theta)\{\bar{A}_{0aj}z(t) + \bar{B}_{0a}u(t)\} + \bar{J}d_2(t) \\ y(t) = \sum_{a=1}^{p} h_a(\vartheta)\bar{C}_{0a}z(t) \end{cases} \tag{17.10}$$

where $z(t) = [x^T(t), w^T(t)]^T, \bar{A}_{0aj} = \begin{bmatrix} A_{0a} + HFE_{1a} & (B_{0a} + HFE_{2a})V_j \\ 0 & W_j \end{bmatrix}, \bar{B}_{0a} = \begin{bmatrix} B_{0a} + HFE_{2a} \\ 0 \end{bmatrix}$, $\bar{C}_{0a} = \begin{bmatrix} C_{0a} & 0 \end{bmatrix}, \bar{J} = \begin{bmatrix} I & 0 \end{bmatrix}^T$.

The composite observer for both $x(t)$ and $w(t)$ is designed as

$$\begin{cases} \dot{\hat{z}}(t) = \sum_{a=1}^{p}\sum_{j=1}^{r} h_a(\vartheta)h_j(\theta)\{\bar{A}_{0aj}\hat{z}(t) + \bar{B}_{0a}u + L(\hat{y}(t) - y(t))\} \\ \hat{y}(t) = \sum_{a=1}^{p} h_a(\vartheta)\bar{C}_{0a}\hat{z}(t) \end{cases} \tag{17.11}$$

where $\hat{z}(t) = [\hat{x}^T(t), \hat{w}^T(t)]^T, L = [L_1^T, L_2^T]^T$ is the observer gain to be determined later. Moreover, the full-state estimation error $\tilde{e} = z(t) - \hat{z}(t)$ can be expressed as

$$\dot{\tilde{e}}(t) = \sum_{a=1}^{p}\sum_{j=1}^{r} h_a(\vartheta)h_j(\theta)(\bar{A}_{0aj} + L\bar{C}_{0a})\tilde{e}(t) + \bar{J}d_2(t) \tag{17.12}$$

The composite-observer-based (COB) PI-type controller with fuzzy rules is refined as

$$u(t) = -\hat{d}_1(t) + \sum_{b=1}^{p} h_b(\vartheta)\left(K_{Pb}\hat{x} + K_{Ib}\int_0^t e(\tau)d\tau\right), K_b = \left[K_{Pb}\ K_{Ib}\right] \tag{17.13}$$

where $\hat{d}_1(t) = \sum_{j=1}^{r} h_j(\theta)\left[0\ V_j\right]\hat{z}(t)$.

Substituting (13) into (9) yields

$$\dot{\bar{x}}(t) = \sum_{a,b=1}^{p}\sum_{j=1}^{r} h_a(\vartheta)h_b(\vartheta)h_j(\theta)\left[(A_a + B_aK_b)\bar{x}(t) + D_{abj}\tilde{e} + Cy_d + Jd_2(t)\right] \tag{17.14}$$

where $D_{abj} = [-B_aK_{Pb}, B_aV_j]$.

Combining the estimation error model (12) with the closed-loop model (14) yields

$$\begin{bmatrix}\dot{\bar{x}}(t)\\ \dot{\tilde{e}}(t)\end{bmatrix} = \sum_{a,b=1}^{p}\sum_{j=1}^{r} h_a(\vartheta)h_b(\vartheta)h_j(\theta)\left\{\begin{bmatrix}A_a + B_aK_b & D_{abj}\\ 0 & \bar{A}_{0aj} + L\bar{C}_{0a}\end{bmatrix}\begin{bmatrix}\bar{x}(t)\\ \tilde{e}(t)\end{bmatrix} + \begin{bmatrix}C\\ 0\end{bmatrix}y_d\right\} \tag{17.15}$$

17.4 Theorem Proof via Convex Optimization Algorithm

Theorem 1 *For the augmented system (15) consisting of PI control input and the disturbance observer, if there exist* $Q_1 = P_1^{-1} > 0$ *and* R_{1b} *satisfying*

$$\Theta_{aa} < 0, a = 1, 2\ldots, p;\ \Theta_{ab} + \Theta_{ba} < 0, a < b, a, b = 1, 2\ldots, p \tag{17.16}$$

and $P_2 > 0$ *and* R_2 *satisfying*

$$\sum_{a=1}^{p}\sum_{j=1}^{r} \Xi_{aj} < 0 \tag{17.17}$$

where

$$\Theta_{ab} = \begin{bmatrix} sym(\tilde{A}_a Q_1 + \tilde{B}_a R_{1b}) & C & J & G_1 H & G_2 H & Q_1 G_1 E_{1a} & R_{1b}^T E_{2a}^T \\ C^T & -\mu_1^2 I & 0 & 0 & 0 & 0 & 0 \\ J^T & 0 & -\mu_3^2 I & 0 & 0 & 0 & 0 \\ H^T G_1^T & 0 & 0 & -\alpha_1^{-1} I & 0 & 0 & 0 \\ H^T G_2^T & 0 & 0 & 0 & -\alpha_2^{-1} I & 0 & 0 \\ E_{1a} G_1^T Q_1 & 0 & 0 & 0 & 0 & -\alpha_1 I & 0 \\ E_{2a} R_{1b} & 0 & 0 & 0 & 0 & 0 & -\alpha_2 I \end{bmatrix} \quad (17.18)$$

$$\Xi_{aj} = \begin{bmatrix} sym(P_2 \tilde{A}_{0aj} + R_2 \bar{C}_{0a}) & P_2 G_3 H & G_3 E_{1a}^T & P_2 G_4 H & G_5^T V_j^T E_{2a}^T & P_2 \bar{J} \\ H^T G_3^T P_2 & -\beta_1^{-1} I & 0 & 0 & 0 & 0 \\ E_{1a} G_3^T & 0 & -\beta_1 I & 0 & 0 & 0 \\ H^T G_4^T P_2 & 0 & 0 & -\beta_2^{-1} I & 0 & 0 \\ E_{2a} G_5 & 0 & 0 & 0 & -\beta_2 I & 0 \\ \bar{J}^T P_2 & 0 & 0 & 0 & 0 & -\mu_2^2 I \end{bmatrix} \quad (17.19)$$

and $\mu_1 > 0, \mu_2 > 0, \mu_3 > 0, \alpha_1 > 0, \alpha_2 > 0, \beta_1 > 0, \beta_2 > 0$ *are known parameter. Then the augmented system (15) under the composite controller (13) is stable and the tracking error e(t) convergent to zero. The gains are given by* $K_b = R_{1b} Q_1^{-1}$ *and* $L = P_2^{-1} R_2$.

Please noted that due to the limitation of the paper length, the corresponding proof of Theorem 1 is omitted.

17.5 Simulation Example

Similarly with [22], the following T-S fuzzy models are introduced to describe the simple airplane plant

$$A_{01} = \begin{bmatrix} -0.833 & 1.000 \\ -2.175 & -1.392 \end{bmatrix}, A_{02} = \begin{bmatrix} -1.134 & 1.000 \\ -4.341 & -2.003 \end{bmatrix}, A_{03} = \begin{bmatrix} -1.644 & 1.000 \\ -22.547 & -3.607 \end{bmatrix}$$

$B_{01} = [-0.1671, -10.9160]^T, B_{02} = [-0.2128, -19.8350]^T, B_{03} = [-0.2110, -32.0813]^T, C_{01} = C_{02} = C_{03} = [1, 1], d_2(t) = [0.1, 0.1]^T sin(0.1\pi t), E_{21} = E_{22} = E_{23} = [0.1, 0.1]^T, H = \begin{bmatrix} 1 & 0 \\ 0 & 1 \end{bmatrix}, F = \begin{bmatrix} sin(t)/2 & 0 \\ 0 & cos(t)/2 \end{bmatrix}, E_{11} = E_{12} = E_{13} = \begin{bmatrix} 0.2 & 0 \\ 0 & 0.2 \end{bmatrix}$.

The member functions are chosen as

$$M_1^a = exp\left(\frac{-(y - s_a)^2}{2\sigma_1^2}\right) / \left[exp(\frac{-(y+1)^2}{2\sigma_1^2}) + exp(\frac{-(y-1)^2}{2\sigma_2^2}) + exp(\frac{-y^2}{2\sigma_3^2}) \right]$$

where $\sigma_1 = \sigma_2 = \sigma_3 = 0.8, s_1 = -1, s_2 = 1, s_3 = 0$.

The nonlinear irregular exogenous disturbance is described by two T-S fuzzy rules, and $W_1 = \begin{bmatrix} -1 & 2 \\ -5 & 0 \end{bmatrix}$, $V_1 = \begin{bmatrix} 0 & 4 \end{bmatrix}$, $W_2 = \begin{bmatrix} 0 & -6 \\ 4 & 0 \end{bmatrix}$, $V_2 = \begin{bmatrix} 0 & 4 \end{bmatrix}$.

The member functions are chosen as

$$A_1^1 = \frac{exp(\frac{-(w_1-1.2)^2}{2\sigma_1^2})}{exp(\frac{-(w_1-1.2)^2}{2\sigma_1^2}) + exp(\frac{-(w_1-1)^2}{2\sigma_2^2})}, A_1^2 = \frac{exp(\frac{-(w_1-1)^2}{2\sigma_2^2})}{exp(\frac{-(w_1-1.2)^2}{2\sigma_1^2}) + exp(\frac{-(w_1-1)^2}{2\sigma_2^2})}$$

where $\sigma_1^2 = 0.5, \sigma_2^2 = 1$.

Supposed that the initial values in augmented system (10)–(11) are taken to be $x_0 = \begin{bmatrix} 2 & 3 \end{bmatrix}^T, \hat{x}_0 = \begin{bmatrix} -2 & -2 \end{bmatrix}^T, w_0 = \begin{bmatrix} 2 & 1 \end{bmatrix}^T, \hat{w}_0 = \begin{bmatrix} 1 & 1 \end{bmatrix}^T$.

The desired tracking objective is design as $y_d = 5$. Figure 17.1 displays the response of nonlinear disturbance and its observation value, which illustrates the tracking performance of the disturbance observer is satisfactory. Figure 17.2 is the trajectory of system output and the good performance dynamic tracking performance can be embodied.

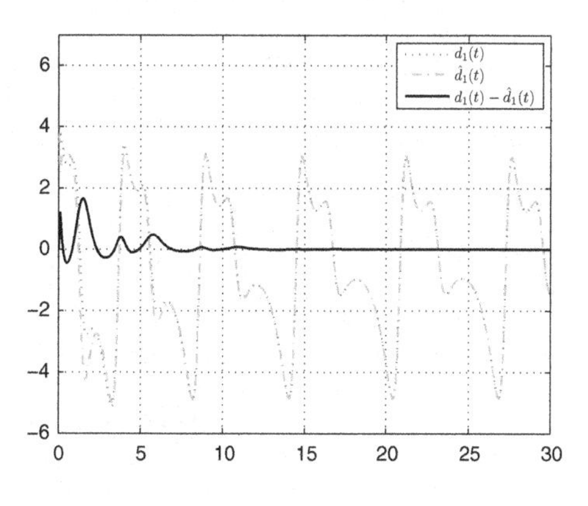

Fig. 17.1 Disturbance and its estimation

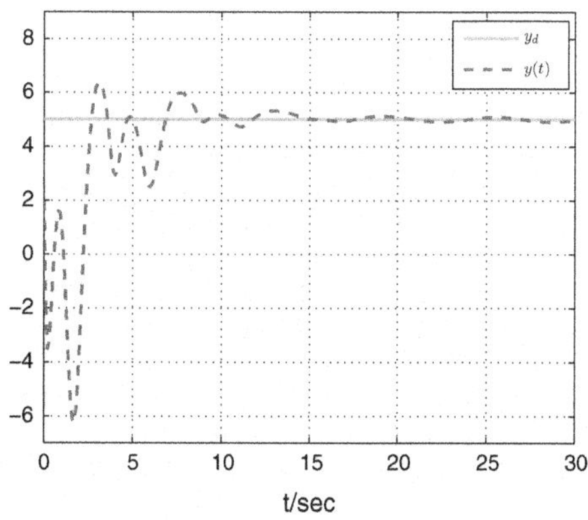

Fig. 17.2 The trajectory of system output

17.6 Conclusion

This paper studies the anti-disturbance tracking control framework for T-S fuzzy models by using T-S disturbance modeling and disturbance observer design. The DOB PI composite controller is designed based on T-S fuzzy models and T-S fuzzy disturbance models simultaneously. As a result, a convex optimization approach is adopted to ensure the augmented closed-loop systems stable and convergence of the tracking error to zero.

References

1. Yang Z, Tsubakihara H (2008) A novel robust nonlinear motion controller with disturbance observer. IEEE Trans Control Syst Technol 16(1):137–147
2. Chen WH, Ballanceand DJ, Gawthrop PJ (2000) A nonlinear disturbance observer for robotic manipulators. IEEE Trans Ind Electron 47(4):932–938
3. Yang J, Li SH, Su JY, Yu XH (2013) Continuous nonsingular terminal sliding mode control for systems with mismatched disturbances. Automatica 49(7):2287–2291
4. Li SH, Xia CJ, Zhou X (2012) Disturbance rejection control method for permanent magnet synchronous motor speed-regulation system. Mechatronics 22(6):706–714
5. Guo L, Chen WH (2005) Disturbance attenuation and rejection for systems with nonlinearity via DOBC approach. Int J. Robust Nonlinear Control 15:109–125
6. Guo L, Cao SY (2013) Anti-disturbance control for systems with multiple disturbances. CRC Press, Boca Raton
7. Liu C, Peng H (2000) Disturbance observer based tracking control. ASME J Dyn Syst Ser Control Meas 122:332–335
8. Guo L, Wen X, Xin X (2010) Hierarchical composite anti-disturbance control for robotic systems using robust disturbance observer. In: Robot intelligence, advanced information and knowledge processing, pp 229–243
9. Kempf C, Kobayashi S (1999) Disturbance observer and feedforward design for a highspeed direct drive positioning table. IEEE Trans Control Syst Technol 7(5):513–526
10. Kim K, Baik I, Moon G, Youn M (1999) A current control for a permanent magnet synchronous motor with a simple disturbance estimation scheme. IEEE Trans Control Syst Technol 7(5):630–633
11. Huang YH, Massner W (1998) A novel disturbance observer design for magnetic hard drive servo system with rotary actuator. IEEE Trans Mechatron 34:1892–1894
12. Wei XJ, Guo L (2009) Composite disturbance-observer-based control and terminal sliding mode control for nonlinear systems with disturbances. Int J Control 82(6):1082–1098
13. Yao XM, Guo L (2013) Composite anti-disturbance control for Markovian jump nonlinear systems via disturbance observer. Automatica 49(8):2538–2545
14. Zhang HB, Shen YY, Feng G (2007) Delay-dependent stability and H_∞ control for a class of fuzzy descriptor systems with timedelay. Fuzzy Sets Syst 160(12):1689–1707
15. Zhang HG, Yang DD, Chai TY (2007) Guaranteed cost networked control for T-S fuzzy systems with time delays. IEEE Trans Syst Man Cybern C 37(2):160–172
16. Yi Y, Zhang TP, Guo L (2009) Multi-objective PID control for non-Gaussian stochastic distribution system based on two-step intelligent models. Sci China-Ser F: Inf Sci 52(10):1754–1765
17. Qiu J, Feng G, Gao H (2011) Nonsynchronized-state estimation of multichannel networked nonlinear systems with multiple packet dropouts via T-S fuzzy-affine dynamic models. IEEE Trans Fuzzy Syst 19(1):75–90

18. Tseng CS (2006) Model reference output feedback fuzzy tracking control design for nonlinear discrete-time systems with time-delay. IEEE Trans Fuzzy Syst 14(1):58–70
19. Zhang K, Jiang B, Shi P (2012) Fault estimation observer design for discrete-time Takagi-Sugeno fuzzy systems based on piecewise Lyapunov functions. IEEE Trans Fuzzy Syst 20(1):192–200
20. Daniel WCH, Niu YG (2007) Robust fuzzy design for nonlinear uncertain stochastic systems via sliding-mode control. IEEE Trans Fuzzy Syst 15(3):350–358
21. Su XJ, Shi P, Wu LG, Nguang SK (2013) Induced L2 filtering of fuzzy stochastic systems with time-varying delays. IEEE Trans Cybern 43(4):1251–1264
22. Xie ZH (2000) Fuzzy controller design and stability analysis for flight control systems. Flight Dyn China 18(2):30–33

Chapter 18
Technology Developments of Micro Fluid Dispensing

Fudong Li, De Xu, Tianping Zhang and Yuequan Yang

Abstract Micro fluid dispensing technology is widely applied in electronics packaging, micro electromechanical system assembly, and biotechnology experiments, in which pl or nl amount of fluid materials (such as solder paste, adhesive, and DNA solution) are delivered controllably for the purpose of conducting, bonding, sealing, etc. This paper reviewed the latest developments as well as advantages and limits of three kinds of micro dispensing technology, which are needle nozzle type, integrated nozzle type and pin transfer type, classified according to the configuration of the nozzle unit. The measuring methods for the micro droplets are also briefly introduced in the article. Our work of dispensing less than 3 pl adhesive in an microassembly task is briefly introduced, and the trends and challenges of micro fluid dispensing are also discussed.

Keywords Micro fluid dispensing · Needle nozzle · Integrated nozzle · Pin transfer

18.1 Introduction

Fluid dispensing technology is widely applied in many fields, such as processes of die attachment and encapsulation in the electronic packaging industry [1], parts joint and structure seal in the micro electromechanical system assembly [2], microinjection and microarray in the biotechnology experiments [3], and so on. As micro manufacturing is developing at a high speed, the scales of the electronic

F. Li (✉) · T. Zhang · Y. Yang
School of Information Engineering, Yangzhou University, Jiangsu 225127, China
e-mail: lfd19850108@126.com

D. Xu
Research Center of Precision Sensing and Control, Institute of Automation Chinese Academy of Sciences, Beijing, China
e-mail: de.xu@ia.ac.cn

Y. Jia et al. (eds.), *Proceedings of the 2015 Chinese Intelligent Systems Conference*, Lecture Notes in Electrical Engineering 360,
DOI 10.1007/978-3-662-48365-7_18

and mechanical devices are dropping dramatically, which induces a growing demand for nl even pl micro dispensing of adhesives and other polymers [4–6].

There is no fixed criterion for distinction between micro dispensing and traditional dispensing, but it is generally acknowledged that the micro dispensing dot diameter is less than 0.25 mm, and the traditional dispensing dot diameter falls between 0.4 and 1.5 mm [7]. As dimensions shrink down, many traditional ways of handling the fluid do not work as well as they do at large scale, because the surface tension and viscous force play much greater roles on micro scale [8].

Traditional fluid dispensing methods are often classified into contact dispensing and non-contact dispensing according to the dispensing process [9]. Since this classification method does not show the inner characteristics of different micro dispensing technologies, in this paper, the existing micro dispensing technologies are classified into 3 groups according to the nozzle unit configuration, which are needle nozzle dispensing, integrated nozzle dispensing and pin transfer dispensing, and this classification method can present a better understanding of the existing micro dispensing technologies. In the following sections, specific micro dispensing methods are introduced, and their advantages and disadvantages are discussed as well, and also, the measuring technologies are briefly introduced.

18.2 Needle Nozzle Dispensing

Traditional needle nozzle dispensing technology is the oldest dispensing form, which has been developed for many decades, and has the widest range of application. Needle nozzle dispensing includes time-pressure dispensing, auger dispensing and piston dispensing, which can also be used for micro dispensing purposes. An obvious characteristic of this kind of dispensing technology is the separation of the nozzle and the actuation. This type of configuration makes it very suitable for some delicate manipulations, since the nozzle is simply a leading tube, and the diameter of the nozzle can be fabricated to as small as 0.5 μm. The needle is a precise locating device for the micro droplets, so the location of the droplets can be manipulated in a highly precise way. These advantages make needle nozzle dispensing technology very suitable for biotechnology, such as cell injection, gene transfer, drug delivery, etc. But it is a time consuming process using needle nozzle equipment to dispense micro fluids, so, it is not very promising in the microchip assembly, which needs to be time effective. Needle nozzle dispensing is suitable for some experiments, especially for the pioneering experiments, which may need pl scale volume, while efficiency is less important.

The typical structure of the needle nozzle dispensing is illustrated in Fig. 18.1.

The needle nozzle micro dispensing device has the same configuration with the traditional one [10], which always includes a syringe, a needle, and driving component (piston, auger, or just air pressure). Changes from traditional dispensing to micro dispensing for the needle nozzle type are the smaller diameter of the needle and higher resolution of the actuation. Such changes may not affect the structure of

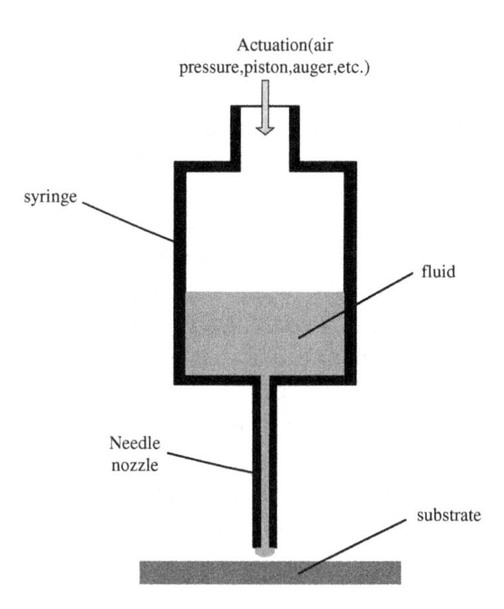

Fig. 18.1 Sketch of the needle nozzle dispensing structure

the equipment in a large scale, but they surely do for the characteristics of the dispensing process, especially for the fluidic behavior of the dispensed micro fluid. Many factors that are neglected in the modeling on large scale, have to be reconsidered, or some approximations in the equations have to be modified [11–14]. This makes the modeling of the micro dispensing process a very difficult issue. And reproducibility is much more difficult to maintain on micro scale, since a slight change in the droplet volume will cause a large deviation percentage.

The actuation for simple nozzle type can be various, like time-pressure, rotational screw [15], piston and so on. It is an interesting phenomenon that the above mentioned actuation types are mostly contact dispensing. That is because the fluid to be dispensed is usually connected with the rest of the fluid in the syringe, and in micro scale, the gravity or the inertial energy of the droplet are not adequate to break the droplet from the rest of the fluid in the needle or syringe since the mass of the micro droplet is too small. The breaking energy usually comes from the contact force when the droplet is in contact with substrate. So in most cases, needle nozzle dispensing is the contact dispensing.

In [16], researchers in Lawrence Livermore National Laboratory uses Femtojet (Eppendorf's micro-injector) and a 10 μm polymide needle tip successfully controled the quantity of the dispensed glue to less than 0.5 ng.

In the field of biology, most cell injection experiments use needle nozzle dispensing method to deliver DNA, protein, drugs, or other materials to the biological cells [17–21]. In [22], Researchers in the Advanced Micro and Nanosystem Laboratory, University of Toronto, Canada, has realized the automated injection of individual zebra fish cells using an optical microscope (SZX12, Olympus), a computer-controlled pico-injector (PLI-100, Harvard Apparatus), and less than 20 μm outer diameter glass needles.

18.3 Integrated Nozzle Dispensing

Integrated nozzle dispensing is the most promising dispensing technique in the electronic assembly industry, and the most studied type, because of its many advantages, which will be elaborated later.

Detailed configurations may vary from example to example, but the main configuration of the integrated nozzle unit mostly stays the same [23–26], which always consists of an actuator, a membrane plate, a fluid chamber, inlet micro channels and a nozzle, as is shown in Fig. 18.2.

There are two main factors that greatly affect the efficiency of the dispensing process for the integrated nozzle type. One is the frequency of the membrane that activates the fluid in the chamber, the other is the filling speed of the chamber. These two factors must align with each other to achieve the most ideal dispensing effect, and either one of them lags behind, the efficiency would be jeopardized.

Various forms of integrated nozzle dispensing technologies have been studied and developed, like thermal actuation, pneumatic actuation, piezoelectric actuation, acoustic actuation [27], electrowetting actuation [28], and even hydrogel actuation, etc. And among these different kinds of actuations, piezoelectric actuation is the most commonly used type. Here are some lately developments and achievements in the integrated nozzle dispensing technologies.

de Heij, Chris Steinert from University of Freiburg (Germany) invented a tunable and highly-parallel picoliter-dispenser [29], which is based on direct displacement of the liquids using an elastomer stamp to simultaneously actuate up to 96 different dosing channels, at a pitch of 500 μm. The droplet volume could be tuned from 150 to 720 pl and droplet speed from 0.2 to 2.8 m/s, using printheads with 50 μm nozzles. Their previous work, a pneumatically actuated printhead with 96 channels at a pitch of 500 μm, was industrialized because of its high throughput production of microarrays. And the improved technology has more advantages comparing to the pneumatic actuated type, and it overcomes the disadvantage of not being able to dispense consistent droplets of fluids with different viscosity.

Hirata and Ishii used buckling diaphragm as activation to dispense micro droplets [30]. Configuration of the equipment is shown in Fig. 18.3a.

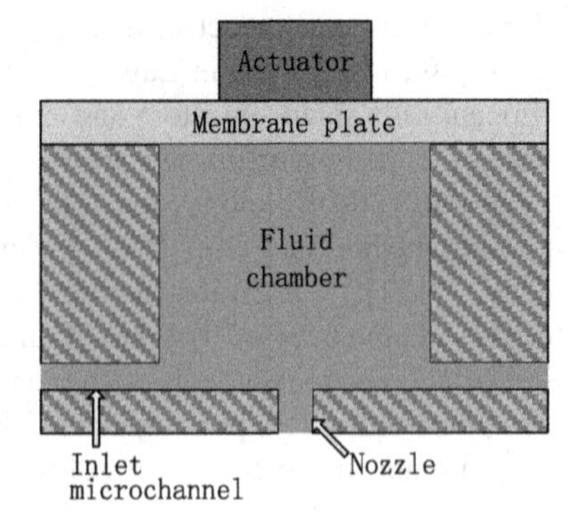

Fig. 18.2 General configuration of the integrated nozzle

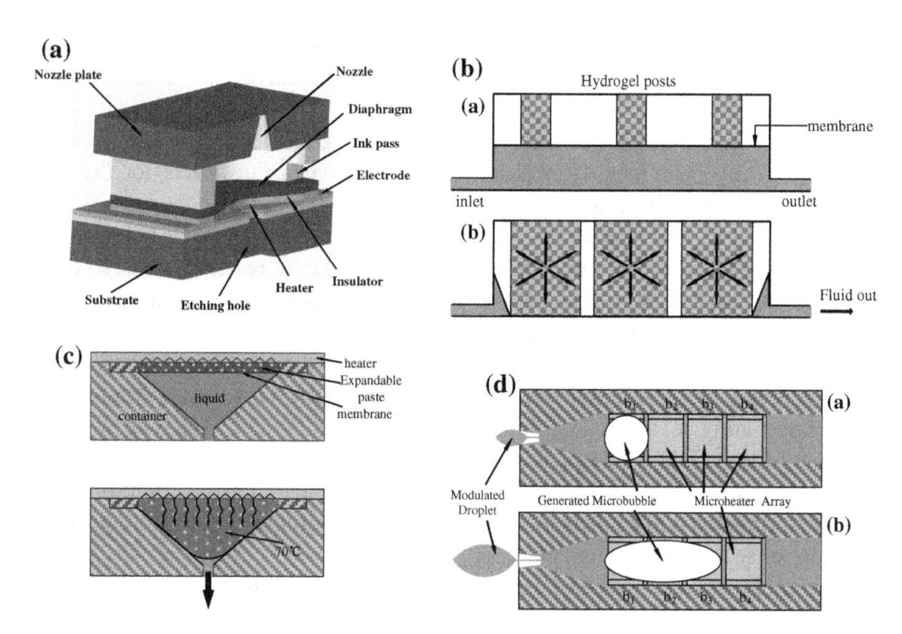

Fig. 18.3 Various integrated nozzle: **a** Schematic view of the ink jet head; **b** Illustration of the hydrogel actuated micro dispensing device; **c** Illustration of the disposable micro dispensing device; **d** Digital microdispensor working principle illustration

The ink jet head has four parts, which are a thick silicon substrate, a silicon dioxide insulator, a nickel heater and a nickel diaphragm. The diaphragm, the spacer and the nozzle plate form the ink chamber. When the electrode is electrified, the nickel heater will generate heat and the diaphragm will be heated. Compressive stress is induced by the heat, which causes the diaphragm to buckle and deflect toward the nozzle plate. Space of the ink chamber is compressed by the deformed diaphragm and the pressure in the chamber will arise. An ink droplet will be ejected through the nozzle under the raised pressure. The developed nozzle head can eject ink droplets at 1.8 kHz frequency with the velocity up to 8 m/s without any bubbles. Droplets with the diameter of 65 μm are achieved with this method.

A hydrogel actuated micro dispensing device was developed in Wisconsin-Madison University [31], whose configuration is shown in Fig. 18.3b.

The device consists of two layers, the top actuation layer, and the bottom reservoir layer. The principle of the hyrogel actuation is that a 2.0 M phosphate buffer initiates the volume change of the hydrogel, which expands under basic conditions and shrinks under acidic conditions. The device can reliably generate a pressure of 3500 N/m^2 with a stroke volume of 32.7 μl.

The advantage of the hydrogel actuation is the simplicity of its constructure, and the easy manipulation of the device. One limitation of the system is that the actuating hydrogels are triggered by local changes in pH, requiring buffer to be injected into the system to expand the hydrogels and dispense the fluid. Such limitation impedes the dispensing frequency and affect the dispensing volume accuracy in a bad way. A new way to actuate the hydrogel would be favorable to solve this problem, like lights or electricity.

Roxhed, Rydholm, etc. From Royal Institute of Technology, Stockholm, Sweden successfully manufactured a disposable micro dispensing device for the purpose of intradermal injections [32]. See Fig. 18.3c.

When the device is heated, the expandable paste expands and the liquid in the container will be squeezed out. The expansion in this case is irreversible, so there won't be any back flow in the process. The device was successfully tested, and it can dispense liquid of a mean volume of 101 μl with a relative deviation 3.2 %.

The integrated nozzle can even be fabricated into digital microdispensor, one example is the microdispensor developed by Kang and Cho, which is actuated by microheater array to dispense volume adjustable droplets [33, 34]. The working principle of the microdispensor is shown in Fig. 18.3d.

As is shown in Fig. 18.3d when different microheaters are activated, different volume of droplets will be dispensed. And 15 different droplet volumes are possible to be dispensed by this equipment.

Integrated nozzle dispensing can act in a very high frequency, which leads to time effective. And it is usually controlled by a circuit board, which is very easy to realize and of high stability. Its configuration allows it to integrate a sensor conveniently, like a capacitor, to measure the dispensed micro droplet in real time. These advantages make the integrated nozzle dispensing highly favorable in the electronic assembly industry.

The disadvantages of the integrated nozzle dispensing comes from its configuration, when the membrane bulks, it won't cover the whole chamber space, and this usually leaves some dead space in the chamber [35]. Because of the existence of the dead space, the accuracy of the dispensed volume will be affected in a bad way. The maintenance of the integrated nozzle dispensing equipments is more trivial, compared with needle nozzle dispensing.

18.4 Pin Transfer Type

A pin transfer applicator usually includes a number of parallel, vertical pins which depend from a movable frame. The pins are arranged in an array that correlates to multiple, space apart locations on the circuit board. In use, the frame is first moved toward a tray containing liquid to be dispensed, and the ends of the pins are simultaneously dipped in the liquid and then retracted, causing a small dot of liquid to adhere to the end of each pin. Next, the frame is moved toward corresponding, pre-selected locations on the circuit board until simultaneous contact is established between each dot of solder and the respective location on the circuit board. Finally, the frame is moved away from the circuit board, whereupon the dots of liquid are transferred from the pins to the circuit board.

Pin transfer dispensing method is not as common as the other two types of dispensing, because of its obvious disadvantages, like difficult to control the transferred droplet volumes, not consistent when transferring liquids of different viscosities, the droplet may evaporate during the transferring process. But is a favorable choice in the electronic assembly process, where time is of very important role. Besides, the pin can even be fabricated individually, and then assembled on a frame in a flexible way, which can meet different configuration requirements.

Pin transfer dispensing can be very effective in some specific application, like delivering solder, or high viscosity liquid, where other dispensing technologies are very limited in these fields.

In [36], a new set of pin transfer dispensing equipment is developed by V&P scientific, inc., which improves the performance of pin transfer dispensing in many ways. First, EDM (Electrical Discharge Machining) is used to cut precise nanoliter slots in the end of these uniform pins so that the volume transferred can be less affected by the height of the source well. Second, the pins are treated with hydrophobic coating to reduce the effect of surface tension of the pin. And also, the pins are cleaned thoroughly after each transfer in a very effective way, which involves dipping the pins into different wash reservoir and blotting them after washing. The carryover can be reduced significantly after cleaning. With these improvements, the pin transfer tool can deliver between 2 and 200 nl of tested compound, with the accuracy better than 5 %.

18.5 Micro Droplets Measuring Methods

To evaluate the performance of the micro dispensing methods, a reliable way to measure the droplets dispensed is a must.

Many highly advanced measuring technologies can be used to determine the volume of the dispensed micro droplets with some changes, like electron microscopy, X-ray, or high resolution balance. When electron microscopy is used in measuring the volume of the droplet, the droplet usually needs to be cured and sprayed with conductive material on its surface. X-ray can also be used for the measurement of the micro droplet when the liquid needs to infiltrate into the crevices of the component, and the X-ray technology seems the only possible way to determine the quantity of the liquid that infiltrate into the crevices.

The above mentioned methods can measure very small droplets, but they can't be used in on-line monitor. New ways of monitoring the dispensed droplets in real time has been developed [37–39], like computer vision and the non-contact capacitive measuring method.

The non-contact capacitive method is a promising way to monitor the volume of the dispensed droplets in real time. This method is based on the principle that when the droplet passes through the two spherical shaped parts formed capacitor, the permittivity of the capacitor will be affected [39]. According to the changes of the voltage of the capacitor, the volume of the dispensed micro droplets can be determined with considerable accuracy. The size of capacitive sensor is very small, so it can be integrated into the nozzle unit, thus, on-line monitoring can be realized.

18.6 Our Research

We have applied micro dispensing technology in a microassembly task of bonding a 10 μm glass tube to a 500 μm sphere. The micro sphere has a 12 μm hole on it, where the glass tube is to be inserted into. Because of the extremely small amount of the adhesive and the precise positioning of the adhesive spot demanded, we chose needle nozzle dispensing technology to fulfill the task. A glass needle with the inner diameter less than 1 μm was used to dispense the adhesive, and an Eppendof FemtoJet to apply air pressure to the glass dispensing needle. Sutter Mp-285 was used to control the precise positioning of the dispensing needle. The exact amount of the adhesive dispensed was achieved by monitoring the diameter of the adhesive spot using a microsopic camera. Ideal adhesive spots with a good consistency were achieved in the microassembly task [40].

18.7 Trends and Challenges

Micro dispensing technologies have developed rapidly in the past decades. nl-pl volume are achieved with considerable accuracy and consistency. Many micro fluid dispensing are achieved by improving the performance of the traditional dispensing technologies, others are developed using new technologies and new dispensing concepts. While integrated nozzle dispensing methods seems very favorable in the industry production lines, needle nozzle dispensing are indispensible for some delicate operation in scientific experiments, and pin transfer dispensing is surely an effective way to deliver solder to the circuit board. Despite the great developments in micro dispensing, there is a serious problem that impedes the technology from further developments, modeling the micro dispensing process. Fluidic behavior of the micro fluid is different from the normal scale fluid, some approximations made on normal scale fluid modeling would cause great error in micro scale, and some factors, like surface tension, wetting angle, would play much more important roles on micro scale. If the precise modeling problem is solved, it will provide very valuable directions on how to improve the micro dispensing equipments and its process.

Acknowledgments This work is supported by National Natural Science Foundation of China (Grant NO.61175111).

References

1. Meyer W (2001) Micro dispensing of adhesives and other polymers, IEEE, pp 35–39
2. Tan KK, Putra AS (2005) Microdispensing system for industrial applications, IEEE, pp 1186–1191
3. Geyl L, Amberg G, van der Wijngaart W, Stemme G (2006) Study of the flight of small liquid droplets through a thin liquid film for picoliter liquid transfer. IEEE, MEMS, Istanbul, pp 24–27
4. Zhang J, Jia H, Zhang J (2010) A fluid dynamic analysis in the chamber and nozzle for a jetting dispenser design. In: 2010 11th international conference on electronic packaging technology and high density packaging
5. McGuire S, Fisher C, Holl M, Meldrum D (2008) A novel pressure-driven piezodispenser for nanoliter volumes. Rev Sci Instrum 79(8):086111
6. Quinones H, Babiarz A, Fang L (2003) Jetting technology: a way of the future in dispensing
7. Li J, Deng G (2004) Technology development and basic theory study of fluid dispensing-a review. In: Proceeding of HDP'04, pp 198–205
8. Hobbs ED, Pisano AP (2003) Micro capillary-force driven fluidic accumulator/pressure source. In: IEEE Transducers'03 the 12th international conference on solid state sensors, actuators and microsystems, Boston, USA, June 2003, pp 155–158
9. Fukuda T, Arai F (2000) Prototyping design and automation of micro/nano manipulation system. In: Proceedings of the 2000 IEEE international conference on robotics and automation. IEEE, San Francisco, USA, pp 192–197
10. Chen XB, Schoenau G, Zhang WJ (2000) Modeling of time-pressure fluid dispensing process. IEEE Trans Electron Packag Manuf 23(4):300–305
11. Fratila D, Palfinger W, Bou S, Almansa A, Mann W et al (2007) A method for measurement and characterization of microdispensing process. In: Proceedings of the 2007 IEEE international symposium on assembly and manufacturing Ann Arbor. IEEE, Michigan, USA, pp 209–214
12. Chen XB, Ke H (2006) Effects of fluid properties on dispensing processes for electronics packaging. IEEE Trans Electron Packag Manuf 29(2):75–82
13. Chen XB, Li MG, Cao N (2009) Modeling of the fluid volume transferred in contact dispensing processes. IEEE Trans Electron Packag Manuf 32(3):133–137
14. Yao Y, Lu S, Liu Y (2011) Numerical simulation of droplet formation in contact micro-liquid dispensing. In: 2011 third international conference on measuring technology and mechatronics automation. IEEE, pp 709–712
15. Peng J, Guiling D (2007) Numerical simulations of 3D flow in the archimedes pump and analysis of its influence on dispensing quality. In: Proceedings of HDP'07, IEEE
16. Takagi M, Saito K, Frederick C, Nikroo A, Cook R (2007) Fabrication and attachment of polyimide fill tubes to plastic NIF capsules
17. Ergenc AF, OlgacN (2007) A new micro injector and an optical sensor. In: Proceedings of the 2007 American control conference Marriott Marquis hotel at Times Square. IEEE, New York, USA, pp 2905–2909
18. Kuncova J, Kallio P (2004) Challenges in capillary pressure microinjection
19. Wang WH, Liu XY, Sun Y (2007) Contact detection in microrobotic manipulation. Int J Robot Res 26(8):821–828
20. Matsuno Y, Nkajima M, Kojima M, Tanaka-Takiguchi Y, Takiguchi K, Kousuke et al (2009) Pico-liter injection control to individual nano-liter solution coated by lipid layer. IEEE, pp 249–254
21. Wang WH, Hewett D, Hann CE, Chase JG, Chen XQ (2008) Machine vision and image processing for automated cell injection. IEEE, pp 309–314
22. Wang WH, Liu XY, Yu S (2009) High-throughput automated injection of individual biological cells. IEEE Trans Autom Sci Eng 6(2):209–219
23. Oeftering R (1999) Acoustic liquid manipulation. In: IEEE ultrasonic symposium, pp 675–678

24. Araz MK, Lal A (2010) Acoustic mixing and chromatography in a PZT driven sillicon microfluidic actuator. IEEE, pp 1111–1114
25. Gaugel T, Bechtel S, Neumann-Rodekirch J (2001) Advanced micro-dispensing system for conductive adhesives. IEEE, pp 40–45
26. Ahamed MJ, Gubarenko SI, Ben-Mrad R, Sullivan P (2010) A piezoactuated driplet-dispensing microfluidic chip. IEEE J Microelectromech Syst 19(1):110–119
27. Strobl CJ, von Guttenberg Z, Wixforth A (2004) Nano- and pico-dispensing of fluids on plannar substrates using SAW. IEEE Trans Ultrason Ferroelectr Freq Control 51 (11):1432–1436
28. Ren H, Fair RB (2002) Micro/nano liter droplet formation and dispensing by capacitance metering and electrowetting actuation, IEEE, pp 369–372
29. de Heij B, Steinert C, Sandmaier H, Zengerle R (2002) A tunable and highly-parallel picoliter-dispenser based on direct liquid displacement. IEEE, pp 706–709
30. Hirata S, Ishii Y, Matoba H, Inui T (1996) An ink-jet head using diaphragm microactuator. IEEE, pp 418–423
31. Eddington DT, Beebe DJ (2002) A hydrogel actuated microdispensing device. In: Proceedings of the second joint EMBS/BMES conference. IEEE, Houston, USA, pp 1824–1825
32. Roxhed N, Rydholm S, Samel B, van der Wijngaart W, Griss P, Stemme G (2004) Low cost device for precise microliter range liquid dispensing. IEEE, pp 326–329
33. Kang TG, Cho YH (2005) A four-bit digital microinjector using microheater array for adjusting the ejected droplet volume. IEEE J Microelectromech Syst 14(5):1031–1038
34. Kang TG, Cho YH (2003) Droplet volume adjustable microinjectors using a microheater array. IEEE, pp 690–693
35. Koltay P, Bohl B, Taoufik S, Steger R, Messner S et al (2003) Dispensing Well Plate (DWP): a highly integrated nanoliter dispensing system. IEEE Transducers'03 the 12th international conference on solid state sensors, actuators and microsystems, Boston, USA, June 2003, pp 16–19
36. Cleveland PH, Koutz PJ (2005) Nanoliter dispensing for uHTS using pin tools. Assay Drug Dev Technol 3(2):213–225
37. Ernst A, Streule W, Zengerle R, Koltay P (2009) Quantitative volume determination of dispensed nanoliter droplets on the fly. IEEE Transducers, Denver, USA, pp 1750–1753
38. Daoura MJ, Meldrum DR (1999) Precise automated control of fluid volumes inside glass capillaries. IEEE J Microelectromech Syst 8(1):71–77
39. Mutschler K, Ernst A, Paust N, Zengerle R, Koltay P (2011) Capacitive detection of nanoliter droplets on the fly-investigation of electric field during droplet formation using CFD-simulation. Transducers'11, Beijing, China, June 2011, pp 430–433
40. Li F, Xu D, Zhang Z, Shi Y (2013) Realization of an automated microassembly task involving micro adhesive bonding. Int J Autom Comput 10(6):545–551

Chapter 19
Fluid Analysis for a PEPA Model

Jie Ding, Xinshan Zhu and Minyi Wang

Abstract It is, as the state space explosion problem indicates, not uncommon that tremendous complexity and size of a system would annoyingly quiver the performance of discrete state-based modeling formalisms. The past few years, however, have inspiringly witnessed a brand new PEPA-based strategy offering a feasible solution against such disturbing puzzle. Via PEPA, a family of ordinary differential equations (ODEs) is figured out as continuous state space approximation. This paper establishes some significant properties for the fluid approximation of a PEPA model, including the existence, uniqueness, boundedness and convergence of the derived ODEs solution.

Keywords Fluid approximations · PEPA · Convergence

19.1 Introduction

Stochastic process algebras (e.g. [1–3]), by authorizing untested parts (such as scalability) of a concurrent system to be parsed ahead of system deployment, have boasted their eminent performance over the past decade. Nevertheless, even those powerful modeling formalisms described above, while facing enormous complexity and size of large scale system applications, appear to be ability-strapped in terms of model construction and analysis. Actually, such problem takes deep root in discrete state

J. Ding
School of Information Engineering, Yangzhou University,
Yangzhou 225127, China

X. Zhu (✉)
School of Electrical Engineering and Automation, Tianjin University,
Tianjin 300072, China
e-mail: xszhu126@126.com

M. Wang
Focus Technology Co., Ltd., Nanjing 210061, China

Y. Jia et al. (eds.), *Proceedings of the 2015 Chinese Intelligent Systems Conference*, Lecture Notes in Electrical Engineering 360,
DOI 10.1007/978-3-662-48365-7_19

approach adopted by stochastic process algebras as well as a host of other formal modeling methodologies.

In the territory of stochastic process algebras, it is a conventional assumption that quantified durations along with activities are random variables subjected to exponential distribution. Thus, there exists little wonder that the quantitative evaluation-based stochastic models, in fact, are Markov chains in most scenarios. While computing steady state probability distribution of some Markov chain, a *state space explosion* problem emerges. And this issue places harsh restriction on system size for model analysis. Recently, a crafty strategy based on the modeling language PEPA (Performance Evaluation Process Algebra, a stochastic process algebra) has succeeded to hedge the risk [4]. In brief, the strategy deduces a set of ODEs as fluid approximation, leading to the evaluation of transient as well as steady-state measurements.

Indeed, fluid approximation of PEPA has already gained fruitful achievements on performance analysis of various large scale systems, e.g. [5–8]. A set of ODEs, however, still boasts substantial potentiality for further exploitation and utilization. This paper elaborates several potential tips via a specific model whose solution converges as time tends toward infinity.

The paper is structured as follows. There are some briefly introduced notions on PEPA presented in Sect. 19.2. In Sect. 19.3, a detailed PEPA model as well as its fluid approximation is described. Furthermore, Sect. 19.4 demonstrates significant properties of the model, i.e., we prove the solution of the derived ODEs is uniquely existent, bounded, as well as convergent to a limit. After showing numerical results in Sect. 19.5, we conclude the paper in Sect. 19.6.

19.2 A Short Introduction of PEPA Language

PEPA language, developed by Hillston in [1], is a high-level model specification formulism for low-level stochastic models. Through this language, systems are described as an interaction of the components engaging in activities. These activities have durations which satisfy exponential distributions. Therefore, each activity can be represented as a pair (α, r). Here α is the action type while r denotes the activity rate. There are some combinators in this language, which will be introduced later. In addition, please refer to [1] for the structured operational semantics. The grammar of this language is listed as follows:

$$S ::= (\alpha, r).S \mid S + S \mid C_S$$
$$P ::= P \underset{L}{\bowtie} P \mid P/L \mid C$$

Here S stands for a *sequential component*, P represents a *model component*, C denotes a constant, while C_S are constants representing sequential components.

In the following, we introduce the combinators in the language.

Prefix: $(\alpha, r).P$: This component is a system behaviour description mechanism. That is, after firing an activity (α, r), the component will behave as P.

Choice: $P + Q$: Competition between P and Q is represented by the component $P + Q$, which means the system behavior is either the component P or Q. A race policy is adopted, i.e., the choice between P and Q depends on whose activity is first finished. The component with the slower one will be discarded.

Constant: $A \stackrel{def}{=} P$: This component gives A the same behaviour of component P, allowing infinite behaviour to be defined by recursive definitions.

Hiding: P/L: For an external observer, hiding makes invisible for any activity with type included in set L. This component behaves as P except for the hidden action type in L.

Cooperation: $P \underset{L}{\bowtie} Q$: The cooperation combinator describes the synchronisation between two components P and Q. The set of action type of the cooperated activities is L. These two components carry out an activity with action type included in L, while perform independently in the case of action type not included. If the set L is empty, i.e., there are no shared activities between P and Q, then $P \underset{\emptyset}{\bowtie} Q$ is simply denoted by $P||Q$, called a **parallel** component.

As mentioned before, each activity has a duration satisfying exponential distribution. By the memoryless property, we can imply that a PEPA model must have Markov property. Hence underlying a PEPA model there is a continuous time Markov chain (CTMC) [1]. Let Q be the infinitesimal generator matrix (usually called *Q- matrix*) of the CTMC. Then by solving the balance equation $\pi Q = 0$ with the unit condition $\sum \pi_i = 1$, we can obtain the steady-state probability distribution π of the CTMC. Performance metrics such as average throughput and utilisation can therefore be derived. Furthermore, expected response time can also be derived through this approach by the Litter's law [10].

However, with the increase of the size of the state space as the scale of the model getting larger, we meet a *state space explosion* problem, i.e., the size is too huge to being feasible to know the exact form of the matrix Q, letting alone solving the algebra equation $\pi Q = 0$. As mentioned before, a method through fluid approximation has been proposed to cope with this problem [4], which leads to a family of ODEs, from which performance measures can be obtained. See an example in the next section.

19.3 Fluid Approximation of a PEPA Model

Let us first consider a simple User-Provider system, which consists of two subsystems: *User* subsystem and *Provider* subsystem. All activities of these two subsystems, i.e., $task_1$ and $task_2$, are cooperated. After $task_1$ is carried out, $User_1$ becomes $User_2$ while $Provider_1$ becomes $Provider_2$ simultaneously. Similarly, $User_2$ becomes $User_1$ while $Provider_2$ becomes $Provider_1$ after firing $task_2$. This User-Provider system is be modelled using the PEPA language and presented below.

PEPA definition for *User*:

$$User_1 \overset{def}{=} (task_1, a).User_2$$
$$User_2 \overset{def}{=} (task_2, b).User_1$$

PEPA definition for *Provider*:

$$Provider_1 \overset{def}{=} (task_1, c).Provider_2$$
$$Provider_2 \overset{def}{=} (task_2, d).Provider_1$$

System equation:

$$\big(User_1[M_1]||User_2[M_2]\big) \underset{\{task_1, task_2\}}{\bowtie} \big(Provider_1[N_1]||Provider_2[N_2]\big)$$

The system equation of the model determines that there are M_i copies of $User_i$ and N_i copies of $Provider_i$ in the initial stage, where $i = 1, 2$. Throughout this paper we assume the parameters to be positive, i.e., $a, b, c, d > 0$. According to the mapping semantics presented in [4, 9], the ODEs derived from this model are as follows:

$$\begin{cases} \dfrac{dx_1}{dt} = -\min\{ax_1, cx_3\} + \min\{bx_2, dx_4\} \\ \dfrac{dx_2}{dt} = \min\{ax_1, cx_3\} - \min\{bx_2, dx_4\} \\ \dfrac{dx_3}{dt} = -\min\{ax_1, cx_3\} + \min\{bx_2, dx_4\} \\ \dfrac{dx_4}{dt} = \min\{ax_1, cx_3\} - \min\{bx_2, dx_4\} \end{cases} \tag{19.3.1}$$

where x_i and x_{j+2} represent the populations of $User_i$ and $Provider_j$ respectively, $i, j = 1, 2$. The following initial condition is concerned, which corresponds to the initial stage of the model:

$$x_i(0) = M_i, \; x_{j+2}(0) = N_j. \tag{19.3.2}$$

Obviously, Eq. (19.3.1) are nonlinear, but linear in piecewise. They are equivalent to

$$\begin{pmatrix} \frac{dx_1}{dt} \\ \frac{dx_2}{dt} \\ \frac{dx_3}{dt} \\ \frac{dx_4}{dt} \end{pmatrix} = I_{\{ax_1 \le cx_3, bx_2 \le dx_4\}} Q_1 \begin{pmatrix} x_1 \\ x_2 \\ x_3 \\ x_4 \end{pmatrix} + I_{\{ax_1 \le cx_3, bx_2 > dx_4\}} Q_2 \begin{pmatrix} x_1 \\ x_2 \\ x_3 \\ x_4 \end{pmatrix}$$
$$+ I_{\{ax_1 > cx_3, bx_2 > dx_4\}} Q_3 \begin{pmatrix} x_1 \\ x_2 \\ x_3 \\ x_4 \end{pmatrix} + I_{\{ax_1 > cx_3, bx_2 \le dx_4\}} Q_4 \begin{pmatrix} x_1 \\ x_2 \\ x_3 \\ x_4 \end{pmatrix}, \tag{19.3.3}$$

where the coefficient matrices Q_i are specified as follows:

$$Q_1 = \left(\begin{array}{cc|cc} -a & b & 0 & 0 \\ a & -b & 0 & 0 \\ \hline -a & b & 0 & 0 \\ a & -b & 0 & 0 \end{array}\right), Q_2 = \left(\begin{array}{cc|cc} -a & 0 & 0 & d \\ a & 0 & 0 & -d \\ \hline -a & 0 & 0 & d \\ a & 0 & 0 & -d \end{array}\right),$$

$$Q_3 = \left(\begin{array}{cc|cc} 0 & 0 & -c & d \\ 0 & 0 & c & -d \\ \hline 0 & 0 & -c & d \\ 0 & 0 & c & -d \end{array}\right), Q_4 = \left(\begin{array}{cc|cc} 0 & b & -c & 0 \\ 0 & -b & c & 0 \\ \hline 0 & b & -c & 0 \\ 0 & -b & c & 0 \end{array}\right).$$

By simple calculation, we know that the matrix Q_1 has eigenvalues 0 (three folds) and $-(a+b)$. For the matrices Q_2, Q_3 and Q_4, they have eigenvalues 0 (three folds) and nonzero eigenvalues $-(a+d), -(c+d), -(b+c)$ respectively. This obversion is important and will be used in the proof of the convergence of the solution, which is shown in the next section.

19.4 Main Results

This section will present some basic characteristics of the fluid approximation for the above User-Provider PEPA model. In particular, analogy with the steady-state probability distribution of the CTMC, we want to see the solution of the ODEs converges to a limit, which will be studied later. But first, if the ODEs have a solution, then the solution must be bounded and positive, as the following prior estimations indicates.

Theorem 1 *If the ODEs (19.3.1) with the initial condition (19.3.2) have a solution* $(x_1, x_3, x_2, x_4)^T$, *and suppose* $M_i, N_i \geq 0$, *then for any time* t, $0 \leq x_i(t) \leq M_1 + M_2$, $0 \leq x_{i+2}(t) \leq N_1 + N_2$ *for* $i = 1, 2$.

Proof First, we assume the initial values to be positive, i.e., $M_i, N_i > 0, i = 1, 2$. We will show that for any $t \geq 0$, $\min_i x_i(t) > 0$. Otherwise, we must have $\min_i x_i(t) \leq 0$ for some time t. Because the initial value of the solution is positive, so we must have a time $t' > 0$ such that $\min_i x_i(t') = 0$. We may also suppose $0 < t' < \infty$ is the first such point, i.e.,

$$t' = \inf \left\{ t > 0 \mid \min_i x_i(t) = 0 \right\}.$$

We may suppose x_1 reaches zero at t' without loss of generality, i.e., $x_1(t') = 0$ and $x_i(t) > 0,\ t \in [0, t')$, for any $i \in \{1, 2, 3, 4\}$.

Notice that if $t \in [0, t']$, then

$$\frac{dx_1(t)}{dt} = -\min\{ax_1, cx_3\} + \min\{bx_2, dx_4\} \geq -\min\{ax_1, cx_3\} \geq -ax_1.$$

Solving the above inequality, we can obtain

$$x_1(t) \geq x_1(0)^{-at}, t \in [0, t']. \tag{19.4.1}$$

Because $x_1(0) = M_1 > 0$, we know that $x_1(t') > 0$. This is contradicted to $x_1(t') = 0$. Therefore, for any $t > 0$, $x_i(t)(i = 1, 2, 3, 4)$ are positive.

Secondly, if $\min\{M_1, M_2, N_1, N_2\} = 0$, then let $x_i^\delta(0) = M_i + \delta$ and $x_{i+2}^\delta(0) = N_i + \delta$, where $\delta > 0$ and $i = 1, 2$. Let $x_i^\delta(t)(i = 1, 2, \cdots, 4)$ be the solution to the ODEs (19.3.1) with the initial condition solution $x_i^\delta(0)$. Then by a similar argument, $x_i^\delta(t) > 0$. Note that the minimum function is Lipschitz continuous. Then by a fundamental inequality (page 14, [11]), we have

$$\left|x_i^\delta(t) - x_i(t)\right| \leq \delta e^{Ct}, \tag{19.4.2}$$

where C is a Lipschitz constant. So we have that for any $t \geq 0$,

$$x_i(t) \geq x_i^\delta(t) - \delta e^{Ct} > -\delta e^{Ct}. \tag{19.4.3}$$

Let $\delta \downarrow 0$ in (19.4.3), then we have $x_i(t) \geq 0$ for any i.

Notice that $\frac{d(x_1+x_2)}{dt} = 0$ and $\frac{d(x_3+x_4)}{dt} = 0$, so $x_1(t) + x_2(t) \equiv M_1 + M_2$ and $x_3(t) + x_4(t) \equiv N_1 + N_2$. Therefore $0 \leq x_i(t) \leq M_1 + M_2$ and $0 \leq x_{i+2}(t) \leq N_1 + N_2$ at any time, $i = 1, 2$. The proof is completed.

In the following, we denote $M = M_1 + M_2$, $N = N_1 + N_2$. By the proof of Theorem 1, if $\{x_i(t)\}_i$ are the solution of the ODEs, then

$$\sum_{i=1}^4 x_i(t) = M + N, \quad \sum_{i=1}^2 x_i(t) - \sum_{i=3}^4 x_i(t) = M - N.$$

Theorem 2 *The ODEs (19.3.1) with the initial condition (19.3.2) have a unique solution* $\mathbf{x}(t) = (x_1, x_3, x_2, x_4)^T$ *in* $[0, \infty)$*. Moreover,*

1. *if* $ax_1(0) \leq cx_3(0)$ *and* $bx_2(0) \leq dx_4(0)$*, then* $\mathbf{x}(t) = e^{tQ_1}\mathbf{x}(0)$*;*
2. *if* $ax_1(0) \leq cx_3(0)$ *and* $bx_2(0) > dx_4(0)$*, then* $\mathbf{x}(t) = e^{tQ_2}\mathbf{x}(0)$*;*
3. *if* $ax_1(0) > cx_3(0)$ *and* $bx_2(0) > dx_4(0)$*, then* $\mathbf{x}(t) = e^{tQ_3}\mathbf{x}(0)$*;*
4. *if* $ax_1(0) > cx_3(0)$ *and* $bx_2(0) \leq dx_4(0)$*, then* $\mathbf{x}(t) = e^{tQ_4}\mathbf{x}(0)$*.*

Proof Notice that the coefficient functions in the ODEs (19.3.1) are globally Lipschtz continuous, so according to the classical theorem (page 14, [11]), the equations (19.3.1) admit a unique global solution. Noticing $\frac{dx_1}{dt} - \frac{dx_3}{dt} = 0$ and $\frac{dx_2}{dt} - \frac{dx_4}{dt} = 0$, so for any t,

$$x_1(t) - x_3(t) = x_1(0) - x_3(0), \quad x_2(t) - x_4(t) = x_2(0) - x_4(0).$$

Therefore, if $ax_1(0) \le cx_3(0)$ and $bx_2(0) \le dx_4(0)$, then for any t,

$$ax_1(t) \le cx_3(t), \quad bx_2(t) \le dx_4(t).$$

So the ODEs (19.3.1) become linear equations $\frac{d\mathbf{x}(t)}{dt} = Q_1\mathbf{x}(t)$, and thus they have the solution $\mathbf{x}(t) = e^{tQ_1}\mathbf{x}(0)$. Similarly, we can prove other conclusions.

As time goes to infinity, it is straightforward to see the solution of the ODEs converges, analogy with the transient probability distribution converging in the corresponding CTMC of the model. This indicates the consistence between the PEPA model and its fluid approximation.

Corollary 1 *The solution of the equations (19.3.1) with the initial condition (19.3.2) tends to a limit as time goes to infinity.*

Proof We first assume $ax_1(0) \le cx_3(0)$ and $bx_2(0) \le dx_4(0)$, then the solution is $\mathbf{x}(t) = e^{tQ_1}\mathbf{x}(0)$ by Theorem 2. Because the eigenvalues of the coefficient matrix Q_1 are either zeros or negative, and the solution is bounded as shown in Theorem 1, we can determine the convergence of the solution by classical theories of differential equation. That is, as time tends to infinity, the solution will converge to a constant

$$\mathbf{x}(\infty)^T = \left(\frac{bM}{a+b}, \frac{aM}{a+b}, \frac{bM}{a+b} + x_3(0) - x_1(0), \frac{aM}{a+b} + x_4(0) - x_2(0)\right).$$

For other cases, we can also similarly determine the convergence and the limits for the solution.

By the proof of Corollary 1, in the first case of Theorem 2, the limit depends on the total population M. In fact, the limit $\mathbf{x}(\infty)$ satisfies $Q_1\mathbf{x}(\infty) = 0$ with the conservation equations: $x_1(\infty) + x_2(\infty) = M$ and

$$x_1(\infty) - x_3(\infty) = x_1(0) - x_3(0), \quad x_2(\infty) - x_4(\infty) = x_2(0) - x_4(0).$$

19.5 Numerical Experiments

This section will present the numerical solution to the ODEs discussed previously. We set the parameters $a = 1, b = 2, c = 3$ and $d = 4$. As we can see from Figs. 19.1, 19.2, 19.3 and 19.4, if we let the initial conditions be $(0, 1, 1, 3)$, $(0, 8, 2, 4)$, $(9, 8, 0, 3)$, $(2, 0, 1, 4)$, the coefficient matrices Q_1, Q_2, Q_3, Q_4 will dominate respectively, and thus the solutions will converge to the corresponding limits. The limits are $\frac{1}{3}(2, 1, 5, 7)$, $\frac{1}{5}(16, 24, 26, 4)$, $\frac{1}{7}(75, 44, 12, 9)$, $\frac{1}{5}(7, 3, 2, 23)$ respectively. This is consistent with Theorem 2 and Corollary 1. These results are organized in Table 19.1.

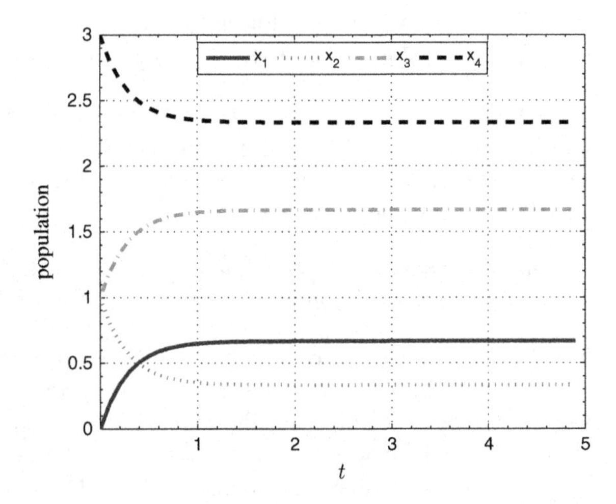

Fig. 19.1 Dominator Q_1, $(M_1, M_2, N_1, N_2) = (0, 1, 1, 3)$

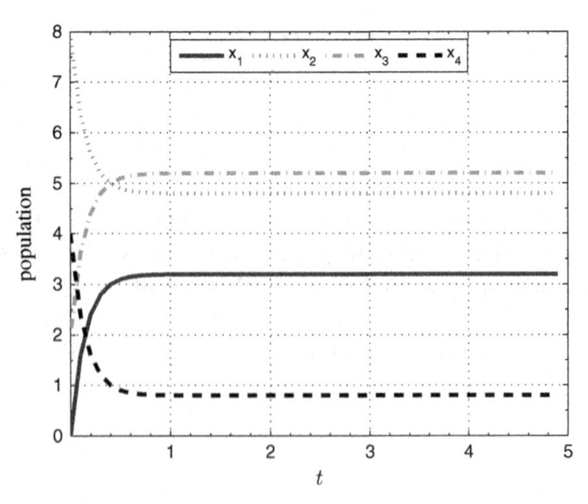

Fig. 19.2 Dominator Q_2, $(M_1, M_2, N_1, N_2) = (0, 8, 2, 4)$

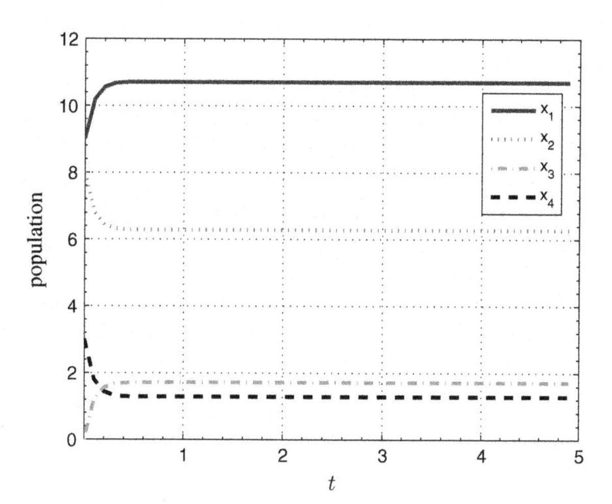

Fig. 19.3 Dominator Q_3, $(M_1, M_2, N_1, N_2) = (9, 8, 0, 3)$

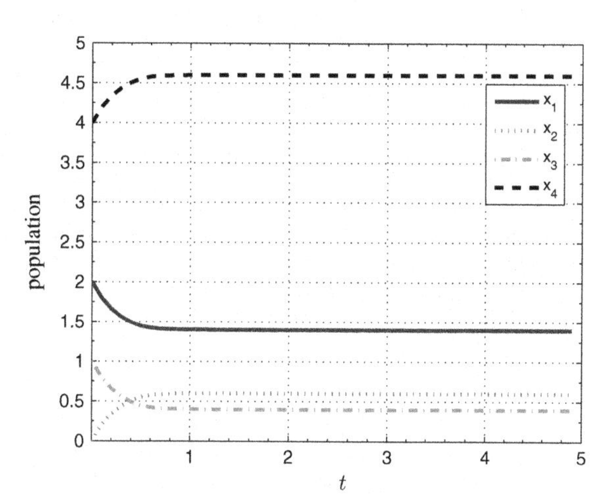

Fig. 19.4 Dominator Q_4, $(M_1, M_2, N_1, N_2) = (2, 0, 1, 4)$

Table 19.1 Numerical experiments

(M_1, M_2, N_1, N_2)	Dominator	Limit	Figure
$(0, 1, 1, 3)$	Q_1	$\frac{1}{3}(2, 1, 5, 7)$	Fig. 19.1
$(0, 8, 2, 4)$	Q_2	$\frac{1}{5}(16, 24, 26, 4)$	Fig. 19.2
$(9, 8, 0, 3)$	Q_3	$\frac{1}{7}(75, 44, 12, 9)$	Fig. 19.3
$(2, 0, 1, 4)$	Q_4	$\frac{1}{5}(7, 3, 2, 23)$	Fig. 19.4

19.6 Conclusions

This paper has pointed out some significant properties on the fluid approximation of a PEPA model. We have exhibited the solution of the derived ODEs is existent, unique, bounded and convergent. Moreover, the limit of the solution is given and analysized. Additionally, the convergence of a ODEs solution with corresponding dominator has been demonstrated via illustrated numerical results.

Acknowledgments The work is supported by the NSF of China under Grant No. 61472343, and the NSF of Jiangsu Province of China under Grants BK2012683 and BK20140492.

References

1. Hillston J (1996) A compositional approach to performance modelling, PhD Thesis, Cambridge University Press
2. Bernardo M, Gorrieri R (1998) A tutorial on EMPA: a theory of concurrent processes with nondeterminism, priorities, probabilities and time. Theor. Comput. Sci. 202:1–54
3. Götz N, Herzog U, Rettelbach M (1992) TIPP—a language for timed processes and performance evaluation. Technical report, 4/92, IMMD7, University of Erlangen-Nörnberg, Germany, November 1992
4. Hillston J (2005) Fluid flow approximation of PEPA models. In: International conference on the quantitative evaluation of systems (QEST'05). IEEE Computer Society
5. Benoit A, Cole M, Gilmore S, Hillston J (2005) Enhancing the effective utilisation of grid clusters by exploiting on-line performability analysis. In: Proceedings of the first internatial workshop on grid performability, May 2005
6. Calder M, Gilmore S, Hillston J (2006) Modelling the influence of RKIP on the ERK signalling pathway using the stochastic process algebra PEPA. In: Lecture Notes in Computer Science, vol 4230, pp 1–23. Springer (2006)
7. Ding J, Hillston J, Laurenson D (2009) Performance modelling of content adaptation for a personal distributed environment. Wirel Pers Commun 48:93–112
8. Ding J, Zhu XS, Li B (2012) On the fluid approximation of a PEPA model. J Comput Inf Syst 8(22):9259–9267
9. Ding J (2015) Structural and fluid analysis of large scale PEPA models—with applications to content adaptation systems. Ph.D. thesis, The Univeristy of Edinburgh (2010). http://www.dcs.ed.ac.uk/pepa/jie-ding-thesis.pdf, Accessed 16 June 2015
10. Clark A, Duguid A, Gilmore S, Hillston J (2008) Espresso, a little coffee. In: Proceedings of the 7th workshop on process algebra and stochastically timed acitvities (2008)
11. Hubbard JH, West BH (1990) Differential equations: a dynamical systems approach (higher-dimensional systems). No. 18 in Texts in Applied Mathematics, Springer, Berlin (1990)

Chapter 20
Movement Curve and Hydrodynamic Analysis of the Four-Joint Biomimetic Robotic Fish

Zhibin Xue, Hai Lin and Qian Zhang

Abstract Focusing on the four-joint robotic fish, based on the fish body wave motion curve equation, deduced a mathematical model of four joint rotation of robotic fish. By setting two different parameters of amplitude coefficient and polarization coefficient, using simulation software MATLAB for different motion curves. For dynamics modeling of four-joint robotic fish provide theoretical basis. Hydrodynamic analysis is done too, the simulation software FLUNT is using for the underwater robotic fish prototype's theoretical analysis. The simulation results have validated its effectiveness, reliability, and scalability and embodied out in the proposed prototype mechanical structure model.

Keywords Robotic fish · Wave motion curve equation · Amplitude coefficient · Polarization coefficient · Hydrodynamic analysis

20.1 Introduction

In nature, some creatures provide scientists with significant enlightenment and inspiration, and robotic fish is a product where one of the inspirations turns into reality. Robotic fish is an imitation of the most successful creature in water—fish. Through relevant strict and careful experiments, good functions of robotic fish can be understood and created. Compared to underwater machine with traditional propeller as the driving force, new-style robotic fish has many advantages, such as higher efficiency, quiet moving and flexible operability. At present, the propulsion

Z. Xue (✉)
School of Chemical Engineering, Qinghai University, Xining 810016, China
e-mail: zbxue_jack@163.com

H. Lin · Q. Zhang
School of Mechanical Engineering, Qinghai University, Xining 810016, China

Y. Jia et al. (eds.), *Proceedings of the 2015 Chinese Intelligent Systems Conference*, Lecture Notes in Electrical Engineering 360,
DOI 10.1007/978-3-662-48365-7_20

mode of carangidae caudal fin is considered to be the best propulsion model in propulsion efficiency. The kind of fish gains propulsion mainly by swinging the tail. Fish body wave motion curve is the most important foundation in the study of swimming process of robotic fish.

The study of fish body wave motion curve of robotic fish is the foundation and the most important part of kinetic study of robotic fish.

20.2 Motion Features of Robotic Fish

The feature of robotic fish movement is decided by different movement of robotic fish's caudal fin. The researchers point out that, the traveling wave from a robotic fish's caudal peduncle to caudal fin is implied in a swimming fish. Gradually increasing amplitude shows the speed of advance of robotic fish. The propulsion model is designed by the carangidae class fish. Propulsive wave curve of the fish starts from its centre of gravity and ends at the caudal fin's rotation axis. According to the conclusion of Lighthill and others, the curve can be described by Eq. (20.1) [1].

$$y_{bc}(x,t) = (f_{c1}x + f_{c2}x^2)\sin(f_k x + f_\omega t) \tag{20.1}$$

Herein, $y_{bc}(x,t)$ is lateral displacement of fish body. x is the displacement along principal axis. f_k is body wave number and $f_k = 2\pi/\lambda$. f_{c1} is the coefficient of amplitude envelope of linear wave. f_{c2} is the coefficient of amplitude envelope of the second wave. f_ω is the frequency of body wave and $f_\omega = 2\pi/T$ (Fig. 20.1).

For four-joint robotic fish, the rotating rule of the end of joints complies with Eq. (20.1). Suppose that the length of joint 1 is l_1, H_1 is the corresponding arc length of sector region formed by the rotation of joint 1.

We can deduce that:

$$H_1 = (f_{c1}l_1 + f_{c2}l_1^2)\sin(f_k l_1 + f_\omega t) \tag{20.2}$$

Suppose θ_1 is the arc length of H_1 and $\theta_1 = \frac{H_1}{l_1} = (f_{c1} + f_{c2}l_1)\sin(f_k l_1 + f_\omega t)$.

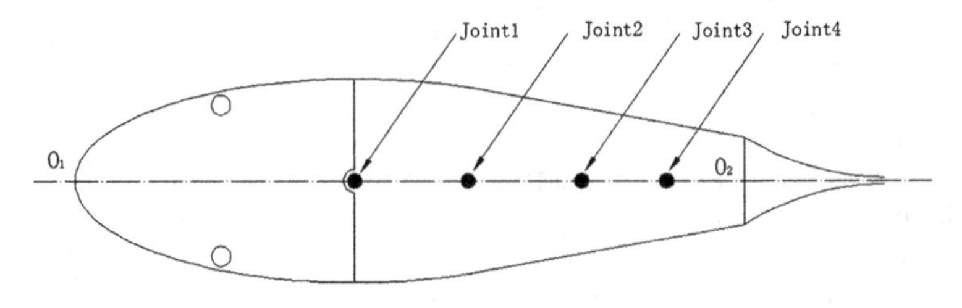

Fig. 20.1 Joints of robotic prototype

So we can conclude that the rotating rule of the first joint follows the rule of sine.

Suppose the length of the second joint is l_2 and H_2 is the arc length of sector region formed by the rotation of joint 2.

So we can conclude that $H_2 = [f_{c1}(l_1+l_2)+f_{c2}(l_1+l_2)^2]\sin[f_k(l_1+l_2)+f_\omega t]$

Form geometrical relationship of the two joints, we can deduce that:

$$\begin{aligned} H_2 - H_1 = l_2\theta_2 \Rightarrow & [f_{c1}(l_1+l_2)+f_{c2}(l_1+l_2)^2]\sin[f_k(l_1+l_2)+f_\omega t] \\ & -(f_{c1}l_1+f_{c2}l_1^2)\sin(f_k l_1+f_\omega t) = l_2\theta_2 \\ & \Rightarrow \theta_2 = \frac{f_{c1}(l_1+l_2)+f_{c2}(l_1+l_2)^2}{l_2}\sin[f_k(l_1+l_2)+f_\omega t] \\ & -\frac{(f_{c1}l_1+f_{c2}l_1^2)}{l_2}\sin(f_k l_1+f_\omega t) \end{aligned} \tag{20.3}$$

Suppose that $A = \frac{f_{c1}(l_1+l_2)+f_{c2}(l_1+l_2)^2}{l_2}$, $B = \frac{(f_{c1}l_1+f_{c2}l_1^2)}{l_2}$, $\delta_1 = f_k(l_1+l_2)$ and $\delta_2 = f_k l_1$, and bring these parameters into formula (20.3), we can deduce that:

$$\begin{aligned} \theta_2 = & A\sin(\delta_1+f_\omega t) - B\sin(\delta_2+f_\omega t) = A[\sin\delta_1\cos(f_\omega t)+\cos\delta_1\sin(f_\omega t)] \\ & -B[\sin\delta_2\cos(f_\omega t)+\cos\delta_2\sin(f_\omega t)] = \cos(f_\omega t)(A\sin\delta_1 - B\sin\delta_2) \\ & +\sin(f_\omega t)(A\sin\delta_1 - B\sin\delta_2) = \cos(f_\omega t)\frac{A\sin\delta_1 - B\sin\delta_2}{\sin\delta} \\ & +\sin(f_\omega t)\frac{(A\sin\delta_1 - B\sin\delta_2)}{\sin\delta}\cos\delta \end{aligned}$$

Hereinto, $\delta = \operatorname{arccot}\frac{A\cos\delta_1 - B\cos\delta_2}{A\sin\delta_1 - B\sin\delta_2}$; $K = \frac{A\sin\delta_1 - B\sin\delta_2}{\sin\delta} \Rightarrow \theta_2 = K\sin(\omega t+\delta)$

Therefore, the rotating rule of the second joint follows the rule of sine. Similarly, the rotating rules of the third and the fourth joints follow the rule of sine as well.

20.3 Mathematical Models of the Rotation of Robotic Fish and MATLAB Simulation

The mathematical model of the first rotating joint of robotic fish is [2]:

$$A_1(t) = \varphi_1 + A_1\sin(2\pi ft) \tag{20.4}$$

Hereinto, φ_1 is the included angle between the rotation of the first joint and axis of symmetry $(O_1 - O_2)$ of fish body;

- A_1 Range of oscillation of rotation of the first joint
- t Rotation duration of the four joints of robotic fish
- f Rotation frequency of the four joints of robotic fish

Suppose that amplitude coefficient d_1 is the ratio of actual amplitude and maximum amplitude, polarization coefficient d_2 is the extent of polarization of joint rotating away from center axis $(O_1 - O_2)$. $A_{1\max}$ is the maximum rotating amplitude of the first joint away from fish body's axis of symmetry. $\varphi_{1\max}$ is the maximum value of rotation of the first joint. Here into,

$$A_1 = d_1 \times A_{1\max}; \varphi_1 = d_2 \times \varphi_{1\max}; A_{1\max} = \varphi_{1\max} + A_1$$
$$\Rightarrow \varphi_{1\max} = A_{1\max} - A_1 = A_{1\max} - d_1 A_{1\max} = (1 - d_1) A_{1\max} \Rightarrow \varphi_1 = d_2 \cdot (1 - d_1) A_{1\max}$$

Bring the above equation into formula (20.4), we can deduce the mathematical model of the rotation of the first joint:

$$A_1(t) = d_2 \cdot (1 - d_1) A_{1\max} + d_1 A_{1\max} \sin(2\pi ft) \tag{20.5}$$

Suppose that $A_{2\max}$ is the maximum rotating amplitude of the second joint with respect to the first joint of robotic fish, and α_1 is the rotating phase difference of the first and the second joints. With formula (20.5), we can deduce that:

The mathematical model of the rotation of the second joint is:

$$A_2(t) = d_2 \cdot (1 - d_1) A_{2\max} + d_1 A_{2\max} \sin(2\pi ft + \alpha_1) \tag{20.6}$$

Suppose that $A_{3\max}$ is the maximum rotating amplitude of the third joint with respect to the second joint of robotic fish, and α_2 is the rotating phase difference of the second and the third joints, $A_{4\max}$ is the maximum rotating amplitude of the fourth joint with respect to the third joint of robotic prototype, α_3 is the rotating phase difference of the robotic prototype's from third to fourth joints. In the similar way, we can deduce that:

The mathematical model of the rotation of the third joint is:

$$A_3(t) = d_2 \cdot (1 - d_1) A_{3\max} + d_1 A_{3\max} \sin(2\pi ft + \alpha_2) \tag{20.7}$$

The mathematical model of the rotation of the fourth joint is:

$$A_4(t) = d_2 \cdot (1 - d_1) A_{4\max} + d_1 A_{4\max} \sin(2\pi ft + \alpha_3) \tag{20.8}$$

Combining (20.5–20.8), we get the equation set of mathematical model of the rotation of robotic fish's each joint.

20.3.1 Analysis of Swimming Style of Robotic Fish

20.3.1.1 Swimming in a Straight Line

First, choose a value for each parameter. $A_{1\max} = 60°$, $A_{2\max} = 50°$, $A_{3\max} = 40°$, $A_{4\max} = 30°$, $\alpha_1 = -30°$, $\alpha_2 = -50°$, $\alpha_3 = -70°$, $f = 1$ Hz. Then use MATLAB for simulation.

The study shows that when the robotic fish moves in a straight line, the rotating axis of symmetry of four joints overlaps the axis $O_1 - O_2$ of robotic fish. Then we can deduce that $d_2 = 0$. Letting $d_1 = 1$, we can get Fig. 20.2 with MATLAB.

20.3.1.2 Making a Turn

When the robotic fish makes a turn, the rotating axis of symmetry of four joints do not overlap the axis $O_1 - O_2$ of robotic fish. The rotating axis sways to the left when the fish turns to left and the situation is the same in the right side. Therefore, suppose that $d_1 = 0.5$, $d_2 = 1$ or $d_2 = -1$.

When $d_1 = 0.5$ and $d_2 = 1$, with simulation software MATLAB we get Fig. 20.3. When $d_1 = 0.5$ and $d_2 = -1$, we get Fig. 20.4.

We come to the following conclusion with the above simulation results:

(1) When amplitude coefficient $d_1 = 0$, the robotic prototype is swimming in straight line. In a certain period, the rotating of robotic fish's joints is symmetrical and the resultant force of lateral forces borne by the robotic fish is zero.
(2) When $d_1 \neq 0$ and $d_2 = \pm 1$, the robotic fish makes a turn. In a certain period, the rotating of robotic fish's joints sways to the left (right) side of the fish. The resultant force of lateral forces borne by the robotic fish is not zero.

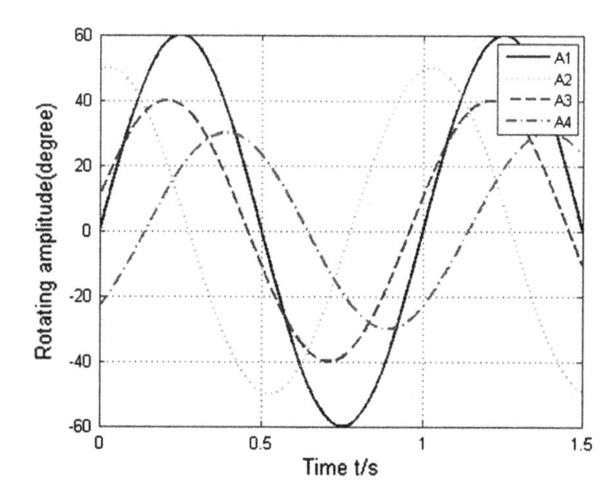

Fig. 20.2 When $d_1 = 1$, $d_2 = 0$, the rotating of four joints robotic prototype is swimming in *straight line*

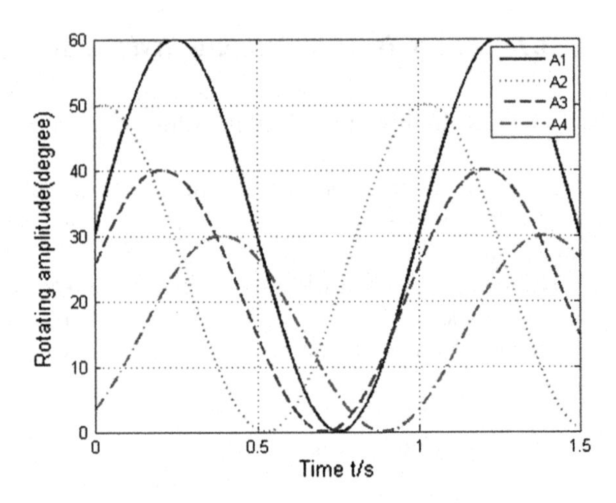

Fig. 20.3 When $d_1 = 0.5$, $d_2 = 1$, the motion of four joints when the robotic fish makes a turn

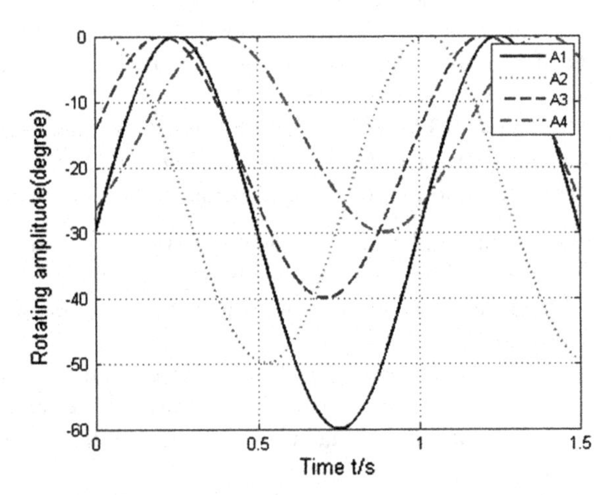

Fig. 20.4 When $d_1 = 0.5$, $d_2 = -1$, the motion of four joints when the robotic fish makes a turn

20.3.2 Example of MATLAB Simulation

20.4 3D Hydrodynamic Simulation Analyses

The four-joint biomimetic robotic fish prototype's SolidWorks 3D mechanical structure as Fig. 20.5 shows.

The Fig. 20.6 shows the maximum pressure on the prototype's mouth, fish belly up and down pressure values appear negative pressure.

Figure 20.7 shows the minimum speed appears on the fish's mouth, the maximum speed appear on the fish's upper and lower abdomen.

Fig. 20.5 The robotic prototype's mechanical structure

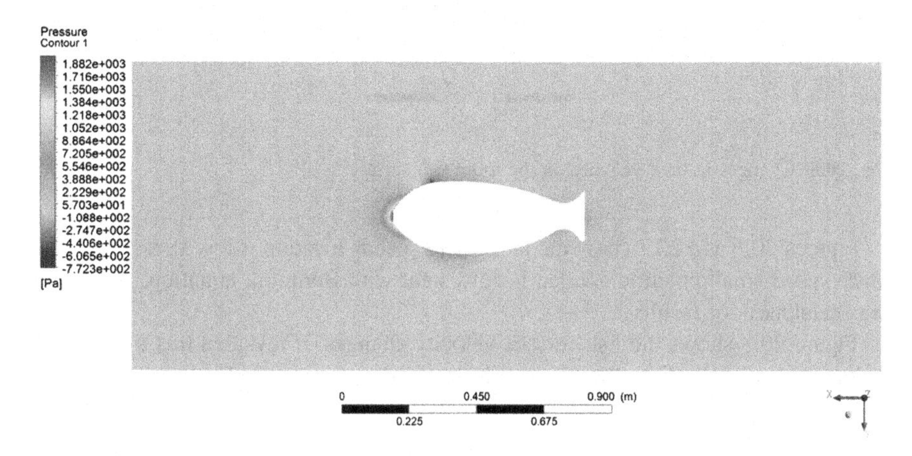

Fig. 20.6 The robotic prototype's pressure contours

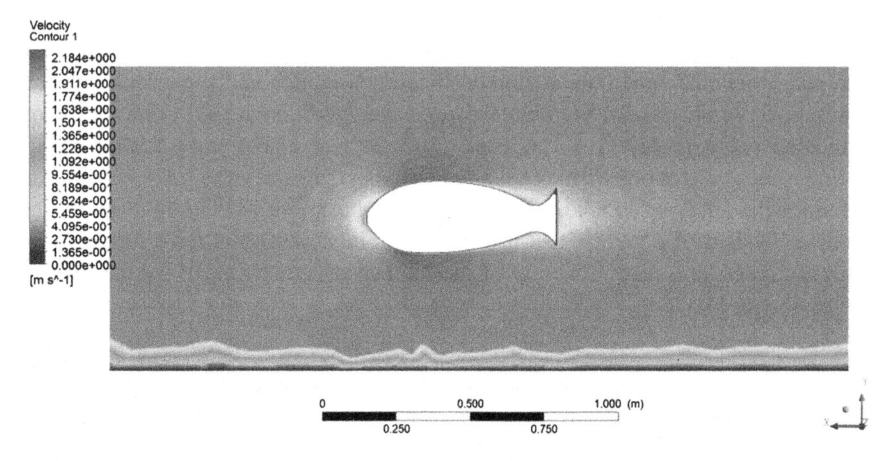

Fig. 20.7 Velocity contours of the robotic fish

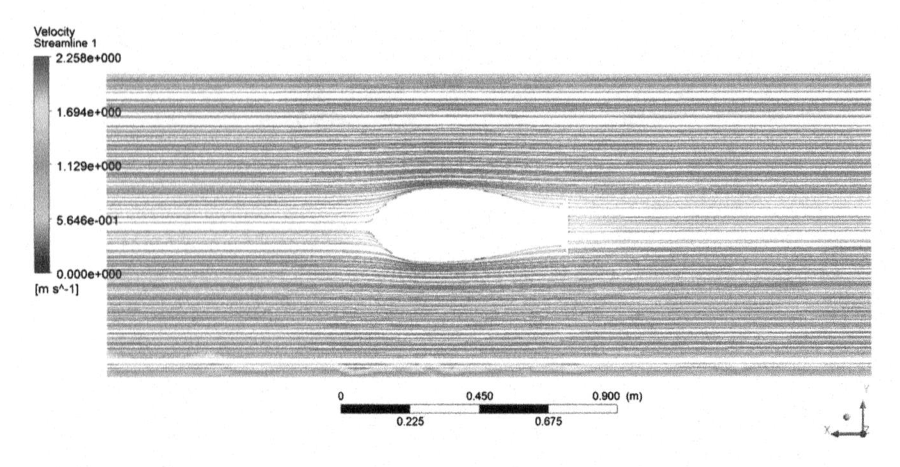

Fig. 20.8 Changes in flow velocity on the robot fish surface

Figures 20.6 and 20.7 show the fact where speeds location is low stress, stressful place speed small position. Which is consistent with Bernoulli equation, verify the reasonableness of results.

Figure 20.8 shows the fish surface velocity changes. It revealed that fish appear to have a streamlined, so that we can have good swimming behavior.

20.5 Conclusions

We established the rotating equations of four joints of the robotic fish through analysis of fish body wave motion curve equation. We discussed the influence of amplitude coefficient and polarization coefficient on motion and swimming impetus of robotic fish respectively. We obtained curves under different value with the aid of simulation software MATLAB as Figs. 20.2, 20.3 and 20.4 show and analyzed motion states of robotic fish. The above studies provide way of thinking for dynamics modeling of four-joint robotic fish in the future [3].The fluid hydrodynamic analysis of the robotic fish based on the Fluent software, have further validated the practicality and reliability of mechanical structural design for the robotic fish prototype [4].

Acknowledgments This work was supported by NSFC of China (Grants No.61165016) and High-level talents training project of Qinghai University (2012-QGC-10). Meanwhile, the authors would also like to thank Intelligence System and Control Laboratory of Qinghai University in China for the support to this work.

References

1. Lighthill MJ (1055) Large-amplitude elongated-body theory of fish locomotion. Proc Roy Soc London Ser B-Biol Sci 1971(179):125–138
2. Junzhi Y, Erkui C, Tanmin W (2003) Progress and analysis on robotic fish study. Control Theor Appl 20(4):485–491
3. Qingping W, Wang S, Tan M, Wang Y (2012) Research development and analysis of biomimetic robotic science. J Syst Sci Math Sci 32(10):1274–1286
4. Guan Z, Gao W, Gu N, Nahavandi S (2010) 3D hydrodynamic analysis of a biomimetic robot fish. In: Proceedings of the 11th international conference on control, automation, robotics and vision (ICARCV2010), Singapore, Dec 2010, pp 793–798

Chapter 21
Classification of Seizure in EEG Signals Based on KPCA and SVM

Weifeng Zhao, Jianfeng Qu, Yi Chai and Jian Tang

Abstract In this study, the electroencephalogram (EEG) signals-analysis experiments were made to classify seizures patients. Principal component analysis (PCA) and kernel principal component analysis (KPCA) were used for the data compression with the (EEG) signals. Classifiers based on support vector machine (SVM)-PCA and SVM-KPCA were designed. The classification performances of four kinds of kernel function were also compared using the same dataset. The results showed that using SVM-KPCA had higher recognition performance than SVM-PCA. Experimental results showed that the algorithm using SVM-KPCA with Gaussian-kernel had better recognition performance than the other three methods.

Keywords Electroencephalogram (EEG) signal · Seizure classification · PCA · KPCA · SVM

21.1 Introduction

PCA and KPCA could be applied to decrease the features in a low dimensional space. Then the projected features are fed into support vector machine (SVM) to distinguish the normal and epileptic. PCA is used extensively for feature extraction which can also lower the dimension [1, 2]. In KPCA [3], kernel function is used to attain randomly the high-order correlation. KPCA could be used to describe either the Gaussian distribution data, or the non-Gaussian distribution data. Some machine

W. Zhao · J. Qu (✉) · Y. Chai · J. Tang
College of Automation, Chongqing University, Chongqing 400044, China
e-mail: qujianfeng@cqu.edu.cn

Y. Chai
Key Laboratory of Power Transmission Equipment and System Security,
College of Automation, Chongqing University, Chongqing 400044, China

Y. Jia et al. (eds.), *Proceedings of the 2015 Chinese Intelligent Systems Conference*, Lecture Notes in Electrical Engineering 360,
DOI 10.1007/978-3-662-48365-7_21

learning classification theories for seizure classification have been proposed in 2010 [4]. Among those theories, support vector machine (SVM) has been proved to be better utilized in classification [4]. SVM was began in applied in the binary classifier. Therefore it was better used to solve the binary classification problems such as epilepsy detection.

The organization of the paper is: Sect. 21.2 illustrates the proposed method from three aspects: PCA, KPCA and SVM for classification. The experimental materials and results are presented in Sect. 21.3. At last, Sect. 21.4 draws a conclusion of these experiments.

21.2 Methodologies

In the paper, an advanced seizure detection approach was proposed based on KPCA and SVM. KPCA was used for personal EEG signals feature extraction and then SVM was used to train the classifier.

21.2.1 Feature Extraction and Reduction

21.2.1.1 Principal Component Analysis

By throwing away axes with small variances, PCA could be used to decrease dimensionality. And this variances can ensure the data matrix projected into its principal components with losing the least information. The set of original samples is assumed to be $\{x_i|x_i = (x_{i1}, x_{i2}, \ldots, x_{ip})^T, i = 1, 2, \ldots, m\}$. ρ represents the number of features, m represents the number of instances and k represents the feature. The "variance accumulation rate" is defined as:

$$\alpha_\iota = \lambda_i / \sum_{j=1}^{p} \lambda_p \tag{21.1}$$

and the former m principal components' cumulative variance rate α_1, $\alpha_2, \ldots \alpha_m$ as $\sum_{i=1}^{m} \lambda_i / \sum_{j=1}^{p} \lambda_p$.

21.2.1.2 Kernel Principal Component Analysis

(a) Kernel PCA is described in [5]. The attribute of KPCA makes it a better way to present the nonlinear data than the linear data. Supposing two vectors $<x', y'>$ are input in the original space. And the two input vectors' mappings in the feature space are $\varphi(x)$ and $\varphi(y)$, and the kernel function $K(x, y)$ is expressed as follows:

$$K(x^{'}, y^{'}) = \langle \varphi(x^{'}),\ \varphi(y^{'}) \rangle \tag{21.2}$$

and in this formula, K is the kernel function.

The computation in space F was defined, and it is associated with the input space through the nonlinear mapping function

$$\psi\colon R^{N} \to G.x \to \psi(x), \tag{21.3}$$

and the formula above is assumed that

$$\sum_{i=1}^{m} \psi(x) \ = 0. \tag{21.4}$$

In order to keep the simplicity, we have assumed the observations are centered, and it can be completed by substituting the matrix K with

$$\tilde{K} = K - L_m K - K L_m + L_m K L_m \tag{21.5}$$

where

$$L_m = \frac{1}{m} \begin{vmatrix} 1 & \cdots & 1 \\ \vdots & \ddots & \vdots \\ 1 & \cdots & 1 \end{vmatrix} \tag{21.6}$$

(b) Kernel Function Selection. The Gaussian kernel equation, polynomial kernel equation, sigmoid kernel equation and linear kernel equation are commonly used, which are defined as follows respectively:

$$k_{\sigma}^{Gaussian}(x, y) \ = \exp\left(\frac{\|x - y\|^{2}}{\sigma^{2}} \right) \tag{21.7}$$

$$k_{\alpha,\beta}^{Polynomial}(x, y) \ = (\langle x, y \rangle + \alpha)^{\beta} \tag{21.8}$$

$$k_{\lambda,\gamma}^{Sigmoid}(x, y) \ = \tanh(\lambda \langle x, y \rangle + \gamma) \tag{21.9}$$

$$k_{\beta}^{Linear} = \beta \langle x, y \rangle \tag{21.10}$$

21.2.1.3 Support Vector Machine for Classification

SVM [6] is built on the foundation of statistical learning theory. By selecting a non-linear mapping, the SVM establishes a hyper-plane in the higher dimensional space with maximum margin. Therefore SVM well known for maximum margin classifier. As the two-class problem, a classifier should be trained with a learning algorithm from a set of samples $(\{(x_i, y_i), i = 1, 2, \ldots N\})$. For each training example x_i, $y_i \in \{1, -1\}$, 1 or−1 is the given label of it. The linear classifier is described as:

$$f(x) = sign[W^T x + w] \tag{21.11}$$

The optical separating hyper-plane need to fulfil the formula below, which is used for preventing the misclassifications:

$$y_i[W^T x + w] \geq 1 - \theta_k, K = 1, \ldots n \tag{21.12}$$

where θ_K is a slack variable.

21.3 Experimentations and Discussion

The EEG datasets is from CHB-MIT Scalp EEG Database [7] which is an open database. And more comprehensive information of the data refers to [8]. In this paper, we choose to use the chb01-26.edf of the database. The file containing 23 EEG signals was selected. Before performing the classification test, the optimal parameters of c and g should be determined. In this work, the grid search method is used to find the optimal parameters of c and g. One thousand groups of the datasets are randomly chosen for training. And then five hundred groups are chosen to test the accuracy of the classifier.

The schematic diagram for SVM-KPCA framework is shown in Fig. 21.1. The experimental EEG signals are shown in Fig. 21.2.

(1) SVM-PCA. Firstly, the selected matrix was sent to PCA processing method for dimensionality reduction. We took the PCs' columns which occupied over 95 % information of the original data. Ten eigenvectors were selected as the

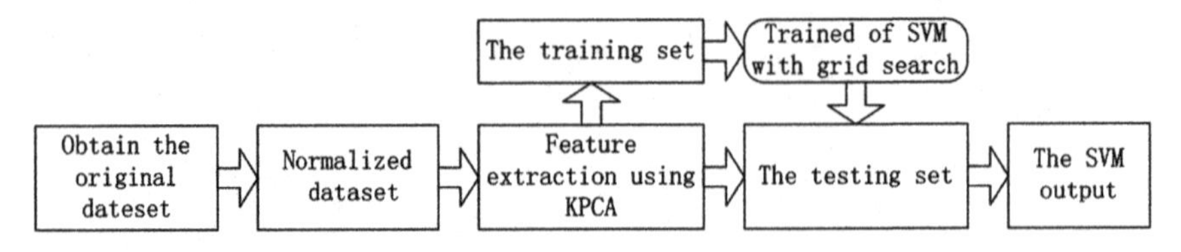

Fig. 21.1 SVM-KPCA framework

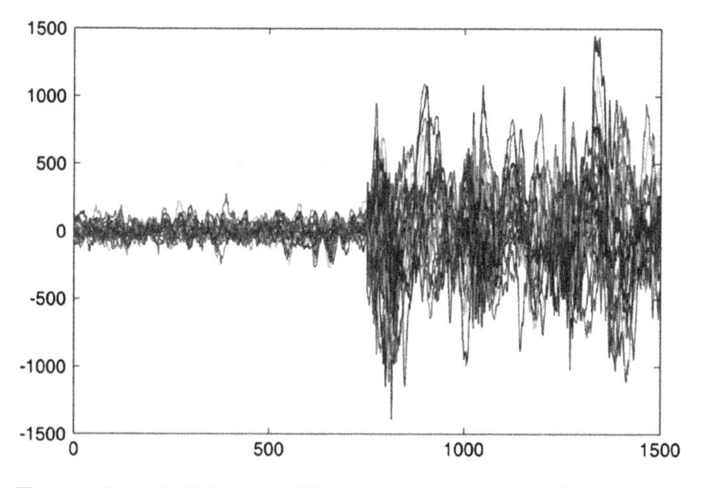

Fig. 21.2 The experimental EEG signals (We select the same number of signals between normal and seizure signals)

Table 21.1 Classification results with SVM-PCA

Feature extraction	Number of eigenvalues	*c*	*g*	Training accuracy (%)	Testing accuracy (%)
PCA	10	0.03125	6.9644	100	98.6

Table 21.2 Accuracy of same eigenvalue number with different kernel function

Function name	Number of eigenvalues	*c*	*g*	Training accuracy (%)	Testing accuracy (%)
Gauss	3	0.5	1	100	100
Linear	3	5.6569	2048	98.7	84.4
Poly	3	0.70711	2048	99.3	89.8
Sigmoid	3	4	8192	97.9	92.4

new input ones. The outcome of inputting the new eigenvectors in classifier is shown in Table 21.1.

(2) SVM-KPCA. Firstly, the input matrix was send to KPCA processing method for dimensionality reduction. The effect of dimensionality reduction on different kernel functions was analyzed.

The outcomes of inputting the new eigenvectors in classifier with different kernel functions are presented in Tables 21.2 and 21.3.

In order to make comparison with PCA, we chose ten eigenvalues with different kernel functions. By comparing Tables 21.1 and 21.4, we found that SVM-KPCA got a better classification result than SVM-PCA.

Table 21.3 Accuracy of same eigenvalue number with different kernel function

Function name	Number of eigenvalues	c	g	Training accuracy (%)	Testing accuracy (%)
Gauss	4	0.5	1	100	100
Linear	4	0.5	128	100	95.8
Poly	4	0.5	1024	99.8	86.8
Sigmoid	4	1	8192	99.8	91.2

Table 21.4 Accuracy of same eigenvalue number with different kernel function

Function name	Number of eigenvalues	c	g	Training accuracy (%)	Testing accuracy (%)
Gauss	10	0.5	2	100	100
Linear	10	0.5	4	100	100
Poly	10	0.5	8	100	100
Sigmoid	10	1	128	100	99.4

21.4 Conclusions

By using SVM-PCA and SVM-KPCA, a new seizure detection approach for the EEG signals are presented in this paper. The results show that KPCA has better performance for processing nonlinear data. The KPCA based kernel function is used to the seizure and non-seizure EEG signals. The features are obtained by computing the new feature vector of EEG signals. We selected three and four features as input to a classifier, which were used for training to obtain the optimal parameters with grid search method. The experimental results indicate that Gauss kernel function can perform better classification result when it classified the seizure and non-seizure EEG signals.

References

1. Wang GW, Zhang C, Zhuang J (2012) An application of classifier combination methods in hand gesture recognition. Math Probl Eng (Article ID 346951):17
2. Cho JH, Lee JM, Choi SW, Lee D et al (2005) Fault identification for process monitoring using kernel principal component analysis. Chem Eng Sci 60(1):10
3. Chen ZG, Ren HD, Du XJ (2008) Minimax probability machine classier with feature extraction by kernel PCA for intrusion detection. In: Wireless Communications, Networking and Mobile Computing, p 4
4. Nandan M, Talathi SS, Myers S et al (2010) Support vector machines for seizure detection in an animal model of chronic epilepsy. J Neural Eng 7(3):13
5. Schölkopf B, Smola A, Muller KR (1996) Nonlinear component analysis as a kernel eigenvalue problem. Neural Comput 10(3):17

6. Li K, Huang WX, Huang ZH (2013) Multi-sensor detected object classification method based on support vector machine. J Zhejiang Univ (Eng Sci) 47:8
7. CHB-MIT Scalp EEG Database. http://physionet.nlm.nih.gov/pn6/chbmit/
8. Shoeb AH Application of machine learning to epileptic seizure onset detection and treatment, Ph.D. thesis, Massachusetts Institute of Technology, no. 9, p 20

Chapter 22
Time-Delay Prediction Method Based on Improved Genetic Algorithm Optimized Echo State Networks

Zhongda Tian and Tong Shi

Abstract In a networked control system, the time-delay has random and nonlinear characteristics, make the stability of system is hard to ensure. It need the controller in system can accurately predict time-delay. So the precise time-delay prediction of networked control system is an important factor in ensuring the stability of the control system. The echo state networks has good predictive ability on nonlinear time series, it is suitable for predict the time-delay. But parameters of echo state networks learning algorithm has a great influence on the prediction accuracy. An improved genetic algorithm is proposed for parameters optimization of echo state networks. The simulation results show that the prediction accuracy of the predictive method in this paper is higher than the conventional predictive model such as auto regressive and moving average (ARMA) model, least square support vector machine (LSSVM) model and Elman neural network.

Keywords Networked control system · Echo state networks · Genetic algorithm · Time-delay · Prediction

22.1 Introduction

The time-delay in the networked control system (NCS) has time-varying and nonlinear characteristics. At the same time, it is also an important parameter affecting the stability of the system [1]. In order to improve the control performance

---

Z. Tian (✉)
School of Information Science and Engineering, Shenyang University of Technology, Shenyang 110870, China
e-mail: tianzhongda@126.com

T. Shi
Department of Humanities, Liaoning Forestry Vocation-Technical College, Shenyang 110101, China

Y. Jia et al. (eds.), *Proceedings of the 2015 Chinese Intelligent Systems Conference*, Lecture Notes in Electrical Engineering 360,
DOI 10.1007/978-3-662-48365-7_22

and quality of the networked control system, accurately measurement, analyze and predict network time-delay is very important to ensure stability of system [2].

Currently, a large number of scholars have proposed many time-delay predictive model of NCS, and made some progresses. AR (auto regressive) and ARMA (auto regressive moving average) model is adopted as predictive model in some literatures [3, 4], AR or ARMA predictive model is suitable for linear time series, but network time-delay has strong nonlinear characteristic, AR or ARMA is not a appropriate model. Neural network can identify nonlinear system, the calculation time is short, so it can be seen as the time-delay prediction model [5, 6]. The shortcoming of neural network is local optimal value problem, it is hard to determine the parameters of network. SVM (support vector machines) has unique advantages for nonlinear, small samples and high dimensional pattern recognition problem, and can be seen as prediction model for network time-delay with strong nonlinearities [7, 8], but it is difficult to determine the parameters of prediction algorithm based on SVM. The time-delay sample of networked control system is analyzed to study the distribution of data is also one of research directions [9, 10], however, predictive method based on statistical analysis is difficult to solve distribution parameters.

In order to enhance prediction performance, a new type of recurrent neural networks-echo state networks (ESN) is chosen as a prediction model. An improved genetic algorithm with better fitness and faster convergence speed is presented for ESN parameters optimization, and achieve better prediction accuracy and effect.

22.2 Echo State Networks

ESN neural network algorithm [11] was put forward in 2001. ESN compared with traditional recurrent neural networks, predictive and modeling ability for nonlinear chaotic system have been greatly improved [12]. Firstly, spectral radius of ESN internal network connection weights matrix can ensure stability of reserve pool recursive network. Secondly, linear regression algorithm is used to train the output weights, the learning algorithm itself is convex optimization, thus avoiding the problem of fall into local minimum value. All these advantages made ESN become an academic research focus. ESN can be applied into a variety of areas.

22.2.1 ESN Learning Algorithm

In a neural network system, input units number as K, internal processing units number as N, and output units as number L. At time t, there is

$$\begin{aligned} \boldsymbol{u}(t) &= [u_1(t), u_2(t), \ldots, u_K(t)]^{\mathrm{T}} \\ \boldsymbol{x}(t) &= [x_1(t), x_2(t), \ldots, x_N(t)]^{\mathrm{T}} \\ \boldsymbol{y}(t &= [y_1(t), y_2(t), \ldots, y_L(t)]^{\mathrm{T}} \end{aligned} \tag{22.1}$$

Compared with normal neural networks, hidden layer of ESN has more neurons constitute a dynamic reservoir (DR) of cycle network. Training samples can connected to the internal unit by $\boldsymbol{W}_{in}$, and achieve the collection of training data, make DR record corresponding information. The linear regression process and DR state vector can minimize the average error of the target output through $\boldsymbol{W}_{out}$. Learning steps of ESN are:

$$\boldsymbol{x}(t+1) = f(\boldsymbol{W}_{in} \cdot \boldsymbol{u}(t+1) + \boldsymbol{W}_{back}\boldsymbol{x}(t)) \tag{22.2}$$

$$\boldsymbol{y}(t+1) = f_{out} \cdot (\boldsymbol{W}_{out} \cdot (\boldsymbol{u}(t+1), \boldsymbol{x}(t+1))) \tag{22.3}$$

22.2.2 Weight Calculation

The actual network output $\tilde{\boldsymbol{y}}(t)$ can be regarded as the estimation value of the expected output value $\boldsymbol{y}(t)$

$$\boldsymbol{y}(t) \approx \hat{\boldsymbol{y}}(t) = \boldsymbol{W}_{out}\boldsymbol{x}(t) \tag{22.4}$$

Computing the weight matrix to meet the minimum mean square error, solving the following objectives:

$$\min \frac{1}{P-m+1} \sum_{n=m}^{P} (\boldsymbol{y}(t) - \boldsymbol{W}_{out}\boldsymbol{x}(t))^2 \tag{22.5}$$

and thus the following equation can be deduced:

$$\boldsymbol{W}_{out} = (\boldsymbol{M}^{-1} \cdot \boldsymbol{T})^T \tag{22.6}$$

22.2.3 Key Parameters of Reservoir

The dynamic reservoir can influence the whole performance of ESN.

(1) *SR*, it is a connection weights of the spectral radius (SR), it means the largest absolute eigenvalue of connection weight matrix $\boldsymbol{W}$, denoted by $\lambda_{\max}$, network is stability when $\lambda_{\max} < 1$.

(2) N, the scale of reservoir, it means the numbers of reservoir neurons. N has relationship with the number of samples, the greater the N, the more accurate description of the given dynamic system, but it will bring over-fitting problem.
(3) IS, it is the input unit scale of reservoirs. It is a multiplying scale factor between input signal connected to the reservoir neurons. It is a certain scaling of the input signal. Generally the stronger of the object nonlinear, the bigger IS.
(4) SD, the sparse degree of dynamic reservoirs. It is a connection parameter between the reservoirs neurons. There is not all neurons connected between the reservoir neurons in a ESN.

22.3 Time-Delay Prediction Model Based on ESN

Assuming the time-delay sequence to be predicted can be expressed as $d(t)$, prediction origin is at time t, prediction step length is h, time-delay sequence prediction problem can be expressed as: for a given $d(t)$, predict future $t+h$ time-delay value. Firstly, it need to establish the mapping of prediction origin of k and prediction step $t+h$. Put $d(n), n \leq t$ follow next equation and convert as:

$$\boldsymbol{D}(t) = [d(t), d(t-1), \ldots, d(t-m)] \tag{22.7}$$

wherein, m is the embedding dimension, in accordance with the ENS algorithm, $\boldsymbol{D}(t)$ as input, $d(t+h)$ as output, the ESN network is training, then the input-output mapping can be determined. The established ESN model can be used for time-delay prediction. Time-delay predictive model based on ESN can be described as the following steps:

Step 1 Determining the length of the training sample and the prediction length, the input and output training set $\{\boldsymbol{D}(t), d(t+h), t = 1, 2, \ldots, p\}$.
Step 2 Determining initialization parameters: IS, SR, N, SD, m, etc.
Step 3 In accordance with the Eqs. (22.5) and (22.6), ESN network is trained and generated the output weight matrix.
Step 4 Generating the input, internal and output according to Eq. (22.1) and predict future h time time-delay value through Eqs. (22.2) and (22.3).

The parameters of the reservoir and the embedding dimension m play an important role on the predictive accuracy of ESN prediction effect. To illustrate the impact of these parameters on the prediction accuracy, 500 Internet network delay data is used for training, 100 group delay data for validation. Figure 22.1 shows the SR range is [0.1, 1), the step size is 0.05. N range is [20, 200), step size is 1. $m = 5$, $IS = 0.1$, $SD = 0.3$, the distribution of root mean square error (RMSE) between prediction time-delay and actual time-delay. Figure 22.2 shows $SR = 0.5$, $N = 30$,

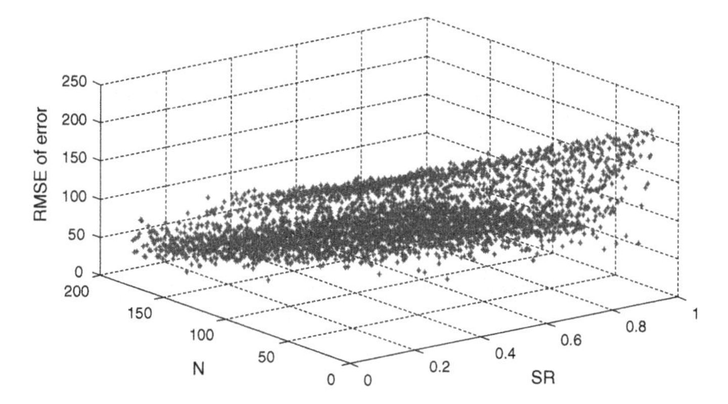

Fig. 22.1 Time-delay RMSE when *SR* and *N* changes

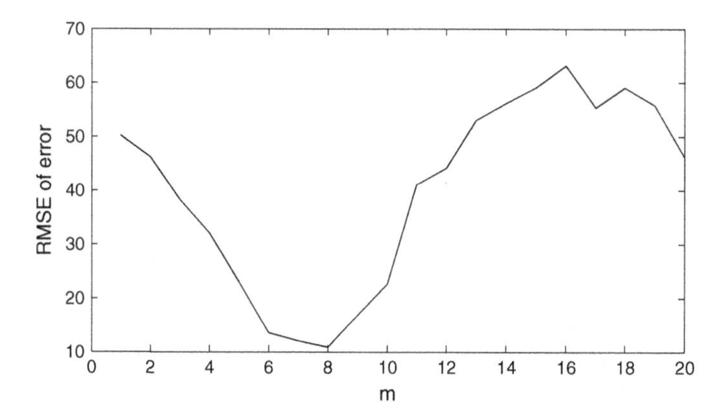

Fig. 22.2 Time-delay RMSE when *m* changes

$IS = 0.1$, $SD = 0.3$, m range is from 1 to 20, the distribution of RMSE between prediction time-delay and actual time-delay. Figure 22.3 shows $SR = 0.5$, $N = 30$, $m = 5$. IS range is [0.01, 1), step size is 0.05. SD range is [0.01, 1), step size is 0.05, the distribution of RMSE between prediction time-delay and actual time-delay.

It can be seen from Figs. 22.1, 22.2 and 22.3, different embedding dimension m and the reservoir parameters will result in different effects for prediction accuracy, how to select the appropriate training parameters is an important problem. In order to find the suitable and optimized parameters, this paper proposes an improved genetic algorithm for parameters optimization, so it can improve the prediction accuracy.

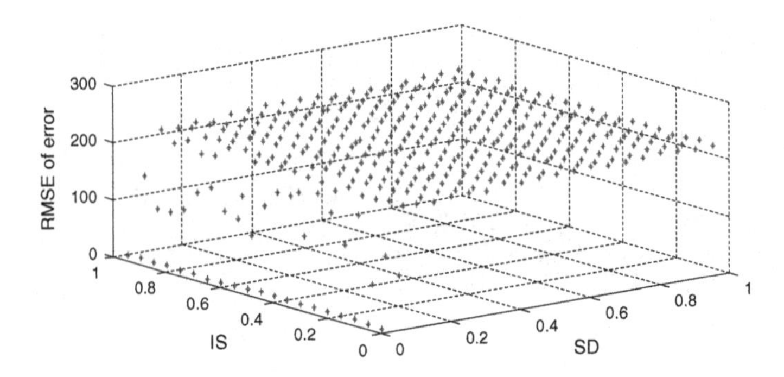

Fig. 22.3 Time-delay RMSE when *SD* and *IS* changes

22.4 Improved Genetic Algorithm

Genetic algorithm uses some theories of Biology, the population will be evolved by natural selection, exchange and mutation. Genetic algorithm has good global search capability, and can search the global optimal value in a short time. But standard genetic algorithm has the shortcomings of results often fall into the local, decoding errors and slow convergence speed, etc. Based on the hierarchical genetic algorithm, an improved genetic algorithm proposed in this manuscript. The adaptive crossover and mutation is used in the optimization process, this improved genetic algorithm has better convergence accuracy and faster convergence speed. Algorithm is designed as follows.

(1) Crossover operation improvement

Because hierarchical chromosome includes both binary code and decimal code genes, so need to be respectively carried out cross operation. Firstly, each individual involved in the cross operation is selected according to the adaptive crossover probability p_c, then the crossover is operated. The single point crossover is used for controlling gene. The each individual of subgroup $P_1(t)$ randomly determine a point of intersection and exchange side of the two individual gene of intersection position. For parameters gene, arithmetic crossover operation performed by the following equation:

$$\begin{cases} a' = (1-u)a + ub \\ b' = (1-u)b + ua \end{cases} \tag{22.8}$$

wherein, u is scale factor, usually a constant from 0 to 1. Because of the above equation u is arbitrarily selected, cross operation has certain blindness, and easy to fall into local optimum, not search in the global scope, thus leading to reduced the accuracy of the algorithm.

The improved crossover operation as the next equations:

$$\begin{cases} a_1^{'} = (1 - P_c)a + P_c b & f_a < f_{avg} \\ a_2^{'} = (1 - P_c)b + P_c a & f_a \geq f_{avg} \end{cases} \tag{22.9}$$

$$\begin{cases} b_1^{'} = \left(1 - P_c^{'}\right)b + P_c^{'} a & f_b < f_{avg} \\ b_2^{'} = P_c^{'} b + \left(1 - P_c^{'}\right)a & f_b \geq f_{avg} \end{cases} \tag{22.10}$$

wherein,

$$P_c = \begin{cases} p_{c1} - \frac{(p_{c1} - p_{c2})(f_a - f_{avg})}{f_{\max} - f_{avg}}, & f_a \geq f_{avg} \\ p_{c1}, & f_a < f_{avg} \end{cases} \tag{22.11}$$

$$P_c^{'} = \begin{cases} p_{c1} - \frac{(p_{c1} - p_{c2})(f_b - f_{avg})}{f_{\max} - f_{avg}}, & f_b \geq f_{avg} \\ p_{c1}, & f_b < f_{avg} \end{cases} \tag{22.12}$$

In the above equations, $p_{c1} = 0.9$, $p_{c2} = 0.6$. As can be seen from the equations, scale factor of crossover operation is changed with adaptive crossover probability of the individuals. So the improved crossover operation not only ensures the global searching ability, but also accelerates the convergence speed of the algorithm.
(2) Mutation operation improvement
Mutation operation should also carried out on the controlling gene and parameters gene. Firstly, individual is selected according to the adaptive mutation probability p_m shown in the next equation, and then mutation operation is conducted on selected individual.

$$p_m = \begin{cases} p_{m1} - \frac{(p_{m1} - p_{m2})(f - f_{avg})}{f_{\max} - f_{avg}}, & f \geq f_{avg} \\ p_{m1}, & f < f_{avg} \end{cases} \tag{22.13}$$

wherein, $p_{m1} = 0.1$, $p_{m2} = 0.001$.

Parametric genes commonly used uniform mutation methods are as follows:

$$x_k^{'} = x_k + u\left(x_k^1 - x_k^2\right) \tag{22.14}$$

wherein, x_k^1 is upper bounds of x_k, and x_k^2 is lower bounds of x_k, u is scale factor that range from 0 to 1. It can be seen that the mutation operation process is random and non-directional. This paper adopt improved mutation operator for individual. Let the individual mutation operate toward best direction and amplitude of the population and individual optimal solution, as shown in the following equations:

$$\Delta x_{k,t+1} = r_1\left(x_{\max,k} - x_{k,t}\right) + r_2\left(x_{\max} - x_{k,t}\right) \tag{22.15}$$

$$x_{k,t+1} = x_{k,t} + \Delta x_{k,t+1} \quad (22.16)$$

In the above equations, $\Delta x_{k,t+1}$ is change range of individual x_k mutate to $t+1$ generation, $x_{k,t}$ is the value of x_k at t generation, $x_{\max,k}$ is the optimal value of x_k in the evolution process, $x_{\max}$ is the optimal value of population, $x_{k,t+1}$ is individual x_k after mutated, r_1 and r_2 is random number from 0 to 1. The individual mutated toward the direction of the optimal value of the population, therefore, the algorithm speed up the convergence speed and enhance convergence precision.

22.5 Time-Delay Prediction Method Based on Improved GA Optimized ESN

Here gives the detail prediction steps of prediction method in this manuscript.

Step 1. The parameters in the prediction model is coded.
Step 2. The initial population is created.
Step 3. The fitness function is chosen as the RMSE of actual and predicted time-delay.
Step 4. The time-delay in a network is collected. These sample will be used to train ESN model. In each evolution process, the fitness value is calculated and recorded.
Step 5. The individual with best fitness value will be chosen as the next generation. The new population is determined by improved genetic algorithm. The best fitness is calculated and recorded.
Step 6. The best parameters of prediction model will be obtained when the algorithm is end.

22.6 Simulation

The parameters to be optimized are m, SR, N, IS, SD, decimal coding is used, encoded bits are 10. The population is 20, the number of evolution iterations is 200, the higher sub-population is controlling gene, encoding as binary, chromosome length is 10. The lower sub populations of parameters gene coded as real number. The range of parameters are, m from 1 to 20, SR from 0.01 to 1, N from 10 to 200, IS from 0.01 to 1, SD from 0.01 to 1. 500 group delay data obtained by time-delay measurement software to train ESN network. The fitness function is taken as the RMSE of actual and predicted time-delay. The parameters optimization results by improved GA are: $m = 8$, $SR = 0.7194$, $N = 45$, $IS = 0.1329$, $SD = 0.4581$. After optimized parameters of ESN model are obtained, 100 measured time-delay sample data is used to verify predictive effect. Predictive value and the actual value of

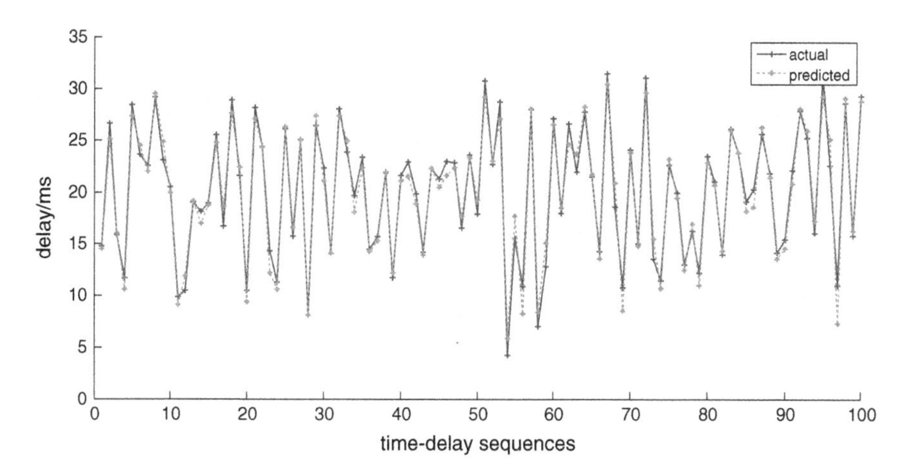

Fig. 22.4 Time-delay contrast curve of this paper method

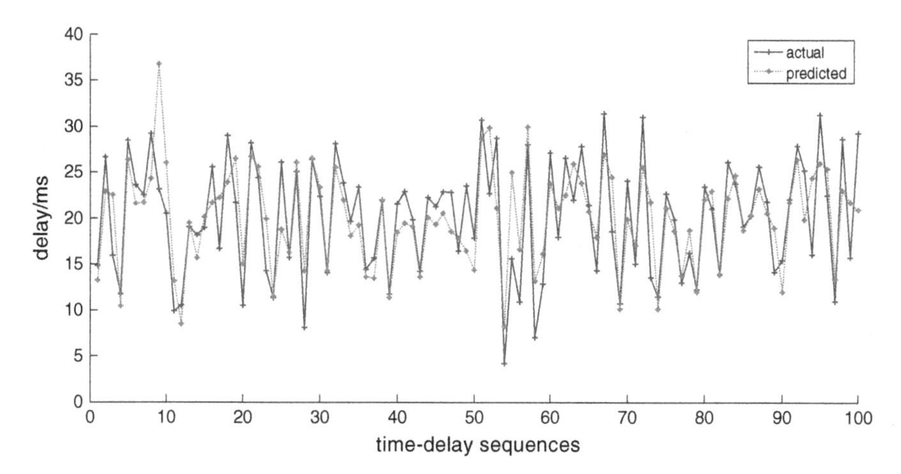

Fig. 22.5 Time-delay contrast curve of ARMA algorithm

contrast curve is shown in Fig. 22.4, where '+' for the actual time-delay, '*' for the prediction time-delay.

In order to verify the superiority of the time-delay predictive method in this paper, the same delay data predicted by ARMA method in literature [4], Elman neural network in literature [6], LSSVM algorithm in literature [7] respectively. The contrast curve of actual and predicted time-delay is shown in Figs. 22.5, 22.6 and 22.7. The order of ARMA is 10, parameters of LSSVM are $\gamma = 1.3895$, $\sigma^2 = 38.24$, the parameters of the Elman neural network include input layer is 10, the middle layer is 20, the output layer of neural network is taken as 1, and maximum iteration number is chosen as 4000. The predictive error distribution histogram chart as shown in Fig. 22.8. In order to further compare prediction effect, the root mean

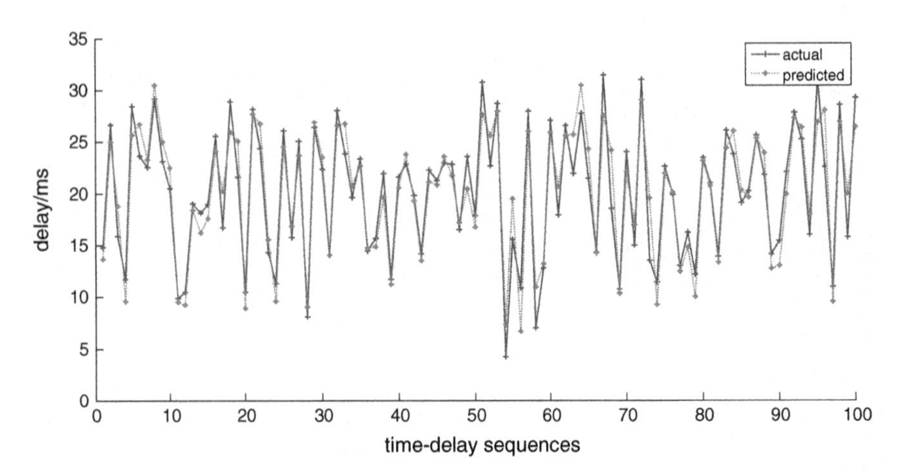

Fig. 22.6 Time-delay contrast curve of LSSVM algorithm

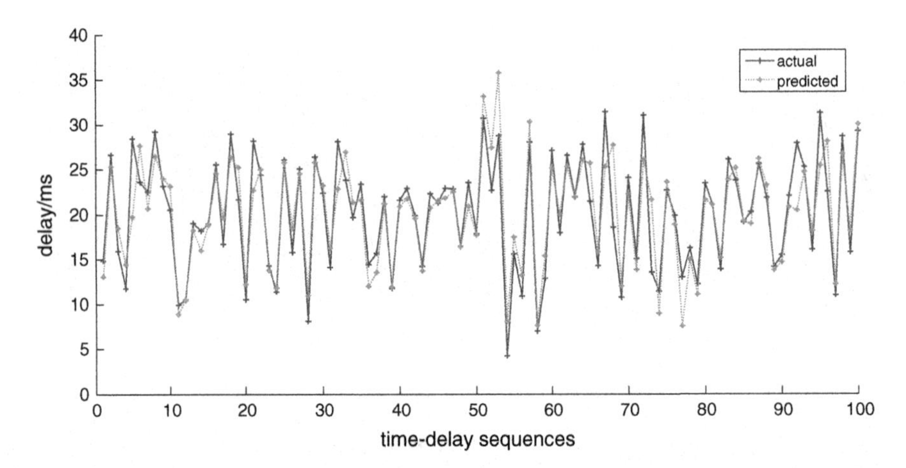

Fig. 22.7 Time-delay contrast curve of Elman neural network

square error (RMSE) and mean absolute percent error (MAPE) of predictive methods are given in Table 22.1.

It can be seen from Figs. 22.4, 22.5, 22.6, 22.7, 22.8 and Table 22.1, predictive effect and accuracy of the paper's method is higher than other methods, and prediction error distribution is more uniform. The main reason of the predictive accuracy improved is echo state network take advantages of traditional neural network, and overcome the shortcomings of the traditional method, has the very good capability of nonlinear approximation. At the same time, parameters of echo state network model are optimized by improved genetic algorithm, the optimal predictive parameters are obtained reasonably, avoid blind setting, thereby the accuracy of prediction improved.

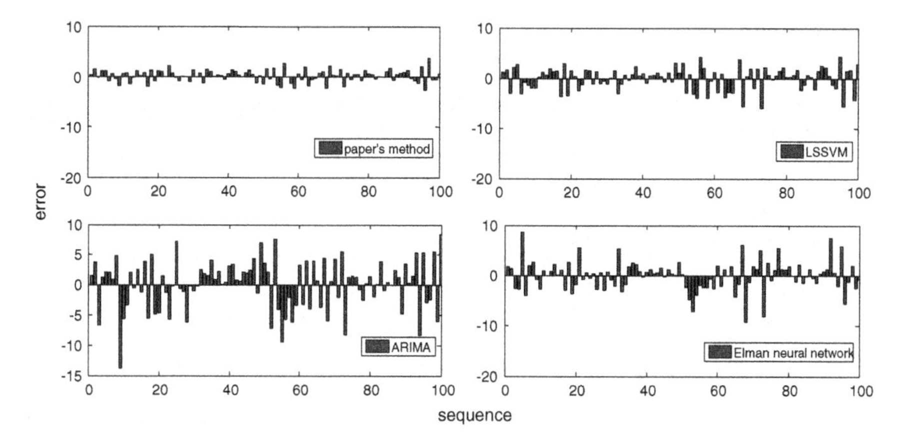

Fig. 22.8 The predictive error distribution

Table 22.1 The predictive effect of several methods

Prediction method	RMSE	MAPE
This paper's method	1.15	0.22
LSSVM	4.91	0.58
Elman	7.43	0.92
ARMA	14.42	0.63

22.7 Conclusion

The precise time-delay prediction is very important for the stability analysis of network control system, but it is difficult to guarantee prediction accuracy because nonlinear and randomness characters of time-delay. In order to enhance prediction effect, a time-delay prediction model based on echo state networks optimized by improved genetic algorithm is proposed in this paper. Echo state networks has good nonlinear modelling and prediction ability, it is a appropriate time-delay prediction model. This paper adopt an improved genetic algorithm to solve the parameters optimization of echo state networks, the optimal predicative parameters are obtained. The simulation verify that the improved genetic algorithm has better convergence performance, and predictive accuracy is higher than the others predictive model.

Acknowledgments This work is a project supported by the Liaoning Province Doctor Startup Fund under Grant 20141070.

References

1. Zhang J, Luo DY (2014) Analysis of time delay and packet dropouts based on state observer for a kind of networked control system. J Netw 9(2):466–473
2. Tian ZD, Gao XW, Li K (2014) A hybrid time-delay prediction method for networked control system. Int J Autom Comput 11(1):19–24
3. Li H, Xiong N, Park JH, Cao Q (2010) Predictive control for vehicular sensor networks based on round-trip time-delay prediction. IET Commun 4(7):801–809
4. Medjiah S, Taleb T, Ahmed T (2014) Sailing over data mules in delay-tolerant networks. IEEE Trans Wirel Commun 31(1):5–13
5. Tian ZD, Gao XW, Li K (2013) Adaptive predictive control of networked control system. J Appl Sci-Electron Inf Eng 31(3):303–308
6. Han CW (2014) Fuzzy neural network-based time delay prediction for networked control systems. Appl Math Inf Sci 8(1):407–413
7. Tian ZD, Gao XW, Li K (2013) Networked control system time-delay prediction method based on KPCA and LSSVM. Syst Eng Electron 35(6):1281–1285
8. Wang T, Wang H, Wang P (2010) Application of support vector machine in networked control system with delayed variable sampling period. In: 2010 3rd IEEE international conference on computer science and information technology, pp 127–130
9. Fu XH, FuX (2011) A predictive algorithm for time delay Internet network. In: 2011 International conference on electronics, communications and control, pp 666–669
10. Zhao JP, Gao XW (2009) Time-delay analysis and estimation of Internet-based robot teleoperation system. Chin Control Decis Conf 2009:4643–4646
11. Jaeger H (2001) The “echo state” approach to analyzing and training recurrent neural networks. GMD Report 148, GMD-German National Research Institute for Computer Science, 2001
12. Cui HY, Feng C, Chai Y (2014) Effect of hybrid circle reservoir injected with wavelet-neurons on performance of echo state network. Neural Netw 57:141–151

Chapter 23
Interactive Speech Recognition Based on Excel Software

Xiujuan Meng and Chaoli Wang

Abstract With the rapid development of modern computer technology, the communication between man and machine is becoming more frequent. A large amount of data is needed to input when people use Microsoft Excel software in the fields of account, office, finance, hospital, etc. This paper presents a recognition method of naming speech as input for Excel table. In this system, we use the characteristics of the name as a basic speech unit by using the Mel Frequency Coefficients (MFCC) as the feature parameters. Moreover, we use the Hidden Markov model (HMM) as the basis to train the acoustic models of this environment. The HMM can ease the mismatch caused by the identification of the test environment and training environment, which can improve the recognition rate further. Finally, experiments show that this system has good recognition and input function. This study establishes the foundation for future development of the method of application system.

Keywords Interactive · Speech recognition · MFCC · HMM

23.1 Introduction

Since speech recognition has no directional limitation and allows the existence of distance, but it can liberate the eyes and hands. So speech recognition can be used to a particular environment, such as a large amount of data to Excel table entry. The vocoder based on speech recognition and synthesis can get very low bit rate, and be transmitted in almost any channel. People can complete communication directly

X. Meng (✉) · C. Wang
Control Science and Engineering Department, University of Shanghai for Science and Technology, Shanghai 200093, China
e-mail: mengxiujuan8421@163.com

Y. Jia et al. (eds.), *Proceedings of the 2015 Chinese Intelligent Systems Conference*, Lecture Notes in Electrical Engineering 360,
DOI 10.1007/978-3-662-48365-7_23

between the two languages by exploiting the machine translation algorithm. This paper mainly discusses the interactive speech recognition based on Excel software. In the process of making an Excel work table, there is a lot of data input, such as sales, wages, performance, etc. For example, in the teaching management system, teachers often need to input the scores of students manually, which is very troublesome. Can we use voice input the results in Excel? In other words, read out the names and records into Microphone, and then the computer automatically inputs each student scores in Excel. Speech recognition has a huge application foreground in many fields, there are many relevant literatures that extend the theory to application, such as [1–3], just to name a few: Chen and Jiang [4] applied the application of speech recognition in medicine. The literature [5] gave the price of agricultural products information acquisition method based on speech recognition. Document [6] discussed the method of using voice recognition of forest road detection method. Reference [7] described the application of speech recognition in the intelligent home furnishing. However, in many previous papers, there are very few results available regarding the special study based on interactive speech recognition input method of Excel software. So it is the first time that discusses such a subject in the topic. The main research focuses speech recognition, office software, and computer science. The development environment of this system mainly uses Visual Studio 2010 as the development platform. Moreover, the platform is exploited to test the recognition process of speech from input to output. In this system, the core part of the speech recognition system is mainly composed of two parts: speech feature extraction, acoustic model and pattern matching. Speech feature parameters are commonly used in Linear Prediction Cestrum Coefficient (LPCC) [8]; MFCC [9]; Accent Sensitive Cestrum Coefficient (ASCC) [10]. Speech recognition model is mainly Dynamic Time Warping (DTW) [11]; HMM [12]; Support Vector Machine (SVM) [13]; neural network [14]; Gauss Mixture Model (GMM) [15]. Identification of acoustic noise is mainly divided into two stages: feature extraction and pattern recognition. The extracted features in recognition should have the ability to distinguish among different names. In this study, we use the MFCC as the feature extraction from the name of the sound in the sample, and use the HMM to classify the different sound model. The algorithm uses a large number of speech data to build the speech model of identification entries. It can extract features from the speech which is recognized, and these features are matched with speech model. By comparing the matching rate, we can get the recognition results. By thronging the establishment of a speech database, we can obtain a robust statistical model, which can improve the recognition rate in various practical situations. This paper selects 20 vocabularies (name) as the speech samples, where these samples were respectively performed by 5 men and 5 women for voice recording and transmission. Experiments show that the system has certain robustness and a good recognition rate.

23.2 The Realization of Speech Recognition System Based on HMM

Speech recognition is a problem of pattern recognition. It is divided into training phase and recognition phase. In the training phase, we use names of the system as the training corpus. The system makes a digital processing for the training corpus, and set up each name of the template or model parameters according to the characteristic parameter reference. In the recognition phase, the system compares characteristic parameters extracted from the speech signal with that obtained in the training process reference model parameters. The system makes the decision according to some similarity criterion, and then gets the recognition result. For names identification, the characteristic parameters of the system are to be compared with the reference model of each name in the training process. And the corresponding reference model which has the minimum differentiation is considered to be the output. For name confirmation, the system compares the feature parameters extracted from the output speech with the reference model parameters, if the distance between them is less than a certain threshold, and then confirm. Figure 23.1 is the block diagram of the speech recognition. It consists of preprocessing, feature extraction, model training, pattern matching and judgment.

23.2.1 The Sound Source and Pretreatment of Samples

23.2.1.1 Sound Recording and Data Acquisition

The name of the recording material used in this study is recorded in the laboratory, and we set up a small speech database to record the voice library names sound 20. To make the HMM model has sufficient data for training, we choose each name 5 men and 5 women respectively for voice recording.

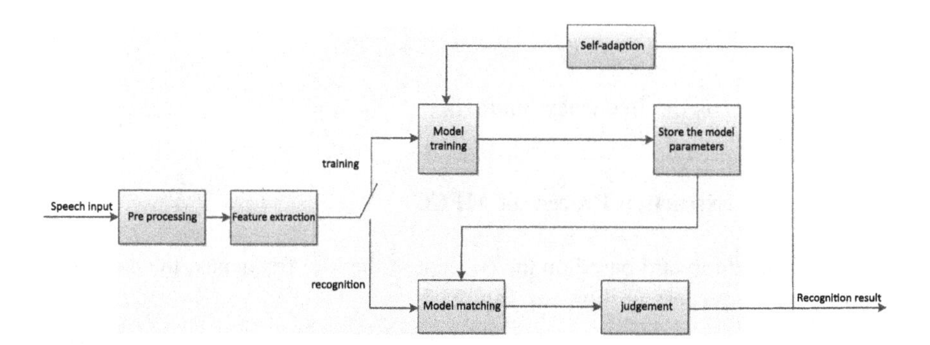

Fig. 23.1 The block diagram of the speech recognition

23.2.1.2 Pre Processing

Assuming all the names sound recording has been digitally quantified, and then the preprocessing includes normalization, pre-emphasis and segmentation.

Normalization: voice signal normalization is each sample divided by the amplitude of the signal's peak value, that:

$$\tilde{X}(n) = x(n)/\max x(n) \quad 0 \leq n < m - 1 \tag{23.1}$$

Among them, $x(n)$ is the original signal, $X(n)$ is the signal after normalization and m is the length of the signal.

23.2.2 MFCC Feature Extraction

Mel cestrum coefficient is a representation of the short-time energy spectrum of the sound. It is the logarithm power spectrum of nonlinear Mel frequency spectra standard linear cosine transform. Mel cestrum coefficient constitutes all the coefficient of the MFC. They originated from a kind of audio cestrum representation. The difference between the Mel cestrum and Cestrum is the frequency band of the uniform distribution of the Mel frequency is more close to the human auditory system.

23.2.2.1 Mel Frequency Campestral Coefficients (MFCC)

The human ear has different perception ability to different frequency of voice. The experiment shows that, when <1000 Hz, perception is linear with the frequency, while above 1000 Hz, perception and frequency are logarithmic relations. Therefore, people put forward the concept of the *Mel* frequency, its meaning is: 1 *Mel* is 1/1000 of the 1000 Hz tone awareness degree. The conversion formula between frequency and Mel frequency is:

$$Mel(f) = 2595 \lg(1 + f/700) \tag{23.2}$$

In the formula, f is the frequency, unit: Hz.

23.2.2.2 The Extraction Process of MFCC

MFCC is putted forward based on the concept of the Mel frequency, the extraction and calculation process is shown in Fig. 23.2.

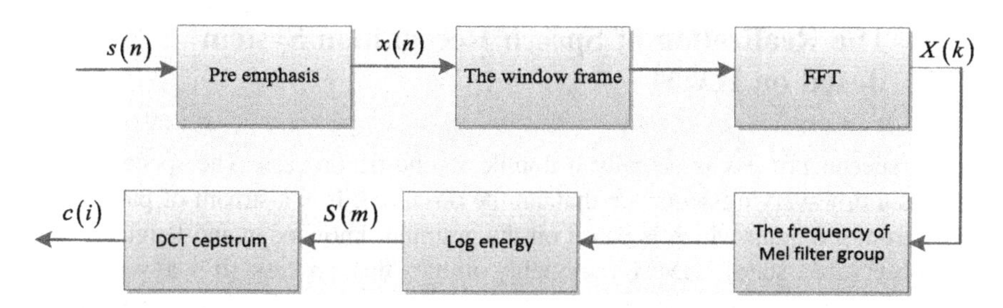

Fig. 23.2 The extraction process of MFCC

① Pre processing. The original speech signals after pre emphasis, framing, add window processing module, we can get the time-domain signal of each speech frame.

② FFT. The time-domain signal $x(n)$ after from 0 to N to form (usually from $N = 512$) sequence, and then through the DFT is obtained after linear spectrum of $X(k)$, the conversion formula is

$$X(k) = \sum_{n=0}^{N-1} x(n)e^{-j2\pi nkN}, 0 \leq n, k \leq N-1 \tag{23.3}$$

③ Mel frequency filter. The linear spectrum of $X(k)$ is obtained by the Mel filter modules Mel frequency spectrum, and logarithmic spectrum is obtained by the processing of the logarithmic spectrum $S(m)$.

④ Demand for energy. In order to make the results of spectral estimation has better robustness to noise and noise, the *Mel* spectrum of the filter through the *Mel* group gets the logarithmic energy. The linear spectrum of $X(k)$ to the logarithmic spectrum $S(m)$ of the total transfer function:

$$S(m) = \ln\left[\sum_{k=0}^{N-1} |X(k)|^2 H_m(k)\right], 0 \leq m < M \tag{23.4}$$

⑤ Discrete cosine transforms. The role of DCT is to transform the output filter campestral coefficients.
The logarithmic spectrum $S(m)$ after the (DCT) transforming cestrum domain can be obtained MFCC of $c(n)$

$$c(n) = \sum_{m=1}^{M-1} S(m) \cos\left[\frac{\pi n(m + 1/2)}{M}\right], 0 \leq m < M \tag{23.5}$$

23.3 The Realization of Speech Recognition System Based on HMM

Man's speech process is actually a double stochastic process. The speech signal itself is a time-varying sequence that can be observed. It is a stream of parameters flows from the brain which is based on the grammar knowledge and language (the non observable state). HMM reasonably mimic this process. It's a very good description of the non-stationary of speech signals and local stability. So it is an ideal model of the speech. From the whole speech, the human voice is a non-stationary random process, but if the whole speech is divided into several short speech signals, these short speech signals can be considered as a stationary process, so we can use the linear method to analyze these short speech signals. If the speech signals hidden Markov model is established, we can use different parameters to identify short time stationary signal, and also can track the conversion between them. So the problem of establishing the model of the speech rate and the acoustic variation is solved.

The speech recognition process is divided into two parts: one is the HMM training process, we can get the HMM speech recognition model, that is, we establish the basic recognition library. The second is the HMM recognition process, we can get speech recognition results from this process.

23.3.1 *HMM Training*

Initialization. HMM model $\lambda = (\pi, A, B)$ for each part parameter initialization

Using the training samples to calculate the forward probability and the backward probability, as shown in Figs. 23.3 and 23.4

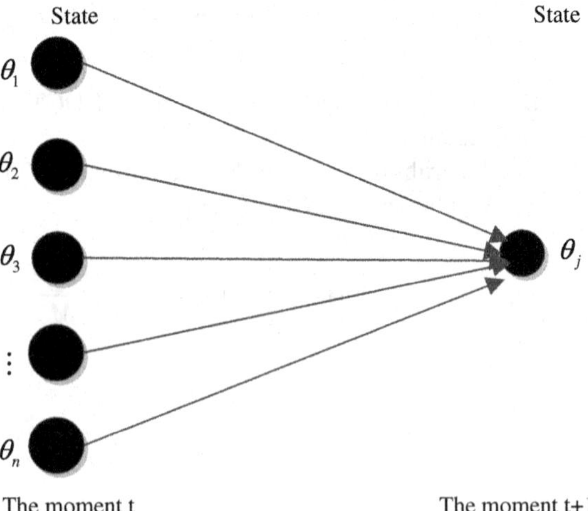

Fig. 23.3 To push the process forward probability

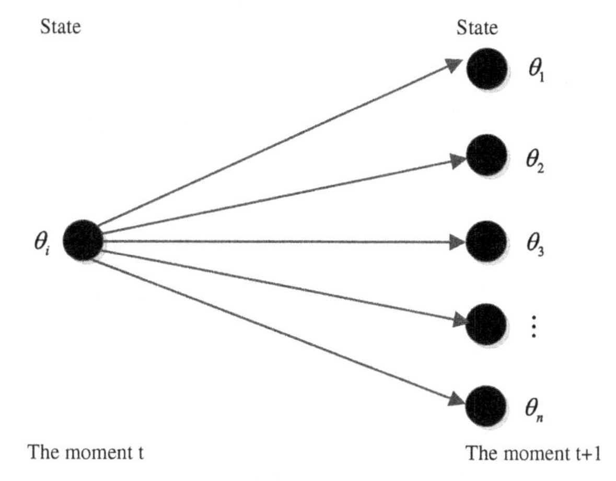

Fig. 23.4 After the probability of recursive process

HMM forward probability: $\alpha_t(i) = P(o_1, o_2, \ldots, o_t, \mathrm{q}_t = i/\lambda)$, it is said in the given observation sequence model under λ, the first t time for $\{o_1, o_2, \ldots, o_t\}$, and in the probability of state i at time t.

HMM after the probability: $\beta_t = P(o_{t+1}, o_{t+2}, \ldots, o_T, \mathrm{q}_t = i/\lambda)$, it is said in the given model under λ, from the time $t + 1$ started to observe the order of this section of the observation sequence $\{o_{t+1}, o_{t+2}, \ldots, o_T\}$, and in the probability of state i at time t.

A new set of parameters of HMM is obtained by the revaluation formula is $\lambda = (\pi, A, B)$.

23.3.1.1 Calculation of the Probability of $P(O/\lambda)$

If the distance between $P(O/\lambda)$ and $P(O/\lambda)$ is too large, then return to step 2, iteratively calculated, until the HMM model parameters are no significant change.

23.3.2 Identification of HMM Model

First, 8000 Hz is carried out on the identify voice, each point of 8 bit quantization. Second, the speech signals inputted must be pre-processed, and judged the starting point of voice roughly. If the recognition of the speech part on the features extraction, so we get the observation sequence o. These steps are similar with training part, so as not to repeat. According to the speech recognition grammar network diagram, Fig. 23.5: regarded as the probability of occurrence of keywords and corpus of roughly equal, $\left(w_{k1} = w_{k2} = \ldots w_{k20} = w_{f1} = w_{f2}\right)$, frame synchronization using Viterbi decoding algorithm according to the best way to match the observation sequence with the reference template sequence, and get the final recognition results.

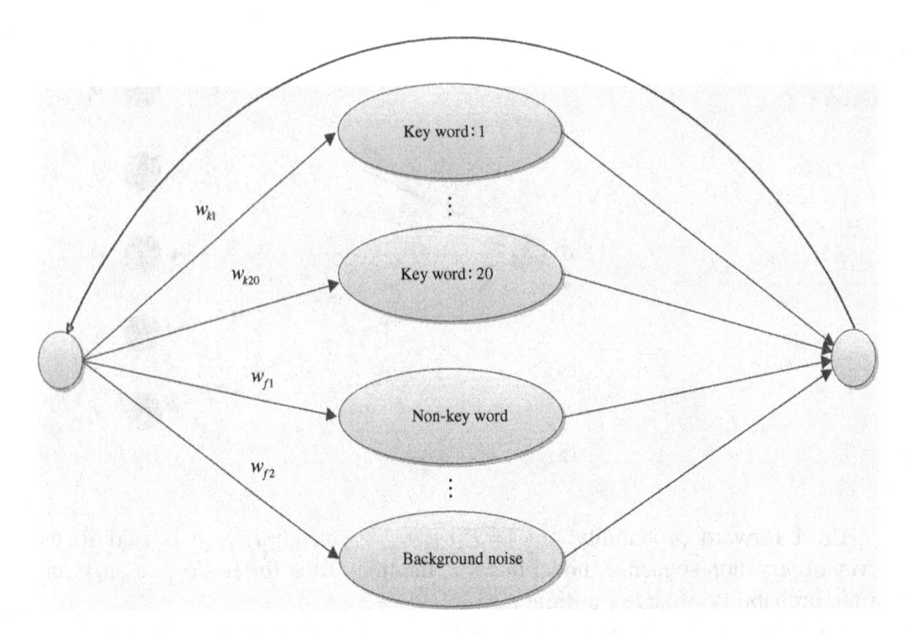

Fig. 23.5 The speech recognition grammar network diagram

23.4 Analysis of Experimental Results

In this paper, the experiments are carried out on the PC machine on Windows XP. And we use VS2010 to build the interface, as shown in Fig. 23.6.

This experiment uses OLE/COM to achieve the operation of the Excel table. Open the Excel interface as shown in Fig. 23.7. The program is divided into two parts. The first part is keywords recognition for speech input: speak the name of the person through the microphone, after realizing the conversion of speech to the text, then locate the name or name corresponding cell. The other part is speech inputting method: after speech locating the cell needed, speech input scores through the microphone. Finally achieve the conversion of speech to text.

Multiple input in this system, experiments show that the recognition rate of this system can reach more than 80 %, then optimized to the speech model of the system: each name records up to 30 times, and the number of states of each word is improved from 10 to 12. And then input in this system repeatedly, we found that the recognition rate has improved by 10–15 %. The experimental results show that: if want to improve the recognition rate of this system, in the speech modeling, each name of the tape in the library of recognition is the more the better, and it may be appropriate to increase the number of states in each word, it will also improve the recognition rate of the model.

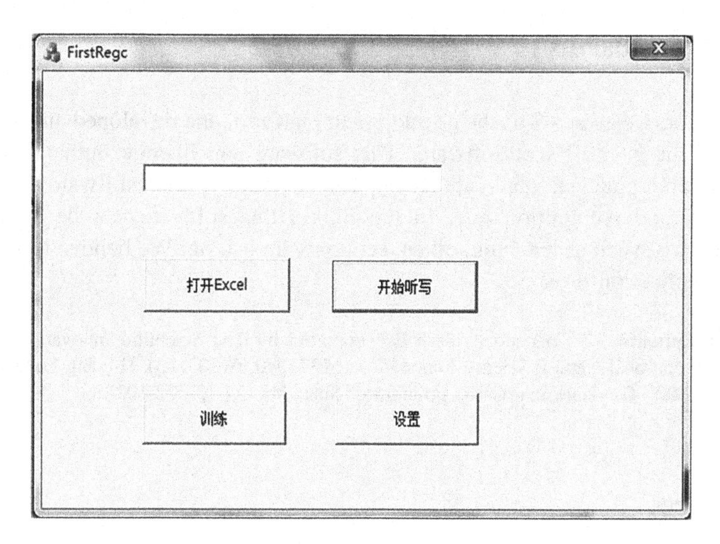

Fig. 23.6 The operation interface

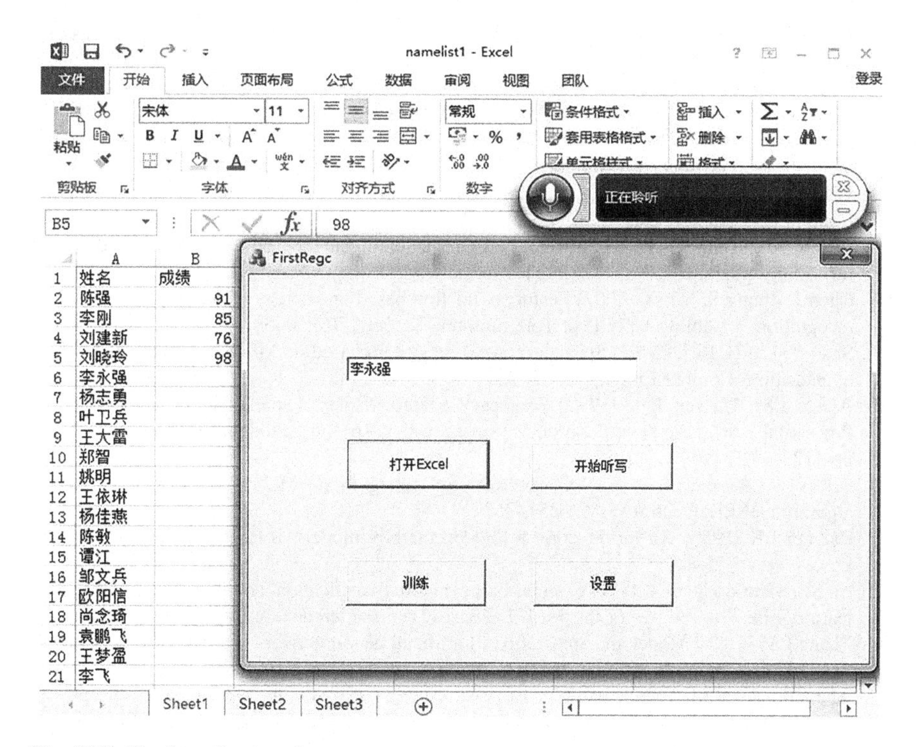

Fig. 23.7 The inputting interface

23.5 Conclusions

In this paper, we used VS as the development platform, and developed an automatic speech input in the Excel software. This software can liberate human eyes and hands in the practical application. Experiment shows, this software has good robustness and recognition rate. In the future, this software can be applied to various fields, such as teaching, office, economy and so on. We believe it will have a good application prospect.

Acknowledgments This paper was partially supported by The Scientific Innovation program (13ZZ115), National Natural Science Foundation (61374040, 61203143), Hujiang Foundation of China (C14002), Graduate Innovation program of Shanghai (54-13-302-102).

References

1. Yan Y (2013) The latest application of linguistic acoustics. Chin J Acoust 35(2):241–247
2. Cai L, Huang D, Cai R (2003) Technology and application of modern speech. Tsinghua University Press, Beijing
3. Kim DS, Lee SY, Kil RM (1999) Auditory processing of speech signals for robot speech recognition in real-world noisy environments. IEEE Trans Speech Audio Process 7(1):55–69
4. Chen S, Jiang Q (1993) Evaluation of speech technology for enhancing performance of man–machine systems. Space Med Med Eng 6(1):31–38
5. JinPu Xu, YePing Zhu (2015) The price of agricultural products information acquisition method based on speech recognition. Sci Agric Sinica 48(3):449–459
6. Wang Z, Tang Y, Han P (2012) Using voice recognition of forest road detection method. Comput Eng Appl 48(30)
7. ZiHao Xu, TengFei Zhang (2012) Design of intelligent speech recognition and home furnishing system based on wireless sensor network. Comput Measur Control 20(1):180–182
8. Chen J, Zhang L, Wu X (2007) Feature extraction based on wavelet packet-LPCC in speaker recognition. J Nanjing Univ Post Telecommun Nat Sci 27(6):54–56
9. Shao Y, Liu B, Li Z (2002) Speaker recognition system based on MFCC and weighted vector quantization. Comput Eng Appl 127–128
10. Arslan LM, Hansen JHL (1997) Frequency characteristics of foreign accented speech. In: Proceedings of international conference on acoustics, speech and signal processing, vol 97, pp 1123–1127 (1997)
11. Myers C, Rabiner L (1981) Connected digit recognition using a level building DTW algorithm. IEEE Trans ASSP 29:351–363
12. Rabiner LR (1986) An introduction to hidden markov models. IEEE Acoust Speech Signal Process Mag 1:5–16
13. Li SZ, Guo-dong G (2000) Content-based audio classification and retrieval using SVM learning. In: Proceedings of the 1st IEEE pacific-rim conference on multimedia
14. Zhang LM (1994) Model and application of artificial neural network. Fudan University Press, Shang Hai, pp 23–123
15. Abu El-Yazeed MF, El Gamal MA, El Ayadi MMH (2007) On the determination of optimal model order for GMM—based text-independent speaker identification. J Appl Sig Process 8:1078–1087

Chapter 24
Decoupling Control of Manipulators Based on Active Disturbance Rejection

Xiaoming Song and Chaoli Wang

Abstract This paper discusses the decoupling control problem of manipulators and proposes a novel control method. This method does not need an accurate model of the robots and only needs input and output information of the robot. The system will be first transformed into a certain integral form by an extensive state observer (ESO) which is used to accurately estimate the systems various states and disturbances inside and outside. Then the decoupling control is investigated by using feed forward compensation. In the end, PD control with gravity compensation and active disturbance rejection control are provided to illustrate effectiveness of the controllers in the simulation.

Keywords Active disturbance rejection control (ADRC) · Manipulator control · Decoupling control

24.1 Introduction

The traditional manipulator control relies heavily on model, and the manipulator system is a multi-input multi-output nonlinear time-varying system with strong coupling and nonlinear dynamics. The system has not only nonlinear characteristics but also unknown uncertainty which is difficult to be modelled exactly. Thus, model-based control methods are not applicable to robot manipulator systems. An et al. [1] and Kelly et al. [11] described PD gravity compensation control for an n degree of freedom (DOF) manipulator. It was capable of achieving the position control, while tuning the parameters was roughly complex. Although the performance improved so much, it was still unsatisfying. Liu [2] designed an adaptive controller that adjusted control law or control parameters in real-time, while online identification of parameters demanded too much computation. Facing non-parametric uncertainties the

X. Song · C. Wang (✉)
Control Science and Engineering Department, University of Shanghai for Science and Technology, Shanghai 200093, China
e-mail: clclwang@126.com

Y. Jia et al. (eds.), *Proceedings of the 2015 Chinese Intelligent Systems Conference*, Lecture Notes in Electrical Engineering 360,
DOI 10.1007/978-3-662-48365-7_24

program was difficult to ensure the system stable or achieve ideal performance. Liu et al. [3] proposed an adaptive genetic algorithm (AGA) to optimize ADRC parameters tuning on a six DOF manipulator, while the region of tuning parameters should be computed before executing programs.

The traditional PID has great difficulty in achieving a desired performance, owing to its multi-variable strong coupling, parameter perturbation and external disturbance input. Facing a set of nonlinear uncertain objects we design a nonlinear controller consisting of three parts: nonlinear tracking differentiator (TD) [4], ESO [5]and nonlinear state error feedback (NLSEF) [6]. The controller does not require a precise mathematical model of the object, and only uses ESO to estimate system states. Observer's input derives from the input and output of the object, which demonstrates perfect decoupling characteristics. After ESO real-time estimates the impact of internal model uncertainties and unknown external disturbance, the nonlinear feedback controller will compensate them in controlling. Even though there is a model uncertainty in system, it will achieve the purpose.

This paper is based on a two DOF manipulator platform. Using ADRC decoupling features constructs a decoupling controller. Then we analyse the dynamic model of the robot. The ARDC controller includes two independent extensive state observers, two tracking differentiators, and two nonlinear state feedback controllers. Finally we use parametric bandwidth adjustment [7] to tune controller parameters. The performances of the proposed controller will provide an applicability in control engineering.

24.2 Background

24.2.1 Structure of Manipulators

For simplicity, assuming manipulators' movement is measurable. Our purpose is to design a controller for controlling the robot manipulator to track the set points. Two DOF manipulator is a typical example of manipulators control, as shown in the Fig. 24.1.

In Fig. 24.1, in accordance with D-H method, we establish a manipulator coordinate system. i, θ_i, d_i, a_i, α_i are the joint label, the joint angle, the joint distance, the link length and the link twist angle of link i respectively. Where $\alpha_i \in (-\pi, \pi]$, $\theta_i \in (-\pi, \pi]$ [8]. Parameters of the manipulator are acquired by D-H method, as shown below (Table 24.1).

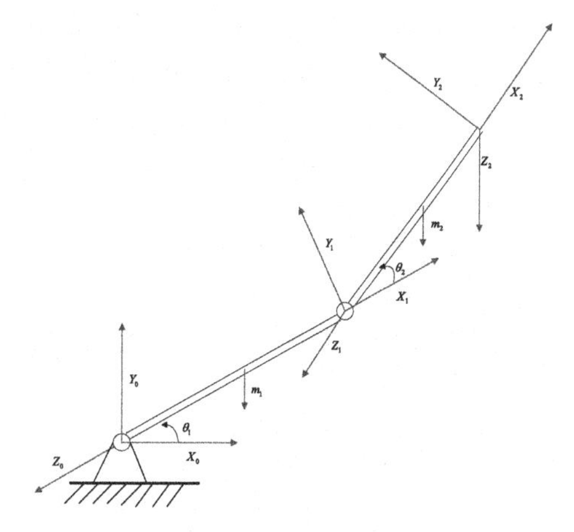

Fig. 24.1 The manipulator structure

Table 24.1 Parameters of two DOF manipulators

i	d_i	a_i	α_i	θ_i
1	0	l_1	θ_1	0
2	0	l_2	θ_2	0

24.2.2 Dynamic Model

It is available for getting the position of the robot manipulator and each position is determined by joint variable p. When considering the uncertainty of the system, the system dynamics model of the manipulator [8] can be deduced by

$$\begin{cases} H(p)\ddot{p} + C(p,\dot{p})\dot{p} + \mathrm{G}(p) = U \\ Y = h(p) \end{cases} \tag{24.1}$$

with

$$\begin{cases} H(p) = \begin{bmatrix} \frac{1}{3}m_2\left(3l_1{}^2 + 3l_1l_2\cos(p_2) + l_2{}^2\right) + \frac{1}{3}m_1l_1{}^2 & \frac{1}{6}m_2l_2\left(2l_2 + 3l_1\cos(p_2)\right) \\ \frac{1}{6}m_2l_2\left(2l_2 + 3l_1\cos(p_2)\right) & \frac{1}{3}m_2l_2{}^2 \end{bmatrix} \\ C(p,\dot{p}) = \begin{bmatrix} -\frac{1}{3}m_2l_1l_2\sin(p_2)\dot{p}_2 & -\frac{1}{2}m_2l_1l_2\sin(p_2)(\dot{p}_1 + \dot{p}_2) \\ \frac{1}{2}m_2l_1l_2\sin(p_2)\dot{p}_1 & 0 \end{bmatrix} \\ G(p) = \begin{bmatrix} \frac{1}{2}g\left(m_1l_1\cos(p_1)\right) + 2m_2l_1\cos(p_1) + m_2l_2(p_1 + p_2) \\ \frac{1}{2}m_2gl_2\cos(p_1 + p_2) \end{bmatrix} \end{cases} \tag{24.2}$$

where $p = [p_1, p_2]^T$. $\dot{p}_1, \dot{p}_2, p_1, p_2$ represent respectively the velocity and the position of the link. Matrix H denotes the inertia matrix. Matrix C indicates the centrifugal-Coriolis matrix. Matrix G refers to gravity effects. m_i, l_i are the weight and length of the link i, respectively. g represents the acceleration of gravity, $U = \left[u_1\ u_2\right]^T$ is the torque input vector.

24.3 Controller Design

24.3.1 Active Disturbance Rejection Control

There are two kinds of disturbances: external disturbances and unknown internal dynamics in model based control. These disturbances require to be modelled exactly while it is not capable in reality engineering domain sometimes [12]. If we define the gravity term and system parameters uncertainties as the entire disturbance ξ, we can rewrite Eq. (24.1) as:

$$\ddot{p} = H(p)^{-1}\left(U - C\left(p, \dot{p}\right)\dot{p} - G\left(p\right)\right) = \xi + b\tau, b \neq 0. \tag{24.3}$$

where $H\left(p\right)$ is reversible in sampling moment [9]. We rewrite Eq. (24.3) in a decoupling structure:

$$\ddot{p}_i = \xi_i + b_i * \tau_{i1}. \tag{24.4}$$

where p_i τ_{i1} ξ_i b_i represent position, control law, disturbances and controller gain of link i, respectively. When the system tracks step signal and square wave signal because the object output is dynamic, there exists a inertia. However, the changes cannot shift quickly, which means that it is impossible to make a slow-changing signal track a rapidly-changing signal [9].

When the error is large, it will increase the controller gain to track quickly, which causes much overshoot and direct impact on the initial system. To overcome those defects we design a tracking differentiator (TD) [4] as shown in formula (24.5) to make the input signal $r_i\left(t\right)$ smoother, to ease the contradiction between less overshoot and quick system response. TD arranges the transition process by using target signal $r_i\left(t\right)$

$$\begin{cases} T_{i1}\left(t+1\right) = T_{i1}\left(t\right) + h * T_{i2}\left(t\right) \\ T_{i2}\left(t+1\right) = T_{i2}\left(t\right) + h * fhan\left(T_{i1}\left(t\right) - r_i\left(t\right), T_{i2}\left(t\right), \delta_i, h_{i0}\right) \end{cases} \tag{24.5}$$

where $T_{i1}\left(t\right)$, $T_{i2}\left(t\right)$ track $r_i\left(t\right)$ and the derivative of $r_i\left(t\right)$, respectively. h is sampling duration. δ_i is the speed factor to track the differential. h_{i0} is filter factor. When the input signal is polluted by noise, increasing filter factor can enhance filtering effect [9]. *fhan* in (24.5) is given by (24.6) and (24.7)

$$\begin{cases} s_{i1}(t) = T_{i1}(t) - r_i(t) \\ s_{i2}(t) = T_{i2}(t) \\ \delta = \delta_i \end{cases} \tag{24.6}$$

$$\begin{cases} d = \delta * h_{i0} \\ d_0 = h_{i0} * d \\ \psi(t) = s_{i1}(t) + h_{i0} * s_{i2}(t) \\ a_0(t) = \sqrt{d^2 + 8 * \delta|\psi(t)|} \\ a_i(t) = \begin{cases} s_{i2}(t) + \frac{(a_0(t)-d)}{2} sign(\psi(t)), |\psi(t)| > d_0 \\ s_{i2}(t) + \frac{\psi(t)}{h_{i0}}, |\psi(t)| \le d_0 \end{cases} \\ fhan = -\begin{cases} \delta * sign(a_i(t)), |a_i(t)| > d \\ \delta * \frac{a_i(t)}{d}, |a_i(t)| \le d. \end{cases} \end{cases} \tag{24.7}$$

The input signal is $r_i(t)$. It is easy to make $T_{i1}(t)$ track $r_i(t)$, and $T_{i2}(t)$ track the derivative of $r_i(t)$ by using TD. We define the internal uncertainties in the robot manipulator system (coupled joint terms) and external disturbances term (gravity term) as an independent extended state. We design an ESO to estimate the velocity, unknown disturbances of link i because ADRC has decoupling characteristics that the observer input is only concerned with output and input of the link.

$$\begin{cases} es_{i1}(t) = \epsilon_{i1}(t) - y_{i1}(t) \\ \epsilon_{i1}(t+1) = \epsilon_{i1}(t) + h * (\epsilon_{i2}(t) - \beta_{i1} * es_{i1}(t)) \\ \epsilon_{i2}(t+1) = \epsilon_{i2}(t) + h * (\epsilon_{i3}(t) - \beta_{i2} * es_{i1}(t) + b_i\tau_{i1}) \\ \epsilon_{i3}(t+1) = \epsilon_{i3}(t) - h\beta_{i3} * es_{i1}(t) \end{cases} \tag{24.8}$$

where y_{i1} is output of the link i. Article [7] proposed a linearizing, parameterized observer and a tuning method to tune $\beta_{i1}, \beta_{i2}, \beta_{i3}$. If the observer's bandwidth is w_{io}, It can utilize pole placement method [7, 10] to tune observer's parameters (i.e. $\beta_{i1} = 3w_{io}, \beta_{i2} = 3w_{io}^2, \beta_{i3} = w_{io}^3$). Therefore $\epsilon_{i1}(t)$, $\epsilon_{i2}(t)$, $\epsilon_{i3}(t)$ estimate $p_i(t)$, $\dot{p}_i(t)$, ξ_i, respectively. $\epsilon_{i3}(t)$ represents the extended state to estimate the disturbance of the link i which can be used in controller design part.

Traditional linear PID control is the most widely used while it is not the most effective combination form. However, the nonlinear PID control will be faster and more accurate under the same gain. It effectively improves the adaptability and robustness of the controller [6]. Controller design based on non-linear state error feedback law (NLSEF) (24.9) is shown below

$$\begin{cases} e_{i1}(t) = T_{i1}(t) - \epsilon_{i1}(t) \\ e_{i2}(t) = T_{i2}(t) - \epsilon_{i2}(t) \\ \tau_{i0}(t) = kp_{i1}fal(e_{i1}(t), \alpha_{i1}, \sigma_1) + kd_{i1}fal(e_{i2}(t), \alpha_{i2}, \sigma_2) \\ \tau_{i1}(t) = \tau_{i0}(k) - \epsilon_{i3}(t)/b_i \end{cases} \tag{24.9}$$

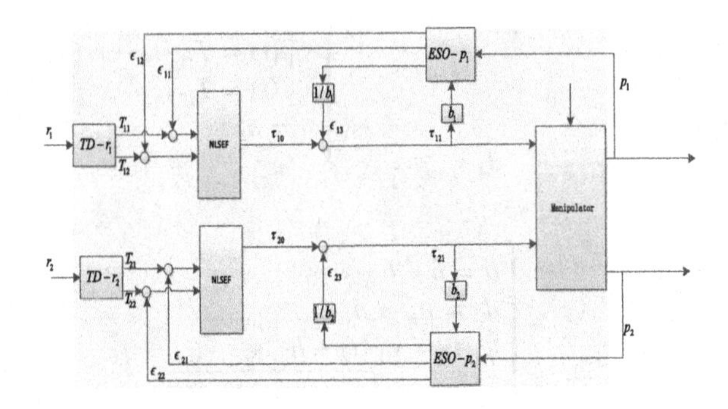

Fig. 24.2 Manipulator structure of ADRC

where T_{i1}, T_{i2} are TD outputs of link i. ϵ_{i1}, ϵ_{i2}, ϵ_{i3} are ESO outputs of link i. with $0 < \alpha_{i1} < 1 < \alpha_{i2}$ [9]. Based on parametric bandwidth controller design methods [7, 9], there are two equations $kp_{i1} = 2\omega_{ic}$, $kd_{i1} = \omega_{ic}{}^2$ in two order system, where ω_{ic} is the bandwidth of the controller. Function of (24.10) is introduced to avoid high-frequency vibration. Power function $|ec|^{\gamma} * sign\,(ec)$ is a saturation function [9] to correct the error. d is an interval length of linear segments

$$fal\,(ec, \gamma, d) = \begin{cases} |ec|^{d} * sign\,(ec)\,, ec > d \\ ec/(d)^{\gamma-1}\,, ec \leq d. \end{cases} \tag{24.10}$$

The entire structure of the controller is shown in Fig. 24.2.

24.4 Simulation Results

To illustrate the effectiveness of the controller, simulation is carried out between PD control with gravity compensation and ADRC on the set-point tracking. h (sampling time) is 0.001 s. We use the step signal of set-points as input signals r_1, r_2. Parameters of the manipulator are chosen as: $m_1 = m_2 = 1\,\text{kg}$, $l_1 = 1\text{m}$, $l_2 = 0.8\,\text{m}$, $g = 9.8\,\text{m/s}^2$. Parameters of PD control with gravity compensation are given $kp_1 = 30$, $kd_1 = 30$, $kp_2 = 30$, $kd_2 = 30$. Transient response curves are shown in Figs. 24.4a and 24.5a. Parameters of the ADRC are given as below: bandwidths of the controller are $\omega_{1c} = 80$, $\omega_{2c} = 100$. Bandwidths of the observer are $\omega_{1o} = 100$, $\omega_{2o} = 50$. Speed factors of tracking differentiator are $\delta_1 = 10, \delta_2 = 10$. Filter factors are $h_{10} = h, h_{20} = h$. Parameters of nonlinear feedback controller: the exponent of the power are $\alpha_{i1} = 3/4$, $\alpha_{i2} = 3/2$. Linear segment interval lengths are $\sigma_1 = 2h, \sigma_2 = 2h$. Controller gains are $b_1 = 1, b_2 = 1$. System responses are shown in Figs. 24.3, 24.4, 24.5. From the comparison, in Fig. 24.3 PD control with gravity compensation has an accurate gravity model, it still has a significant overshoot.

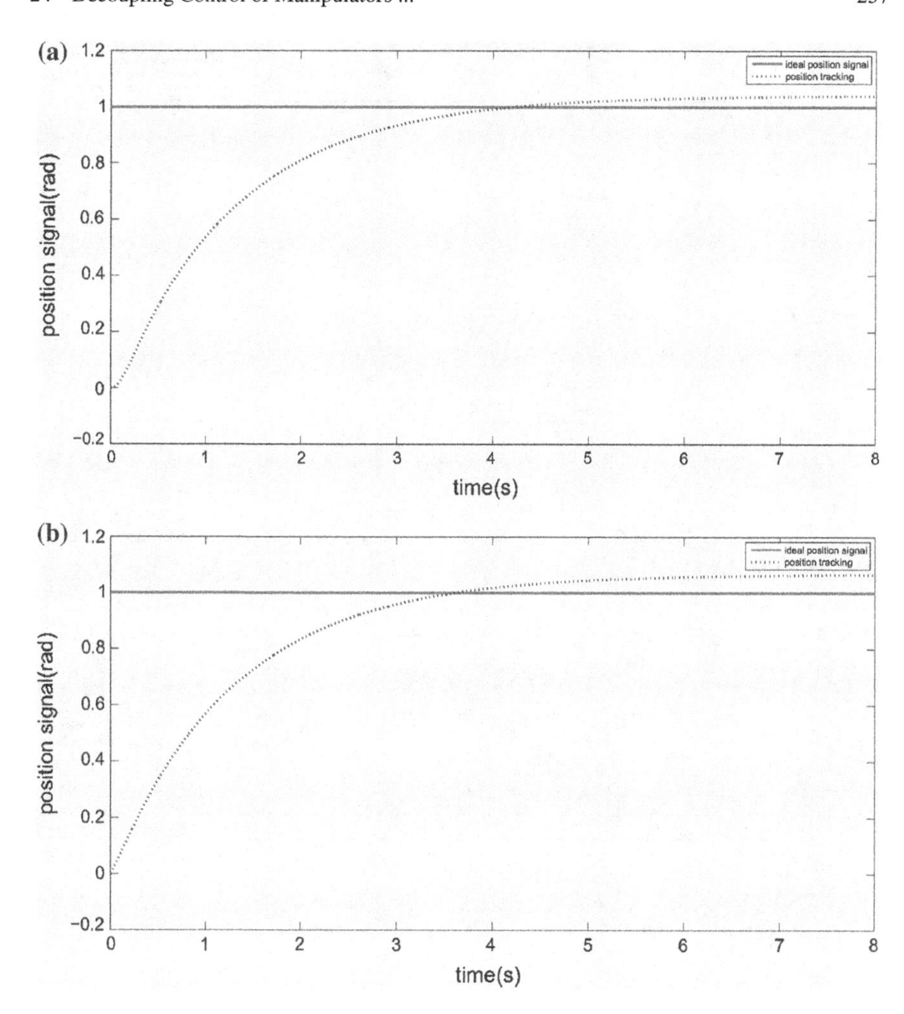

Fig. 24.3 System responses for PD control with gravity compensation, **a** link 1, **b** link 2

However, the transition response of ADRC achieves a perfect coincidence to the designed curve in Figs. 24.4a and 24.5a. Not only has it a better quickness but also it can suppress the overshoot of the system. The performance is caused by efficiently estimating the system states and disturbances by ESO, then the nonlinear feedback controller cancels the disturbances (Figs. 24.4b and 24.5b). Moreover, the reason why ADRC indicates a superior performance is that it has a proper transition process.

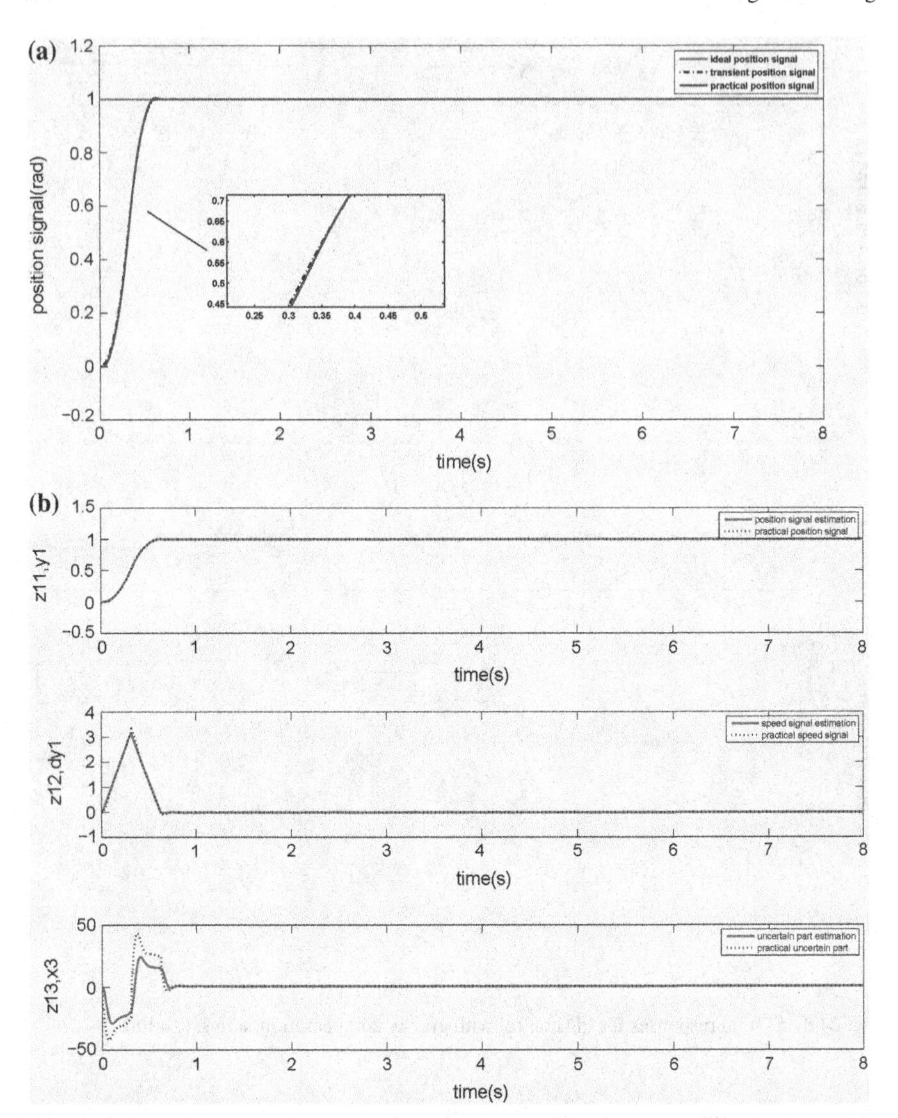

Fig. 24.4 System responses for the active disturbance rejection control of the link 1. **a** Position details of link 1. **b** States details of link 1

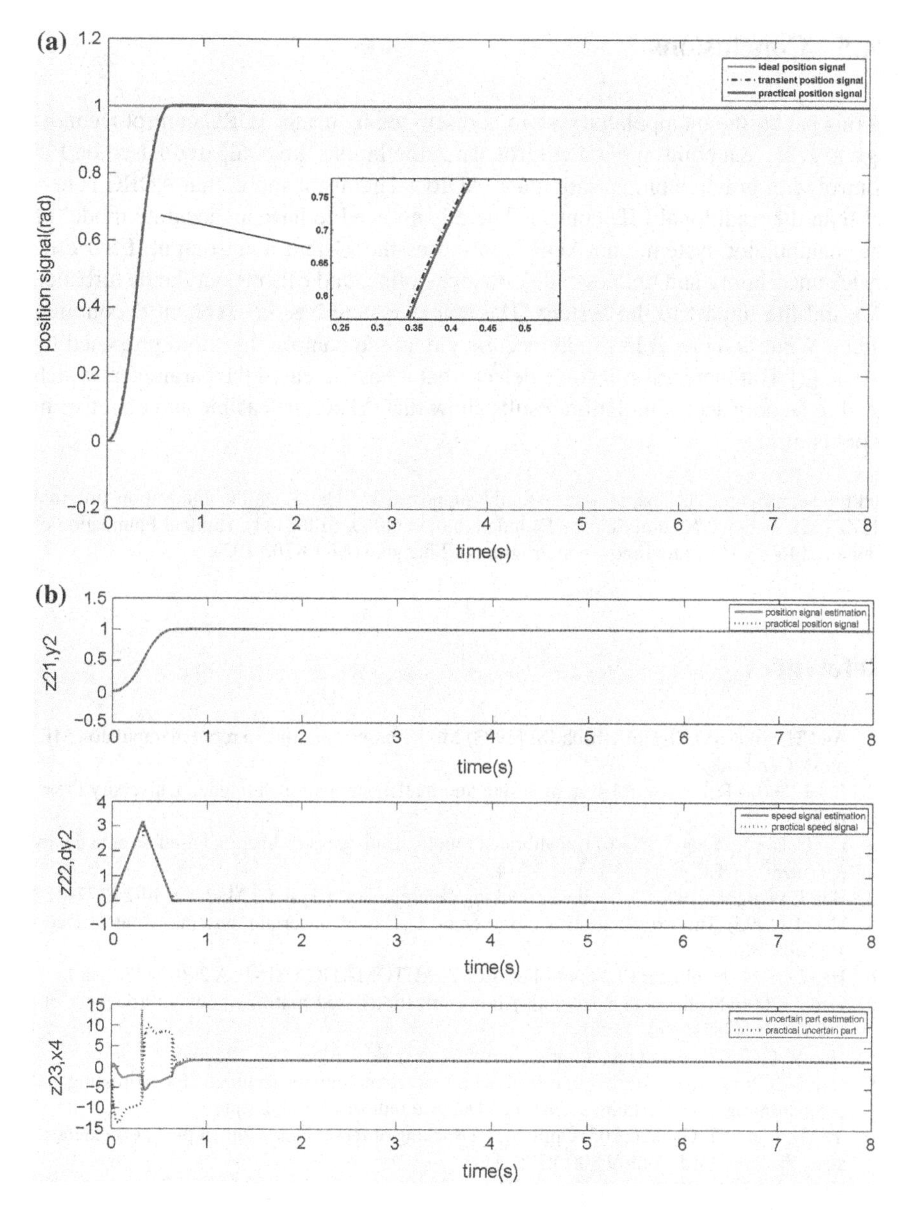

Fig. 24.5 System responses for the active disturbance rejection control of the link 2. **a** Position details of link 2. **b** States details of link 2

24.5 Conclusions

In this paper, the manipulator system is discussed by using ADRC control technology to achieve a point to point control. The simulations are compared based on PD control with gravity compensation and ADRC. The result shows that ADRC is better than the traditional PID control. There is no need to have an accurate model of the manipulator system, and ADRC only uses the system input-output. ESO estimates uncertainty and unknown disturbance online, and cancels out the disturbance to avoid the impact to the system. The whole system has an excellent decoupling effect. What is more, it has more accuracy than the control algorithm proposed by article [2]. But there are still some defects that it has increased the parameters which need to be adjusted. Simulation results show that ADRC is feasible and effective in robot control.

Acknowledgments This paper was partially supported by The Scientific Innovation program (13ZZ115), National Natural Science Foundation (61374040, 61203143), Hujiang Foundation of China (C14002), Graduate Innovation Program of Shanghai (54-13-302-102).

References

1. An CH, Atkeson CG, Hollerbach JM (1988) Model-based control of a robot manipulator. MIT press, Cambridge
2. Liu J (2008) Robot control system design and matlab simulation. Tsinghua University Press, Beijing
3. Liu D, Liu X, Yang Y (2006) Uncalibrated robotic hand-eye coordination based on genetically optimized ADRC. Robot 28(5):510–514
4. Han J, Wang W (1994) A nonlinear tracking differentiator. J Syst Sci Math Sci 14(2):177–183
5. Han J (1995) The extended state observer of a class of uncertain systems. Control Decis 10(1):85–88
6. Han J (1994) Nonlinear PID controller. ACTA AUTOMATICA SINICA 20(04):487–490
7. Gao Z (2006) Scaling and bandwidth-parameterization based controller tuning. Proc Am Control Conf 6:4989–4996
8. Huo W (2005) Robot dynamics and control. Higher Education Press, Beijing
9. Han J (2008) Active disturbance rejection control technology the technique for estimating and compensating the uncertainties. National Defense Industry Press, Beijing
10. Yoo D, Yau SST, Gao Z (2007) Optimal fast tracking observer bandwidth of the linear extended state observer. Int J Control 80(1):102–111
11. Kelly R, Davila VS, Lora A (2005) PD control with gravity compensation. Control of robot manipulators in joint space. Springer, London, pp 157–169
12. Zhao S, Zheng Q, Gao Z (2012) On model-free accommodation of actuator nonlinearities. World Congress on Intelligent Control and Automation. pp. 2897–2902

Chapter 25
Sensor Fault Detection of Double Tank Control System Based on Principal Component Analysis

Hailian Du, Yubin Liu, Wenxia Du and Xiaojing Fan

Abstract Production process system is a dynamic process, so whether the dynamic process' sensor is faulted or not is determined through the method of various sensor data acquisition and analysis. The double water tank data processing and fault diagnosis model was established according to the basic method of principal component analysis theory and its application research in the field of fault diagnosis. The test data was input into the model, so whether there was a failure was determined by comparing thresholds, and which sensor and what kind of fault are determined. The effectiveness was proved by the simulation result.

Keywords Principal component analysis · Sensor · Fault detection

25.1 Introduction

With the development of modern industry, more and more industrial automation involves more widely, and has also made remarkable progress in recent decade. Process control is closely related to the daily life such as in textile, ceramics, food, papermaking, and also covering power, chemical industry, light industry, aerospace and petroleum and so on, which are very important industrial departments of the national economy. In the actual production, more and more high even close to the limit parameter is set in the production. The product quality and the higher production efficiency has been the purpose of people. The stability of the production system and security of the production process has been paid more attention.

H. Du (✉) · W. Du · X. Fan
Electrical Department of Hebei, Normal University, Shijiazhuang 050024, China
e-mail: duhailian@126.com

Y. Liu
The Department of Computer Science, Tangshan Normal University,
Tangshan 063000, China

Y. Jia et al. (eds.), *Proceedings of the 2015 Chinese Intelligent Systems Conference*, Lecture Notes in Electrical Engineering 360,
DOI 10.1007/978-3-662-48365-7_25

When the sensor is malfunctioned or abnormal, the output data may be unreliable or incomplete, which results to insufficient computing data. Contacting the actual industrial process, it is difficult to establish a precise analytical model of complex system. The principal component analysis (PCA) method was processed and studied in this article, which was a method of fault diagnosis for sensors. Although there is no accurate model for fault judgment, but it is suited for industrial production process.

In order to overcome the diagnosis difficulties, a large number of fault detection methods have been developed. Multivariate statistical process methods such as principal component analysis (PCA) [1] and partial least squares (PLS) [2] have been implemented more frequently in the fault detection and diagnosis schemes [3–5]. Yoo et al. [6] used PCA model and a sensor validation model to generate a reconstructed sensor value which has demonstrated its enhanced monitoring performance. Hua et al. [7] presented a diagnostic analysis and implementation study based on PCA method for the fuel cell system, which was successfully implemented for detecting a fuel cell stack sensor network failure in the fuel cell bus fleet. Moon et al. [8] developed an operational map for the objective diagnosis of the process operating states of a municipal wastewater treatment plant, in which a K-means clustering analysis and a Fisher's linear discriminant analysis were applied.

25.2 Principal Component Analysis

Principal component analysis is a multivariate statistical method, which was put forward by Karl Pearson who was studying on how to find the best straight line and plane fitting. The earliest theory can be traced back to the beginning of the twentieth Century, after several decades, PCA was further modified and innovated and promoted by Hotelling. The scientific research personnel calculated the PCA from the different aspects, and applied to the different field [9–12].

25.2.1 The Basic Idea of Principal Component Analysis

n is sampling point, and m is measured variable, which constitute a matrix of $n*m$. The maximum number of principal component is m. In general, the number of principal component is a, which is less than m. The principal component is strictly linear irrelevant variables, which can represent the amount of information that the m measurement variable expression. In the PCA method, how much the amount of information can be provided is described by the cumulative contribution rate.

The original data vector $\mathbf{X} = (x_1, x_2, \ldots, x_m)^T$, there are m measurement variables of $\mathbf{X}$. It is worth noting that in the practical industrial production, measurement unit of each measurement variables may not be identical. Therefore, the original data matrix must be processed standard. Using the formula $\mathbf{X} = [\mathbf{X} - \mathbf{I}u^T]D_\sigma^{-\frac{1}{2}}$, in the formula, $\mathbf{I}$ is a full rank matrix with all elements of 1, $\boldsymbol{u} = (u_1, u_2, \ldots, u_m)^T$ is the mean vector of the original data matrix $\mathbf{X}$, $D_\sigma = \mathrm{diag}(\sigma_1^2, \sigma_2^2, \ldots \sigma_m^2)$ is the covariance matrix of the original data matrix $\mathbf{X}$.

The process of standardization is the foundation, after the establishment of good foundation the principal components can be computed in the next step. Now assume that $\mathbf{X}$ is the normalized data matrix from the original data matrix, and is composed of n sampling samples, namely $\mathbf{X}$ is $n \times m$ matrix.

The correlation coefficient matrix $\mathbf{R}$ is solved from the standardized data matrix $\mathbf{X}$. $\mathbf{R}$ can be calculated according to the formula $\mathbf{R} = (r_{ij})^{m \times m} = \mathbf{X}^T\mathbf{X}/n$. Obviously $\mathbf{R}$ is positive definite matrix, and the covariance matrix of $\mathbf{X}$. After coefficient of correlation matrix $\mathbf{R}$ is gotten, the non negative eigenvalue and the corresponding eigenvector are obtained according to the relevant knowledge of linear algebra. The eigenvalue is arranged from big to small as $\lambda_1 \geq \lambda_2 \geq \cdots \lambda_m \geq 0$, the corresponding eigenvector is denoted as $\boldsymbol{p_1}, \boldsymbol{p_2}, \ldots, \boldsymbol{p_m}$.

After calculating the eigenvalue and eigenvector, the principal component $\boldsymbol{t}_i (i = 1, 2, \ldots, m)$ can be expressed as follows:

$$\boldsymbol{t}_i = \boldsymbol{p}_i^T \boldsymbol{x} = \boldsymbol{p}_{i1}\boldsymbol{x}_1 + \boldsymbol{p}_{i2}\boldsymbol{x}_2 + \cdots + \boldsymbol{p}_{im}\boldsymbol{x}_m \tag{25.1}$$

In the equation, $\boldsymbol{t}_i (i = 1, 2, \ldots, m)$ is the ith principal component of $\mathbf{X}$; $\boldsymbol{p}_i^T$ is the principal component coefficient vector of $\mathbf{X}$, also known as the load vector. Where $\boldsymbol{t}_i$ and $\boldsymbol{t}_j$ are not related, namely cov $(\boldsymbol{t}_i, \boldsymbol{t}_j) = 0$, $(i \neq j)$.

As previously mentioned, the cumulative contribution rate is used to calculate the number of principal components value of a by the PCA method. The contribution of $\boldsymbol{t}_i$ is δ_i, which is the proportion of correlation coefficient matrix $\mathbf{R}$ value in the entire eigenvalue sum. η_a is the cumulative contribution rate of the former a principal component, namely the sum of contribution rate of the former a principal component.

$$\delta_i = \frac{\lambda_i}{\sum_{j=1}^{m} \lambda_j} \tag{25.2}$$

$$\eta_a = \sum_{i=1}^{a} \delta_i = \frac{\sum_{i=1}^{a} \lambda_i}{\sum_{i=1}^{m} \lambda_i} \tag{25.3}$$

The rate of contribution is refers to as each principal component account for the share of the total amount of information, the share of former a principal component is gotten by accumulated. If the information included in a principal components contain all the information or $a = m$, the η_a is 100 %. Generally, the cumulative contribution rate $\eta_a \geq 85$ %.

According to the above calculation method of principal component analysis method, after obtained the number of principal components of a, the data matrix $\mathbf{X}$ will be divided into two parts, as follows:

$$\mathbf{X}=\widehat{\mathbf{X}}+\widetilde{\mathbf{X}}=\mathbf{T}\mathbf{P}^{T}+\widetilde{\mathbf{T}}\widetilde{\mathbf{P}}^{T} \tag{25.4}$$

$\mathbf{T}$ and $\mathbf{P}$ are respectively a principal components score matrix and principal components load matrix, the column vector $\mathbf{P}$ respectively is former a eigenvector of $\mathbf{R}$. $\mathbf{P}=[p_1,p_2,\ldots,p_a]$; $\tilde{\mathbf{T}}$ and $\tilde{\mathbf{P}}$ were residual score matrix and residual load matrix, one of the columns of $\tilde{\mathbf{P}}$ are the rest of $(m-a)$ eigenvector, $\tilde{\mathbf{P}}=[p_{a+1},p_{a+2},\ldots,p_m]$.

25.2.2 Squared Prediction Error SPE

After the establishment of a data model, the new measurement data can be analyzed and judged by the use of the data model. $\mathbf{x}$ is a new data matrix for the real-time measurement,where $\mathbf{x}$ is the data matrix, which is normalized by the same method, $\mathbf{x}$ can be decomposed into two parts, then

$$x=\widehat{x}+\widetilde{x} \tag{25.5}$$

In the equation, $\hat{x}=\mathbf{P}\mathbf{P}^{T}x=\mathbf{C}x$, $\mathbf{C}$ is the projection matrix in principal component subspace (PCS), that is, $\hat{x}$ is its projection. $\widetilde{\mathbf{x}}=\widetilde{\mathbf{P}}\widetilde{\mathbf{P}}^{T}\boldsymbol{x}=(\mathbf{I}-\mathbf{C})x=\widetilde{\mathbf{C}}\boldsymbol{x}$, $\boldsymbol{I}$ is the unit matrix, $\boldsymbol{C}$ is the projection matrix of (RS), that is, $\tilde{x}$ is its projection0. describes most of the information of all data,residual subspace expresses the measured noise and fault. Principal component subspace and the residual subspace are two orthogonal complementary spaces.

Abnormal or fault is diagnosed through the analysis of SPE statistic data of the new real time measurement. The SPE statistic is the square prediction error, whose detection performance is superior. SPE is the scalar deviation from the PCA model in the residual space of the measurement values. Under normal condition, SPE $\leq Q_\alpha$, on the other hand, when the abnormal or there is a failure, SPE $> Q_\alpha$. Obviously, the control limit of SPE is represented by the Q_α, the calculation formula is as follows:

$$Q_\alpha=\theta_1\left[\frac{C_\alpha\sqrt{2\theta_2 h_0^2}}{\theta_1}+1+\frac{\theta_2 h_0(h_0-1)}{\theta_1^2}\right]^{\frac{1}{h_0}} \tag{25.6}$$

In the equation, $\theta_i=\sum_{j=a+1}^{m}\lambda_j^i$, $i=1,2,3$; $h_0=1-\frac{2\theta_1\theta_3}{3\theta_2^2}$; a is the number of principal components, m is the number of measurement variables, λ_j ($j=a+1$, $a+2,\ldots,m$) is the covariance matrix of the original variable data matrix normalized by big features to order value, C_α is the normal distribution of confidence for the statistics of α.

The calculation formula of SPE statistics is as follows:

$$\text{SPE} = ||\tilde{x}||^2 = \mathbf{X}_i(\mathbf{I}-\mathbf{C})\mathbf{X}_i^T = \mathbf{X}_i\left(\mathbf{I}-\mathbf{P}\mathbf{P}^T\right)\mathbf{X}_i^t, \quad (i=1,2,\ldots,\mathrm{n}) \tag{25.7}$$

In the equation, i is the ith sample point, $\boldsymbol{I}$ is the full rank matrix with elements of 1.

25.3 The Fault Diagnosis Realization of Double Tank

In the actual industrial production, there are abnormal or fault in the running process of the system. If the correct judgment can be made, the fault diagnosis is realized. In general, the establishment of dynamic mathematical model of controlled system provides the basis for the fault detecting.

Sensors are divided into many different types according to the applicable working environment. The sensor fault can be divided into two main types. Commonly one type is known as the soft fault such as fault of bias, fault of drift and fault of precision grade lower. Another type is called hard fault such as complete failure. The sensor fault of double tank belonged to soft fault.

In the industrial production, there are many types of sensors. The principal component analysis is used to analyze the sensor fault, which uses only a few of the linear variable indicators to summarized most of the information. The feasibility of principal component analysis in multi-sensor fault detection is investigated through the double tank controlled system. The sensor data of double tank was detected as the following.

The unit of each sensor data used the international standard unit. The PCA method was used to construct the model and analyze the data matrix of normal working conditions (Fig. 25.1).

25.3.1 *The Bias Fault*

The bias fault refers to the deviation constant, which produced a fixed difference value between the real value and the measured value. Because the effective value of each sensor is different, so the hypothesis deviation value is 30 % of the measured value. Through the calculation and analysis of main element analysis, the curve of SPE statistics is gotten in the following figure.

As shown in Fig. 25.2, when the double tank was in a stable working state, the change of SPE statistics curve was little. When the deviation fault was occurred, the SPE statistic curve increased obviously, after trouble was cleared, SPE statistic curve returned to normal level.

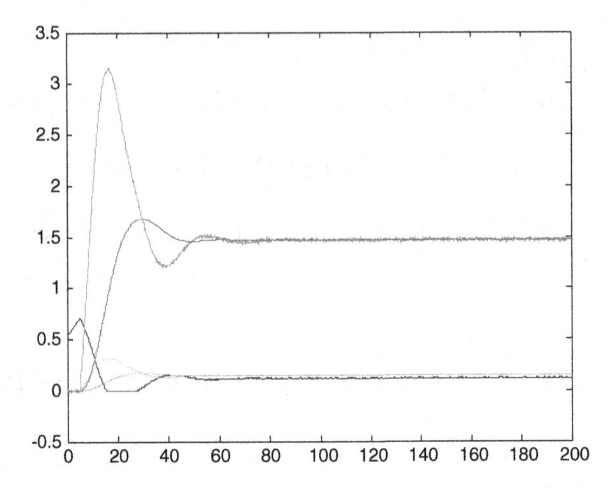

Fig. 25.1 The sensor data of the double water tank

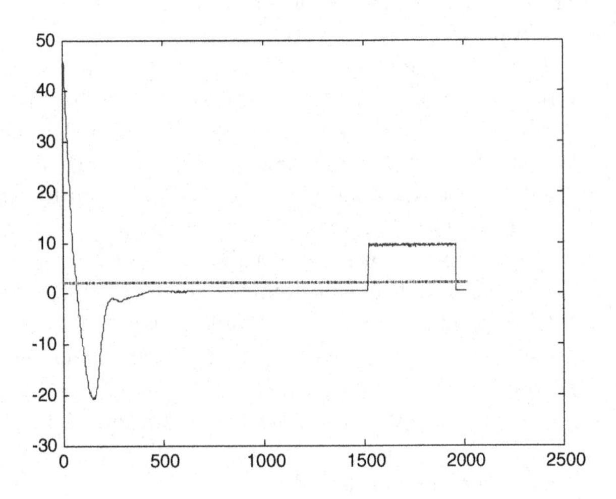

Fig. 25.2 The constant deviation fault of the data

25.3.2 *Drift Failure*

The drift fault is defined that the size of the fault is linear changed according to the change of time. Assuming at the time point 150 the drift fault was occurred,. After calculation, curve of SPE statistics was gotten in the following figure.

As shown in Fig. 25.3, when the double tank was in the stable working state, SPE statistics curve changed little. When the drift failure happened, SPE statistic curve increased obviously. After the trouble was cleared, SPE statistic curve decreased to the normal state.

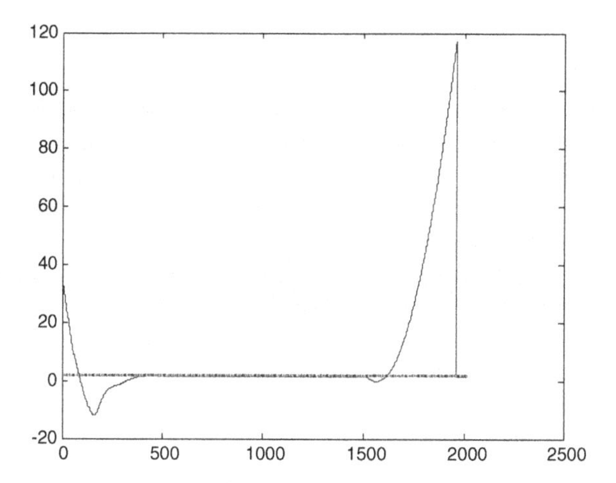

Fig. 25.3 The drift fault

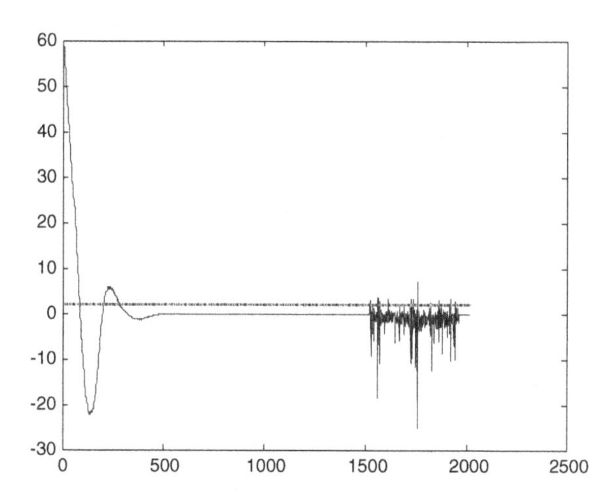

Fig. 25.4 The fault of precision decrease

25.3.3 The Fault of Reduced Precision

The fault is defined that the measured average value remains unchanged but the measurement variance changes, which is similar to the increased free noise variance. After calculation, the SPE statistics curve is in the following figure.

As shown in Fig. 25.4, when the double tank was in the stable working state, the SPE statistics curve was changed little. When the fault of reduced precision was happened, the SPE statistic curve fluctuated obviously, after the trouble was cleared, the SPE statistic curve returned to the normal state.

25.4 Conclusion

It is difficult to establish a precise analytical model of the complex controlled system. Without the precise model, the general method is needed to realize the fault diagnosis for the industrial production process, multivariate statistics is one of them. Because it is possible to appear bias fault judgment results, it is important for the further fault diagnosis. The PCA of multivariate statistical analysis was taken as the basic methods of the data. The double tank was taken as a controlled object to create the data and simulated fault. According to the material balance transfer function of double tank water system, the offline data model for the original data was established under the normal condition, and the SPE statistic value was calculated. At last the compared result of the new monitoring data and the historical data showed the PCA method was effective.

Acknowledgement This paper is supported by fund projects: the national natural science funds projects (code: 60974063), the nature science foundation of Hebei province(code: F2014205115), Hebei province department of education (code: Z2011141), Hebei province department of science and technology (code: 11215650), Hebei normal university scientific research fund (code: L2011Q10).

References

1. Jackson JE, Mudholkar GS (1979) Control procedures for residuals associated with principal component analysis. Technometrics 21:341–349
2. Geladi P, Kowalski BR (1986) Partial least-squares regression: a tutorial. Anal Chimica Acta 185:1–17
3. Wang SW, Cui JP (2005) Sensor-fault detection, diagnosis and estimation for centrifugal chiller systems using principal–component analysis method. Appl Energy 82:197–213
4. Du ZM, Jin XQ, Wu LZ (2007) Fault detection, diagnosis based on improved PCA with JAA method in VAV systems. Build Environ 42:3221–3232
5. Du ZM, Jin XQ (2009) Multiple fault diagnosis for sensors in air handing unit using Fisher discriminant analysis. Energy Conversation Manag 49:3654–3665
6. Yoo CK, Villez K, Van Hulle S, Vanrolleghem PA (2008) Enhanced process monitoring for wastewater treatment systems. Environmetrics 19:602–617
7. Hua JF, Li JQ, Ouyang MG, Lu LG, Xu LF (2011) Proton exchange membrane fuel cell system diagnosis based on the multivariate statistical method. Int J Hydrogen Energy 36:9896–9905
8. Moon TS, Kim YJ, Kim JR, Cha JH, Kim DH, Kim CW (2009) Identification of process operating state with operational map inmunicipal wastewater treatment plant. Environ Manag 90:772–778
9. Tamura Masayuki, Tsujita Shinsuke (2007) A study on the number of principal components and sensitivity of fault detection using PCA. Comput Chem Eng 31:1035–1046
10. Hailian Du, Wang Zhanfeng, Lv Feng, Wenxia Du (2011) Fault detection for multi-sensor based on PCA. J Nanjing Univ Aeronaut Astronaut 43:138–141
11. Alkaya Alkan, Eker İlyas (2011) Variance sensitive adaptive threshold-based PCA method for fault detection with experimental application. ISA Trans 50:287–302
12. BinShams MA, Budman HM, Duever TA (2011) Fault detection, identification and diagnosis using CUSUM based PCA. Chem Eng Sci 66:4488–4498

Chapter 26
Steel Liquid Level Tracking Via Iterative Learning with Extended Error Information

Yunzhong Song

Abstract This paper aims to improve the steel liquid level control quality via iterative learning control (ILC) with extended error information. The ILC is one kind of type P iterative learning control, and besides the forgetting factor and the on-off switching action, error information was further extended on account of introduction of the just past and the second past cycles error signals. Results demonstrated that, the control quality can still be improved even under the model uncertainties, periodic bulging disturbances, the measuring noises, as well as the input signal error, the state error and the output error can be guaranteed to be ultimately bounded. Simulation results were provided to clarify the suggested idea further.

Keywords Iterative learning control with extended error information · Just past error information · Molten steel level · On-off switching action · Forgetting factor

26.1 Introduction

Steel liquid level is an quite important parameter to be controlled in continuous casting process, since the influence of the periodic bulging disturbances, the single PID controller can not meet the high standard requirements of the steel liquid level control [1–3]. Because the ILC is pertinent to the periodic manipulation, it is introduced to be combined with the already existing PID controller to improve the steel liquid level quality, and the reported results are promising [4, 5]. After the examination of the ILC scheme used in steel liquid level control, we have noticed that only just past error information is used, and this can limit the steel liquid level control improvement further. Considering the available memory module embedded in the ILC, and the price of the memory hardware is not so high, we try to borrow the second error information to the ILC controller. In order to keep the balance of the just past error

Y. Song (✉)
School of Electrical Engineering and Automation, Henan Polytechnic University, Jiaozuo 454003, China
e-mail: songhpu@126.com

Y. Jia et al. (eds.), *Proceedings of the 2015 Chinese Intelligent Systems Conference*, Lecture Notes in Electrical Engineering 360,
DOI 10.1007/978-3-662-48365-7_26

information and the second past information, an forgetting factor is designed to make the tradeoff between them.

Our contributions lie in two folds, one is the introduction of the second past error information for the ILC, and the second one is that the strict proof of the convergent properties of the proposed scheme.

26.2 The Main Results

26.2.1 System Description

The steel liquid level model can be expressed as

$$\begin{cases} x_i(k+1) = x_i(k) + B(x_i(k), u_i(k), k) + \beta_i(k) + d_i(k), \\ y_i(k) = x_i(k) + n_i(k). \end{cases} \tag{26.1}$$

where i is the iterative number, and k is the discrete time, $k \in \mathbb{N}$. And, $x_i(k) \in R^n, u_i(k) \in R^r, y_i(k) \in R^m, \beta_i(k) \in R^p, d_i(k) \in R^p, n_i(k) \in R^m$ demonstrate system states, controller inputs, system outputs, model uncertainties and periodic burgling disturbances.

26.2.2 The ITL Controller

The ILC controller is designed as

$$\begin{aligned} u_{i+1}(k) = & s_{i1}(k)[(1-p)u_i(k) + L(k)(\gamma e_i(k) \\ & + (1-\gamma)e_{i-1}(k))] + s_{i2}(k)[u_{i+1}(k-1)]. \end{aligned} \tag{26.2}$$

During the ith iterate, the tracking error is modeled as $e_i(k) = y_d(k) - y_i(k)$, $s_{i1}(k) = 0.5[\text{sgn}(T_v - e_{i+1}(k) + 1)], s_{i2}(k) = -0.5[\text{sgn}(T_v - e_{i+1}(k) - 1)]$, $L(k)$ is the gain factor and $L(k) \leq b_L$; γ is the proportional coefficient and it is assigned as 0.7; p is defined as forgetting factor, and it is designed as $0 < p < 1$. When the steel liquid level is below the threshold value, under auspices of the on-off action, the P-type learning law with forgetting factor is used; if the steel liquid level is bigger than the assigned threshold value, the controller input will main the error information of the just past error and the second past error. and the PID controller is the fundamental controller, it is employed to complete the basic control action, the system structure can be described as Fig. 26.1.

The parameters of PID controller is selected as $K_p = 1, K_i = 0.1, K_d = 0.1$. and $B(x_i(k), u_i(k), k)$ represents $(\Delta T/A)(\sqrt{2ghSG(F_{fb}(u_i(k)))} - Q_{out}).\overline{SG(\cdot)}$ is the linear overlay area.

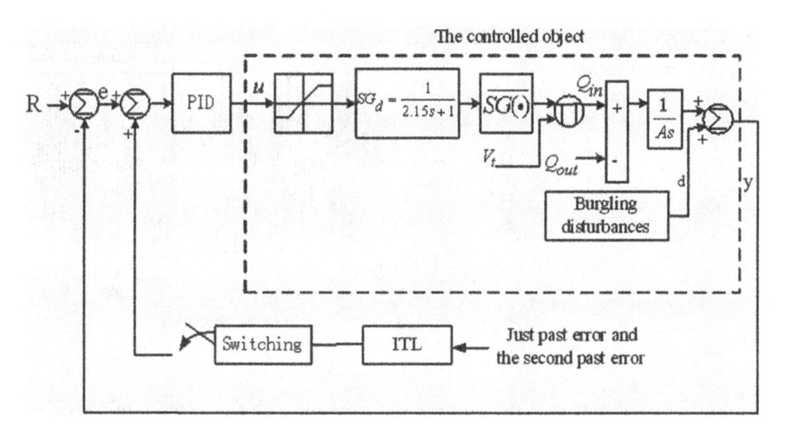

Fig. 26.1 Block diagram for the proposed molten steel level control system

26.2.3 The Simulation Results

In case of validity verification of the suggested control strategy,simulation under Matlab environment is provided. The steel liquid level is assigned as 80 mm, the disturbances which represent the burgling variations are comprised of two sinusoid signals, their peak values are designed as 10.5 mm, and their frequencies are designed between 0.120 and 0.140 Hz. $L(k)$ is chosen as 0.6, the forgetting factor p is assigned as 0.3. The proportional coefficient is selected as 0.7. The simulation span covers 8 cycles under the burgling disturbances.

Through the simulation, it is not difficult to know that when only the PID controller is possible, the liquid level error can be approached to 13 mm, it is unexpected huge. When the suggested scheme is extra introduced, under the next coming 4 cycles, the bad effect of the burgling disturbances are inhibited effectively, and on account of that, the high quality steel liquid level control is expected. And the corresponding result is demonstrated as Fig. 26.2. And the affiliated switching parameter together with the controlled output are demonstrated as Fig. 26.3.

Fig. 26.2 Simulation results of PID and the new proposed control scheme

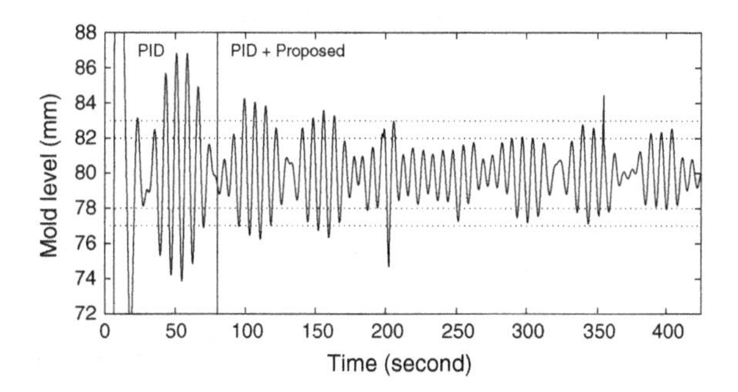

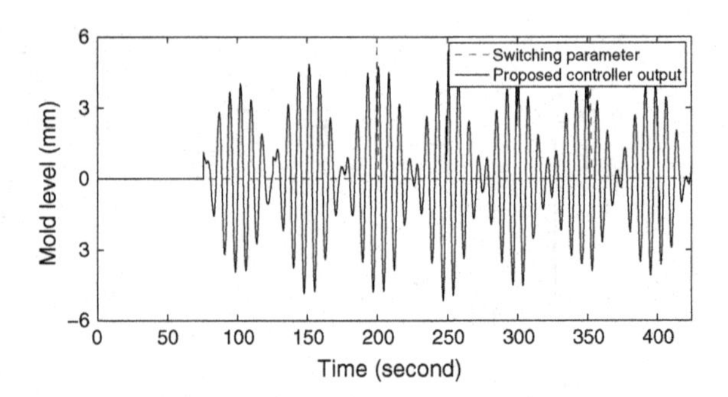

Fig. 26.3 The affiliated switching parameter together with the controlled output

It is clearly demonstrated that under the auspices of the introduction of the just past error and the second past error information in to ILC, together with the fundamental controller PID, the high quality molten liquid level control can be expected.

26.2.4 The Convergent Analysis of the ILC

Definition 1 $b_s = \max\limits_{1 \le i < \infty} \max\limits_{1 \le k \le n} ||(1 - s_{i1}(k))(1 - p)||, b_{s_1} = \max\limits_{1 \le i < \infty} \max\limits_{1 \le k \le n} ||s_{i1}(k)||, b_{s_2} = \max\limits_{1 \le i < \infty} \max\limits_{1 \le k \le n} ||s_{i2}(k)||$.

Assumption 1 $\max\limits_{1 \le k \le n} ||u_d(k)|| \le b_{u_d}$.

Assumption 2
$$\max_{1 \le i < \infty} \max_{1 \le k \le n} ||\beta_i(k)|| \le b_\beta,$$
$$\max_{1 \le i < \infty} \max_{1 \le k \le n} ||d_i(k)|| \le b_d,$$
$$\max_{1 \le i < \infty} \max_{1 \le k \le n} ||n_i(k)|| \le b_n.$$

Assumption 3 $||x_d(0) - x_i(0)|| \le b_{x_0}$.

Assumption 4 $||s_{i1} + (1 - p)|| + b_{S_1} b_L c_B \le \rho < 1$.

Theorem 1 *Under the suggested ITL law (26.2), the system (26.1) gives bounded output series.*

Proof

$$\begin{aligned}
\tilde{x}_i(k+1) &= x_d(k+1) - x_i(k+1) \\
&= [x_d(k) + B(x_d(k), u_d(k), k)] - [x_i(k) + B(x_i(k), u_i(k), k) + \beta_i(k) + d_1(k)] \\
&= \tilde{x}_i(k) + B(x_d(k), u_d(k), k) - B(x_i(k), u_i(k), k) - \beta_i(k) - d_1(k).
\end{aligned}$$

and then, we have

$$||\tilde{x}_i(k+1)|| \le ||\tilde{x}_i(k)|| + c_B||\tilde{x}_i(k)|| + c_B||\tilde{u}_i(k)|| + b_\beta + b_d \tag{26.3}$$

It is easy to conclude that

$$||\tilde{x}_i(k+1)|| \le (1+c_B)||\tilde{x}_i(k)|| + c_B||\tilde{u}_i(k)|| + b_\beta + b_d.$$

and

$$||\tilde{x}_i(k)|| \le \sum_{j=0}^{k-1}(1+c_B)^{k-1-j}[c_B||\tilde{u}_i(j)|| + b_\beta + b_d] + (1+c_B)^k b_{x_0}.$$

For control input, we have

$$\begin{aligned}
\tilde{u}_{i+1}(k) &= u_d(k) - u_{i+1}(k)\\
&= u_d(k) - s_{i1}(k)[(1-p)u_i(k) + L(k)e_i(k+1)] - s_{i2}(k)u_{i+1}(k-1)\\
&= u_d(k) - s_{i1}(k)(1-p)u_i(k) - s_{i1}(k)(1-p)u_d(k) + s_{i1}(k)(1-p)u_d(k)\\
&- s_{i1}(k)L(k)e_i(k+1) - s_{i2}(k)u_{i+1}(k-1)\\
&= s_{i1}(k)(1-p)\tilde{u}_i(k) + u_d(k) - s_{i1}(k)(1-p)u_d(k) - s_{i2}(k)u_{i+1}(k-1)\\
&- s_{i1}(k)L(k)[y_d(k+1) - y_i(k+1)]\\
&= s_{i1}(k)(1-p)\tilde{u}_i(k) + (1 - s_{i1}(k)(1-p))u_d(k) - s_{i2}(k)u_{i+1}(k-1)\\
&- s_{i1}(k)L(k)[y_d(k+1) - y_i(k+1)]\\
&= s_{i1}(k)(1-p)\tilde{u}_i(k) + (1 - s_{i1}(k)(1-p))u_d(k) - s_{i2}(k)u_{i+1}(k-1)\\
&- s_{i1}(k)L(k)[x_d(k+1) - x_i(k+1) - n_i(k)]\\
&= s_{i1}(k)(1-p)\tilde{u}_i(k) + (1 - s_{i1}(k)(1-p))u_d(k)\\
&- s_{i1}(k)L(k)[x_d(k) + B(x_d(k), u_d(k), k) - x_i(k) - B(x_i(k), u_i(k), k)\\
&- \beta_i(k) - d_1(k) - n_i(k)] - s_{i2}(k)u_{i+1}(k-1)\\
&+ s_{i2}(k)u_d(k-1) - s_{i2}(k)u_d(k-1).
\end{aligned} \tag{26.4}$$

The equation can be turned into

$$\begin{aligned}
&||\tilde{u}_{i+1}(k)||\left(\tfrac{1}{\alpha}\right)^k \le \rho||\tilde{u}_i(k)||\left(\tfrac{1}{\alpha}\right)^k + h_1 b_{x_0}\left(\tfrac{1+c_B}{\alpha}\right)^k + b_1\left(\tfrac{1}{\alpha}\right)^k\\
&+\left(\tfrac{h_1}{\alpha}\right)\sum_{j=0}^{k-1}\left(\tfrac{1+c_B}{\alpha}\right)^{k-1-j}\left[c_B||\tilde{u}_i(j)||\left(\tfrac{1}{\alpha}\right)^j + (b_\beta + b_d)\left(\tfrac{1}{\alpha}\right)^j\right]\\
&+b_{s_2}\left(\tfrac{1}{\alpha}\right)||\tilde{u}_{i+1}(k-1)||\left(\tfrac{1}{\alpha}\right)^{k-1}
\end{aligned} \tag{26.5}$$

and α is chosen according to

$$\alpha > \max[1, b_\beta + b_d].$$

And we can have

$$
\begin{aligned}
||\tilde{u}_{i+1}||_\alpha &\le \rho||\tilde{u}_i||_\alpha + h_1 b_{x_0}\left(\tfrac{1+c_B}{\alpha}\right)^k + b_1\left(\tfrac{1}{\alpha}\right)^k \\
&+ (b_B||\tilde{u}_i||_\alpha + b_\beta + b_d)\left(\tfrac{h_1}{\alpha}\right)\sum_{j=0}^{k-1}\left(\tfrac{1+c_B}{\alpha}\right)^{k-1-j} + \left(\tfrac{b_{s_2}}{\alpha}\right)||\tilde{u}_{i+1}||_\alpha \\
&\le \rho||\tilde{u}_i||_\alpha + h_1 b_{x_0}\left(\tfrac{1+c_B}{\alpha}\right)^k + b_1\left(\tfrac{1}{\alpha}\right)^k \\
&+ (b_B||\tilde{u}_i||_\alpha + b_\beta + b_d)\tfrac{h_1[1-((1+c_B)/\alpha)^k]}{\alpha-(1+c_B)} + \left(\tfrac{b_{s_2}}{\alpha}\right)||\tilde{u}_{i+1}||_\alpha \\
&= \left(\tfrac{\alpha}{\alpha-b_{s_2}}\right)\left(\rho + b_B h_1\tfrac{1-((1+c_B)/\alpha)^k}{\alpha-(1+c_B)}\right)||\tilde{u}_i||_\alpha + \left(\tfrac{\alpha h_1}{\alpha-b_{s_2}}\right)\left(\tfrac{1+c_B}{\alpha}\right)^k b_{x_0} \\
&+ \left(\tfrac{\alpha b_1}{\alpha-b_{s_2}}\right)\left(\tfrac{1}{\alpha}\right)^k + \tfrac{\alpha(b_\beta+b_d)h_1[1-((1+c_B)/\alpha)^k]}{(\alpha-b_{s_2})(\alpha-(1+c_B))}
\end{aligned}
\tag{26.6}
$$

and for

$$
\frac{1}{\alpha}\sum_{j=0}^{k-1}\left(\frac{1+c_B}{\alpha}\right)^{k-1-j} = \frac{((1+c_B)/\alpha)^{k-1}((\alpha/(1+c_B))^k-1)}{\alpha(\alpha/(1+c_B)-1)} = \frac{1-((1+c_B)/\alpha)^k}{\alpha-(1+c_B)}
$$

so we can have

$$
||\tilde{u}_{i+1}||_\alpha \le \tilde{\rho}||\tilde{u}_i||_\alpha + \varepsilon. \tag{26.7}
$$

where

$$
\begin{aligned}
\tilde{\rho} &= \left(\tfrac{\alpha}{\alpha-b_{s_2}}\right)\left(\rho + b_B h_1\tfrac{1-((1+c_B)/\alpha)^k}{\alpha-(1+c_B)}\right), \\
\varepsilon &= \left(\tfrac{\alpha h_1}{\alpha-b_{s_2}}\right)\left(\tfrac{1+c_B}{\alpha}\right)^k b_{x_0} + \left(\tfrac{\alpha b_1}{\alpha-b_{s_2}}\right)\left(\tfrac{1}{\alpha}\right)^k + \tfrac{\alpha(b_\beta+b_d)h_1[1-((1+c_B)/\alpha)^k]}{(\alpha-b_{s_2})(\alpha-(1+c_B))}.
\end{aligned}
\tag{26.8}
$$

so, we can have

$$
||\tilde{u}_{i+1}||_\alpha \le \hat{\rho}^i||\tilde{u}_i||_\alpha + \varepsilon\sum_{j=0}^{i-1}\hat{\rho}^j = \hat{\rho}^i||\tilde{u}_i||_\alpha + (1-\frac{\hat{\rho}^i}{1-\hat{\rho}}). \tag{26.9}
$$

If the value of α is chosen big enough, then we can have $\rho < 1$; and then we can conclude that

$$
\lim_{i\to\infty}||\tilde{u}_i||_\alpha \le \frac{\varepsilon}{1-\hat{\rho}}
$$

and accordingly, we can do the same kind of manipulation on the states of the system, so we can have

$$||\tilde{x}_i(k)||\left(\frac{1}{\alpha}\right)^k \leq \left(\frac{1}{\alpha}\right)\sum_{j=0}^{k-1}\left(\frac{1+c_B}{\alpha}\right)^{k-1-j}[c_B||\tilde{u}_i(j)||\left(\frac{1}{\alpha}\right)^j + (b_\beta + b_d)\left(\frac{1}{\alpha}\right)^j] + \left(\frac{1+c_B}{\alpha}\right)^k b_{x_0}. \tag{26.10}$$

and we can have

$$\begin{aligned} ||\tilde{x}_i||_\alpha &\leq (c_B||\tilde{u}_i||_\alpha + b_\beta + b_d)\frac{1-((1+c_B)/\alpha)^k}{\alpha-(1+c_B)} + \left(\frac{1+c_B}{\alpha}\right)^k b_{x_0} \\ &= c_B\frac{1-((1+c_B)/\alpha)^k}{\alpha-(1+c_B)}||\tilde{u}_i||_\alpha + (b_\beta + b_d)\frac{1-((1+c_B)/\alpha)^k}{\alpha-(1+c_B)} + \left(\frac{1+c_B}{\alpha}\right)^k b_{x_0}. \end{aligned} \tag{26.11}$$

which means

$$\begin{aligned} \lim_{i\to\infty}||\tilde{x}_i||_\alpha \leq c_B\frac{1-((1+c_B)/\alpha)^k}{\alpha-(1+c_B)}\frac{\varepsilon}{1-\hat{\rho}} + \\ (b_\beta + b_d)\frac{1-((1+c_B)/\alpha)^k}{\alpha-(1+c_B)} + \left(\frac{1+c_B}{\alpha}\right)^k b_{x_0}. \end{aligned} \tag{26.12}$$

As to the output of the system, we can have

$$\tilde{y}_i(k) = y_d(k) - y_i(k) = x_d(k) - [x_i(k) + n_i(k)] = \tilde{x}_i(k) - n_i(k). \tag{26.13}$$

and with the helps of the following transform used for control input and system states

$$||\tilde{y}_i(k)||\left({}^1/_\alpha\right)^k = ||\tilde{y}_i(k)||_\alpha$$

we can have

$$||\tilde{y}_i||_\alpha \leq ||\tilde{x}_i||_\alpha + \left(\frac{1}{\alpha}\right)b_n$$

considering the results we have gotten for the system states, we can have

$$\begin{aligned} &\lim_{i\to\infty}||\tilde{y}_i||_\alpha \leq c_B\frac{1-((1+c_B)/\alpha)^k}{\alpha-(1+c_B)}\frac{\varepsilon}{1-\hat{\rho}} + (b_\beta + b_d)\frac{1-((1+c_B)/\alpha)^k}{\alpha-(1+c_B)} \\ &+\left(\frac{1+c_B}{\alpha}\right)^k b_{x_0} + \left(\frac{1}{\alpha}\right)b_n \end{aligned}$$

and that is the end of the proof.

26.3 Conclusions

The deep information of the operating plant is the valuable source to improve the quality of the control, since the history information like error including the disturbances, and such like. To fully utilize these information effectively, the characteristics of the plant and its environments must be considered at first. To the molten steel liquid level, the burgling disturbances are the slow sinusoid signals, the ILC, which is born to deal with the repeated motion, can be borrowed to deal with that kind of disturbances. So, with the fundamental action of PID, together with the proposed ILC, and especially the extended error information transferred into the control loop, the expected high quality control of molten liquid level is made possible, the idea explained here is much powerful for its quite general characteristics, we will try to explore this idea even further in the near coming future.

Acknowledgments The author would like to express his appreciation for his graduate student *Xu Zhao*, for his laborious work towards this paper. This work is also supported in part by National Science Foundation of China (61340041 and 61374079), and The Project-sponsored by SRF for ROCS, SEM to Yunzhong Song.

References

1. Barron MA, Aguilar R, Gonzalez J, Melendez E (1998) Model-based control of mold level in a continuous steel caster under model uncertainties. Control Eng Pract 6(3):191–196
2. Keyser R (1991) Improved mould-level control in a continuous steel casting line. Control Eng Pract 5(2):231–237
3. Lee D, Kueon Y, Lee S (2003) High performance hybrid mold level controller for thin slab caster. Control Eng Pract 12(3):275–281
4. You B, Kim M, Dukman L et al (2011) Iterative learning control of molten steel level in a continuous casting process. Control Eng Pract 19(3):234–242
5. Chen Y, Wen C, Sun M (1997) A robust high-order P-type iterative learning controller using current iterative tracking error. Int J Control 68(1):331–342

Chapter 27
Analysis of Quantization Noise Spectrum in Signal Reconstruction

Su Xu, Hongpeng Yin and Yi Chai

Abstract Quantization is an essential but often ignored part of the realization of compressive sampling (CS), and the analysis of quantization noise arise from CS is incomplete and not sufficient until now. The quantization noise is generated from quantizing CS values by a uniform quantizer under ideal and noise conditions. And also, the auto correlation function and power spectrum have been derived. It is concluded that the quantization noise is always uncorrelated with the input signals, the quantization noise is white and the spectrum is white noise spectrum. On this basis, we analyze the reconstruction error introduced by quantization noise quantitatively and give the upper and lower bounds of reconstruction error. Simulation results validate the validity of the analysis for further.

Keywords Compressed sensing · Quantization noise · Reconstruction error

27.1 Introduction

Compressive sampling (CS) [1–3] is a novel theory of signal requisition and gets a lot of attention over the past few years. It introduces the methods of incoherent linear measurement and nonlinear reconstruction to require sparse signal in a lower

S. Xu (✉) · H. Yin · Y. Chai
College of Automation, Chongqing University, Chongqing 400030, China
e-mail: xusu44@ctbu.edu.cn

S. Xu · H. Yin
Key Laboratory of Dependable Service Computing in Cyber Physical Society, Ministry of Education, Chongqing 400030, China

S. Xu
College of Computer Science and Information Engineering, Chongqing Technology and Business University, Chongqing 400067, China

S. Xu
Chongqing Engineering Laboratory for Detection, Control and Integrated System, Chongqing Technology and Business University, Chongqing 400067, China

Y. Jia et al. (eds.), *Proceedings of the 2015 Chinese Intelligent Systems Conference*, Lecture Notes in Electrical Engineering 360,
DOI 10.1007/978-3-662-48365-7_27

sampling rate. As the corresponding device of CS theory, analog-to-information converter (AIC) [4] plays a very important role in promoting CS theory to practical development. The quantization is an essential but often ignored part of AIC, because of unsufficient research. Recently, the analysis of quantization error after CS has been focused on [5–14, 15, 16]. Some literatures concern on the way to reduce the quantization error during CS, e.g., adopting nonuniform quantization method [5], designing new automatic gain control methods [6, 7] and so on. Also, some concern on designing new reconstruction algorithm to reduce quantization error, under the condition of quantization noise. For example, an optimum quantization strategy which can get the minimum reconstruction error is proposed in [8], a method of reconstructing signal from the measured value of nonuniform quantization is addressed in [9], Furthermore, the mean errors of these two algorithms are calculated in [10]. Under the condition of non-uniformly quantized compressed sensing, weighted fidelity in quantization noise is discussed in [11]. In addition, the upper error bounds of basis pursuit (BP) and subspace pursuit (SP) reconstruction are given in [12, 13]. The sensitivity function defines the optimal quantizer to a quantized compressed sensing model in [14]. The average distortion introduced by quantizing compressive sensing measurements in [15]. A simple and effective automatic gain control strategy for CS is to amplify the signal by the finite range quantizer overflows at a pre-determined rate [16].

However, the analysis of CS's quantization noise is not sufficient. Comparing with the condition of traditional Nyquist theory, the shape of CS's quantization noise spectrum and the condition when it is white noise spectrum are still not clear and waiting to be solved. Aiming that, this paper analyzes the CS's quantization noise spectrum when its quantizer is uniform, and quantitatively derives the upper and lower bounds of reconstruction error under the condition of quantization noise.

The remainder of this paper is organized as follows. Section 27.2 briefly introduces the CS's basic knowledge. Section 27.3 analyzes the quantization noise spectrum under the ideal and noise situations. In Sect. 27.4, the reconstruction error is analyzed. And Sect. 27.5 shows the results of simulations and experiments. The paper is concluded in Sect. 27.6 finally.

27.2 CS's Basic Knowledge

From the viewpoint of mathematics, if the signal in some basis is sparse or compressible [1], CS noncoherent measurement procession can be expressed as:

$$y = \mathbf{\Phi} x \tag{27.1}$$

Here the sparse signal $x \in \mathbb{R}^N$, measurement vector $y \in \mathbb{R}^M$, measurement matrix $\mathbf{\Phi} \in \mathbb{R}^{N \times M}$ and it is noncoherent to the recovery signals y [1]. For a sparse signal

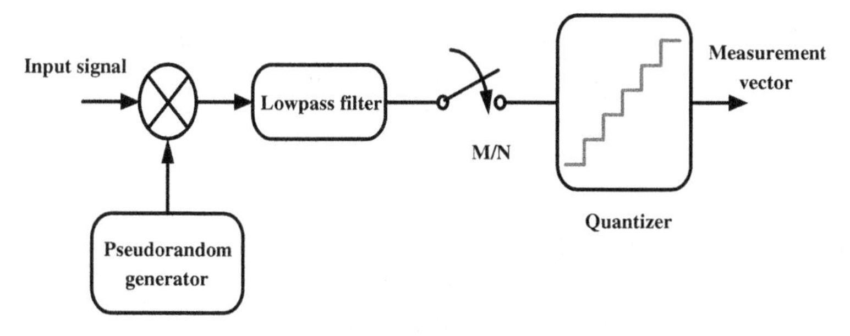

Fig. 27.1 AIC structure based on random demodulation

with K order, if $M \geq O(K\log(N/K))$ and $\mathbf{\Phi}$ meets the restricted isometry property (RIP) with K order, i.e.:

$$(1-\delta)\|x\|_2^2 \leq \|\mathbf{\Phi}x\|_2^2 \leq (1+\delta)\|x\|_2^2 \quad (27.2)$$

It can be exactly or highly accurately reconstructed y from a small set of incomplete measurements x by the restricted isometry constant $\delta \in (0,1)$. So many existing methods for ill-posed linear inverse problems can be applied to the CS recovery. The reconstruction algorithms mainly include BP [12], SP [13], linear optimization (L-OPT) [17], orthogonal matching pursuit (OMP) [18], Regularized Orthogonal Matching Pursuit (ROMP) [19], Stage wise Orthogonal Matching Pursuit (StOMP) [20], Compressive Sampling Matching Pursuit, (CoSaMP) [21], Optimized Orthogonal Matching Pursuit (OOMP) [22].

As the concrete implement of above measurement process, AIC is the key to practical application of CS theory. A useful AIC frame is designed by means of wideband pseudorandom demodulator and low-rate sampler in [4], which is shown in Fig. 27.1. Obviously, whether the sampling device of traditional Nyquist sampling theory, or the sampling device of CS theory, quantization is an indispensable key point during the sampling process.

27.3 The Quantization Noise Spectrum

The condition of quantization noise spectrum can be divided into the ideal and noise situations. Ignoring the noise influence, (27.1) can be transformed into:

$$y_i = \sum_{j=1}^{N} \phi_{ij} x_j \quad (27.3)$$

If random measurement matrix $\mathbf{\Phi}$ in which elements ϕ_{ij} satisfy Gaussian matrix with mean is 0 and variance is $\frac{1}{M}$, (27.3) can be expressed as:

$$y_i = \sum_{j=1}^{N} \phi_{i,j} x_j = \sum_{j=1}^{N} S_{ij} \tag{27.4}$$

where the mean of $\tilde{S}_{ij}$ can easily deduce the result (27.5). And the variance is (27.6).

$$E\left[S_{i,j}\right] = E\left[\phi_{i,j} x_j\right] = x_j E\left[\phi_{i,j}\right] = 0 \tag{27.5}$$

$$D\left[S_{i,j}\right] = E\left[S_{i,j}^2\right] = x_j^2 E\left[\phi_{i,j}^2\right] = \frac{x_j^2}{M} \tag{27.6}$$

When N increases, from Lyapunov fundamental theorem, $\tilde{y}_i$ approximate to a random Gaussian with the mean is 0 and the variance is $\sum_{j=1}^{N} \frac{x_j^2}{M} = \frac{\|X\|_2^2}{M}$. Using the uniform quantizer, the amplitude of the input signal subjects to the Gaussian distribution with mean is 0, and then the autocorrelation function $R_e(m)$ of the quantization noise is as follows:

$$R_e(m) = \frac{\Delta^2}{2\pi^2} \sum_{k=1}^{\infty} \frac{1}{k^2} \exp\left[-4\pi^2 \frac{\sigma^2}{\Delta^2} k^2 \left(1 - r_y(m)\right)\right] \tag{27.7}$$

When Δ is the width of the quantized steps and $r_y(m) = \frac{R_y(m)}{R_y(0)}$ is the normalized autocorrelation function from the input signal, the autocorrelation function $R_y(m)$ can be expressed as:

$$\begin{aligned} R_y(m) &= E[y_i y_{i+m}] \\ &= \mathrm{E}\left[\sum_{j=1}^{N} \phi_{i,j} x_j \sum_{k=1}^{N} \phi_{i+m,\mathrm{k}} x_k\right] \\ &= \mathrm{E}\left[\sum_{j=1}^{N} \sum_{k=1}^{N} \phi_{i,j} \phi_{i+m,\mathrm{k}} x_j x_k\right] \\ &= \sum_{j=1}^{N} \sum_{k=1}^{N} \mathrm{E}\left[\phi_{i,j} \phi_{i+m,\mathrm{k}}\right] x_j x_k \end{aligned} \tag{27.8}$$

When $j = k$, $m = 0$ $R_y(0) = \frac{\|X\|_2^2}{M}$ and $r_y(m) = \begin{cases} 1, & \mathrm{m} = 0 \\ 0, & \text{else} \end{cases}$, (27.7) can be transformed into:

$$R_e(m) = \begin{cases} \frac{\Delta^2}{2\pi^2} \sum_{k=1}^{\infty} \frac{1}{k^2}, & \mathrm{m} = 0 \\ \frac{\Delta^2}{2\pi^2} \sum_{k=1}^{\infty} \frac{1}{k^2} \exp\left[-4\pi^2 \frac{\sigma^2}{\Delta^2} k^2\right], & \text{else} \end{cases} \tag{27.9}$$

The number of quantized steps $\frac{\sigma}{\Delta} \geq 1$, if $m \neq 0$ and $\frac{\sigma}{\Delta} = 1$, (27.9) becomes:

$$R_e(m) = \frac{\Delta^2}{2\pi^2}\left[\frac{e^{-4\pi^2}}{1} + \frac{e^{-16\pi^2}}{4} + \frac{e^{-36\pi^2}}{9} + \cdots\right] \tag{27.10}$$

Because of $e^{-4\pi^2} \approx 7 \times 10^{-18}$ $e^{-16\pi^2} \approx 2 \times 10^{-69}$, etc., $R_e(m) \approx 0$, thus $m = 0$, $R_e(0) = \frac{\Delta^2}{2\pi^2} \sum_{k=1}^{\infty} \frac{1}{k^2} = \frac{\Delta^2}{12}$. Quantization noise power spectrum is the autocorrelation function of Fourier transform, due to the approximation as a function of power spectrum, so the quantization noise power spectrum is the white noise spectrum.

Consider the quantized error is inevitable, when the noise signal n in which elements n_j subject to Gauss white noise with the mean is 0 and variance is δ^2, (27.3) can be expressed as:

$$\tilde{y}_i = \sum_{j=1}^{N} \phi_{ij}(x_j + n_j) = \sum_{j=1}^{N} \tilde{S}_{ij} \tag{27.11}$$

For the same reason, the mean of $\tilde{S}_{ij}$ is (27.12) and the variance is (27.13).

$$E[\tilde{S}_{i,j}] = E[\phi_{ij}x_j + \phi_{ij}n_j] = E[\phi_{ij}]E[x_j + n_j] = 0 \tag{27.12}$$

$$D[\tilde{S}_{ij}] = E\left[\tilde{S}_{ij}^2\right] = E\left[\phi_{ij}^2\right]E\left[(x_j + n_j)^2\right] = \frac{x_j^2 + \sigma^2}{M} \tag{27.13}$$

Also, it is proved that $\tilde{y}_i$ approximate to a random Gaussian with the mean is 0 and the variance is $\sum_{j=1}^{N} \frac{x_j^2 + \sigma^2}{M} = \sum_{j=1}^{N} \frac{\|X\|_2^2 + N\sigma^2}{M}$. The autocorrelation function of $\tilde{Y}$ can be expressed as:

$$\begin{aligned} R_{\tilde{y}}(m) &= E[\tilde{y}_i\tilde{y}_{i+m}] \\ &= \sum_{j=1}^{N}\sum_{k=1}^{N} E[\phi_{ij}\phi_{i+m,k}]E[(x_k + n_k)(x_j + n_j)] \end{aligned} \tag{27.14}$$

When $j = k$ and $m = 0$, $R_{\tilde{y}}(0) = \frac{\|X\|_2^2 + N\sigma^2}{M}$ is equivalent to the variance of $\tilde{y}_i$. That is the case of noise, the quantization error of compress sampling spectrum can be approximated as white noise spectrum, and the quantization noise variance is $\frac{\Delta^2}{12}$.

27.4 The Reconstruction Error

The actual quantization process of compressed sample can be expressed as:

$$Y_q = \mathbf{\Phi} X + N' \tag{27.15}$$

where $N' \in \mathbb{R}^M$ denotes the quantization error which is equal with the Gaussian white noise of the mean is 0 and the variance is $\frac{\Delta^2}{12}$. Let X be the reconstructed signal based on the noisy measurements Y_q. Then the reconstruction distortion is defined as $\|X - \hat{X}\|_2^2$ and the upper boundary is $\|X - \hat{X}\|_2^2 \leq c^2 \|N'\|_2^2 = \frac{c^2 \Delta^2}{12}$.

For the BP method, the parameter c_{bp} can be expressed as:

$$c_{bp} = \frac{4}{\sqrt{3 - 3\delta_{4K}} - \sqrt{1 + \delta_{4K}}} \tag{27.16}$$

While the parameter c_{sp} of the SP method can be also expressed as:

$$c_{sp} = \frac{1 + \delta_{3K} + \delta_{3K}^2}{\delta_{3K}(1 - \delta_{3K})} \tag{27.17}$$

Assume the location T of the sparse signal has been accurate estimation; the sparse signal corresponding to the columns of the measurement matrix $\mathbf{\Phi}_{\mathrm{T}}$. Then $\widehat{X}$ can be recovery by the following expression:

$$\hat{X} = \left(\mathbf{\Phi}_T^* \mathbf{\Phi}_T\right)^{-1} \mathbf{\Phi}_T Y_q \tag{27.18}$$

The lower boundary of the reconstruction error can be expressed as:

$$\begin{aligned} \|X - \hat{X}\|_2^2 &\geq \left(\frac{\sqrt{1 - \delta_K}}{1 + \delta_K}\right)^2 \|Y_q - Y\|_2^2 \\ &= \frac{1 - \delta_K}{(1 + \delta_K)^2} \|N'\|_2^2 \\ &= \frac{\Delta^2 (1 - \delta_K)}{12 (1 + \delta_K)^2} \end{aligned} \tag{27.19}$$

If $c_{lb} = \frac{\sqrt{1 - \delta_K}}{1 + \delta_K}$, the boundaries can be obtained by the reconstruction error caused by the quantization error of the compressed sampling values is as follows:

$$\frac{\Delta^2}{12} c_{lb}^2 \leq \|X - \hat{X}\|_2^2 \leq \frac{\Delta^2}{12} c^2 \tag{27.20}$$

27.5 Simulations and Experiments

To illustrate the effectiveness of proposed method for reconstructing CS's quantization noise spectrum, random Gaussian matrix and the coefficient of sparse signal and normalize the CS value before quantization are employed. All the testify experiment are done using the MATLAB 7.0 and XP operation system in the laptop with 1 GB RAM and 1.60 GHZ AthlonTM Processor.

If $N = 256, M = 128, K = 6$, CS's quantization noise spectrum is shown in Fig. 27.2. From top to bottom, the quantization steps are 1bit, 4bits, 8bits and 12bits, respectively. As is shown, CS's quantization noise spectrum is white, all the while.

The reconstruction error curves by utilizing measurement values after quantization are shown in Fig. 27.3. It compares the reconstruction performance of BP

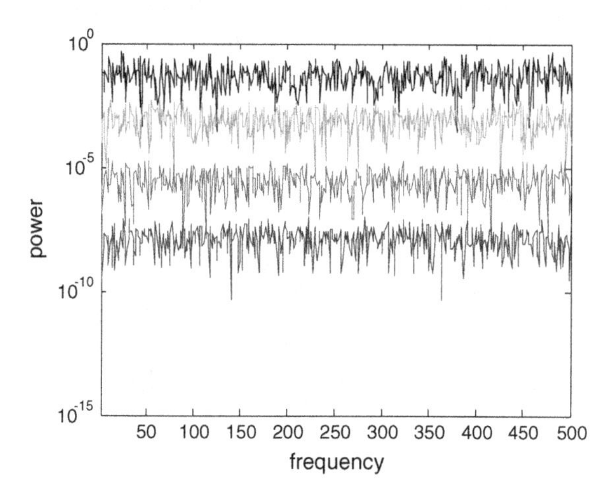

Fig. 27.2 CS's quantization noise spectrum

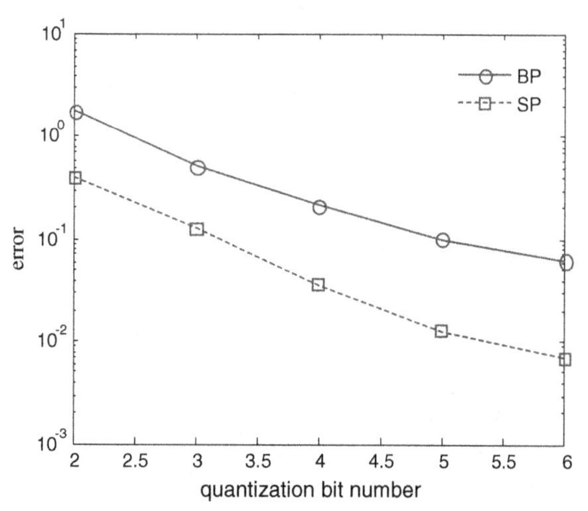

Fig. 27.3 Reconstruction error curves

and SP algorithm with different quantization bits number. From Fig. 27.3, both of these two algorithms' reconstruction errors decrease as the quantization bit number increases. The reason is that, increasing quantization bit number can reduce the width of quantization step. Consequently, the upper and lower bounds of reconstruction error given by (27.20) become smaller.

27.6 Analysis of Experimental Results

In this work, we take uniform quantizer to quantize the CS values and analyze the quantization noise, under the ideal and noise situations. The conclusion is that, the whiten function of measurement process can eliminate the correlation of measurement values. As a result, the CS's quantization noise spectrum is white, all the time. Based on these, the bounds of reconstruction error arising from quantization are proposed in the paper. Experimental results can prove the proposed bound is feasible.

References

1. Candès EJ (2006) Compressive sampling. In: Proceedings oh the international congress of mathematicians: Madrid, 22–30 Aug 2006, invited lectures, pp 1433–1452
2. Donoho DL (2006) Compressed sensing. IEEE Trans Inf Theory 52(4):1289–1306
3. Candès EJ, Romberg J, Tao T (2006) Robust uncertainty principles: exact signal reconstruction from highly incomplete frequency information. IEEE Trans Inf Theory 52 (2):489–509
4. Laska JN, Kirolos S, Duarte MF et al (2007) Theory and implementation of an analog-to-information converter using random demodulation. In: IEEE International symposium on circuits and systems, 2007. ISCAS 2007. IEEE, pp 1959–1962
5. Goyal VK, Fletcher AK, Rangan S (2008) Compressive sampling and lossy compression. Sig Process Mag IEEE 25(2):48–56
6. Sun JZ, Goyal VK (2009) Quantization for compressed sensing reconstruction. In: SAMPTA'09, International conference on sampling theory and applications, vol 2, pp 3–6
7. Laska JN, Boufounos PT, Davenport MA et al (2011) Democracy in action: quantization, saturation, and compressive sensing. Appl Comput Harmonic Anal 31(3):429–443
8. Zymnis A, Boyd S, Candes E (2010) Compressed sensing with quantized measurements. IEEE Signal Process Lett 17(2):149–152
9. Jacques L, Hammond DK, Fadili JM (2011) Dequantizing compressed sensing: when oversampling and non-Gaussian constraints combine. IEEE Trans Inf Theory 57(1):559–571
10. Dai W, Pham HV, Milenkovic O (2009) A comparative study of quantized compressive sensing schemes. In: Proceedings of the 2009 IEEE international conference on symposium on information theory-volume 1. IEEE Press, pp 11–15
11. Jacques L, Hammond D K, Fadili J (2011) Weighted fidelity in non-uniformly quantized compressed sensing. In: 2011 18th IEEE International Conference on Image Processing (ICIP). IEEE, pp 1921–1924
12. Chen SS, Donoho DL, Saunders MA (1998) Atomic decomposition by basis pursuit. SIAM J Sci Comput 20(1):33–61

13. Dai W, Milenkovic O (2009) Subspace pursuit for compressive sensing signal reconstruction. IEEE Trans Inf Theory 55(5):2230–2249
14. Sun JZ, Goyal VK (2009) Quantization for compressed sensing reconstruction. In: SAMPTA'09, International Conference on Sampling Theory and Applications, vol 2, pp 3–6

Chapter 28
Attitude Control for Rigid Satellite Under Actuator Constraint

Aihua Zhang, Haitao Meng and Zhiqiang Zhang

Abstract An attitude controller is proposed via employing backstepping control technique, and being represented by modified Rodriguez parameters. A general dynamic attitude model of satellites is deduced, along with a general model of actuator dynamics which can describe presumably all actuators for space application. External disturbances and actuator constraints are explicitly addressed. The control performance is proved in the numerical simulation experiences at last.

Keywords Attitude control · Backstepping control · Actuator constraint

28.1 Introduction

The external disturbance rejection issue of satellite attitude control is always the interest point for the researchers, although the satellite attitude stabilization control ideas have discussed in Refs. [1–3]. The reason comes from the difficulty of simple control structures design to complete satellite attitude control subject to external disturbance. The rapid development of space technology has brought more benefits to the space progress, e.g. a higher accuracy of attitude control, higher efficient and lower energy consumption under severe environment. Similar issues have been paid attention in the industrial process, e.g., Yin et al. show their remarkable performance of fault diagnosis or prediction in their recent work which worked pretty well in this field [4–8]. To meet these requirements in space field, many researches put their eyes on this mature technique applied in industry.

Focusing the issue discussed above, a novel attitude control for rigid satellite under actuator constraint is proposed in this paper. A general dynamical model of satellites is deduced along with a general model of actuator dynamics to design the controller. The proposed controller govern the attitude and the angular velocity of

A. Zhang (✉) · H. Meng · Z. Zhang
College of Engineering, Bohai University, Jinzhou 121013, China
e-mail: jsxinxi_zah@163.com

Y. Jia et al. (eds.), *Proceedings of the 2015 Chinese Intelligent Systems Conference*, Lecture Notes in Electrical Engineering 360,
DOI 10.1007/978-3-662-48365-7_28

the satellite to be uniformly ultimately bounded. The main contribution of this work in comparison with the existing attitude stabilization approaches is that, actuator dynamics and actuator constraints are simultaneously addressed.

Notation: Define $\Re$ to be the set of real numbers, define $\Re_+$ to be the set of positive real numbers, firstly. The set of m is defined as $\Re^{m\times n}$ obtained by n real matrices, and the $n\times n$ identity matrix is denoted by I_n. For any matrix $\boldsymbol{A}\in\Re^{m\times n}$, $\boldsymbol{A}^{\mathrm{T}}$ is used to denote its transpose, $\boldsymbol{A}^{\dagger}$ is also defined as its pseudo inverse matrix if $\boldsymbol{A}$ is full-row rank. The symbol $\|\cdot\|$ represents the Euclidean norm or matrix norm. If a vector $\boldsymbol{x}=[x_1\, x_2\ldots x_n]^{\mathrm{T}}\in\Re^n$ is defined first, then a vector-valued saturation function as $\boldsymbol{Sat}(x,\ell_{\max})=[\mathrm{sat}(x_1)\ \mathrm{sat}(x_2)\ldots\mathrm{sat}(x_n)]^{\mathrm{T}}$, $\ell_{\max}\in\Re_+$, and $\mathrm{sat}(x_i)=$ $\begin{cases}\ell_{\max}, & \text{if } x_i>\ell_{\max}\\ \bar{x}_i, & \text{if } -\ell_{\max}\le x_i\le\ell_{\max},\quad i=1,2,\ldots,n.\\ -\ell_{\max}, & \text{if } x_i<-\ell_{\max}\end{cases}$

28.2 Mathematical Model

The kinematics of a rigid Satellite is expressed as follows by using the MRPs [9]:

$$\dot{\boldsymbol{\sigma}}=\frac{1}{4}[(1-\boldsymbol{\sigma}^{\mathrm{T}}\boldsymbol{\sigma})\boldsymbol{I}_3+2\boldsymbol{\sigma}^{\times}+2\boldsymbol{\sigma}\boldsymbol{\sigma}^{\mathrm{T}}]\boldsymbol{\omega}=\boldsymbol{G}(\boldsymbol{\sigma})\boldsymbol{\omega} \tag{28.1}$$

where $\boldsymbol{G}(\boldsymbol{\sigma})\in\Re^{3\times3}$ in Eq. (28.1) is subject to:

$$\boldsymbol{G}^{\mathrm{T}}(\boldsymbol{\sigma})\boldsymbol{G}(\boldsymbol{\sigma})=\left(\frac{1+\boldsymbol{\sigma}^{\mathrm{T}}\boldsymbol{\sigma}}{4}\right)^2\boldsymbol{I}_3 \tag{28.2}$$

The satellite dynamical model here employed from Ref. [10]

$$\boldsymbol{J}\dot{\boldsymbol{\omega}}=-\boldsymbol{\omega}\times\boldsymbol{J}\boldsymbol{\omega}+\boldsymbol{u}+\boldsymbol{d} \tag{28.3}$$

where $\boldsymbol{u}$ is defined by

$$\boldsymbol{u}=\boldsymbol{D}\boldsymbol{\tau}_a \tag{28.4}$$

and $\boldsymbol{\tau}_a$ is defined by

$$\boldsymbol{T}\dot{\boldsymbol{\tau}}_a+\boldsymbol{\tau}_a=\boldsymbol{\tau}_c \tag{28.5}$$

where $\boldsymbol{D}\in\Re^{3\times N}$ represents the distribution matrix of actuator (available and full-row rank for a given Satellite), $\boldsymbol{J}\in\Re^{3\times3}$ represents the positive-definite, $\boldsymbol{\tau}_a=[\tau_{a1}\ \tau_{a2}\ldots\tau_{aN}]^{\mathrm{T}}\in\Re^N$ represents actual control inputs vector, $\boldsymbol{\tau}_c=[\tau_{c1}\ \tau_{c2}\ldots\tau_{cN}]^{\mathrm{T}}\in\Re^N$ represents the desired actuator inputs vector, and $\boldsymbol{T}=\boldsymbol{T}^{\mathrm{T}}\in\Re^{N\times N}$ represents diagonal square matrix subject to time constants T_i, $\boldsymbol{d} \in \Re^3$ represents external disturbance vector.

Assumption 1 Suppose there is a scalar $d_{\max} \in \Re_+$ such that $\|\boldsymbol{d}\| \le d_{\max}$.

Remark 1 Reference [11] proposed the method of avoiding singularity problem if the satellite dynamic model defined via MRPs. Adhering to the idea here, the satellite attitude rotation is done via employed MRPs $\boldsymbol{\sigma}$ with its shadow counterpart $\boldsymbol{\sigma}^s$.

28.3 Controller Design

For easy to prove the controller, we define some variables firstly: $\boldsymbol{x}_1 = \boldsymbol{\sigma}$, $\boldsymbol{x}_2 = \boldsymbol{\omega}$, and $\boldsymbol{x}_3 = \boldsymbol{\tau}_a$. Accordingly, the Satellite attitude model Eqs. (28.1) and (28.3) with actuator dynamics Eqs. (28.4) and (28.5) can be rewritten as:

$$\dot{\boldsymbol{x}}_1 = \boldsymbol{G}(\boldsymbol{x}_1)\boldsymbol{x}_2 \tag{28.6}$$

$$\boldsymbol{J}\dot{\boldsymbol{x}}_2 = -\boldsymbol{x}_2^{\times}\boldsymbol{J}\boldsymbol{x}_2 + \boldsymbol{D}\boldsymbol{x}_3 + \boldsymbol{d} \tag{28.7}$$

$$\boldsymbol{T}\dot{\boldsymbol{x}}_3 = -\boldsymbol{x}_3 + \boldsymbol{\tau}_c \tag{28.8}$$

Let us introduce a change of coordinates based on standard backstepping procedures:

$$z_1 = \boldsymbol{x}_1 \tag{28.9}$$

$$z_2 = \boldsymbol{x}_2 - \boldsymbol{\alpha}_1 \tag{28.10}$$

$$z_3 = \boldsymbol{x}_3 - \boldsymbol{\alpha}_2 \tag{28.11}$$

where $\boldsymbol{\alpha}_1 \in \Re^3$ and $\boldsymbol{\alpha}_2 \in \Re^N$ represent virtual control inputs, and which would be designed in the next part.

Based on the above discussion and analysis, the satellite attitude controller design can be proved detailed as follows.

Step 1. We set the control variable $\boldsymbol{x}_2$ defined in Eq. (28.6) firstly, then the time-derivative of z_1 is can be obtained via

$$\dot{z}_1 = \dot{\boldsymbol{x}}_1 = \boldsymbol{G}(z_1)\boldsymbol{x}_2 \tag{28.12}$$

For the purpose of guarantee $\lim_{t\to\infty} z_1(t) = 0$, here a $\boldsymbol{\alpha}_1$ defined virtual control is proposed. A Lyapunouv candidate function is defined by $V_1 = \frac{1}{2}z_1^{\mathrm{T}}z_1$, and design the virtual control $\boldsymbol{\alpha}_1$ as $\boldsymbol{\alpha}_1 = -k_1\boldsymbol{G}^{\mathrm{T}}(z_1)z_1$, where $k_1 \in \Re_+$ is a scalar. Then, applying Eqs. (28.2) and (28.10) yields

$$\begin{aligned}\dot{V}_1 &= z_1^{\mathrm{T}} \boldsymbol{G}(z_1) \boldsymbol{x}_2 \\ &= -k_1 z_1^{\mathrm{T}} \boldsymbol{G}(z_1) \boldsymbol{G}^{\mathrm{T}}(z_1) z_1 + z_1^{\mathrm{T}} \boldsymbol{G}(z_1) z_2 \\ &\leq -\frac{k_1}{16} \|z_1\|^2 + z_1^{\mathrm{T}} \boldsymbol{G}(z_1) z_2\end{aligned} \tag{28.13}$$

where the inequality $\boldsymbol{G}^{\mathrm{T}}(\boldsymbol{\sigma})\boldsymbol{G}(\boldsymbol{\sigma}) \geq (\frac{1}{16})\boldsymbol{I}_3$ is used. Hence, if $z_2 = \boldsymbol{0}$, then $\dot{V}_1 \leq -\frac{k_1}{16}\|z_1\|^2$. That leads to $\lim_{t\to\infty} z_1(t) = \boldsymbol{0}$.

Step 2. The second error should be recognized, considering the time-derivative of z_2 defined in Eq. (28.10), then we have

$$\boldsymbol{J}\dot{z}_2 = \boldsymbol{J}\dot{\boldsymbol{x}}_2 - \boldsymbol{J}\dot{\boldsymbol{\alpha}}_1 = -\boldsymbol{x}_2^{\mathrm{T}}\boldsymbol{\omega}^{\times}\boldsymbol{J}\boldsymbol{x}_2 + \boldsymbol{D}\boldsymbol{x}_3 + \boldsymbol{d} - \boldsymbol{J}\dot{\boldsymbol{\alpha}}_1 \tag{28.14}$$

Choose another Lyapunov candidate function $V_2 = V_1 + \frac{1}{2}z_2^{\mathrm{T}}\boldsymbol{J}z_2$, and design the virtual control law $\boldsymbol{\alpha}_2$ as

$$\boldsymbol{\alpha}_2 = \boldsymbol{D}^{\dagger}[\boldsymbol{x}_2^{\times}\boldsymbol{J}\boldsymbol{x}_2 - \boldsymbol{G}^{\mathrm{T}}(z_1)z_1 - k_2 z_2 + \boldsymbol{J}\dot{\boldsymbol{\alpha}}_1] \tag{28.15}$$

where $k_2 \in \mathfrak{R}_+$ is a constant. Differentiating both sides of V_2 and inserting Eqs. (28.11) and (28.15) yields

$$\begin{aligned}\dot{V}_2 &= \dot{V}_1 z_2^{\mathrm{T}}\boldsymbol{J}\dot{z}_2 \\ &\leq -\frac{k_1}{16}\|z_1\|^2 + z_1^{\mathrm{T}}\boldsymbol{G}(z_1)z_2 + z_2^{\mathrm{T}}[-\boldsymbol{x}_2^{\times}\boldsymbol{J}\boldsymbol{x}_2 + \boldsymbol{D}(z_3 + \boldsymbol{\alpha}_2) + \boldsymbol{d} - \boldsymbol{J}\dot{\boldsymbol{\alpha}}_1] \\ &= -\frac{k_1}{16}\|z_1\|^2 - k_2\|z_2\|^2 + z_2^{\mathrm{T}}\boldsymbol{D}z_3 + z_2^{\mathrm{T}}\boldsymbol{d}\end{aligned} \tag{28.16}$$

Theorem 1 *Consider the satellite attitude control system described by Eqs.* (28.1) *and* (28.3) *of a rigid satellite subject to external disturbances, actuator dynamics Eqs.* (28.4) *and* (28.5), *and actuator constraint, design the controller as:*

$$\boldsymbol{\tau}_c = \boldsymbol{Sat}(\boldsymbol{\tau}_I, \tau_{\max}) \tag{28.17}$$

with the input $\boldsymbol{\tau}_1 \in \mathfrak{R}^N$ of the controller synthesized as

$$\boldsymbol{\tau}_1 = \boldsymbol{\alpha}_2 - k_3 z_3 + \boldsymbol{T}\dot{\boldsymbol{\alpha}}_2 - \boldsymbol{D}^{\mathrm{T}} z_2 \tag{28.18}$$

where $k_3 \in \mathfrak{R}_+$ and $k_4 \in \mathfrak{R}_+$ are control gains, and $z_a \in \mathfrak{R}^3$ is the output of an auxiliary system given by:

$$\dot{z}_a = \begin{cases} -Kz_a - (\frac{1}{4\delta_1} + \frac{1}{4\delta_2})\frac{\|\Delta\boldsymbol{\tau}\|^2}{\|z_a\|^2} z_a - \Delta\boldsymbol{\tau}, & \|z_a\| \geq e \\ 0 & \|z_a\| < e \end{cases}. \tag{28.19}$$

where $\Delta\boldsymbol{\tau} = \boldsymbol{\tau}_c - \boldsymbol{\tau}_l$, $K \in \Re_+$ is a scalar, $z_a \in \Re^N$ is the state of auxiliary system, $e \in \Re_+$, $\delta_1 \in \Re_+$, and $\delta_2 \in \Re_+$ are constants. For any given scalar $\delta_3 \in \Re_+$, assume the control gains are defined as follow

$$k_2 - \delta_3 > 0 \tag{28.20}$$

$$k_3 + 1 - \delta_2 > 0 \tag{28.21}$$

$$K - \delta_1 > 0 \tag{28.22}$$

Therefore, the satellite attitude control system is stable and uniformly bounded. and that is $|\tau_{ai}| \le \tau_{\max}$, which denotes the position saturation of the actual control $\boldsymbol{\tau}_{ai}$ is overly prevented for all $t \ge 0$ and $i = 1, 2, \ldots, N$, the problem of actuator constraint is thus solved. Moreover, for the satellite attitude and velocity, there exists a scalar $t_0 \in \Re_+$ such that $\|\boldsymbol{\sigma}(t)\| \le \varepsilon_1$ and $\|\boldsymbol{\omega}(t)\| \le \varepsilon_2$ for all $t \ge t_0$.

Proof For the space limit, the proof part is omitted here. □

28.4 Simulation Results

In this section the properties of the proposed attitude stabilization methodology is evaluated through numerical simulations. A rigid satellite currently developed is numerically simulated. The physical parameters and the initial attitude of the satellite are all selected from Ref. [9]. The control gains for the controller Eq. (28.17) are chosen as $k_1 = 0.2$, $k_2 = 1.5$, $k_3 = 1.5$, $\delta_1 = 0.2$, $\delta_2 = 0.2$, $\delta_3 = 0.01$ and $K = 1.5$.

In the simulation part, a large constant external disturbance torque is defined by $\boldsymbol{d}(t) = [0.02\ \ -0.01\ 0.01]^{\mathrm{T}}$. When the designed control approach is acted on the satellite attitude stabilization control, the control results are shown in Figs. 28.1 and 28.2. The attitude control can be found from the simulation.

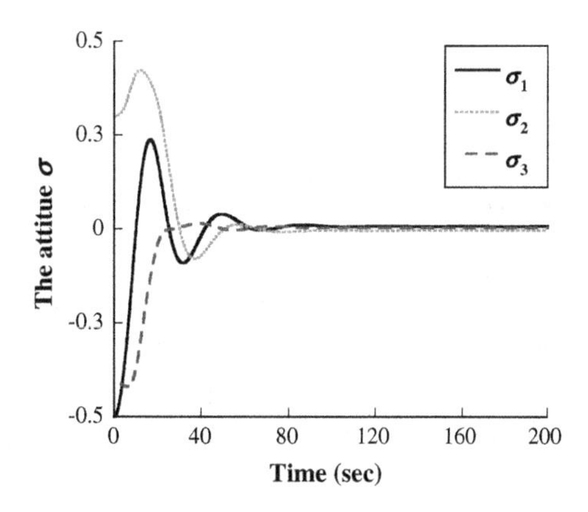

Fig. 28.1 The $\boldsymbol{\sigma}$ in the presence of constant external disturbances

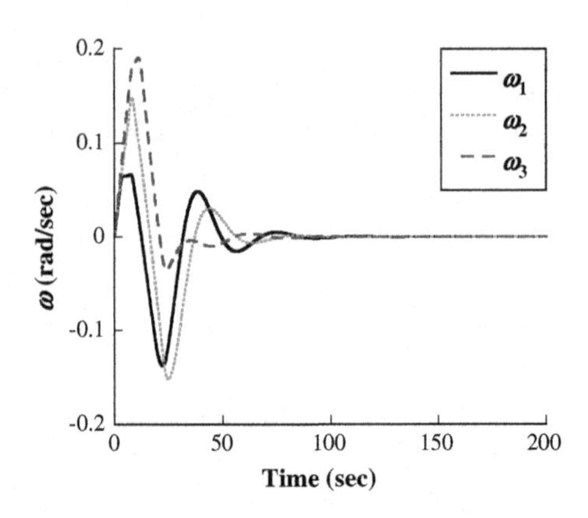

Fig. 28.2 The ω in the presence of constant external disturbances

28.5 Conclusions

A novel backstepping based nonlinear controller was developed to accomplish attitude stabilization control of rigid satellite. External disturbances and actuator constraints were solved. Moreover, the satellite attitude control, the angular velocity control were uniformly ultimately bounded under the proposed controller. Simulation results were presented to demonstrate the control performance of the controller.

References

1. Yoon H, Agrawal BN (2009) Adaptive control of uncertain Hamiltonian multi-input multi-output systems: with application to spacecraft control. IEEE Trans Control Syst Technol 17:900–906
2. Seo D, Akella MR (2007) Separation property for the rigid body attitude tracking control problem. J Guidance Control Dyn 30:1569–1576
3. Seo D, Akella MR (2008) High-performance spacecraft adaptive attitude-tracking control through attracting-manifold design. J Guidance Control Dyn 31:884–891
4. Plestan F, Bregeault V, Glumineau A, Shtessel Y, Moulay E (2012) Advances in high order and adaptive sliding mode control–theory and applications. In: Sliding modes after the first decade of the 21st century. Springer, 2012, pp 465–492
5. Sidi MJ (1997) Spacecraft dynamics and control: a practical engineering approach, vol 7. Cambridge University Press, Cambridge
6. Sekara TB, Matausek MR (2009) Optimization of PID controller based on maximization of the proportional gain under constraints on robustness and sensitivity to measurement noise. IEEE Trans Autom Control 54:184–189
7. Zou A-M, Kumar KD, Hou Z-G (2010) Quaternion-based adaptive output feedback attitude control of spacecraft using Chebyshev neural networks. IEEE Trans Neural Netw 21:1457–1471
8. Yin S, Ding S, Xie X, Luo H (2014) A review on basic data-driven approaches for industrial process monitoring. IEEE Trans Industr Electron 61:6418–6428

9. Crassidis JL, Markley FL (1996) Sliding mode control using modified Rodrigues parameters. J Guidance Control Dyn 19:1381–1383
10. Sidi MJ (1997) Spacecraft dynamics and control. Cambridge University Press, Cambridge
11. Schaub H, Akella MR, Junkins JL (2001) Adaptive control of nonlinear attitude motions realizing linear closed loop dynamics. J Guidance Control Dyn 24:95–100

Chapter 29
Observer-Based Adaptive Consensus for Multi-agent Systems with Nonlinear Dynamics

Heyang Wang and Lin Li

Abstract Distributed consensus problem is investigated for Lipschitz nonlinear multi-agent systems (MASs). Under the assumption that the states of the multiple agents are unmeasured, nonlinear observer for each agent is designed. Based on these observers, a distributed protocol is proposed, in which the coupling weights between adjacent agents are time-varying and can automatically change according to the designed adaptive law. Lyapunov-Krasovskii functional is constructed to analyses the consensus problem of the MASs under the proposed distributed adaptive protocol. By using free-weighting matrix approach, sufficient conditions that can ensure consensus are given. Finally, numerical example is presented to illustrate our result.

Keywords Consensus · Multi-agent systems · Adaptive protocol · Observer

29.1 Introduction

During the past few years, consensus problem of MASs has received more and more attention due to its broader range applications, such as flocking control [1], tracking control [2], unmanned aerial vehicle formation, underwater robot coordinated search and rescue, etc. The consensus problem is that the single agent often in some amount (velocity, position) tends to be consistent with other agents. Over the past decade, research on consensus problem has achieved numerous meaningful results. For example, average consensus problem was studied in [3]. Based on the results obtained in [3], relaxed conditions were obtained in [4]. The H_∞ consensus for MASs with delay was investigated in [5]. More results about the consensus problems can be found in [6–12].

H. Wang · L. Li (✉)
Department of Control Science and Engineering, University of Shanghai for Science and Technology, Shanghai 200093, China
e-mail: lilin0211@163.com

Y. Jia et al. (eds.), *Proceedings of the 2015 Chinese Intelligent Systems Conference*, Lecture Notes in Electrical Engineering 360,
DOI 10.1007/978-3-662-48365-7_29

It is noteworthy that the protocols in the aforementioned literatures are all based on the actual neighbors' states. However, the states of agents are often difficult to be measured in practical applications. In this case, distributed control protocols based on distributed observers are usually need to be considered. The leader-following consensus problem was investigated in [12, 13], where only partial states of the agents can be measured.

On the other hand, when the nonlinear dynamics of a multi-agent system are considered, the consensus problem is more challenging, and the design of adaptive law can solve these problems effectively [13, 14]. However, most of the research results of MASs with nonlinear dynamics have no consideration of the affect of delay on the convergence of the system. Here, the consensus problem of MASs with nonlinear dynamics under a fixed undirected connected topology and constant communication time delay is in focus. Firstly, distributed full-order observers are designed to estimate the states of agents. And then, using the observed information, a distributed adaptive consensus protocol is proposed. Furthermore, sufficient condition for the existence of this protocol is derived. Finally, numerical simulation shows the effectiveness of our obtained method.

29.2 Problem Statement and Preliminaries

Consider a weighted undirected graph $\mathcal{G} = (\mathcal{V}, \mathcal{E}, \mathcal{A})$. Thereinto, $\mathcal{V}$ represents the set of nodes with $\mathcal{V} = \{v_1, \ldots, v_N\}$, and $\mathcal{E}$ is the set of undirected edges, $\mathcal{A} = [a_{ij}]$ is the weighted adjacency matrix where $a_{ii} = 0, a_{ij} \geq 0$. An undirected edge a_{ij} is denoted by the pair of nodes (v_i, v_j), in which $(v_i, v_j) \in \mathcal{E} \Longleftrightarrow a_{ij} > 0$, while $a_{ij} = 0$. Then in-degree of node v_i is defined by $\mathcal{D}_{in}(i) = \sum_{j\in N_i} a_{ij}, i, j = 1, 2, \ldots N$. In undirected graph, the out-degree is equal to the in-degree. The Laplacian matrix is defined as $L = [l_{ij}]_{N\times N} = \Delta - \mathcal{A}$, in which $\Delta = \mathrm{diag}\{\mathcal{D}_{in}(1), \ldots, \mathcal{D}_{in}(N)\}$, $l_{ii} = \sum_{j=1, j\neq i}^{N} a_{ij}$, $l_{ij} = -a_{ij} (i \neq j)$.

Consider a multi-agent system consisting of N agents, and the dynamic of the ith agent is

$$\begin{cases} \dot{\eta}_i(t) = A\eta_i(t) + f(\eta_i) + Bu_i(t) \\ y_i(t) = C\eta_i(t), i = 1, 2, \ldots, N \end{cases} \tag{29.1}$$

where $\eta_i \in \mathbb{R}^n$ is state, $y_i \in \mathbb{R}^q$ is the measured output, and $u_i \in \mathbb{R}^p$ is control input. A, B, C are known real matrices. The nonlinear function $f(\eta_i)$ satisfies the following Lipschitz condition:

$$\|f(m) - f(n)\| \leq \mu \|m - n\|, \tag{29.2}$$

where μ is a known real constant.

For system (29.1), consider the following observer

$$\begin{cases} \dot{\hat{\eta}}_i(t) = A\hat{\eta}_i(t) + f(\hat{\eta}_i) + Bu_i(t) + M(y_i(t) - \hat{y}_i(t)) \\ \hat{y}_i(t) = C\hat{\eta}_i(t), i = 1, 2, \ldots, N \end{cases} \tag{29.3}$$

where M is the observer feedback matrix, $\hat{\eta}_i$ is the estimation of state η_i, $\hat{y}_i$ is the output of the observer.

Based on the above observer (29.3), consider the protocol as follows:

$$u_i(t) = K \sum_{j=1}^{N} c_{ij}a_{ij}(\hat{\eta}_i(t-\tau) - \hat{\eta}_j(t-\tau)), \tag{29.4}$$

with the adaptive law

$$\dot{c}_{ij} = 2\kappa_{ij}a_{ij}[(\hat{\eta}_i(t) - \bar{\hat{\eta}})^T \Gamma(\hat{\eta}_i(t-\tau) - \hat{\eta}_j(t-\tau))], i, j = 1, 2, \ldots, N \tag{29.5}$$

where $\bar{\hat{\eta}}$ is the average value of $\hat{\eta}_i$, $\Gamma \in \mathbb{R}^{n\times n}$, $K \in \mathbb{R}^{p\times n}$ are the determined feedback gain matrices. The positive constants $\kappa_{ij} = \kappa_{ji}$, c_{ij} is the time-varying coupling weight of agent i and j satisfying $c_{ij}(0) = c_{ji}(0)$.

Substituting (29.4) into (29.1) and (29.3) yields,

$$\begin{cases} \dot{\eta}_i(t) = & A\eta_i(t) + f(\eta_i) + BK\sum_{j=1}^{N} c_{ij}a_{ij}(\hat{\eta}_i(t-\tau) - \hat{\eta}_j(t-\tau)) \\ \dot{\hat{\eta}}_i(t) = & A\hat{\eta}_i(t) + f(\hat{\eta}_i) + BK\sum_{j=1}^{N} c_{ij}a_{ij}(\hat{\eta}_i(t-\tau) - \hat{\eta}_j(t-\tau)) \\ & + MC(\eta_i(t) - \hat{\eta}_i(t)). \end{cases} \tag{29.6}$$

Let $e_{1i}(t) = \eta_i(t) - \hat{\eta}_i(t)$ be the state estimation error, and $e_{2i}(t) = \hat{\eta}_i(t) - \frac{1}{N}\sum_{j=1}^{N}\hat{\eta}_j(t)$ be the consensus error with $i, j = 1, 2, \ldots, N$.

It follows from (29.5) and (29.6) that

$$\begin{cases} \dot{e}_{1i}(t) = & (A - MC)e_{1i}(t) + f(\eta_i) - f(\hat{\eta}_i) \\ \dot{e}_{2i}(t) = & Ae_{2i}(t) + f(\hat{\eta}_i) - \frac{1}{N}\sum_{j=1}^{N} f(\hat{\eta}_j) + BK\sum_{j=1}^{N}(\tilde{c}_{ij} + \beta)a_{ij}(e_{2i}(t-\tau) \\ & - e_{2j}(t-\tau)) + MC(e_{1i}(t) - \frac{1}{N}\sum_{j=1}^{N} e_{1j}(t)) \end{cases} \tag{29.7}$$

$$\dot{\tilde{c}}_{ij} = 2\kappa_{ij}a_{ij}[e_{2i}^T(t)\Gamma(e_{2i}(t-\tau) - e_{2j}(t-\tau))], i, j = 1, 2, \ldots, N, \tag{29.8}$$

where the constant $\beta > 0$, and satisfies $\tilde{c}_{ij} = c_{ij} - \beta$.

The objective of this paper is to design a distributed adaptive protocol (29.4) such that the states of system (29.1) achieve consensus asymptotically. That is, for any initial conditions $\eta_i(0)$,

$$\lim_{t\to\infty}(\eta_i(t) - \eta_j(t)) = 0, i, j = 1, 2, \ldots, N.$$

Remark 1 For any $i,j = 1, 2, \ldots, N$, if $\lim_{t\to\infty} e_{2i}(t) = 0$, then $\lim_{t\to\infty} \hat{\eta}_i = \lim_{t\to\infty} \hat{\eta}_j$. That is the observer states $\hat{\eta}_i$ reach consensus. Similarly, $\lim_{t\to\infty} e_{1i}(t) = 0$ can guarantee $\lim_{t\to\infty} \eta_i = \lim_{t\to\infty} \hat{\eta}_i$. Thus, when $\lim_{t\to\infty} e_{2i}(t) = 0$ and $\lim_{t\to\infty} e_{1i}(t) = 0$ are satisfied simultaneously, all the agent states η_i also can reach consensus.

29.3 Main Results

Theorem 1 *Given positive scalars β and μ, if there exist positive definite matrices P, Q, and R, and matrices $Y = [Y_1\ Y_2]^T$ such that*

$$\Theta < 0 \tag{29.9}$$

where

$$\Theta = \begin{bmatrix} I_N \otimes \chi & \Lambda \otimes HCP & I_N \otimes Q & I_N \otimes Q & (1,5) & (1,6) & 0 \\ * & I_N \otimes \psi & 0 & 0 & 0 & 0 & I_N \otimes P \\ * & * & -I & 0 & 0 & 0 & 0 \\ * & * & * & -I_N \otimes R & 0 & 0 & 0 \\ * & * & * & * & (5,5) & -I_N \otimes Y_2^T & 0 \\ * & * & * & * & * & (6,6) & 0 \\ * & * & * & * & * & * & -I \end{bmatrix}$$

$$\chi = AQ + QA^T + \gamma^2 I + Y_1 + Y_1^T,\ \psi = P(A - MC)^T + (A - MC)P + \mu^2 I$$

$$\Lambda = \left[\frac{1}{N}\right]_{N\times N} - I_N,\ (1,5) = -\beta\mathcal{L} \otimes BB^T - I_N \otimes Y_1,$$

$$(5,5) = -I_N \otimes (R - 2Q), (1,6) = -I_N \otimes (Y_1 - Y_2^T),\ (6,6) = -I_N \otimes (Y_2 + Y_2^T),$$

Then, choosing $\Gamma = Q^{-1}BB^TQ^{-1}$, $K = -B^TQ^{-1}$, system (29.1) can achieve consensus under the protocol (29.4). Here, the signal $[a]_{n\times n}$ denotes a matrix that belongs to $\mathbb{R}^{n\times n}$, and all the elements of this matrix is a.

Proof Define the following Lyapunov-Krasovskii functional candidate by

$$\begin{aligned} V(t) &= \sum_{i=1}^{N} e_{1i}^T(t)P^{-1}e_{1i}(t) + \sum_{i=1}^{N} e_{2i}^T(t)Q^{-1}e_{2i}(t) \\ &\quad + \sum_{i=1}^{N} \int_{t-\tau}^{t} e_{2i}^T(t)R^{-1}e_{2i}(t)dt + \sum_{i=1}^{N}\sum_{j=1,j\neq i}^{N} \frac{\tilde{c}_{ij}^2}{2\kappa_{ij}}. \end{aligned}$$

Taking the time derivative of Lyapunov functional $V(t)$, we get

$$\begin{aligned}
\dot{V}(t) =& 2\sum_{i=1}^{N} e_{1i}^T(t)P^{-1}\dot{e}_{1i}(t) + \sum_{i=1}^{N}\sum_{j=1,j\neq i}^{N} \frac{\tilde{c}_{ij}^2}{2\kappa_{ij}}\dot{c}_{ij} + 2\sum_{i=1}^{N} e_{2i}^T(t)Q^{-1}\dot{e}_{2i}(t) \\
&+ \sum_{i=1}^{N}[e_{2i}^T(t)R^{-1}e_{2i}(t) - e_{2i}^T(t-\tau)R^{-1}e_{2i}(t-\tau)] \\
=& 2\sum_{i=1}^{N} e_{1i}^T(t)P^{-1}[(A-MC)e_{1i}(t) + f(\eta_i) - f(\hat{\eta}_i)] + 2\sum_{i=1}^{N} e_{2i}^T(t)Q^{-1}[Ae_{2i}(t) \\
&+ f(\hat{\eta}_i) - \frac{1}{N}\sum_{j=1}^{N} f(\hat{\eta}_j) + \sum_{j=1}^{N}(\tilde{c}_{ij}+\beta)a_{ij}BK(e_{2i}(t-\tau) - e_{2j}(t-\tau)) \\
&+ MC(e_{1i}(t) - \frac{1}{N}\sum_{j=1}^{N} e_{1j}(t))] + \sum_{i=1}^{N}\sum_{j=1,j\neq i}^{N} \tilde{c}_{ij}a_{ij}[e_{2i}^T(t)\Gamma(e_{2i}(t-\tau) \\
&- e_{2j}(t-\tau)) + (e_{2i}(t-\tau) - e_{2j}(t-\tau))^T\Gamma e_{2i}(t)] + \sum_{i=1}^{N}[e_{2i}^T(t)R^{-1}e_{2i}(t) \\
&- e_{2i}^T(t-\tau)R^{-1}e_{2i}(t-\tau)]
\end{aligned}$$

$$\begin{aligned}
=& 2\sum_{i=1}^{N} e_{1i}^T(t)P^{-1}[(A-MC)e_{1i}(t) + f(\eta_i) - f(\hat{\eta}_i)] + 2\sum_{i=1}^{N} e_{2i}^T(t)Q^{-1}Ae_{2i}(t) \\
&- 2\beta\sum_{i=1}^{N}\sum_{j=1}^{N} \ell_{ij}e_{2i}^T(t)Q^{-1}BB^TQ^{-1}e_{2j}(t-\tau) + 2\sum_{i=1}^{N} e_{2i}^T(t)Q^{-1}[f(\hat{\eta}_i) \\
&- f(\bar{\eta}) + f(\bar{\eta}) - \frac{1}{N}\sum_{j=1}^{N} f(\hat{\eta}_j)] + 2\sum_{i=1}^{N} e_{2i}^T(t)Q^{-1}MC(e_{1i}(t) - \frac{1}{N}\sum_{j=1}^{N} e_{1j}(t)) \\
&+ \sum_{i=1}^{N}[e_{2i}^T(t)R^{-1}e_{2i}(t) - e_{2i}^T(t-\tau)R^{-1}e_{2i}(t-\tau)],
\end{aligned} \tag{29.10}$$

According to the Lipschitz condition (29.2), we have

$$\begin{aligned}
&2e_{1i}^T(t)P^{-1}[f(\eta_i) - f(\hat{\eta}_i)] \leq 2\mu\|P^{-1}e_{1i}(t)\|\|e_{1i}(t)\| \leq e_{1i}^T(t)(\mu^2(P^{-1})^2 + I)e_{1i}(t) \\
&2\sum_{i=1}^{N} e_{2i}^T(t)Q^{-1}[f(\hat{\eta}_i) - f(\bar{\eta})] \leq e_{2i}^T(t)(\mu^2(Q^{-1})^2 + I)e_{2i}(t),
\end{aligned} \tag{29.11}$$

Since $\sum_{i=1}^{N} e_{2i}^T(t) = 0$, we can get $\sum_{i=1}^{N} e_{2i}^T(t)Q^{-1}[f(\bar{\eta}) - \frac{1}{N}\sum_{j=1}^{N} f(\hat{\eta}_j)] = 0$.

Let $\tilde{e}_{2i}(t) = Q^{-1}e_2(t), \tilde{e}_{2i}(t-\tau) = Q^{-1}e_{2i}(t-\tau)$. For any matrix Y of the form $Y = [Y_{1i}\ \ Y_{2i}]^T, i = 1, 2, \ldots, N$, we obtain the following equation.

$$\sum_{i=1}^{N} 2[\tilde{e}_{2i}^T(t)Y_{1i} + \int_{t-\tau}^{t} \tilde{e}_{2i}^T(s)Y_{2i}ds][\tilde{e}_{2i}(t) - \tilde{e}_{2i}(t-\tau) - \int_{t-\tau}^{t} \tilde{e}_{2i}(s)ds] = 0. \quad (29.12)$$

In virtue of (29.11) and (29.12), the equation (29.10) can be written as

$$\begin{aligned}
\dot{V}(t) \le\ & 2\sum_{i=1}^{N} e_{1i}^T(t)P^{-1}(A - MC)e_{1i}(t) + \sum_{i=1}^{N} e_{1i}^T(t)(\mu^2(P^{-1})^2 + I)e_{1i}(t) \\
& + \sum_{i=1}^{N} \tilde{e}_{2i}^T(t)[AQ + QA^T + Y_{1i} + \mu^2 I + Q^2 + QR^{-1}Q + Y_{1i} + Y_{1i}^T)\tilde{e}_{2i}(t) \\
& - 2\sum_{j=1}^{N}(\beta\ell_{ij}BB^T + Y_{1i})\tilde{e}_{2i}(t-\tau) - 2(Y_{1i} - Y_{2i}^T)\int_{t-\tau}^{t} \tilde{e}_{2i}(s)ds \\
& + 2MCe_{1i}(t) - \frac{1}{N}\sum_{j=1}^{N} 2MCe_{1j}(t)] - \sum_{j=1}^{N} \tilde{e}_{2i}^T(t-\tau)QR^{-1}Q\tilde{e}_{2i}(t-\tau) \\
& - 2[\tilde{e}_{2i}^T(t-\tau)Y_{2i}^T + \int_{t-\tau}^{t} \tilde{e}_{2i}^T(s)Y_{2i}ds]\int_{t-\tau}^{t} \tilde{e}_{2i}(s)ds.
\end{aligned} \quad (29.13)$$

Let $e_1(t) = [e_{11}^T(t), e_{12}^T(t), \ldots, e_{1N}^T(t)]^T$, $\tilde{e}_2(t) = [\tilde{e}_{21}^T(t), \tilde{e}_{22}^T(t), \ldots, \tilde{e}_{2N}^T(t)]^T$, and $\tilde{e}_2(t-\tau) = [\tilde{e}_{21}^T(t-\tau), \tilde{e}_{22}^T(t-\tau), \ldots, \tilde{e}_{2N}^T(t-\tau)]^T$, then the inequality (29.13) can be rewritten as:

$$\begin{aligned}
\dot{V}(t) \le\ & e_1^T(t)I_N \otimes [(A - MC)^T P^{-1} + P^{-1}(A - MC) + (\mu^2(P^{-1})^2 + I)]e_1(t) \\
& + \tilde{e}_2^T(t)[I_N \otimes (AQ + QA^T + \mu^2 I + Q^2 + QR^{-1}Q + Y_1 + Y_1^T)\tilde{e}_2(t) \\
& - 2(\beta\mathcal{L} \otimes BB^T - I_N \otimes Y_1)\tilde{e}_2(t-\tau) - 2I_N \otimes (Y_1 - Y_2^T)\int_{t-\tau}^{t} \tilde{e}_2(s)ds \\
& - 2(I_N \otimes MC + E \otimes MC)e_1(t)] - 2[\tilde{e}_2^T(t-\tau)I_N \otimes Y_2^T \\
& + \int_{t-\tau}^{t} \tilde{e}_2^T(s)I_N \otimes Y_2 ds]\int_{t-\tau}^{t} \tilde{e}_2(s)ds \\
& - \tilde{e}_2^T(t-\tau)I_N \otimes QR^{-1}Q\tilde{e}_2(t-\tau).
\end{aligned} \quad (29.14)$$

In the following presentation, it is necessary to introduce a matrix inequality (see [15]): For any matrices $H > 0$ and U of appropriate dimensions, we have $U^T H^{-1} U \ge U + U^T - H$. Then, let $\xi(t) = (e_1^T(t), \tilde{e}_2^T(t), \tilde{e}_2^T(t-\tau), \int_0^t \tilde{e}_2^T(s)ds)^T$, we can obtain

$$\begin{aligned}
\dot{V}(t) \le\ & \xi^T(t)\begin{bmatrix} I_N \otimes \psi & (1,2) & 0 & 0 \\ * & I_N \otimes \varphi & (2,3) & -I_N \otimes (Y_1 - Y_2^T) \\ * & * & -I_N \otimes (R - 2Q) & -I_N \otimes Y_2^T \\ * & * & * & -I_N \otimes (Y_2 + Y_2^T) \end{bmatrix}\xi(t) \\
=\ & \xi^T(t)\Phi\xi(t)
\end{aligned}$$

where $E = [\frac{1}{N}]_{N\times N}$, and

$$\varphi = AQ + QA^T + \mu^2 I + Y_1 + Y_1^T,\ \psi = P(A - MC)^T + (A - MC)P + \mu^2 I,$$
$$(1,2) = E \otimes C^T M^T - I_N \otimes C^T M^T, (2,3) = -\beta \mathcal{L} \otimes BB^T - I_N \otimes Y_1,$$

From Schur formula, $\Phi < 0$ is equivalent to the inequality (29.9) in Theorem 1. Thus, the estimation error and the synchronization error of the system (29.7) are asymptotically convergent to zero. Hence, under the distributed adaptive protocol (29.4), the system (29.1) can reach consensus. The proof is completed.

29.4 Simulation Examples

An illustrative example will be given to show the effectiveness of our results. Consider an undirected graph $\mathcal{G}$ including four agents. The elements of the adjacency matrix $\mathcal{A}$ are $a_{12} = a_{21} = a_{23} = a_{32} = a_{34} = a_{43} = a_{14} = a_{41} = 1.3$. Consider the system (29.1) with $f(x_i) = [0\ \ 0\ \ 0\ \ -0.1\sin(x_{i4})]^T$, and

$$A = \begin{bmatrix} -1 & 1 & 0 & 0 \\ 0 & -48.6 & -12.5 & 48.6 \\ 0 & 0 & 0 & 10 \\ 1.95 & 0 & -1.95 & 0 \end{bmatrix},\ B = \begin{bmatrix} 1 \\ 1 \\ 1 \\ 1 \end{bmatrix},\ C = \begin{bmatrix} 0.7 & -0.2 & 0.1 & 0.4 \\ 0.8 & 0.6 & 0.4 & 0.3 \\ 0.23 & 0.25 & 0.18 & 0.4 \\ 0.9 & 0.7 & 0.86 & 0.55 \end{bmatrix}.$$

Obviously, $f(x_i)$ satisfies the condition (29.2) with a constant $\mu = 0.1$. Let $\tau = 0.1$, $\beta = 1$, $\kappa_{ij} = 1, i,j = 1,2,\ldots,N$. The initial states of system (29.1) are $x_{1j}(0) = [2\ -1\ -6\ -3]^T$, $x_{2j}(0) = [4\ \ 2\ -5\ \ 3]^T$, $x_{3j}(0) = [3\ \ 5\ \ 2\ -2]^T$, $x_{4j}(0) = [1\ -3\ \ 4\ \ 4]^T$. Solving the inequality (29.9) gives the feedback gain matrices in (29.3), (29.4) and (29.5) as $K = [-13.9381\ -1.1194\ \ 3.9368\ -46.5433]$, and

$$\Gamma = \begin{bmatrix} 194.3 & 15.6 & -54.9 & 648.7 \\ 15.6 & 1.3 & -4.4 & 52.1 \\ -54.9 & -4.4 & 15.5 & -183.2 \\ 648.7 & 52.1 & -183.2 & 2166.3 \end{bmatrix},\ M = \begin{bmatrix} 236.84 & 276.36 & -318.82 & -89.35 \\ -307.56 & 322.84 & 258.18 & -140.18 \\ -52.36 & -520.45 & -369 & 597.65 \\ 79.31 & -214.42 & 752.54 & -62.32 \end{bmatrix}.$$

The state trajectories of nonlinear multi-agent system (29.1) are shown in Fig. 29.1. From Fig. 29.1, we can see that the states of system (29.1) achieve consensus under the protocol (29.4).

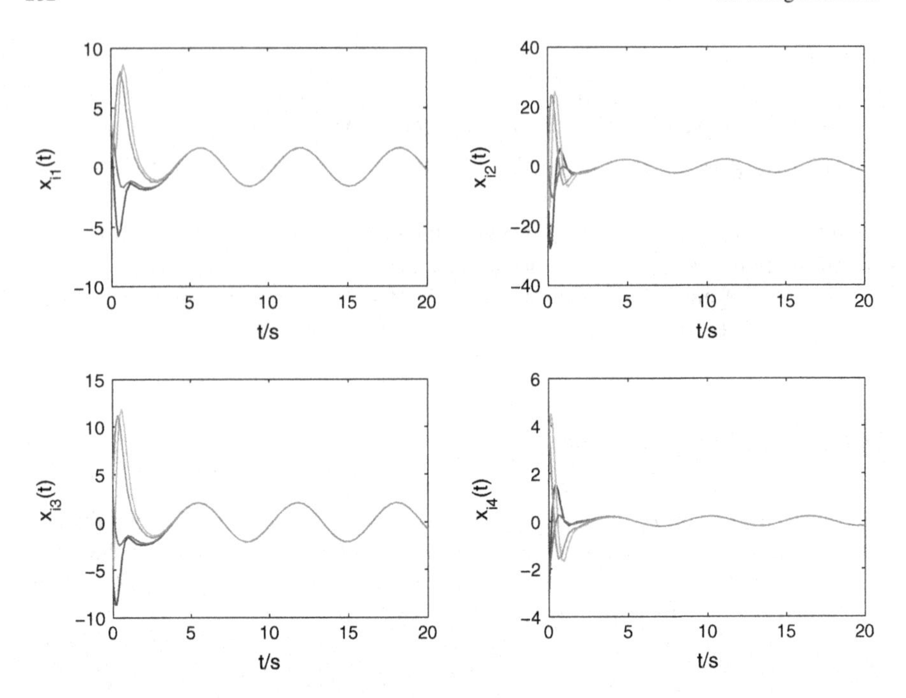

Fig. 29.1 States of the multi-agent system (29.1)

29.5 Conclusions

The distributed adaptive consensus problem of Lipschitz nonlinear multi agent systems has been investigated in this paper. Considering the states of agents be unmeasured, full-order observer for each agent is design to estimate the actual state of agent. And then, a observer-based distributed adaptive protocol is proposed and designed. Some sufficient condition is obtained such that the Lipschitz nonlinear multi agent systems achieve consensus. Simulation example is illustrated to demonstrate the effectiveness of proposed result.

Acknowledgments This paper was supported by National Natural Science Foundation (61203143), and Hujiang Foundation of China (C14002).

References

1. Su H, Wang X, Lin Z (2009) Flocking of multi-agents with a virtual leader. Control Theory Appl 54(2):293–307
2. Wang Y, Yan W (2013) Consensus formation tracking control of multiple autonomous underwater vehicle system. J Guid Control Dyn 30(3):380–384

3. Zhang T, Yu H (2010) Average consensus for directed networks of multi-agent with time-varying delay. In: Advances in swarm intelligence, Beijing, Chian, pp. 723–730
4. Ren W, Beard R (2005) Consensus seeking in multi-agent systems under dynamically changing interaction topologies. IEEE Trans Autom Control 50(5):655–661
5. Lin P, Jia Y (2008) Distributed robust H_∞ consensus control in directed networks of agents with time-delay. Syst Control Lett 57(8):643–653
6. Olfati-Saber R, Murray RM (2004) Consensus problems in networks of agents with switching topology and time-delays. IEEE Trans Autom Control 49(9):1520–1533
7. Ren W, Beard R, Atkins E (2005) A survey of consensus problem in multi-agent coordination. In: Proceedings of the american control conference, pp 1859–1864. IEEE, Portland
8. Olfati-Saber R, Fax JA, Murry RM (2007) Consensus and cooperation in networked multi-agent systems. Proc IEEE 95(1):215–233
9. Li Z, Duan Z, Chen G, Huang L (2010) Consensus of multi-agent systems and synchronization of complex networks: a unified viewpoint. Circuits Syst I Regul Pap IEEE Trans 57(1):213–224
10. Seo JH, Shim H, Back J (2009) Consensus of high-order linear systems using dynamic output feedback compensator: low gain approach. Automatica 45(11):2659–2664
11. DeLellis P, Bernardo M, Garofalo F (2009) Novel decentralized adaptive strategies for the synchronization of complex networks. Automatica 45(5):1312–1319
12. Hong Y, Wang X, Jiang Z (2013) Distributed output regulation of leader-follower multi-agent systems. Int J Robust Nonlinear Control 23(1):48–66
13. Hu J, Cao J, Yu J et al (2014) Consensus of nonlinear multi-agent systems with observer-based protocols. Syst Control Lett 72:71–79
14. Li Z, Ren W, Liu X, Fu M (2013) Consensus of multi-agent systems with general linear and Lipschitz nonlinear dynamics using distributed adaptive protocols. IEEE Trans Autom Control 58(7):1786–1791
15. De Oliveira MC, Bernussou J, Geromel JC (1999) A new discrete-time robust stability condition. Syst Control Lett 37(4):261–265

Chapter 30
Research on Asphalt Gas Concentration Control System Based on Fuzzy-PID Control

Man Feng, Weicun Zhang and Yuzhen Zhang

Abstract The asphalt gas concentration control system is characterized by its long-time delay, large inertia object and varying control parameter, too complicated to be applied in traditional PID control. In this paper, we use fuzzy-PID controller to control the asphalt gas concentration control system. The fuzzy-PID controller is designed based on fuzzy adaptive PID control principle; the PID parameters are adjusted according to error e and error change rate ec. Fuzzy-PID controller has more excellent performances than the traditional PID controller in the gas concentration control from the simulation results comparisons.

Keyword Asphalt gas concentration control · Fuzzy-PID · MATLAB

30.1 Introduction

For a large number of harmful gases generated from the production of asphalt waterproofing coiled material, China Academy of Building Research has developed a set of asphalt gas treatment device [1]. At present, the device commonly used manual or traditional PID controller to control the gas concentration. The features of traditional PID are that it has simple structure and it is stable and reliable. But for the asphalt gas concentration control system, if we use traditional PID controller, we cannot get a good achievement, because it has low control resolution, long rising time and large overshooting [2, 3].

In this paper, we design a fuzzy-PID controller for gas concentration control system based on plenty of experience of practical projects. We combine the fuzzy logic and traditional PID control to improve the robustness as well as control

M. Feng (✉) · W. Zhang · Y. Zhang
School of Automation and Electrical Engineering, University of Science and Technology Beijing, Beijing 100083, China
e-mail: xlshdhl@126.com

Y. Jia et al. (eds.), *Proceedings of the 2015 Chinese Intelligent Systems Conference*, Lecture Notes in Electrical Engineering 360,
DOI 10.1007/978-3-662-48365-7_30

resolution of the gas concentration control system [4]. The gas concentration mathematical model is built based on the object of control system, and then simulated by MATLAB [5]. After that, we compare the simulation results.

30.2 Asphalt Gas Treatment Device and Its Work Principles

A typical asphalt gas treatment device is shown as Fig. 30.1:

Asphalt gas mainly comes from the heating mixing process of the dispensing equipment and oiled, spreading process of the molding equipment. The exhaust gas produced by dispensing equipment is affected by temperature, while the exhaust gas of molding equipment is related to the production line speed.

Firstly, the high-temperature gas produced by batching and forming system is collected to the gathering system and then mixed with the air of a certain concentration with the action of the induced draft fan. After being cooled by the scrubbing system, the asphalt gas is purified by the electrostatic system. Where the electric tar filter can monitor the air volume and adjust the electric fields intensity. Then the purified gas odor is removed by the odor removal system. The electrochemical generator monitors the system exhaust capacity and adjusts the corona intensity in order to control the yield of the active matter. Finally, the standard purified gas is released into the atmosphere through the pipe [1, 6].

The inlet flue gas concentration control of the electrostatic system generally operates through the traditional PID controller. Traditional PID controller is based on the feedback signal obtaining from gas concentration sensor, adjusting the frequency of the inverter automatically by using the PID arithmetic to change the induced draft fan air volume and achieve the purpose of controlling the gas concentration.

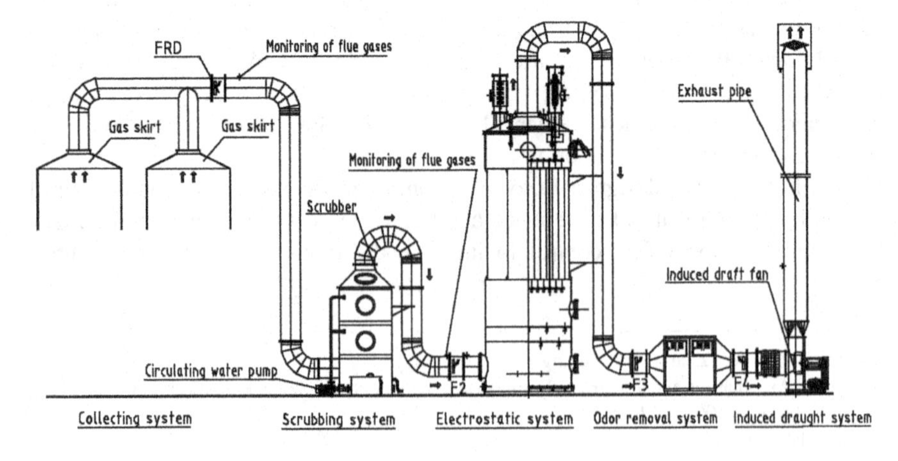

Fig. 30.1 Asphalt gas treatment device

30.3 The Fuzzy-PID Controller

Fuzzy control is established on the basis of human thought fuzziness and relays on fuzzy linguistic variables, fuzzy set theory and fuzzy logical reasoning. We use fuzzy control system in fuzzy controller to replace the traditional controller. The fuzzy-PID controller has two inputs, they are deviation e and deviation rate ec; the two inputs can adjust k_p, k_i, k_d online, put the modified Δk_p, Δk_i, Δk_d into the traditional PID controller, then the system calculates the amount of $u(k)$ by using formula (30.1) to fulfill the requirements of fuzzy-PID gas concentration control [5, 7].

$$u(k) = k_p e(k) + k_i \sum_{j=0}^{k} e(k) + k_d[e(k) - e(k-1)] \tag{30.1}$$

$$e = r(t) - c(t) \tag{30.2}$$

$$ec = \frac{de}{dt} = \frac{e(t) - e(t-T)}{T} \tag{30.3}$$

$r(t)$ is given value of t moment; $c(t)$ is measurement value of t moment.

The schematic diagram of fuzzy-PID control system of the gas concentration system is shown as Fig. 30.2.

30.4 Design of Fuzzy-PID Controller

In order to meet the requirements of the asphalt gas concentration control system, we need two inputs and three outputs, the input variables are the deviation e and the deviation rate ec, and the output variables are k_p, k_i and k_d. The deviation e and deviation rate ec variation range is defined as the range of fuzzy sets. The range of e and ec is $[-3, 3]$, k_d is within $[-3, 3]$, the domain of k_p is $[-0.3, 0.3]$, k_i is within $[-0.06, 0.06]$. The input variables and output variables are described by seven

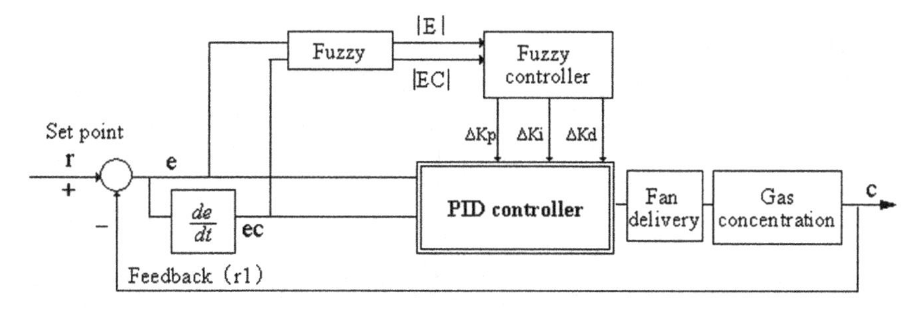

Fig. 30.2 The schematic diagram of fuzzy-PID control system

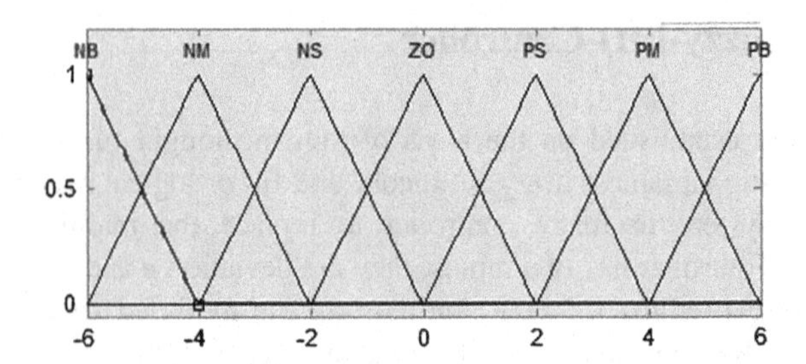

Fig. 30.3 The membership function curves of e, ec, k_p, k_i, k_d

segments fuzzy subject, which is {NB, NM, NS, ZO, PS, PM, PB} [8]. We choose trigonometric functions as the membership functions of the variables. And Fig. 30.3 shows the membership function curves of e, ec, k_p, k_i and k_d.

When regulating the parameters of fuzzy-PID controller, the following principles should be followed:

When $|e|$ is large, in order to make the system's response fast and avoid the differential over saturation, we should take a greater k_p and a smaller k_d. And the value of k_i should be zero to prevent integral saturation and large overshoot.

When $|e|$ and $|ec|$ are in the middle size, in order to make the overshoot of the system smaller, we should take a smaller k_p. In this case, we should make the value of k_i and k_d suitable for this system.

When $|e|$ is small, in order to make the system has a good steady performance, we should increase the value of k_i and k_d.

When $|ec|$ is large, we should take a smaller k_p and a larger k_i [9] (Tables 30.1, 30.2, 30.3).

From the control rule tables above, we can clearly know the relationship between e, ec and k_p, k_i, k_d. And we can get 49 fuzzy control rule sets which can be described in following language:

If e = NB, ec = NB then k_p = PB, k_i = NB, k_d = PS

If e = NB, ec = NS then k_p = PM, k_i = NM, k_d = NB

……

If e = PB, ec = PB then k_p = NB, k_i = PB, k_d = PB

Figure 30.4 shows the fuzzy control rule represented in MATLAB:

Table 30.1 Fuzzy control rule table of k_p

e / ec	NB	NM	NS	ZO	PS	PM	PB
NB	PB	PB	PM	PM	PS	PS	ZO
NM	PB	PB	PM	PM	PS	ZO	ZO
NS	PM	PM	PM	PS	ZO	NS	NM
ZO	PM	PS	PS	ZO	NS	NM	NM
PS	PS	PS	ZO	NS	NS	NM	NM
PM	ZO	ZO	NS	NM	NM	NM	NB
PB	ZO	NS	NS	NM	NM	NB	NB

Table 30.2 Fuzzy control rule table of k_i

e / ec	NB	NM	NS	ZO	PS	PM	PB
NB	NB	NB	NB	NM	NM	ZO	ZO
NM	NB	NB	NM	NM	NS	ZO	ZO
NS	NM	NM	NS	NS	ZO	PS	PS
ZO	NM	NS	NS	ZO	PS	PS	PM
PS	NS	NS	ZO	PS	PS	PM	PM
PM	ZO	ZO	PS	PM	PM	PB	PB
PB	ZO	ZO	PS	PM	PB	PB	PB

Table 30.3 Fuzzy control rule table of k_d

e / ec	NB	NM	NS	ZO	PS	PM	PB
NB	PS	PS	ZO	ZO	ZO	PB	PB
NM	NS	NS	NS	NS	ZO	NS	PM
NS	NB	NB	NM	NS	ZO	PS	PM
ZO	NB	NM	NM	NS	ZO	PS	PM
PS	NB	NM	NS	NS	ZO	PS	PS
PM	NM	NS	NS	NS	ZO	PS	PS
PB	PS	ZO	ZO	ZO	ZO	PB	PB

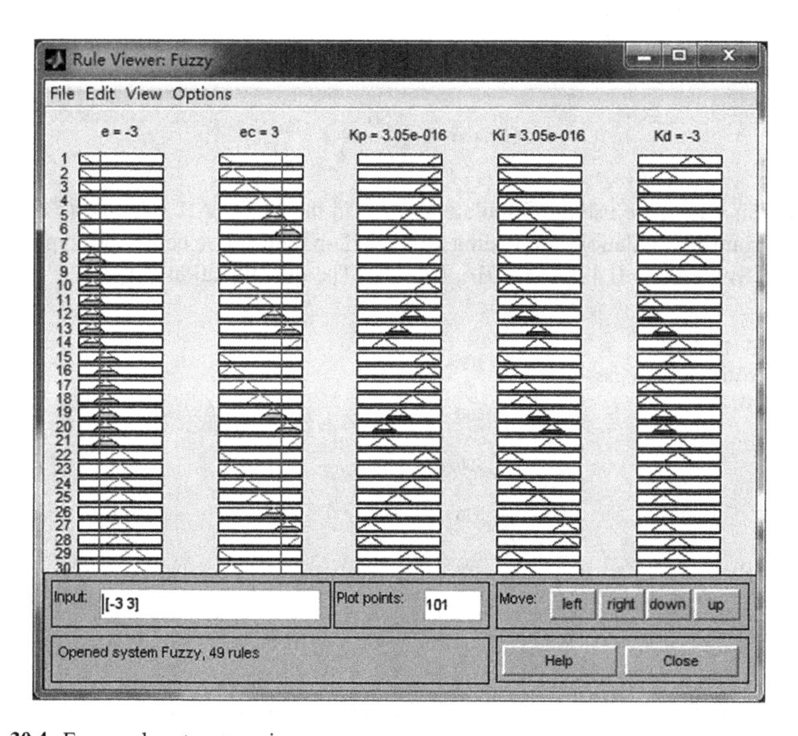

Fig. 30.4 Fuzzy rule sets expression

30.5 Simulation

According to the characteristics of the controlled object and through accessing to a lot of documents of gas concentration control, we can choose a first-order inertia link plus time delay model as asphalt gas concentration mathematical model [2, 7, 10, 11]. Because it is relatively simple and it can represent the control object approximately. So the transfer function of the gas concentration control is:

$$G(s) = \frac{K}{Ts + 1} e^{-\tau s} \tag{30.4}$$

K is amplification factor, T is time constant and τ is lag time.

According to this model we can get the gas concentration mathematical model of the control system, so accurate parameter is necessary. Firstly, we should obtain the dynamic characteristic of the gas concentration. Based on the dynamic characteristic, K, T and τ can be calculated. The lag time τ is 18 s on the basis of the actual measurement. According to the data from locale, Fortran application is used to program these data by the means of Chebyshev curve fitting method. Then we can get the dynamic characteristic curve of the gas concentration by using Origin application (Fig. 30.5).

The approximate values of three parameters which are obtained by calculating are shown as follows:

$K \approx 2.56$, $T \approx 15.27s$, $\tau \approx 18s$. So the asphalt gas concentration mathematical model is:

$$G(s) = \frac{2.56}{15.27s + 1} e^{-18s} \tag{30.5}$$

In this paper, we use the Simulate toolbox in the MATLAB software to comply fuzzy control simulation. According to the PID-parameter selection principles, the parameters are $k_p = 0.42$, $k_i = 0.016$, $k_d = 2.7$. The quantification factor of e is 0.01,

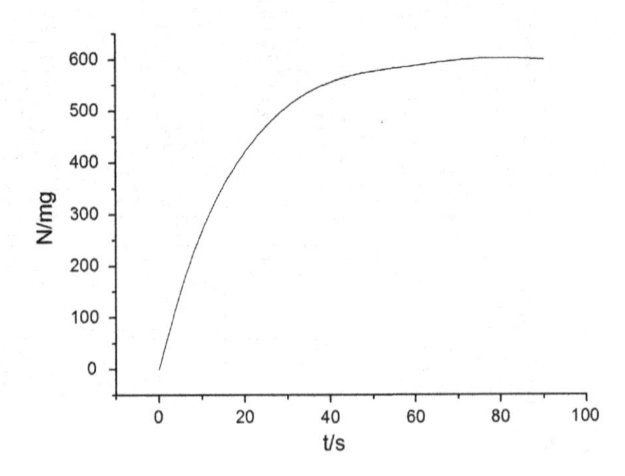

Fig. 30.5 Dynamic characteristic curve of gas concentration

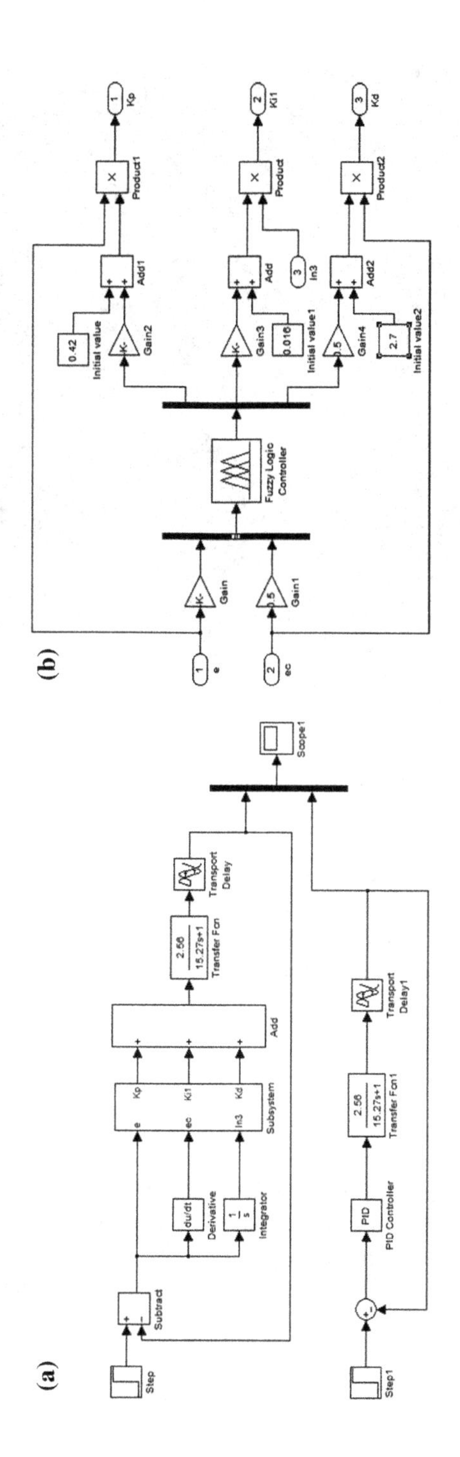

Fig. 30.6 Simulation models of fuzzy-PID control system

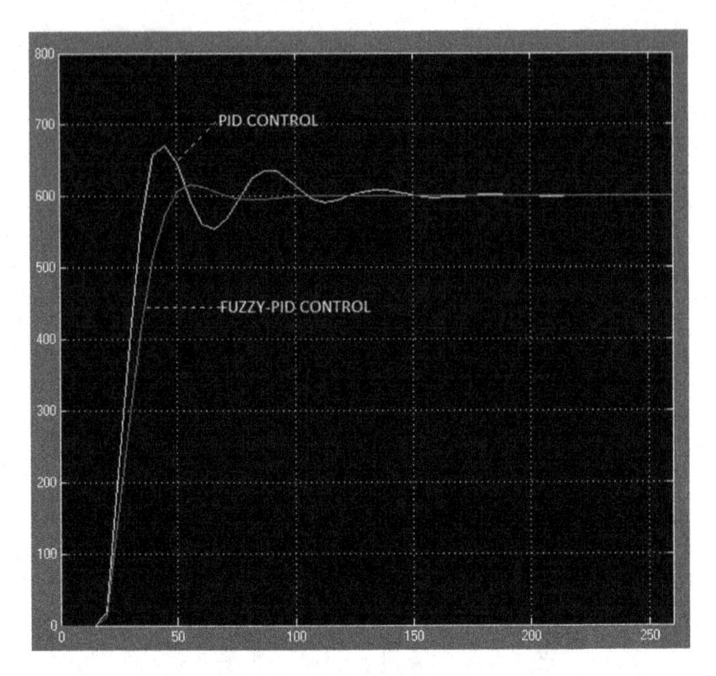

Fig. 30.7 System simulation results

the quantification factor of *ec* is 0.5, the proportionality of k_p is 0.15, the proportionality of k_i is 0.001, the proportionality of k_d is 0.5. The simulation models are shown in Fig. 30.6a, b.

Figure 30.7 shows the system simulation results.

It can be seen from Fig. 30.7 that the traditional PID controller has longer adjusting time, larger overshoot and worse steady state accuracy, while the fuzzy-PID controller has better following performance, shorter adjusting time, smaller overshoot and better steady state accuracy.

30.6 Conclusions

In this paper, we use fuzzy-PID controller to control the asphalt gas concentration. The system simulation results show that the fuzzy-PID control system has better following performance, shorter adjusting time, smaller overshoot and better steady state accuracy. So the fuzzy-PID control is suitable for the asphalt gas concentration control system.

References

1. Guanfei Tian, Shengli Feng, Yong Zhang (2015) R&D of exhausted fumed decontamination system of polymer modified asphalt waterproofing sheet assembly [J]. China Build Waterproofing 4:31–38
2. Fang Huang, Jianqiang Gao, Xuehui Zhang (2013) Application of Fuzzy PID in waste gas treatment system for casting room [J]. Foundry Technol 4:489–491
3. Yuanshao Qiao, Yanlong Cao (2008) Design and simulation of self-adaptive Fuzzy PID controller used in soybean milk concentration control system [J]. Modular Mach tool Autom Manuf Tech 11:56–59
4. Lemke VN, Dezhao W (1985) Fuzzy PID supervisor [J]. In: 24th IEEE CDC Paper 9 (15):602–608
5. Ma F (2014) An improved fuzzy PID control algorithm applied in liquid mixing system [J]. In: International conference on information and automation 7:587–591
6. Aiyu Zheng, Guifen Wang, Shengli Feng (2012) Bitumen fume low temperature plasma cleaning technology [J]. China Build Waterproofing 1:20–23
7. Huan Y, Jianjing S (2010) The application of fuzzy-PID in control system of THJ-2 temperature experimental device [J]. In: International conference on computer design and application 3:101–105
8. Xia Y, Qingding G, Qiang L (2005) Research on fuzzy PID composite control based on AC servo system [J]. Servo control 5:37–39
9. Driankov D, Hellendoorn H, Reinfrank M (1996) An Introduction to Fuzzy Control, 2nd edn. Springer, Berlin, pp32–96
10. Lizheng S, Jun G, Jianhua W (2010) Pulp concentration control by PID with BP neural network in the production of light weight cardboard [J]. In: Computer-Aided industrial design & Conceptual design(CAIDCD), 2:1217–1220
11. Hui Chen, Dongwei Qiao, Jianghua Li (2012) Research on irrigation area intelligent measurement system based on Fuzzy PID control [J]. Measure control Technol 31:253–256

Chapter 31
On the Stability of Linear Active Disturbance Rejection Control: Virtual Equivalent System Approach

Wei Wei, Weicun Zhang, Donghai Li and Yuqiong Zhang

Abstract Active disturbance rejection control is a unique control approach which could provide nice performance and need little knowledge of physical processes/plants. In order to analyze the stability of linear active disturbance rejection control (LADRC) by a direct and simple way, virtual equivalent system (VES) technique is adopted. By VES, global asymptotically stable with known process/plant dynamics and bounded input and bounded output stable with unknown process/plant dynamics are analyzed. The stability of LADRC for general single input single output nonlinear systems subject to dynamical and external uncertainties is analyzed from a brand-new viewpoint, which may be also helpful for stability analysis of other LADRC based system.

Keyword Virtual equivalent system · Linear active disturbance rejection control · Stability

31.1 Introduction

Modern control theory, a state-space model based control theory, have fruitful results in past decades. Its basic framework can be characterized by describing a physical process/plant with a mathematical model first; secondly, setting objects to be another mathematical model, which may be a differential equation or a cost

W. Wei (✉)
School of Computer and Information Engineering, Beijing Technology and Business University, Beijing 100048, China
e-mail: weiweizdh@126.com

W. Zhang
School of Automation and Electrical Engineering, University of Science and Technology Beijing, Beijing 100083, China

D. Li · Y. Zhang
School of Thermal Engineering, Tsinghua University, Beijing 100084, China

Y. Jia et al. (eds.), *Proceedings of the 2015 Chinese Intelligent Systems Conference*, Lecture Notes in Electrical Engineering 360,
DOI 10.1007/978-3-662-48365-7_31

function to be minimized; and then a control algorithm will be designed to meet the desired objectives; last but not least, a rigorous proof of the closed-loop system should be provided. Modern control, obviously, makes an assumption that the dynamics of the physical processes/plants can be described mathematically, and the mathematical model is the basis of the modern control. For mathematical rigor, modern control provides a valuable approach to analyze why and how feedback control works. However, physical processes/plants are always in the state of flux, discrepancies between model and physical process/plant, will degrades performance or even makes the system unstable. But the goal of control system is to keep system output be consistent even if uncertainties exist in processes/plants dynamics. Robust control, which is a typical algorithm to deal with uncertainties in modern control theory, has been pointed out that robust control problem is a paradox which may not be resolvable according to Gödel's incompleteness theorem by Prof. Han in [1]. While, PID, a commonly used control algorithm by practitioners/engineers in industry, focuses on and aims at decreasing the tracking errors, other than the mathematical model of physical process/plant, which will be more effective when uncertainties exist. For modern control, the reliance on faithful models and deductive reasoning is being questioned.

In order to find out what is the essence of control theory, Prof. Han first proposed his way of thinking about control theory in [2], and along his unique thinking, active disturbance rejection control (ADRC) is proposed [3–5], and it is the first time that his unique ideas were systematically introduced into English in [5]. As we can't know exactly about the processes/plants, and if we generalize the disturbance to be any discrepancies between the physical process/plant and the mathematical model, disturbances and uncertainties are synonymous. In this sense, the essence of feedback control will be disturbance rejection. In the viewpoint of ADRC, nominal model is the cascade of integrators, and any discrepancies between the physical process/plant and the nominal model is regarded as generalized disturbance. Contrary to all existing conventions, the exact expression of such generalized disturbance is not required or even necessary for the design of ADRC. What we need is the estimated value of the generalized disturbance, which can be obtained by the key part of ADRC, i.e. extended state observer (ESO). Additionally, according to an observer survey [6], other than most published state and disturbance observers, ESO just need the order of system and the estimate value of coefficient of control input. By estimating and cancelling the generalized disturbance timely and actively, ADRC is able to achieve nice control performance for complex, time-varying, and uncertain nonlinear processes/plants.

For Prof. Han's unique way of thinking, and the existing theory-practice gap in control, an increasing number of researchers and engineers have recognized that the essence of control is disturbance rejection, and there are a plenty of literatures on ADRC and its applications [7–10]. Along Prof. Han's disturbance rejection paradigm, Prof. Zhiqiang Gao has proposed linear active disturbance rejection control (LADRC) and bandwidth-parameterized tuning approach [11], which is more easier to tune and also has nice closed-loop performance. What's more, there are also many literatures discussing on how to improve the performance of ADRC/ESO, a survey on this topic can be found in [12]. At the same time, the stability of the closed-loop system is also a central problem for ADRC/LADRC. For a long time, as ADRC/LADRC is reckoned with the nature of engineering, i.e. reality and uncertainties, not based on a concrete model, the stability proof problem is always a challenge for its designers. However, ADRC/LADRC designers have made significant progress on its stability analysis. Prof. Gao gave out a basic stability analysis on LADRC [13]. A more detailed analysis has been given out for nonlinear time-varying plants which are largely unknown [14]. Particularly, asymptotic stability can be obtained when the dynamics of plant is completely known. Prof. Gao and his cooperators extend such analysis approach to multi-input and multi-output (MIMO) system [15]. By decomposing of the original system into a relatively slow subsystem and a relatively fast subsystem, the singular perturbation method is also utilized to analyze the closed-loop system, and exponentially stable is obtained in [16]. For the stability of LADRC, Prof. Chen and his students also gave out their stability results [17]. Recently, for ADRC, there are also favorable results proposed by Prof. Guo [18–21].

As a matter of fact, the stability analysis could not be obscure. Virtual equivalent system (VES), proposed by Prof. Zhang [22], is the very approach that has quite clear physical explanations. From the viewpoint of VES, controller, in general, is synthesized for the on-line estimated models, but applied to the physical processes/plants, i.e. there is a mismatch between the process/plant and its controller. It will be helpful if we can find a suitable way to compensate for such mismatch with a proper signal. VES, in this sense, is an artificial system equivalent to the physical system from the input-output point of view. By doing so, as we will see later, the analysis of stability becomes more direct and simple. In this paper, we consider the LADRC for general single input single output nonlinear systems subject to dynamical and external uncertainties. By means of VES concept, we can decompose the LADRC system into two simple control structures. Conditions that guarantee LADRC to achieve closed-loop system stability, disturbance rejection and reference tracking, will be established.

31.2 Brief Introduction of LADRC

Consider a generally nonlinear time-varying dynamic system with single-input, u, and single-output y

$$y^{(n)}(t) = f(y^{(n-1)}(t), \ldots, y'(t), y(t), w(t)) + bu(t) \tag{31.1}$$

where w is the external disturbance and b is a given constant, u is the control input, y is the system output. $f(y^{(n-1)}(t), \ldots, y'(t), y(t), w(t))$ is the nonlinear time-varying dynamics of the plant, which is unknown. Assuming f is differentiable, and denoting $\dot{x}_{n+1} = \dot{f} = h$, we have

$$\begin{cases} \dot{x}_1 = x_2 \\ \dot{x}_2 = x_3 \\ \quad \vdots \\ \dot{x}_n = x_{n+1} + bu(t) \\ \dot{x}_{n+1} = h(\boldsymbol{x}, w) \\ y = x_1 \end{cases} \tag{31.2}$$

where $\boldsymbol{x} = [x_1, x_2, \ldots, x_n, x_{n+1}]^T \in \boldsymbol{R}^{n+1}, u, y \in \boldsymbol{R}$ are state, input and output of the system, respectively. Linear extended state observer (LESO) can be designed as

$$\begin{cases} \dot{\hat{x}}_1 = \hat{x}_2 + \beta_1(x_1 - \hat{x}_1) \\ \dot{\hat{x}}_2 = \hat{x}_3 + \beta_2(x_1 - \hat{x}_1) \\ \quad \vdots \\ \dot{\hat{x}}_n = \hat{x}_{n+1} + \beta_n(x_1 - \hat{x}_1) + bu(t) \\ \dot{\hat{x}}_{n+1} = \beta_{n+1}(x_1 - \hat{x}_1) \end{cases} \tag{31.3}$$

For (31.2) and (31.3), we have compact forms, respectively,

$$\dot{\boldsymbol{x}}(t) = \boldsymbol{A}\boldsymbol{x}(t) + \boldsymbol{B}u(t) + \boldsymbol{E}h(\boldsymbol{x}, w) \tag{31.4}$$

where,

$$\boldsymbol{A} = \begin{bmatrix} 0 & 1 & 0 & 0 & \cdots & 0 \\ 0 & 0 & 1 & 0 & \cdots & 0 \\ 0 & 0 & 0 & 1 & 0 & 0 \\ & & & \ddots & & \\ 0 & 0 & \cdots & 0 & 0 & 1 \\ 0 & 0 & 0 & \cdots & \cdots & 0 \end{bmatrix}_{(n+1)\times(n+1)}, \quad \boldsymbol{B} = \begin{bmatrix} 0 \\ 0 \\ \vdots \\ 0 \\ b \\ 0 \end{bmatrix}_{(n+1)\times 1}, \quad \boldsymbol{E} = \begin{bmatrix} 0 \\ 0 \\ \vdots \\ 0 \\ 0 \\ 1 \end{bmatrix}_{(n+1)\times 1}$$

LESO can be written as

$$\dot{\hat{x}} = A\hat{x} + \boldsymbol{B}u + \boldsymbol{L}e \tag{31.5}$$

where $\hat{\boldsymbol{x}} = [\hat{x}_1, \hat{x}_2, \ldots, \hat{x}_n, \hat{x}_{n+1}]^T \in \boldsymbol{R}^{n+1}, e = x_1 - \hat{x}_1$, and

$$\boldsymbol{A} = \begin{bmatrix} 0 & 1 & 0 & \cdots & 0 \\ & & & \ddots & \\ 0 & 0 & \cdots & 0 & 1 \\ 0 & 0 & 0 & \cdots & 0 \end{bmatrix}_{(n+1)\times(n+1)}, \quad \boldsymbol{B} = \begin{bmatrix} 0 \\ \vdots \\ b \\ 0 \end{bmatrix}_{(n+1)\times 1}, \quad \boldsymbol{L} = \begin{bmatrix} \beta_1 \\ \beta_2 \\ \vdots \\ \beta_{n+1} \end{bmatrix}_{(n+1)\times 1}.$$

The block diagram of the corresponding control system is given in Fig. 31.1.

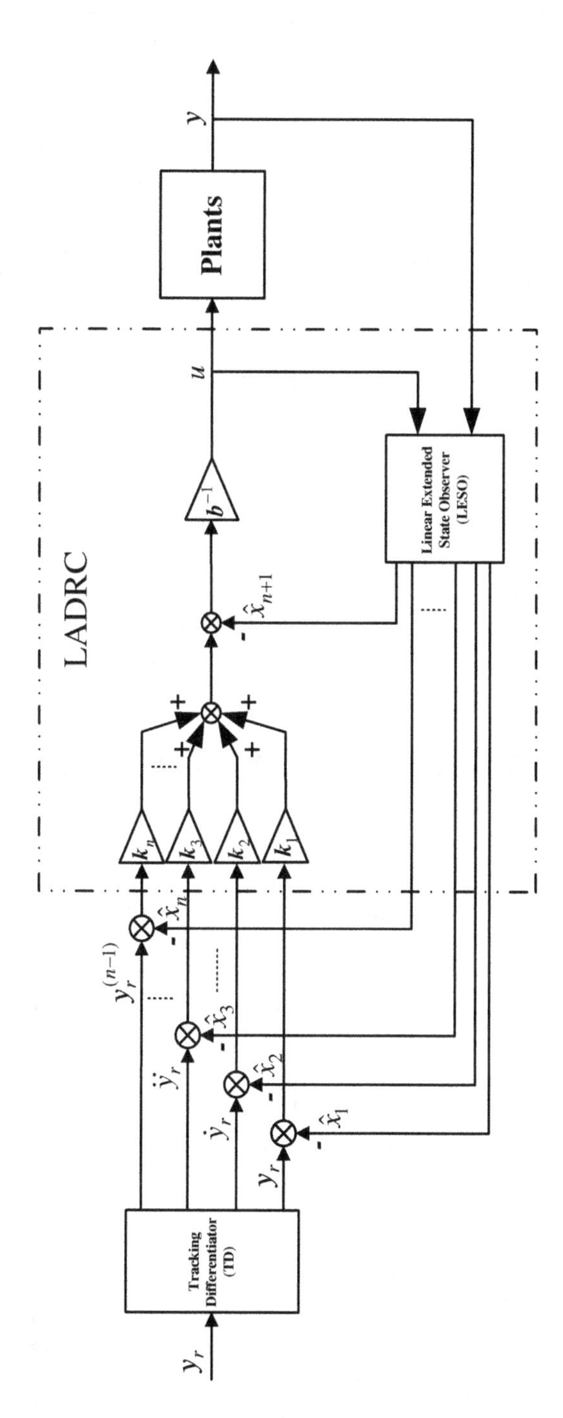

Fig. 31.1 Block diagram of LADRC system

From Fig. 31.1 we may see clearly that control input u of the plant is:

$$u = b^{-1}[k_1(y_r - \hat{x}_1) + k_2(\dot{y}_r - \hat{x}_2) + k_3(\ddot{y}_r - \hat{x}_3) + \ldots + k_n(y_r^{(n-1)} - \hat{x}_n) - \hat{x}_{n+1}] \tag{31.6}$$

Since $y = x_1$, the error of LESO will be rewritten as $e = y - \hat{x}_1$. Define tracking errors as $e_{k1} = y_r - x_1, e_{k2} = \dot{y}_r - x_2, \ldots, e_{n-1} = y_r^{n-2} - x_{n-1}, e_n = y_r^{n-1} - x_n$. $\tilde{x}_1 = x_1 - \hat{x}_1, \tilde{x}_2 = x_2 - \hat{x}_2, \ldots, \tilde{x}_n = x_n - \hat{x}_n, \tilde{x}_{n+1} = x_{n+1} - \hat{x}_{n+1}$ are errors of LESO.

Let $\boldsymbol{e}_k = [e_{k1}, e_{k2}, \ldots, e_{n-1}, e_n]^T \in \boldsymbol{R}^n, \tilde{\boldsymbol{x}} = [\tilde{x}_1, \tilde{x}_2, \ldots, \tilde{x}_n, \tilde{x}_{n+1}]^T \in \boldsymbol{R}^{n+1}$, tracking error dynamic system is

$$\dot{\boldsymbol{e}}_k = \boldsymbol{A}_k \boldsymbol{e}_k + \boldsymbol{B}_k \tilde{\boldsymbol{x}} \tag{31.7}$$

where

$$\boldsymbol{A}_k = \begin{bmatrix} 0 & 1 & 0 & \cdots & 0 \\ 0 & 0 & 1 & 0 & 0 \\ & & \ddots & & \\ 0 & 0 & \cdots & 0 & 1 \\ -k_1 & -k_2 & \cdots & -k_{n-1} & -k_n \end{bmatrix}, \quad \boldsymbol{B}_k = \begin{bmatrix} 0 & 0 & 0 & \cdots & 0 \\ 0 & 0 & 0 & 0 & 0 \\ & & \ddots & & \\ 0 & 0 & \cdots & 0 & 0 \\ -k_1 & -k_2 & \cdots & -k_n & -1 \end{bmatrix}$$

31.3 Stability Analysis for LADRC Via Virtual Equivalent System Approach

If there is no error between one of the LESO outputs, $\hat{x}_1$, and the system output, y, i.e. $y = \hat{x}_1$, the control of system equals to the control of LESO. In this sense, we may view LESO as a model of process/plant. When there is an error between $\hat{x}_1$ and y, we have $e = y - \hat{x}_1$, then $y = e + \hat{x}_1$. In other words, when $e = 0, y = \hat{x}_1$, and when $e \neq 0, y = e + \hat{x}_1$. Accordingly, the VES of closed-loop LADRC system shown in Fig. 31.1 can be drawn in Fig. 31.2.

For VES of closed-loop LADRC system, as shown in Fig. 31.2, we can see clearly that there are two input signals for the closed-loop system. One is y_r, and the

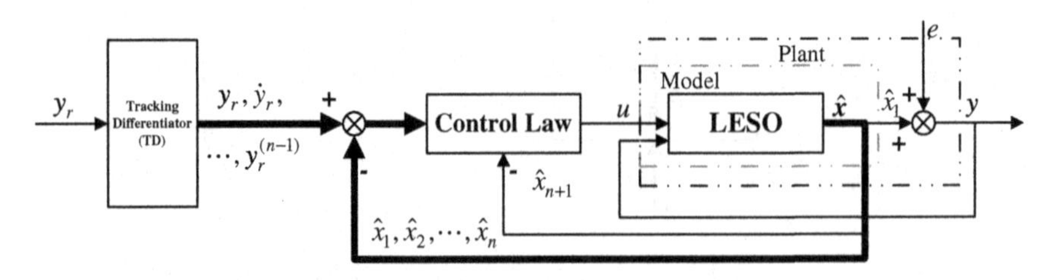

Fig. 31.2 Virtual equivalent system for the closed-loop LADRC system

other is e. Decompose it into two subsystems (see Figs. 31.3 and 31.4, respectively).

According to the superposition principle of linear system, we have:

$$y = y_{\mathrm{I}} + y_{\mathrm{II}} \tag{31.8}$$

y_{II} converges to zero, when e converges to zero, as long as y_{I} tracks y_r gradually, system output will track y_r gradually. y_{II} is bounded, when e is bounded, as y_{I} tracks y_r gradually, tracking error of y and y_r is bounded.

For subsystem I, $y_{\mathrm{I}} = \hat{x}_1$, there are no errors among outputs of LESO and system. LESO is equivalent to the controlled plant, the closed-loop system is

$$\begin{cases} \dot{\hat{\boldsymbol{x}}} = \boldsymbol{A}\hat{\boldsymbol{x}} + \boldsymbol{B}u \\ u = b^{-1}[k_1(y_r - \hat{x}_1) + k_2(\dot{y}_r - \hat{x}_2) + k_3(\ddot{y}_r - \hat{x}_3) + \cdots + k_n(y_r^{(n-1)} - \hat{x}_n) - \hat{x}_{n+1}] \end{cases} \tag{31.9}$$

Since $\hat{\boldsymbol{x}} = \boldsymbol{y}_{\mathbf{I}}$, we have:

$$y_{\mathrm{I}}^{(n)} = k_1(y_r - y_{\mathrm{I}}) + k_2(\dot{y}_r - \dot{y}_{\mathrm{I}}) + k_3(\ddot{y}_r - \ddot{y}_{\mathrm{I}}) + \cdots + k_n(y_r^{(n-1)} - y_{\mathrm{I}}^{(n-1)}) \tag{31.10}$$

Laplace transform is performed for both sides of Eq. (31.9), we have:

$$(s^n + k_n s^{n-1} + \cdots + k_3 s^2 + k_2 s + k_1)y_{\mathrm{I}}(s) = (k_n s^{n-1} + \cdots + k_3 s^2 + k_2 s + k_1)y_r(s)$$

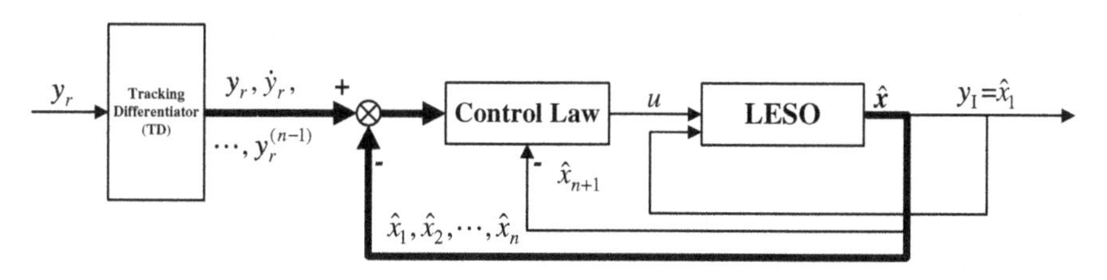

Fig. 31.3 Subsystem I of VES for the closed-loop LADRC system

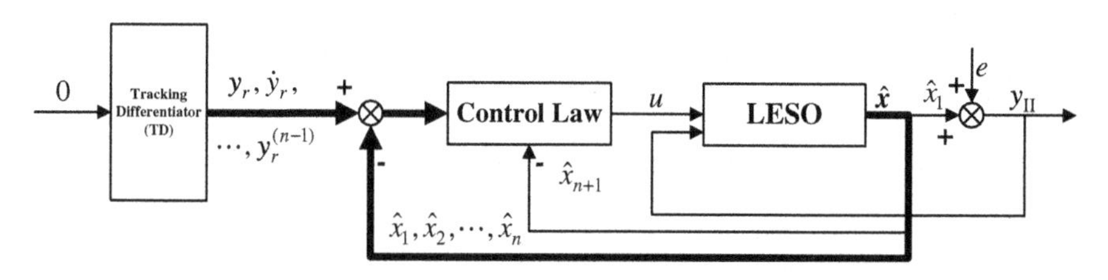

Fig. 31.4 Subsystem II of VES for the closed-loop LADRC system

Then the transfer function between desired output and actual output is

$$G_{y_I y_r}(s) = \frac{y_I(s)}{y_r(s)} = \frac{k_n s^{n-1} + \cdots + k_3 s^2 + k_2 s + k_1}{s^n + k_n s^{n-1} + \cdots + k_3 s^2 + k_2 s + k_1} \tag{31.11}$$

Therefore, for the closed-loop subsystem I of VES, if positive control gains $k_1, k_2, \ldots, k_n$ are selected to make characteristic polynomial $s^n + k_n s^{n-1} + \cdots + k_1$ be Hurwitz, subsystem I of VES is global asymptotically stable. At the same time, tracking error dynamic system (31.7) is becoming:

$$\dot{\boldsymbol{e}}_k = \boldsymbol{A}_k \boldsymbol{e}_k \tag{31.12}$$

Therefore, for tracking error dynamic system (31.7), positive control gains $k_1, k_2, \ldots, k_n$ can be selected such that $\boldsymbol{A}_k$ is Hurwitz, then tracking error dynamic of subsystem I is global asymptotically stable, i.e. $\lim_{t\to\infty} \boldsymbol{e}_k = 0$. We have $\lim_{t\to\infty}(y_r - x_1) = \lim_{t\to\infty}(y_r - y_I) = 0$.

For subsystem II of VES, $y_{II} = \hat{x}_1 + e$. As $y_r, \dot{y}_r, \ldots y_r^{(n-1)} = 0$, we have the control signal of LADRC

$$u = b^{-1}[-k_1\hat{x}_1 - k_2\hat{x}_2 - k_3\hat{x}_3 + \cdots - k_n\hat{x}_n - \hat{x}_{n+1}] \tag{31.13}$$

Substitute (31.13) into (31.3), we have the closed-loop system

$$\dot{\hat{\boldsymbol{x}}} = \boldsymbol{A}_b \hat{\boldsymbol{x}} + \boldsymbol{L}e \tag{31.14}$$

where $\boldsymbol{A}_b = \begin{bmatrix} 0 & 1 & 0 & \cdots & 0 \\ & & \ddots & & \\ -k_1 & -k_2 & \cdots & -k_n & 0 \\ 0 & 0 & 0 & \cdots & 0 \end{bmatrix}_{(n+1)\times(n+1)}$, $\boldsymbol{L} = \begin{bmatrix} \beta_1 \\ \beta_2 \\ \vdots \\ \beta_{n+1} \end{bmatrix}_{(n+1)\times 1}$.

Control positive gains $k_1, k_2, \ldots, k_n$ are selected to make $\boldsymbol{A}_b$ be Hurwitz. Next, we will consider two cases, one is that the plant model is known well, and the other is that the dynamics of plant is not known.

31.3.1 Given Model of the Plant

$$\begin{cases} \dot{\hat{x}}_1 = \hat{x}_2 + \beta_1(x_1 - \hat{x}_1) \\ \dot{\hat{x}}_2 = \hat{x}_3 + \beta_2(x_1 - \hat{x}_1) \\ \quad\vdots \\ \dot{\hat{x}}_n = \hat{x}_{n+1} + \beta_n(x_1 - \hat{x}_1) + bu(t) \\ \dot{\hat{x}}_{n+1} = \beta_{n+1}(x_1 - \hat{x}_1) + h(\hat{\boldsymbol{x}}, \omega) \end{cases} \tag{31.15}$$

Here function h is known. The characteristic polynomial of (31.15) is $\lambda(s) = s^{n+1} + \beta_1 s^n + \cdots + \beta_n s + \beta_{n+1}$. Let $\lambda(s) = (s + \omega_o)^{n+1}$, where $\beta_i = \frac{(n+1)!}{i!(n+1-i)!}\omega_o^i$, $i = 1, 2, \ldots, (n+1)$. Let $\alpha_i = \frac{(n+1)!}{i!(n+1-i)!}$, then $\beta_i = \alpha_i \omega_o^i$. Subtracting (31.15) from (31.2), we have:

$$\begin{cases} \dot{\tilde{x}}_1 = \tilde{x}_2 - \beta_1 \tilde{x}_1 \\ \dot{\tilde{x}}_2 = \tilde{x}_3 - \beta_2 \tilde{x}_1 \\ \quad \vdots \\ \dot{\tilde{x}}_n = \tilde{x}_{n+1} - \beta_{\mathrm{n}} \tilde{x}_1 \\ \dot{\tilde{x}}_{n+1} = h(\boldsymbol{x}, \omega) - h(\hat{\boldsymbol{x}}, \omega) - \beta_{\mathrm{n}+1} \tilde{x}_1 \end{cases} \tag{31.16}$$

i.e.

$$\begin{cases} \dot{\tilde{x}}_1 = \tilde{x}_2 - \alpha_1 \omega_o \tilde{x}_1 \\ \dot{\tilde{x}}_2 = \tilde{x}_3 - \alpha_2 \omega_o^2 \tilde{x}_1 \\ \quad \vdots \\ \dot{\tilde{x}}_n = \tilde{x}_{n+1} - \alpha_n \omega_o^n \tilde{x}_1 \\ \dot{\tilde{x}}_{n+1} = h(\boldsymbol{x}, \omega) - h(\hat{\boldsymbol{x}}, \omega) - \alpha_{n+1} \omega_o^{n+1} \tilde{x}_1 \end{cases} \tag{31.17}$$

Let $\varepsilon_i = \frac{\tilde{x}_i}{\omega_o^{i-1}}$, then we can rewritten (31.17) as

$$\begin{bmatrix} \dot{\varepsilon}_1 \\ \dot{\varepsilon}_2 \\ \vdots \\ \dot{\varepsilon}_{n-1} \\ \dot{\varepsilon}_n \\ \dot{\varepsilon}_{n+1} \end{bmatrix} = \omega_o \begin{bmatrix} -\alpha_1 & 1 & 0 & 0 & \cdots & 0 \\ -\alpha_2 & 0 & 1 & 0 & \cdots & 0 \\ \vdots & & \ddots & & & \vdots \\ -\alpha_{n-1} & 0 & \cdots & 0 & 1 & 0 \\ -\alpha_n & 0 & \cdots & 0 & 0 & 1 \\ -\alpha_{n+1} & 0 & \cdots & 0 & 0 & 0 \end{bmatrix} \begin{bmatrix} \varepsilon_1 \\ \varepsilon_2 \\ \vdots \\ \varepsilon_{n-1} \\ \varepsilon_n \\ \varepsilon_{n+1} \end{bmatrix} + \begin{bmatrix} 0 \\ 0 \\ \vdots \\ 0 \\ 0 \\ 1 \end{bmatrix} \frac{h(\boldsymbol{x}, \omega) - h(\hat{\boldsymbol{x}}, \omega)}{\omega_o^n}$$

If we let $\varepsilon = (\varepsilon_1, \varepsilon_2, \ldots \varepsilon_{n-1}, \varepsilon_n, \varepsilon_{n+1})^T$,

$$\boldsymbol{A}_\varepsilon = \begin{bmatrix} -\alpha_1 & 1 & 0 & 0 & \cdots & 0 \\ -\alpha_2 & 0 & 1 & 0 & \cdots & 0 \\ \vdots & & \ddots & & & \vdots \\ -\alpha_{n-1} & 0 & \cdots & 0 & 1 & 0 \\ -\alpha_n & 0 & \cdots & 0 & 0 & 1 \\ -\alpha_{n+1} & 0 & \cdots & 0 & 0 & 0 \end{bmatrix}_{(n+1)\times(n+1)} \qquad \boldsymbol{B}_\varepsilon = \begin{bmatrix} 0 \\ 0 \\ \vdots \\ 0 \\ 0 \\ 1 \end{bmatrix}_{(n+1)\times 1}$$

then we have

$$\dot{\boldsymbol{\varepsilon}} = \omega_o \boldsymbol{A}_\varepsilon \boldsymbol{\varepsilon} + \boldsymbol{B}_\varepsilon \frac{h(\boldsymbol{x}, \omega) - h(\hat{\boldsymbol{x}}, \omega)}{\omega_o^n} \tag{31.18}$$

Here $\boldsymbol{A}_\varepsilon$ is Hurwitz for α_i, $i = 1, 2, \ldots, (n+1)$ chosen above.

Lemma 1 [23] *If* $\lambda < 0$, $\|u(\tau)\| \to 0$, *then* $\int_0^t e^{\lambda(t-\tau)} \|u(\tau)\| d\tau \to 0$.

Lemma 2 [14] *Assuming* $h(\boldsymbol{x}, \omega)$ *is globally Lipschitz with respect to x, there exist constants* ω_o, c, *when* $\omega_o > c$, $\lim_{t\to\infty} \tilde{\boldsymbol{x}}(t) = 0$.

Lemma 3 *For system* $\dot{\chi}(t) = \boldsymbol{M}\chi(t) + \boldsymbol{\xi}(t)$, *where* $\chi(t) = [\chi_1(t), \chi_2(t), \ldots, \chi_{n+1}(t)]^T$, $\xi = [\xi_1(t), \xi_2(t), \ldots \xi_{n+1}(t)]^T$, $\boldsymbol{M} \in R^{(n+1)\times(n+1)}$, *if* $\boldsymbol{M}$ *is Hurwitz and* $\lim_{t\to\infty} \|\xi(t)\| = 0$, *then* $\lim_{t\to\infty} \|\chi(t)\| = 0$.

Theorem 1 *Assuming that* $h(\boldsymbol{x}, \omega)$ *is globally Lipschitz with respect to x, there exist constants* ω_o *and c, when* $\omega_o > c$, *we can choose control gains* $k_1, k_2, \ldots, k_n$ *to be positive and Hurwitz, such that closed-loop system* (31.2), (31.6), *and* (31.15) *is global asymptotically stable, and the tracking errors converge to zero gradually.*

31.3.2 Plant Dynamics Largely Unknown

If the plant dynamics represented by f is mostly unknown, (31.3) is the LESO. Estimate errors of LESO can be rewritten as

$$\begin{cases} \dot{\tilde{x}}_1 = \tilde{x}_2 - \alpha_1 \omega_o \tilde{x}_1 \\ \dot{\tilde{x}}_2 = \tilde{x}_3 - \alpha_2 \omega_o^2 \tilde{x}_1 \\ \vdots \\ \dot{\tilde{x}}_n = \tilde{x}_{n+1} - \alpha_n \omega_o^n \tilde{x}_1 \\ \dot{\tilde{x}}_{n+1} = h(\boldsymbol{x}, \omega) - \alpha_{n+1} \omega_o^{n+1} \tilde{x}_1 \end{cases} \tag{31.19}$$

Equation (31.18) becomes

$$\dot{\boldsymbol{\varepsilon}} = \omega_o \boldsymbol{A}_\varepsilon \boldsymbol{\varepsilon} + \boldsymbol{B}_\varepsilon \frac{h(\boldsymbol{x}, \omega)}{\omega_o^n} \tag{31.20}$$

Lemma 4 *Assuming that* $h(\boldsymbol{x}, \omega)$ *is bounded, i.e. there exists a positive constant* M_1, *such that* $|h(\boldsymbol{x}, \omega)| \le M_1$. *For LESO, there exists another positive constant* M_2, *such that* $\|\tilde{\boldsymbol{x}}(t)\| \le M_2$.

Lemma 5 *For system* $\dot{\chi}(t) = M\chi(t) + \xi(t)$, *where* $\chi(t) = [\chi_1(t), \chi_2(t), \ldots, \chi_{n+1}(t)]^T, \xi = [\xi_1(t), \xi_2(t), \ldots, \xi_{n+1}(t)]^T, M \in R^{(n+1)\times(n+1)}$. $\|\chi(t)\| \leq M_\chi$, *if* M *is Hurwitz and* $\|\xi(t)\| \leq M_\xi$.

Theorem 2 *Assuming that* $h(x, \omega)$ *is bounded. There exist positive control gains* $k_1, k_2, \ldots, k_n$, *which make characteristic polynomial* $s^n + k_n s^{n-1} + \cdots + k_1$ *be Hurwitz, such that closed-loop system* (31.2), (31.3), *and* (31.6) *is bounded input and bounded output (BIBO) stable, and the tracking error is bounded.*

31.4 Concluding Remarks

In this paper, by virtual equivalent system technique, we have analyzed the stability of LADRC for single input single output nonlinear systems. VES technique, a tool utilized to analyze the stability and convergence more direct and simple, is utilized. Conditions for global gradually stable with known process/plant dynamics and bounded input and bounded output stable with unknown process/plant dynamics are obtained. By VES, a brand-new viewpoint for stability analysis, we have a clearer picture for stability analysis of LADRC.

Acknowledgments This work is supported by National Natural Science Foundation of China (61403006), Beijing Natural Science Foundation (4132005), The Importation and Development of High-Caliber Talents Project of Beijing Municipal Institutions (YETP1449), and Project of Scientific and Technological Innovation Platform (PXM2015_014213_000063).

References

1. Han J (1999) Robustness of control system and the gödel's incomplete theorem. Control Theor Appl 16(1):149–155. Institute of Systems Science, Chinese Academy of Sciences. (In Chinese)
2. Han J (1989) Control theory, is it a model analysis approach or a direct control approach? J Syst Sci Math Sci 9(4):328–335 (In Chinese)
3. Han J (1995) A class of extended state observers for uncertain systems. Control Decis 10 (1):85–88 (In Chinese)
4. Han J (1998) Auto-disturbance rejection control and its applications. Control Decis 13:19–23 (In Chinese)
5. Gao Z, Huang Y, Han J (2001) An alternative paradigm for control system design decision and control. In: Proceedings of the 40th IEEE Conference on IEEE, vol 5, pp 4578–4585
6. Radke A, Gao Z (2006) A survey of state and disturbance observers for practitioners. In: American Control Conference IEEE, pp 5183–5188
7. Huang Y, Zhang W (2002) Development of active disturbance rejection controller. Control Theor Appl 19(4):485–492 (In Chinese)
8. Yi H, Wenchao X, Chunzhe Z (2011) Active disturbance rejection control: methodology and theoretical analysis. J Syst Sci Math Sci 9:1111–1129 (In Chinese)
9. Yi H, Wenchao X (2012) Active disturbance rejection control: methodology, applications and theoretical analysis. J Syst Sci Math Sci 10:1287–1307 (In Chinese)

10. Huang Y, Xue W (2014) Active disturbance rejection control: methodology and theoretical analysis. ISA Trans 53(4):963–976
11. Gao Z (2003) Scaling and bandwidth-parameterization based controller tuning. In: Proceedings of American Control Conference, pp 4989–4996
12. Madoński R, Herman P (2015) Survey on methods of increasing the efficiency of extended state disturbance observers. ISA Trans 56:18–27
13. Gao Z (2006) Active disturbance rejection control: a paradigm shift in feedback control system design. In: American Control Conference, p 7
14. Zheng Q, Gao L, Gao Z (2007) On stability analysis of active disturbance rejection control for nonlinear time-varying plants with unknown dynamics. In: Proceedings of IEEE Conf. Decision Control, pp 3501–3506
15. Zheng Q, Chen Z, Gao Z (2009) A practical approach to disturbance decoupling control. Control Eng Pract 17(9):1016–1025
16. Zhou W, Shao S, Gao Z (2009) A stability study of the active disturbance rejection control problem by a singular perturbation approach. Appl Math Sci 3(10):491–508
17. Zengqiang C, Mingwei S, Ruiguang Y (2013) On the stability of linear active disturbance rejection control. Acta Autom Sin 39:574–580 (In Chinese)
18. Guo BZ, Zhao Z (2011) On the convergence of an extended state observer for nonlinear systems with uncertainty. Syst Control Lett 60(6):420–430
19. Guo BZ, Zhao Z L (2012) On convergence of nonlinear active disturbance rejection for SISO systems. In: Control and Decision Conference (CCDC), 2012 24th Chinese. IEEE, pp 3507–3512
20. Guo BZ, Zhao ZL (2012) On convergence of non-linear extended state observer for multi-input multi-output systems with uncertainty. IET Control Theor Appl 6(15):2375–2386
21. Guo BZ, Zhao ZL (2013) On convergence of the nonlinear active disturbance rejection control for MIMO systems. SIAM J Control Optim 51(2):1727–1757
22. Zhang W (2010) On the stability and convergence of self-tuning control–virtual equivalent system approach. Int J Control 83(5):879–896
23. Zhang W, Liu J, Guangda H (2015) Stability analysis of robust multiple model adaptive control systems. Acta Autom Sin 41:113–121

Chapter 32
A Dual Camera System for Moving Objects Close-up Tracking

Peng Chen, Yuexiang Cao, Peiliang Deng and Hongpeng Yin

Abstract A single camera cannot provide a broad perspective and provide the details of monitoring targets at the same time. This paper design a dual camera system based on a wide-angle camera used to provide a wide-angle visual field and a close-up camera used to provide the details of moving targets. The real-time video data are firstly acquired based on DirectShow. Secondly, a Gaussian mixture model and Kalman filter are used to detect and track the moving targets, respectively. In order to realize the collaborative relationship among the dual camera, the conversion model among coordinate systems is established. Finally, the PID algorithm is used to driver the close-up camera to make the target locate in the center of its visual field. Experimental results demonstrate the effectiveness of the proposed method.

Keywords Dual camera · Gaussian mixture · Kalman filter · Correlation · PID algorithm

32.1 Introduction

Nowadays, video surveillance is widely used in public places such as train stations, airports and so on. However a single camera cannot get a wide-angle visual field and the details of monitoring targets at the same time. As to a wide-angle camera, it can provide a large visual field, but it can only capture little details of monitoring targets which result the poor performance of feature detection. On the contrary, a close-up camera can get the details, but its visual field is small. To overcome the

P. Chen · Y. Cao · P. Deng · H. Yin (✉)
School of Automation, Chongqing University, Chongqing 400044, China
e-mail: yinhongpeng@gmail.com

H. Yin
Key Laboratory of Dependable Servise Computing in Cyber Physical Society, Ministry of Education, Chongqing 400030, China

Y. Jia et al. (eds.), *Proceedings of the 2015 Chinese Intelligent Systems Conference*, Lecture Notes in Electrical Engineering 360,
DOI 10.1007/978-3-662-48365-7_32

limitation of a single camera, there have been many studies involving the multi-camera collaboration. Compared with the single camera system, the algorithm of the multi-camera system is more complicated. But it can increase the information interaction among cameras and capture more details.

Sato et al. [1] proposed a distributed vision agent model based on several monocular cameras for tracking multiple objects expanding the monitoring vision efficiently. Meyer et al. [2] proposed an approach which used the combined evaluation of the two cameras to obtain the 3D information of reduce the monitoring vision. However, their research mainly focused on real-time tracking objects. Recently, Hoi et al. [3] proposed a face recognition system based on one PTZ camera and two static cameras to increase detection distance. Compared to the conventional camera system, it improved the instance range from 5 to 10 m. However, it can only realize the tracking and face recognize of one target. Lain et al. [4] described a sensor based on an omnidirectional camera and a PTZ camera for continuously monitoring the environment. However, the omnidirectional camera cannot provide a large visual field and the distortion is serious.

In this paper, an approach of dual camera correlation is proposed. A wide-angle camera is used to detect the moving targets from its wide-angle visual field, and a close-up is used to get the details of the moving target of interest. This algorithm overcomes the contradiction between breadth and accuracy. Compared to the multi-camera system, it also saves the cost. Firstly, the video data are captured from the dual camera based on DirectShow. Secondly, the background subtraction based on Gaussian mixture model and the moving targets tracking based on Kalman filter is used to obtain the moving targets' basic parameters including coordinates, speed and so on. Thirdly, a conversion model among coordinate systems is established to realize the mapping of the same object and correlation of the dual camera. Finally, to make the target located in the center of close-up camera visual field and get its close-up information, the control command is sent to the close-up camera based on PID.

32.2 The Proposed Approach

32.2.1 The Description of the System Framework

The close tracking of the moving objects based on a dual camera system mainly including the acquisition of real-time video data, the detection and tracking of the moving objects and dual camera coordination control. Figure 32.1 shows the whole structure of the system, and Fig. 32.2 shows the flow diagram of the algorithm.

Shown as Fig. 32.1, the whole system mainly consists of the close-up camera, the wide-angle camera, the video capture camera and the PC. The monitoring equipment captures the video data from the dual camera, and sends it to the video capture card based on the BNC cable. Then the PC reads the video data from the video capture card. After the image post-processing such as the detection and

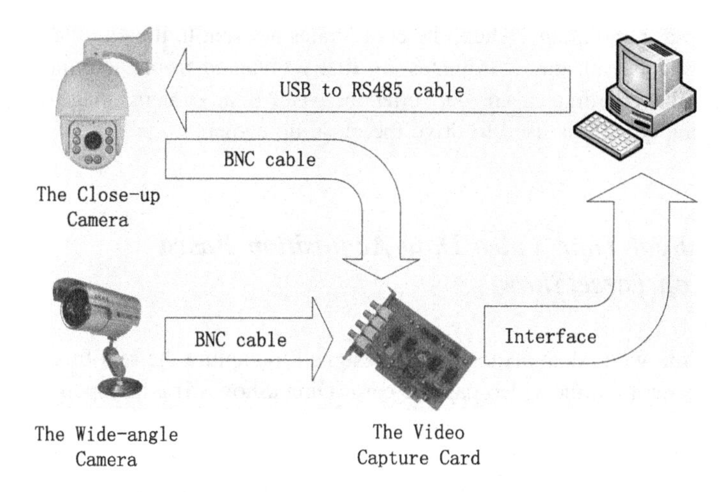

Fig. 32.1 The whole structure of the system

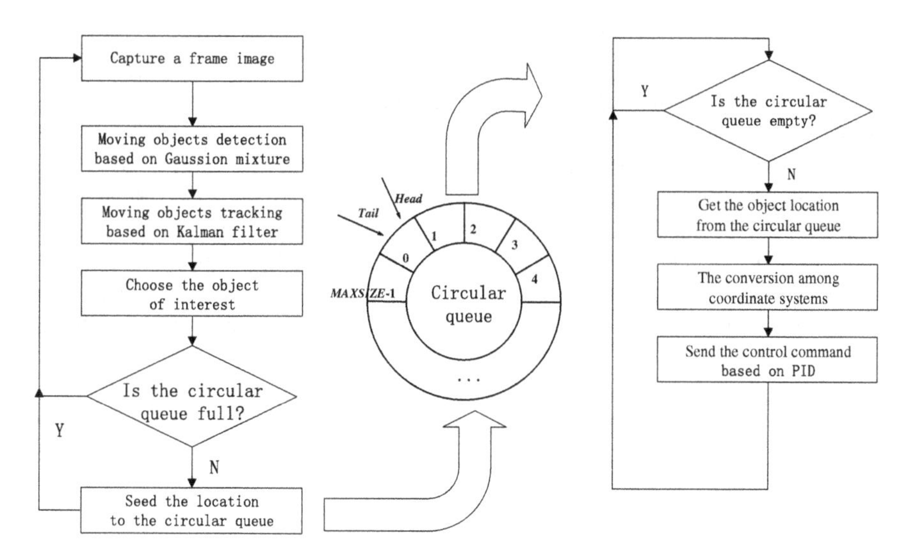

Fig. 32.2 The flow diagram of the algorithm

tracking of the moving objects, the PC sends the control command to the close-up camera via the USB to RS485 cable. Finally the object of interest will locate in the center of visual field.

Shown as Fig. 32.2, there are two threads and a circular queue. One thread is mainly used to get the coordinates of the moving objects in the wide-angle camera coordinate system. The other is mainly used to send control command to the close-up camera. The circular queue is used to keep the coordinates of the moving targets. As to the first thread, the moving targets are detected and tracked whenever

PC capture a frame image. Then, the coordinates are sent to the circular queue. As to the second thread, the coordinates are firstly obtained from the circular queue. Secondly, the coordinates are converted between dual camera image coordinate system. Finally, PID is used to drive the close-up camera.

32.2.2 Real-Time Video Data Acquisition Based on DirectShow

The real-time video data acquisition is designed to capture the real-time data from the dual camera via the video capture card. DirectShow [5] is an open application frame which can capture multimedia stream from WDM driver capture card and VFW driver capture card. It supports a wide range of formats [6], including ASF, WMV, MPEG, AVI, and so on. Figure 32.3 shows the real-time video data acquisition Filter Graph which can be divided into three parts: Source Filter, Transform Filter and Render Filter. Each filter function is as follows:

Source Filter: it is mainly responsible for capturing multimedia data from data source. Here, Crossbar obtains real-time data from cameras. Capture Filter which is the filter of WDM driver capture card. It captures real-time data from Crossbar.

Transform Filter: it is mainly responsible for the multimedia data flow transfer and conversion, for example, encode, decode and so on. Here, Smart Tee divides the video data into two ways. One way which finally goes to Video Renderer is used to preview video in the specified window. And the other is used to preserve video data in the specified directory. Among them, AVI Decompressor and Xvid MPEG-4 Codec are used to encode and decode data flow respectively, AVI Mux synthetic data flow in the form of avi, and SampleGrabber capture the current frame image from data flow.

Render Filter: it is mainly responsible for the output of multimedia data flow. As like in Fig. 32.3, one video stream is stored in the file, the other is sent to a window on the desktop.

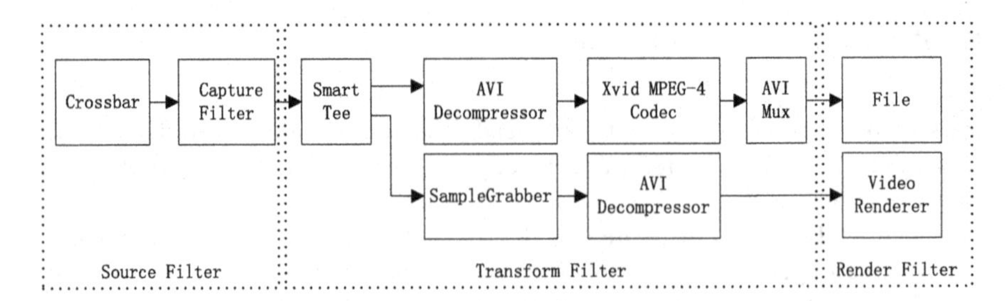

Fig. 32.3 The real-time video data acquisition filter graph

32.2.3 Moving Objects Detection Based on Gaussian Mixture

The result of the real-time video data acquisition is the current frame image. Separate the moving targets from the environment is the key to obtain the basic state information. Gaussian mixture model is a typical background modeling algorithm which can quickly adapt to changing environment. As to each pixel in the scene, it is modelled by a mixture of K Gaussian distributions [7]. With the increase of K, Gaussian mixture model is more adapt to environmental changes [8]. But it also increases the time complexity. As to the pixel x_t, its probability distribution at time t is:

$$P(x_t) = \sum_{i=1}^{K} w_{t,i}\eta(x_t, \mu_{t,i}, \textstyle\sum_{t,i}) \tag{32.1}$$

where $w_{t,i}$, $\mu_{t,i}$ and $\sum_{t,i}$ are the weight, mean and covariance matrix of the ith Gaussian distributions, respectively. Under the assumption that the values among the pixels are independent, the covariance matrix $\sum_{t,i}$ is:

$$\textstyle\sum_{t,i} = (\sigma_{t,i})^2 E \tag{32.2}$$

where $\sigma_{t,i}$ is the standard deviation of the ith Gaussian distributions at time t and E is a unit matrix. The ith Gaussian distributions at time t is represented by:

$$\eta(x_t, \mu_{t,i}, \textstyle\sum_{t,i}) = \dfrac{1}{(2\pi)^{n/2}|\sum_{t,i}|^{1/2}} e^{\frac{1}{2}(x_t - \mu_{t,i})^T \sum_{t,i}^{-1}(x_t - \mu_{t,i})} \tag{32.3}$$

The moving objects detection steps are shown as follows:

STEP 1. Establish the ROI area to reduce the foreground noises' effect
STEP 2. Initialize the Gaussian mixture model using the first frame image
STEP 3. Update the basic parameters including the weight $w_{t,i}$, the mean value $\mu_{t,i}$ and the standard deviation $\sigma_{t,i}$ for each frame image
STEP 4. Distinguish foreground and background
STEP 5. Separate the moving objects from foreground base on the aspect ratios and positions of the blobs

The results of the moving objects detection algorithm are the locations of the moving objects.

32.2.4 Moving Objects Tracking Based on Kalman Filter

The moving objects tracking is the key to gain the current state information of the tracking objects. The Kalman filter [9] is widely used in single or multiple objects tracking, which can estimate the state of dynamical system described by the state vector prediction equation and the filtering equation [10] as follows:

$$x_t = Ax_{t-1} + B\mu_t + w_t \quad (32.4)$$

$$z_t = Hx_t + v_t \quad (32.5)$$

where x_t is the system state vectors at time t, u_t is system input vector at time t, w_t is the process vector at time t, z_t is the system measurement vector at time t, v_t is the measurement noise vector at time t. A, B and H are the state dynamic matrix, the system control input matrix and the measurement matrix, respectively.

The moving objects tracking steps are shown as follows:

STEP 1. Define the basic parameters of Kalman filter such as the system state vector x_t, the system measurement vector z_t and so on

STEP 2. Initialize the parameters defined in step 1 using the moving objects' data of the first frame image

STEP 3. Predict the state of the tracking objects base on the Kalman equations whenever the moving objects are detected from the current frame image

STEP 4. Update the state of the tracking objects by matching the predicted state in step 3 with the actual state detected from the current frame image

STEP 5. Under the assumption of uniform motion between two frame images, the location of the object with the maximum displacement is sent to the circular queue as like in Fig. 32.2

As to each group of the moving objects detected from the current frame image, the moving objects tracking algorithm update the state information of the tracking objects.

32.2.5 Dual Camera Correlation

The results of the detection and tracking of the moving objects are the current coordinates of the objects. However, the object of interest needs to locate in the center of close-up camera visual field through the horizontal movement and vertical movement of the close-up camera. It is key to establish the conversion model among coordinate systems. Figure 32.4 shows the geometry model of the dual camera based on the basic camera imaging model.

Shown as Fig. 32.4, $|OA|$ and $|OB|$ are the focal lengths of dual cameras respectively, and the moving object is at the point of C. Thus its pixel coordinate

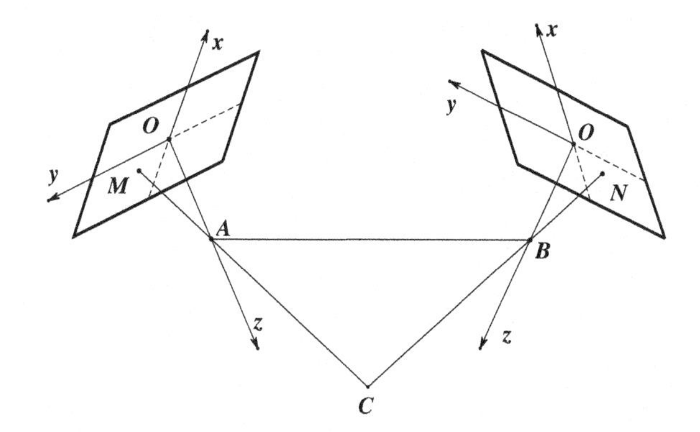

Fig. 32.4 The geometry model of dual camera

relative to the wide-angle camera's image coordinate system is $M(x_M, y_M)$ and its pixel coordinate relative to the close-up camera's image coordinate system is $N(x_N, y_N)$. In actual, the focal lengths of the dual camera are nearly equal, the locations are also very close, that is,$|OA| = |OB|$ and $|AB|$ near 0. There is a small angle θ between the two optical axes, Fig. 32.5 shows it.

Shown as Fig. 32.5, O_1 and O_2 are the lens center of the dual camera respectively. Under the initial condition, the y_1-axis and the y_2-axis are parallel, the other two axis angles are very small. Here, the angles are ignored, that is, the coordinates of the point M and point N in the two image coordinate systems are nearly equal.

Supposing the point M is projected to the point M_x and point M_y on the two axis. To move the point M to the origin, that is $\angle O_2M_xC = 0$ and $\angle O_2M_yC = 0$. Therefore, we can adjust the vertical movement and vertical movement of the close-up camera to move the moving object to the center of visual field, shown as Fig. 32.6.

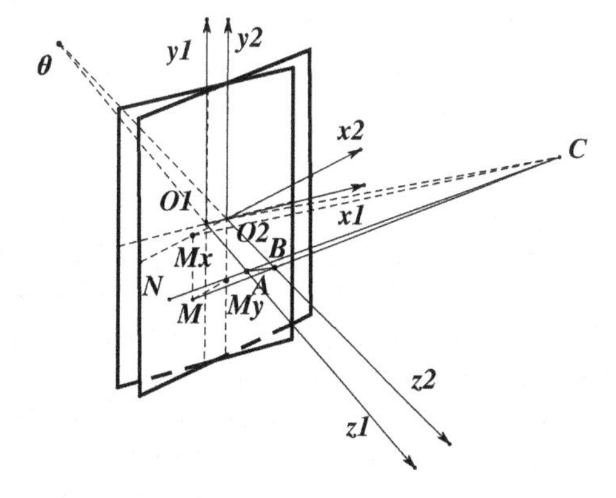

Fig. 32.5 The geometry model of the actual system

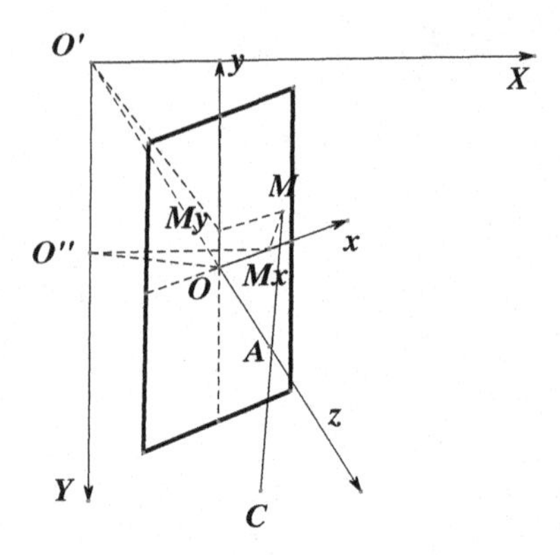

Fig. 32.6 The geometry model of the close-up camera

The horizontal rotation angle and vertical rotation angle are $\angle M_y O' O$ and $\angle M_x O'' O$ which can be calculated by:

$$\angle M_y O' O = \arctan \frac{|OM_y|}{|OO'|} = \arctan \frac{|OM_y| \sin \angle OO'O''}{|OO''|} \quad (32.6)$$

$$\angle M_x O'' O = \arctan \frac{|OM_x|}{|OO''|} \quad (32.7)$$

where $|OO''|$ is a constant, $\angle OO'O''$ is the angle between the close-up camera's optical axis and the vertical direction under the initial conditions.

As to each object, its pixel coordinate relative to the wide-angle camera's image coordinate system can be converted to the horizontal rotation angle and vertical rotation angle of the close-up camera under the initial conditions. As to any moving objects, the angle deviation can be calculated based on the current angles.

32.2.6 Control Strategies for the Close-up Camera Based on PID

After the conversion between the coordinate with the angles, the control command needs to be sent to the close-up camera to drive it. On the one hand, based on the deviation to the center of visual field, regulate the movement speed and zoom ratio are the key to make the object located in its center of visual field. That is, the proportional control is necessary.

On the other hand, the close-up camera may could not catch the moving objects because of the time asynchronous between the image processing and the send of control command. That is, the integral control is necessary to eliminate the error.

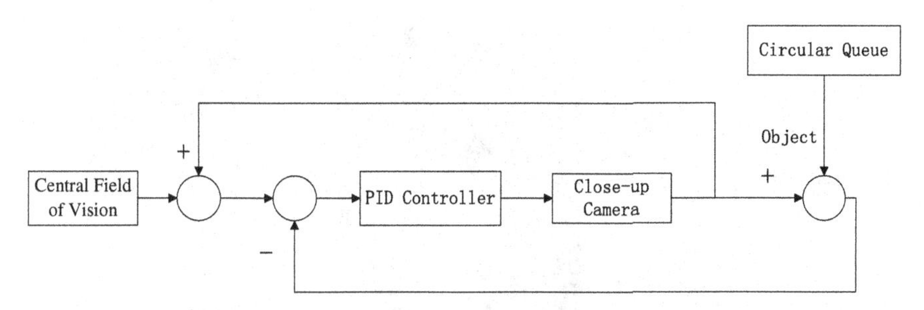

Fig. 32.7 The flow chart of the control algorithm

PID algorithm is one of the most widely used control algorithms which described by:

$$u(t) = K_p e(t) + K_i \int_0^t e(\tau)d\tau + K_d \frac{d}{dt} e(t) \tag{32.8}$$

where $e(t)$ is the deviation signal, $u(t)$ is the output signal. K_p, K_i and K_d are the proportional coefficient, the integral coefficient and the differential coefficient.

In our system, there are three PID controllers [11]: one for controlling the horizontal movement, one for controlling the vertical movement and the last for controlling the zoom. Figure 32.7 shows the flow diagram of the control algorithm.

Shown as Fig. 32.7, the deviation is the rotation angle and the zoom radio relative to the current center of the close-up camera visual field which is updated whenever the close-up camera move. After an object is inputted, the deviation signal is sent to PID controller to make the object locate in the center of the close-up camera visual field.

32.3 Experiment and Results

To illustrate the effectiveness of the proposed approach, an experiment is conducted at our laboratory. Shown as Fig. 32.1, our experiment mainly consist of a wide-angle camera (a WOSHIDA 776F6Z) which can offer a large range of pan angles ($-180° \rightarrow 180°$) and tilt angles ($-92° \rightarrow +2°$), a close-up camera (a DASKIOS DS-HY8007Z12-39), a video capture card (a 10MOONS VC4000) and a PC. Microsoft Visual Studio 2010 under Windows 7 was used for the simulations. The scene is selected at our laboratory.

Figure 32.8 shows the results of the experiment. In this figure, the frame numbers 11, 16, 21, 30 and 35 are presented. In Fig. 32.8a, the blue external rectangles represent the detection and tracking of the moving objects. In Fig. 32.8b, the white pixels represent the foreground of the moving objects. And Fig. 32.8c shows the close-up image which captured from the close-up camera.

Fig. 32.8 The results of the experiment

Shown as Fig. 32.8, when the people move from right to left and turn back in the wide-angle camera visual field, the close-up camera always follows the moving object based on our algorithm. It can be observed that the people's center of gravity always near the center of the close-up camera visual field.

32.4 Conclusion and Discussion

In this paper, a close tracking of moving objects algorithm is proposed. This algorithm balance the contradictory between the wide-angle camera and the close-up camera, and realize the correlation between these two kinds of cameras. Firstly, the background subtraction based on Gaussian mixture model is adopted for the moving objects detection of the wide-angle camera visual field. Secondly, the Kalman filter is used to track the moving objects. Thirdly, a conversion model is established to convert the coordinate to the horizontal and vertical angle. Finally, PID is used to drive the close-up camera to make the object locate in its center of visual field. Experiment results demonstrated the effectiveness of the proposed approach.

Light and other environmental factors affect the performance of the detection and tracking of the moving targets. The difference between the movement speed of the close-up camera and the image processing affects the close-up tracking of the moving objects. A fast-moving object even cannot be tracked because of it. This further research could consider the complex environment and high speed servo.

Acknowledgements We would like to thank the supports by National Natural Science Foundation of China (61203321), China Postdoctoral Science Foundation (2012M521676), China Central Universities Foundation (106112015CDJXY170003), Chongqing Special Funding in Postdoctoral Scientific Research Project (XM2013007), Chongqing Funding in Postgraduate Research Innovation Project (CYB14023) and the National College Students' Innovative Experiment Project (201410611053).

References

1. Sato K, Maeda T, Kato H, Inokuchi S (1994) CAD-Based Object Tracking with Distributed Monocular Camera Security Monitoring[R]. Second CAD-Based Vision Workshop, Champion, pp 291–297
2. Meyer M, Hotter M, Ohmacht H (1998) New options in video surveillance applications using multiple views of a scene[C]. In: International Carnahan conference on security technology, Alexandria, VA, pp 216–219
3. Choi HC, Park U, Jain AK (2010) PTZ camera assisted face acquisition, tracking and recognition[C]. In: Fourth IEEE international conference on biometrics, Washington, DC, pp 1–6
4. Lain B, Ghidoni S, Menegatti E (2011) Robust object tracking in a dual camera sensor[C]. In: 4th International conference on human system interactions (HSI), Yokohama, pp 150–157
5. Xiao Z, Liang J (2006) Design and implementation of network based video monitoring system using DirectShow [J]. Comput Eng Appl 12(1):1–5
6. Manhong L, Xingming S, Gaobo Y (2006) The design and implementation of video transmit filter based on DirectShow. Sci Technol Eng Beijing 6(15):2281–2285
7. Stauffer C, Grimson W.E.L (1999) Adaptive background mixture models for real-time tracking[C]. In: IEEE Computer society conference, Fort Collins, CO, pp 246–252
8. Luo J, Zhu J (2010) Adaptive Gaussian mixture model based on feedback mechanism[C]. In: International conference on computer design and applications (ICCDA), Qinhuangdao, pp 177–181
9. Chen G, Chui CK (2011) A modified adaptive Kalman filter for real-time applications[C]. In: 4th International conference on human system interactions (HSI), Yokohama, pp 150–157
10. Tafti AD, Sadati N (2008) Novel adaptive Kalman filtering and fuzzy track fusion approach for real time applications[C]. In: 3rd IEEE Conference on industrial electronics and applications, Singapore, pp 120–125
11. Zhang L, Xu K, Yu S, Fu R, Xu Y (2010) An effective approach for active tracking with a PTZ camera[C]. In: IEEE International conference on robotics and biomimetics (ROBIO), Tianjin, pp 1768–1773

Chapter 33
Fractional Modeling and Controlling of the Assistant Motor in Electric Power Steering System

Chaoqiang Sheng, Chao Chen, Zhaoli Xie and Kai Huang

Abstract The inductances' nonlinear characteristic in the assistant motor can affect the steady-state and dynamic performance of motor, therefore it also relates with assist torque of Electric Powering Steering (EPS) system closely. This paper constructs an accurately optimized model based on the theory of fractional calculus. This reference model takes the advantage of fractional orders which can improve robustness to the nonlinear of mechanical system. Based on the reference model, the fractional order controller $PI^{\lambda}D^{\mu}$ is designed. Several experiments show the proposed control strategy is effective and stable, furthermore the EPS system using the control strategy has a good anti-interference performance in the presence of nonlinear factors.

Keywords EPS · Assistant motor · Fractional calculus · Fractional order controller

33.1 Introduction

In recent years, the EPS system has taken the place of the conventional hydraulic system gradually because its lower in fuel consumption, higher efficiency, and steering feel tenability and so on [1].

The EPS system is easily affected by nonlinear factors, such as internal fraction and external disturbances, so simplified linear models of EPS system controlled by the conventional linear controllers may have a bad performance and instability in presence of nonlinear factors [2]. To solve this problem, so recent researches, such as fuzzy control [3, 4], neural network control [5–7] and a sliding mode control [8–10] are applied to EPS systems. However, they have not considered the assistant motor's inductances' nonlinearity effect on the EPS system. As motor's inductances not only

C. Sheng · C. Chen (✉) · Z. Xie · K. Huang
School of Automation Chongqing University, Chongqing 400044, China
e-mail: nicechenmail@163.com

Y. Jia et al. (eds.), *Proceedings of the 2015 Chinese Intelligent Systems Conference*, Lecture Notes in Electrical Engineering 360,
DOI 10.1007/978-3-662-48365-7_33

affect average torque, but also relate with torque ripple and phase commutation closely [11, 12], so they may result in system instability. Based on fractional calculus, the fractional order model is more accurate to describe the real physical world [13–16], and it has better robustness to the nonlinear of mechanical system [17]. So the fractional calculus is used to motor's modeling in this paper, and it has better robustness than traditional integer order model. Fractional order controller is designed to improve the dynamic and static performance of the fractional order system.

The motor's fractional order electric dynamic model is presented in this paper so that the characteristics of the actual circuit can be described accurately. It is optimal state feedback control based on $PI^{\lambda}D^{\mu}$ which is the control theory of fractional calculus (λ and μ are real order integral and derivative respectively). The fractional order model improves the accuracy of traditional integer order model and it is taken as the reference model of the EPS system to achieve a new a control strategy.

The key contributions of this paper are summarized as follows: (33.1) The motor's fractional order model is established and served as the reference model of the EPS system. The proposed reference model has a better robustness and stability than the traditional integer order model in the presence of nonlinear factors; (33.2) Fractional order controller $PI^{\lambda}D^{\mu}$ is designed to control the fractional order system. It can perform better in robustness and simulation than the traditional control theory [18].

33.2 Nonlinear Dynamic Model of the EPS System

33.2.1 Fractional Order Model of the Assistant Motor

The EPS nonlinear dynamic model is established by the relation between the tire/road contact forces, the DC motor and the mechanism of steering system [19]:

$$J_c\ddot{\theta}_c = -K_c\theta_c - B_c\dot{\theta}_c + K_c\frac{\theta_m}{N} - F_c sign(\dot{\theta}_c) + T_d \tag{33.1}$$

$$J_{eq}\ddot{\theta}_m = K_c\frac{\theta_c}{N} - \left(\frac{K_c}{N^2} + \frac{K_r R_p^2}{N^2}\right)\theta_m - B_{eq}\dot{\theta}_m + K_t I_m - F_m sign(\dot{\theta}_m) - \frac{R_p}{N}F_r \tag{33.2}$$

$$U = R_m I_m + L_m \dot{I}_m + K_t\dot{\theta}_m \tag{33.3}$$

Fractional order is used to the integral motor's optimization design for nonlinear inductances. The motor's dynamic equation is transformed into the following form after introduction of fractional order:

$$U = R_m I_m + L_m D^{\alpha}\dot{I}_m + K_t D^{\beta}\dot{\theta}_m \tag{33.4}$$

where $-1 < \alpha, \beta < 1$, $T_r = R_p F_r$, $T_a = NK_t I_m$.

$$J_{eq} = J_m + \frac{R_p^2}{N^2} M_r,\ B_{eq} = B_m + \frac{R_p^2}{N^2} B_r,\ \theta_m = N\frac{x_r}{R_p},\ T_c = K_c\left(\theta_c - \frac{\theta_m}{N}\right).$$

After simultaneous (33.1) (33.2) (33.4), with taking the nonlinear effect on the system into account, the fractional order nonlinear dynamic model of the EPS system can be presented in state-space form as follows:

$$\begin{cases} \dot{x} = Ax + B_1 u_1 + B_2 u_2 + F \\ y = Cx \end{cases} \tag{33.5}$$

where $x = [\theta_c\ \dot{\theta}_c\ \theta_m\ \dot{\theta}_m\ I_m]^T$, $u_1 = U$, $u_2 = [T_d\ T_r]^T$, $y = [\theta_c\ \theta_m]^T$.

$$A = \begin{bmatrix} 0 & 1 & 0 & 0 & 0 \\ a_{21} & a_{22} & a_{23} & 0 & 0 \\ 0 & 0 & 0 & 1 & 0 \\ a_{41} & 0 & a_{43} & a_{44} & a_{45} \\ 0 & 0 & 0 & a_{54} & a_{55} \end{bmatrix},\ B_1 = \begin{bmatrix} 0 \\ 0 \\ 0 \\ 0 \\ b_{51} \end{bmatrix},\ B_2 = \begin{bmatrix} 0 & 0 \\ b_{21} & 0 \\ 0 & 0 \\ 0 & b_{42} \\ 0 & 0 \end{bmatrix},\ C = \begin{bmatrix} 1 & 0 & 0 & 0 & 0 \\ 0 & 0 & 1 & 0 & 0 \end{bmatrix},\ F = \begin{pmatrix} 0 \\ f_{21} \\ 0 \\ f_{31} \\ 0 \end{pmatrix},$$

$$f_{21} = -\frac{F_c}{J_c} sign(\dot{\theta}_c),\ f_{31} = -\frac{F_m}{J_{eq}} sign(\dot{\theta}_m),\ a_{21} = \frac{-K_c}{J_c},\ a_{22} = \frac{-B_c}{J_c},\ a_{23} = \frac{K_c}{J_c},$$

$$a_{41} = \frac{K_c}{J_{eq}N},\ a_{43} = \frac{-\left(R_p^2 K_r + K_c\right)}{J_{eq}N^2},\ a_{44} = \frac{-B_{eq}}{J_{eq}},\ a_{45} = \frac{K_t}{J_{eq}},\ b_{42} = \frac{-1}{J_{eq}N},\ b_{51} = \frac{1}{L_m} D^{-\alpha}.$$

where J_m is motor moment of inertia, θ_c is steering wheel angle, J_c is steering column moment of inertia, T_d is driver torque, T_c is steering torque, θ_m is motor angle, T_m is Motor electromagnetic torque, T_a is assist torque, B_m is motor shaft viscous damping, M_r is mass of the rack, B_r is Viscous damping of the rack, x_r is the rack position, R_p is steering column pinion radius, B_c is steering column viscous damping, R_m is motor resistance, L_m is motor inductances, I_m is motor current, K_t is motor torque, voltage constant, K_r is tire spring rate, N is motor gear ratio.

33.2.2 Vehicle Model

The F_r in the Eq. (33.2) represents road reaction force, the force which acts on the rack can be generated by using the vehicle model [20]:

$$\dot{\beta} = \frac{1}{MV}\left\{ -(C_f + C_r)\beta - \left\{ MV - \frac{1}{V}\left(l_f C_f - l_r C_r\right) \right\}\gamma + C_f \delta_f \right\} \tag{33.6}$$

$$\dot{\gamma} = \frac{1}{I_Z}\left\{ -(l_f C_f - l_r C_r)\beta - \frac{1}{V}\left(l_f^2 C_f + l_r^2 C_r\right)\gamma + l_f C_f \delta_f \right\} \tag{33.7}$$

The front-wheel sideslip is given by

$$\alpha_f = \delta_f - \left(\beta + \frac{l_f}{V}\gamma\right) \tag{33.8}$$

The self-aligning torque

$$T_{sat} = (T_p + T_m)F_{yf} = (T_p + T_m)C_f \alpha_f \tag{33.9}$$

Therefore the reaction force F_r which acts on the rack is defined as:

$$F_r = \frac{T_{sat}}{R_p} = \frac{(T_p + T_m)C_f\left[\delta_f - \left(\beta + \frac{l_f}{V}\gamma\right)\right]}{R_p} \tag{33.10}$$

where β is sideslip angle, γ is the yaw rate, T_m is the mechanical trail, C_f and C_r are the cornering stiffness coefficient, l_r is the chassis length of rear, T_p is the pneumatic trail, l_f is the chassis length of front, N_v is the steering system ratio, I_z is the moment of vehicle inertia. And the front steer angle δ_f is given by $\delta_f = \frac{\theta_m}{N_v}$.

33.3 Fractional Controlling

Fractional calculus operator s^α can't be calculated in Matlab directly, so it can be got by algorithm fitting of Oustaloup filter:

$$s^\alpha \approx K \coprod_{K=-N}^{N} \frac{S + w_k'}{S + w_k} \tag{33.11}$$

In order to make the operator to achieve optimal approximation within the frequency bands based on Oustaloup filter [21], adding another filter G_l before the Oustaloup filter G_c, so the fractional operator can be expressed:

$$s^\alpha \approx G_l \times G_c \tag{33.12}$$

And form of the filter G_l can be expressed:

$$G_l = \frac{a_1 s^2 + a_2 s + a_3}{b_1 s^2 + b_2 s + b_1} \tag{33.13}$$

The coefficients of G_l can be determined by parameter optimization. In order to enhance the approximation accuracy of phase frequency and amplitude frequency within frequency bands, taking the error as the optimization performance index:

$$J = \int_{w_b}^{w_h} (\lambda|M_1 - M_2| + (1-\lambda)|P_1 - P_2|)dw \tag{33.14}$$

where M_1 and P_1 are the practical amplitude frequency and phase frequency respectively; M_2 and P_2 are the approximation amplitude frequency and phase frequency respectively; w_h and w_b are the frequency range; taking $\lambda = 0.5$ which is weight factor. The parameters of the filter G_l should be optimized to make J minimum value.

According to the dynamic equation, for the convenience of fractional dynamic modeling in Matlab, the Eq. (33.4) can be changed into following form:

$$U = R_m I_m + L_m D^{\alpha} I_m + K_t D^{\beta} \dot{\theta}_m \tag{33.15}$$

where $0 < \alpha < 2, -1 < \beta < 1$.

According to the Eq. (33.3), the motor's integer order model in simulink is showed in the Fig. 33.1, while the motor's fractional order model is showed according to the Eq. (33.15) in the Fig. 33.2.

The assistant control strategy of the EPS system adopts motor's current feedback control, and fractional order controller $PI^{\lambda}D^{u}$ is used to control fractional order system. Combining integer order vehicle model, steering column model, torque sensor model, rack model with motor's fractional order model, then a new control strategy can be got showing in block diagram of Fig. 33.3.

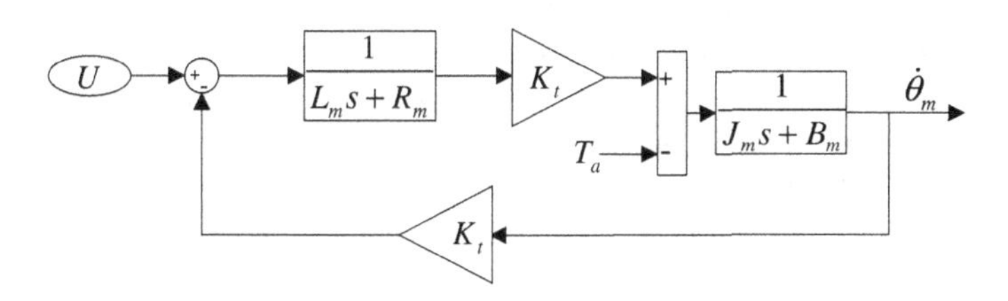

Fig. 33.1 The integer order model of motor

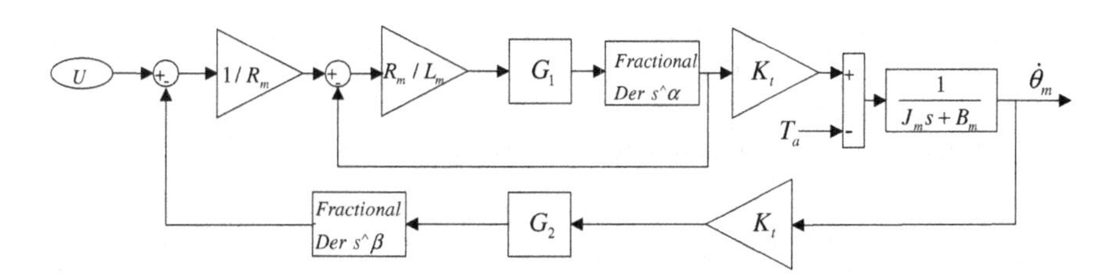

Fig. 33.2 The fractional order model of motor

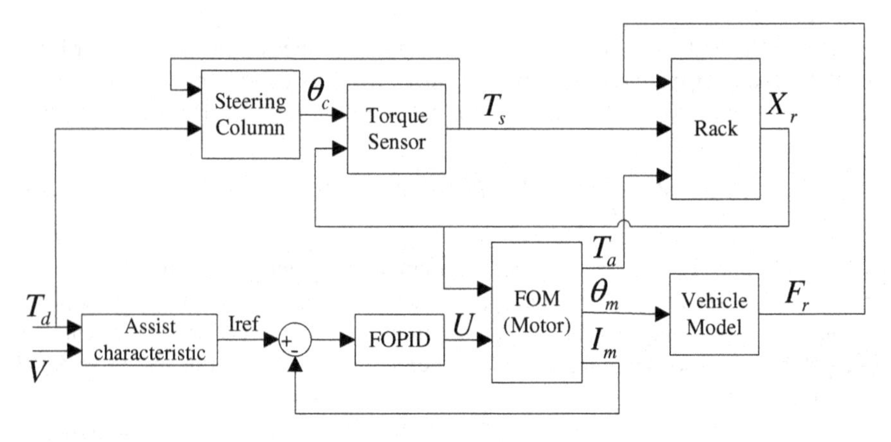

Fig. 33.3 The fractional control diagram of the EPS system

33.4 Simulation Results

The fractional order model of the EPS system is established in this paper, and the fractional order $\alpha = 0.013$, $\beta = 0.086$ can be obtained by optimizing the motor's parameters. The simulation parameters of the dynamic model are introduced in reference [19].

Assuming that the driver's input torque T_d is the ideal sine signal (amplitude is 5 Nm and frequency is 1 HZ). As it can be seen in Fig. 33.4, as the existence of nonlinear inductances, the assistant torque T_a' appears torque ripple under the integer order model, while the assistant torque T_a under the fractional order model has a good tracking performance and stability. The simulation means that both the motor's fractional order modeling and the new control strategy are effective.

The stability of assist power is also an important index to represent stability of the whole system. So the fractional order controller $PI^{\lambda}D^{u}$ (FOPID) is compared with the traditional integral order PID (IOPID) controller when the input is at a fixed steering wheel torque. The parameters of controller: $K_p = 0.1$, $K_I = 50$, $\lambda = 0.85$, $K_D = 0$. As it can been observed, the motor current transient response is showed in Fig. 33.5.

(a) Using the same fractional order controller to control integral order model (IOM) and fractional order model (FOM), the simulations illustrate that IOM has a big overshoot and reaches steady state in a long time.
(b) Using the same fractional order model (FOM) controlled by IOPID and FOPID respectively, the simulations illustrate that the system controlled by FOPID has no overshoot and reaches steady state in a short time.

Therefore we can summarize that the fractional order system controlled by the fractional order $PI^{\lambda}D^{u}$ controller will have a good system performance.

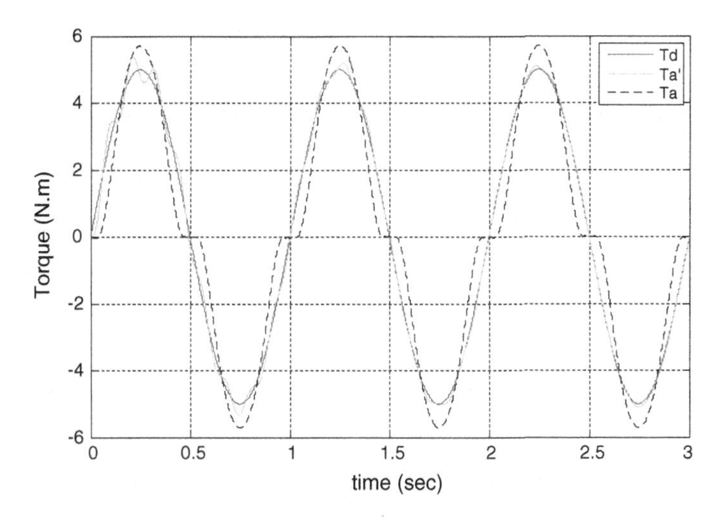

Fig. 33.4 The effectiveness of fractional order model

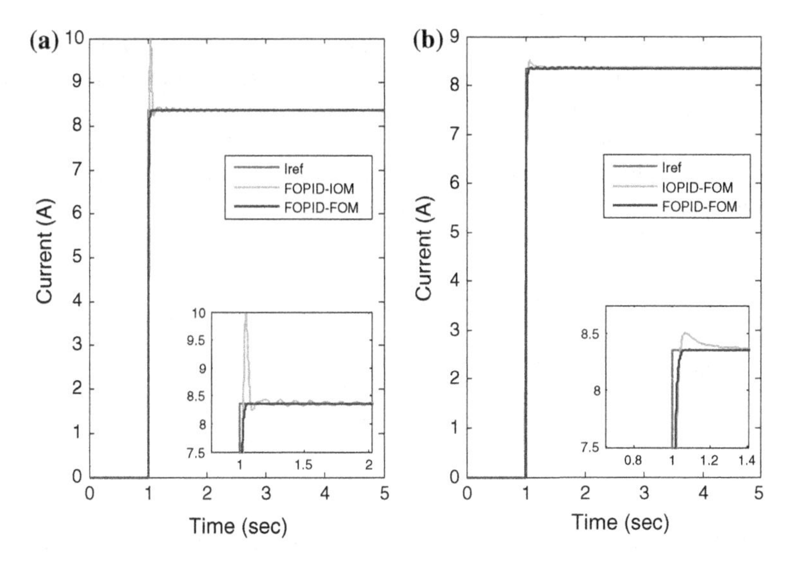

Fig. 33.5 The contrast of step response: **a** Compared IOM with FOM, **b** Compared IOPID with FOPID

The EPS system can be disturbed by the external disturbance torque easily which may cause motor current not able to track the target current accurately. Therefore the controller's anti-disturbance is taken into account. This work adds the frequency 10HZ and 30HZ signals respectively as the external disturbance sine signal. Where Iref represents the motor's reference output current, FO-Im represents motor current controlled by FOPID and IO-Im represents motor currents controlled by IOPID.

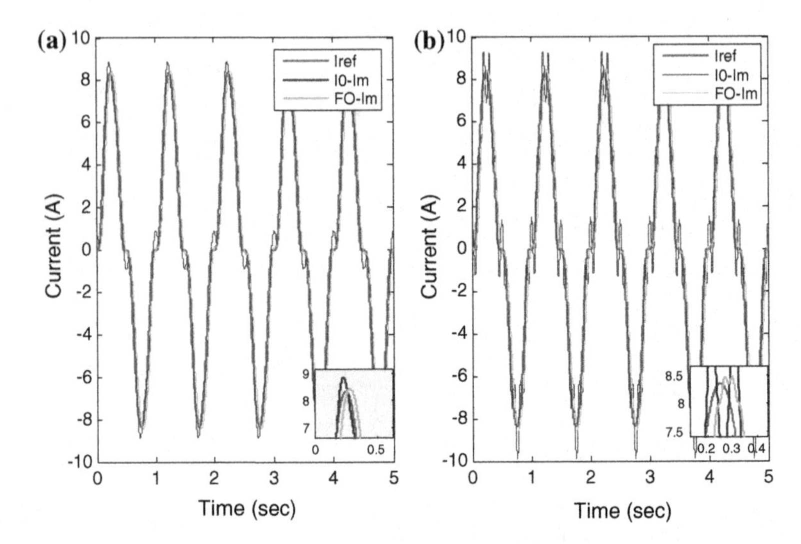

Fig. 33.6 The contrast of anti-disturbance: **a** Disturbance at 10 HZ, **b** Disturbance at 30 HZ

As it can be seen as shown in Fig. 33.6: Both (a) and (b) show that FO-Im has a better anti-disturbance than the traditional PID control when disturbance at 10 or 30 HZ.

In order to test the robustness to parameter uncertainties, another experiment which varies the parameters values (J_c, B_c, K_t, R_m, L_m) with 10 % is performed. The curves after the change are almost coincided with the curves before the change as shown in Fig. 33.7. The results mean that the proposed control strategy has a good performance even with parameter uncertainties.

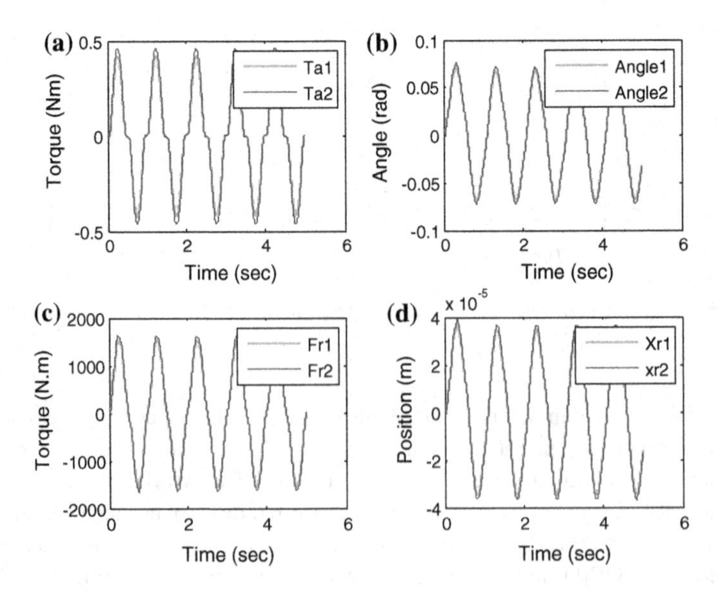

Fig. 33.7 Robustness, with 10 % variation of the parameters: **a** Assist torque, **b** Motor angle, **c** Road reaction torque, **d** Rack position

33.5 Conclusion

The assistant motor's fractional order reference model is established, and the fractional order controller is designed to control the fractional order system that can make the system show better performance. In addition, this paper gives a method which uses Oustloup filter to realize the fractional order system of the EPS system, and its fractional order space-state expression is also showed in this paper. The experiments illustrate that EPS system have a good performance, like good anti-disturbance and robustness, under the fractional order $PI^{\lambda}D^{u}$ control in the presence of nonlinear factors. The feasibility and reliability of the EPS fractional order system have been verified in this paper sufficiently.

Although the fractional order model has better robustness to nonlinear factors, it is just still in the theoretical state in this paper. How to realize the fractional order model in engineering practice is our follow-up study focus. This paper aims at proposing a method to establish the fractional order plant, hoping to provide some opinions and ideas for the fractional order modeling of the EPS system.

References

1. Parmar M, Hung JY (2004) A sensorless optimal control system for an automotive electric power assist steering system [J]. IEEE Trans Industr Electron 51(2):290
2. Saifia D (2015) Fuzzy control for electric power steering system with assist motor current input constraints [J]. J Franklin Inst 352(2):562–576
3. Xin L, Xueping Z, Jie C (2009) Controller design for electric power steering system using T-S fuzzy model approach [J]. Int J Autom Comput 6(2):198–203
4. Messoussi WE, Pages O, Hajjaji AE (2007) Four-wheel steering vehicle control using Takagi-Sugeno fuzzy models [J]. In: Proceedings of IEEE international fuzzy systems conference, pp 1–6
5. Lin F-J, Hung Y-C, Ruan K-C (2014) An intelligent second order sliding mode control for an electric power steering system using a wavelet fuzzy neural network [J]. IEEE Trans Fuzzy Syst 22(6):1598–1611
6. Yingchih H, Faajeng L, Jonqchin H, Jinkuan C, Kaichun R (2015) Wavelet fuzzy neural network with asymmetric membership function controller for electric power steering system via improved differential evolution [J]. IEEE Trans Power Electron 30(4):2350–2362
7. Zang H, Liu M et al (2007) Fuzzy neural network PID control for electric power steering system [J]. In: IEEE International conference on automation and logistics, pp 643–648
8. Marouf A, Sentouh C, Djemai M, Pudlo P (2011) Control of an electric power assisted steering using reference model [J]. In: Proceedings of the 50th IEEE conference on decision and control and european control conference, pp 6684–6690
9. Marouf A, Sentouh C, Djemai M, Pudlo P (2011) Control of electric power-assisted steering system using sliding-mode control [J]. In: Proceedings of international ieee conference on intelligent transportation system, pp 107–112
10. Chen B-C, Hsu W, Huang S-J (2008) Sliding-mode return control of electric power steering [J]. In: Society of automotive engineers, Paper 2008-01-0499
11. Meiqin M, Shaobo Z (2013) Analysis and Optimal design of winding inductances for permanent magnet brushless DC motor [J]. Electr Drive 43(2):19–21 (in Chinese)

12. Jiang X, Li W, Li W, Zhu H (2009) A novel measuring method of PMSM nonlinear incremental inductancess [J]. In: sixth international conference on measuring technology and mechatronics automation, IEEE, pp 170–173
13. Westerlund S, Ekstam L (1994) Capacitor theory [J]. IEEE Trans Dielectr Electr Insul 1 (5):826–839
14. Torvik PJ, Bagley RL (1984) On the appearance of the fractional derivative in the behavior of real materials [J]. Trans Am Soc Mech Eng 51(2):294–298
15. Xue D, Zhao C, Chen Y (2006) A modified approximation method of fractional order system [J]. In: IEEE International Conference on Mechatronics and Automation, IEEE, pp 1043–1048
16. Zhao C, Xue D, Chen Y (2005) A fractional order PID tuning algorithm for a class of fractional order plants [J]. In: IEEE International Conference Mechatronics and Automation, PP 216–221
17. Chunna Z, Zhao Y, Xiangde Z (2009) Simulation research on fractional order controllers with integer order controllers [j]. J Syst Simul 21(3):768–771 (in Chinese)
18. Wei G, Jinghong W, Wangping Z (2010) Fractional-order PID dynamic matrix control algorithm based on time domain [J]. In: world congress on intelligent control & automation, pp 208–212
19. Marouf A, Djemai M, Sentouh C et al (2012) A new control strategy of an electric-power-assisted steering system [J]. IEEE Trans Veh Technol 61(8):3574–3589
20. Marouf A, Sentouh C, Djemai M, Pudlo P (2011) Control of an electric power assisted steering system using sliding mode control [J]. In: The 14th IEEE conference on intelligent transportation systems, IEEE, pp 107–112
21. Naiming Q, Changmao Q, Wei W (2010) Optimal Oustaloup digital realization of fractional order systems [J]. Control Decision 25(10):1598–1600 (in Chinese)

Chapter 34
Optical Mouse Sensor-Based Laser Spot Tracking for HCI Input

Mingxiao He, Quanzhou Wang, Xuemei Guo and Guoli Wang

Abstract When facing with the mid-air interactive tasks in a wide range, the laser spot motion sensing technique can be as a information input mode of human-computer interaction(HCI). This paper explores the use of optical mouse sensors for building a laser spot tracking system, which can be used as a motion-based HCI device. Our work is focused on the characterization of the laser speckle sensing by optical mouse sensors. Based on the laser speckle displacement measurement capability of optical mouse sensors, we demonstrate that the low-cost optical mouse sensor can be used to record the motion of laser spot as a precise, fast and compact sensing method. To make a prototype system for demonstration, we describe a kind of deployment method for building optical mouse sensor array and propose a weighted method to fuse the raw data of multi-optical mouse sensor array. In experimental testing, the ISO standard tests for HCI input device were used for evaluating the efficiency of HCI input by optical mouse sensor-based laser spot tracking. A paradigm that using laser pointer and our laser spot tracking system to complete the dynamic hand gesture recognition task is also given in this paper.

Keywords Optical mouse sensor · Laser speckle · Laser spot tracking · Human-computer interaction · Sensor deployment · Data fusion

34.1 Introduction

Nowadays with an increasing number of intelligent devices, more and more HCI scenes can be found in our life. Projection display scene is a typical example, but the traditional keyboard and mouse limit the interaction flexibility of projection

M. He · Q. Wang · X. Guo · G. Wang (✉)
School of Information Science and Technology Sun Yat-Sen University,
Guangzhou 510006, China
e-mail: isswgl@mail.sysu.edu.cn

G. Wang
SYSU-CMU Shunde International Joint Research Institute, Foshan 5280000, China

Y. Jia et al. (eds.), *Proceedings of the 2015 Chinese Intelligent Systems Conference*, Lecture Notes in Electrical Engineering 360,
DOI 10.1007/978-3-662-48365-7_34

system. What's important in HCI is that we have to figure out the intention of operator. Motion behavior is a natural cue to present human intention. When using projection display to make a presentation, laser pointer helps presenter interact with the audiences. With the development of computer vision HCI with laser-pen in projection display become possible. The laser spot can be detected and tracked by using a camera towards the projection curtain [1]. But the weaknesses of computer vision method are obvious. The frame rate of common cameras limits the speed of laser spot motion-based interaction. The algorithm complexity is high and computer vision can be affected by realtime lighting. What's more, dealing with the image distortion [2] also introduces additional computations and computational errors.

However, when some researchers turn their eyes on the optical mouse sensor, they find that the low-cost sensor chip, which has embedded firmware and a CMOS camera, employs low level image processing and with low energy consumption. The mouse sensor chip is made for traditional optical mouse to measure horizontal 2D displacement of human hand. But the flexibility of optical mouses are sorely limited, conventional optical mouses need to be placed quite near to a work surface. Though Robert Ross [3] and our previous work [4] try to assemble the sensor chip with larger focal length lens to enlarge the reliable working interval between the optical mouse sensor and its work surface, the prototype devices are just fit for robotic device positioning. Due to the configuration restriction, these prototypes can not be used in projection system interaction for the limited working distance. When using an external laser as a direct illumination source, T.W. Ng [5] and D. Font [6] have found that the motion of laser speckle can be measured by the optical mouse sensor with desirable accuracy. And A. Olwal [7] has made a HCI device with a single mouse sensor chip and a laser speckle projector. His prototype can recognise the laser speckle motion trajectory and control the external devices according to the recognition. But this prototype just recognise simple hand gestures and the effective area is small for interaction.

In this paper, a optical mouse sensor-based laser spot tracking system with low computation cost and desirable accuracy has been developed. Multi-optical mouse sensors are employed for giving a large interaction area. We give a data fusion paradigm to show that our system can be easily expanded by adding optical mouse sensors with low computational cost. This system can be easily embedded in the traditional projection system for implementing HCI input applications.

34.2 System Implementation

This laser spot tracking system for HCI input has a number of optical mouse sensors, which are the basic parts for sensing the motion of laser spot. Every sensor chip has the embedded optical flow algorithm firmware for speckle image flow processing. Then the measurements data from all the sensors will be fused by using a microcontroller and then sent to a computer as input information.

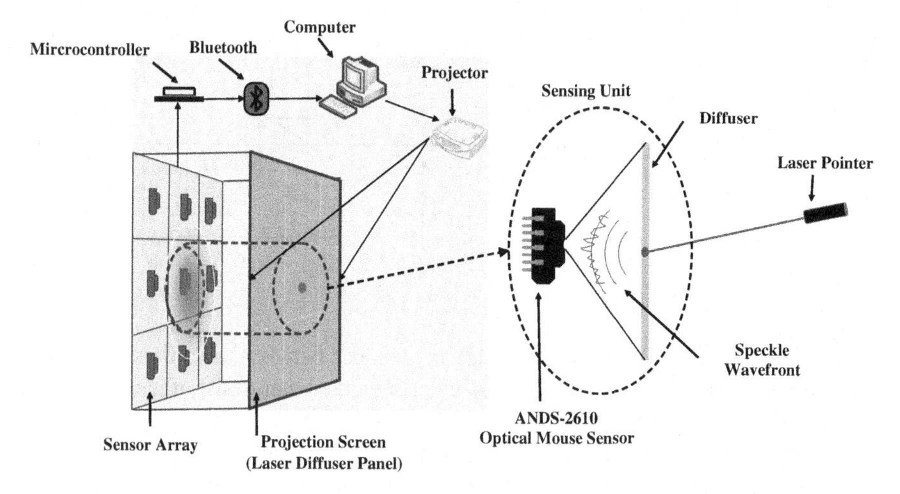

Fig. 34.1 The system framework

34.2.1 System Framework

Figure 34.1 shows the framework of a laser spot tracking system which is developed for projection HCI. This system can be embedded in the traditional projection equipment. Here a Freescale microcontroller (xs128) is used for fusing data from the optical sensor array. And the laser spot motion data which imply for interaction input information can be sent to an external pc via a bluetooth serial port module (HC-06). More than one optical mouse sensor are used for expanding the limited sensing area of a single sensor.

34.2.2 Speckle Sensing

For gaining reliable displacement measurements of the laser spot motion, the law of speckle motion in the image plane of the system shown in needs to be considered in detail. Another important factor is the deployment of the sensors, there should not be any sensing dead zone of the interaction area.

Laser Speckle Correlation

The laser speckle has 2 important statistical properties. In one case, as the laser spot move on the diffuse panel surface, the speckles move as a whole on the observation plane and their shape remains unchanged. This property can be called a "translation" of speckles. On the contrary if the shape of speckles changes while speckle motion, this property is called "boiling" of speckles [8]. Obviously, the optical flow method will lose efficacy when the speckles boiling cause there are no connections between the speckle images.

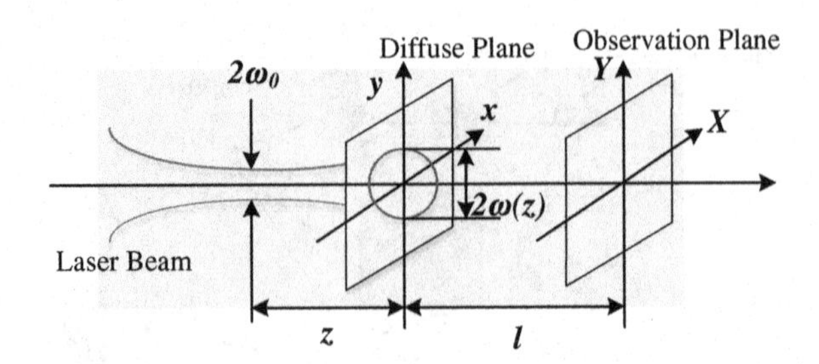

Fig. 34.2 The speckle formation and observation coordinates

In the coordinates shown by Fig. 34.2, the distance between the diffuse plane and the observation plane is l, z is the distance between the beam waist of gaussian laser and the diffuse plane, ω_0 is the radius of beam waist, ω is the radius of spot on the diffuse surface. Supposing the velocity of laser spot on the diffuse plane is v and the speckle translation velocity on the observation is V_s. Research [9] shows that the relation between v and V_s is given by

$$V_s = (1 + \frac{l}{\rho})v, \tag{34.1}$$

where ρ is the radius of curvature of the wave front, which is given by

$$\rho = z(1 + \frac{a^2}{z^2}), \tag{34.2}$$

where

$$a = \frac{\pi}{\lambda}\omega_0^2. \tag{34.3}$$

Parameter a is the area within the scope of beam waist along the optical axis. An evaluation factor η of speckle property is also given in [13] by

$$\eta = (\frac{(l+z)z}{a} + a)\frac{1}{l}. \tag{34.4}$$

When $\eta = 0$, the dynamic speckles are boiling. If $|\eta| \le 1$, the boiling property of speckles plays a major role. If $|\eta| > 1$, the translation property of speckles plays a major role. So we need make the deployment of sensor array and diffuser panel in Fig. 34.2 to satisfy the speckles translation condition.

When $|\eta| \le 1$, there 4 kinds of circumstances:
If $z \ge a > 0$ (beam divergence),

$$\frac{a^2 + z^2}{a - z} \le l \le -\frac{a^2 + z^2}{a + z}. \tag{34.5}$$

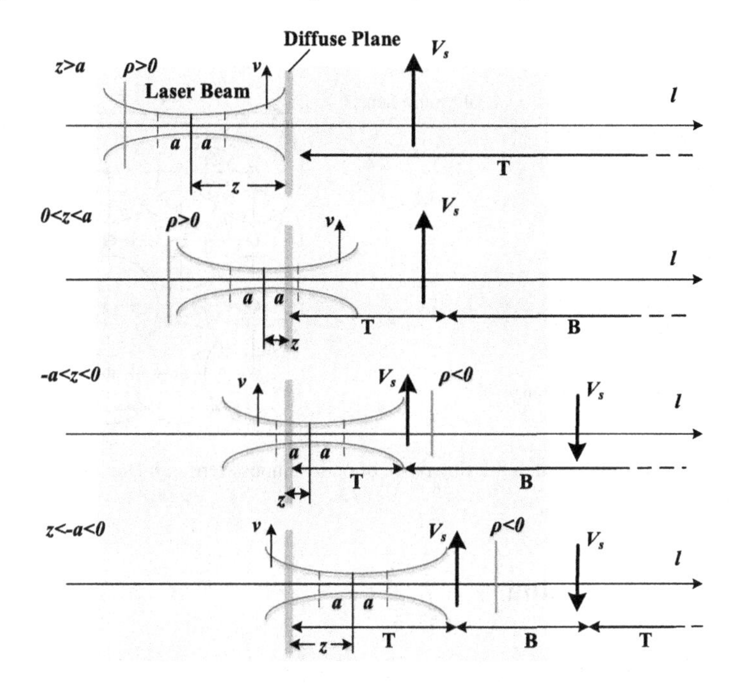

Fig. 34.3 The motion status of dynamic speckles in free space

If $0 < z < a$,

$$l \leq -\frac{a^2 + z^2}{a + z}, \frac{a^2 + z^2}{a - z} \leq l. \tag{34.6}$$

If $-a < z < 0$, same as Eq. (34.6).
If $z \leq -a < 0$ (beam convergence), same as Eq. (34.5). Here we should notice that l can not be negative. Figure 34.3 shows the motion status of dynamic speckles under all conditions.

Sensor Array Deployment

It is obvious that densely-deployed sensor array can eliminate the dead zones. But this will lead a waste of resources. An efficient way has been proposed as shown in Fig. 34.5. Ignoring the FOV(field of view) of the sensor chip, when the radius of speckle image is r, the sensing range of an optical mouse sensor can be modeled as a circular area with radius r, as shown in Fig. 34.4a. So when deploying the optical mouse sensors, the interval between two sensors in sensor array can be as large as $\sqrt{2}r$ with no dead zone in the sensing area. The sensing area of a typical 4×4 sensor array is illustrated by a blue square in Fig. 34.4b.

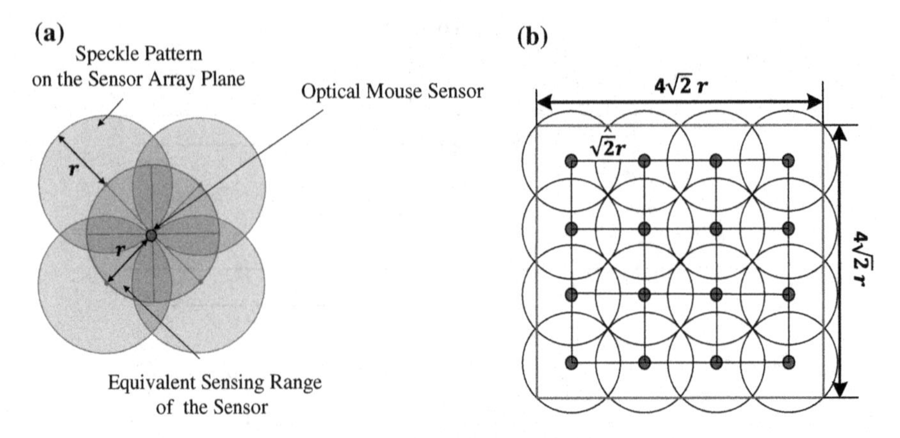

Fig. 34.4 Sensor deployment. **a** Sensing range of optical mouse sensor. **b** Deployment of a 4 × 4 sensor array

34.3 Data Processing

In our laser spot tracking system, every optical mouse sensor output the relative displacement increments when the speckle is exposed to the photosensitive element. At every moment, not all sensors of the sensor array can be activated by the speckle. To figure out the time-spatial relationship of the data provided by the valid sensors at each time, a data fusion method has been proposed for providing reliable laser spot motion measurements. Besides, the measurement noise elimination algorithms for laser spot motion measurement are also proposed.

34.3.1 Multisensor Data Fusion

For a single sensor, when the motion of the speckles is captured, the 2D displacement increments can be read out from the sensor's registers (Delta_X and Delta_Y) time by time. Assuming that the starting point of the speckle motion trajectory is (0,0), the position of the laser spot in time *t* can be calculated by Eq. (34.7):

$$\begin{cases} x_t = x_{t-1} + \Delta x_t \\ y_t = y_{t-1} + \Delta y_t \end{cases} . \tag{34.7}$$

where Δx_t and Δy_t can be read out from the registers of the sensor chip in time *t*. For evaluating the effectiveness of the measurements, the optical mouse sensor also provides a parameter of image quality [10]. The value of this parameter is stored in the SQUAL register. The sensor works better with high quality image flow captured. Meanwhile, the SQUAL value has a great contribution for sensor array data fusion.

The interval between two adjacent sensors in the array is shorter than the diameter of the laser speckle, show as Fig. 34.4b. So more than one sensor can sense the speckle motion at a time. If only a single sensor is selected to provide motion measurement, some useful information is abandoned. To improve the utilization of the sensor array as well as enhance the reliability of the measurements, a weighted data fusion method is used.

$$\Delta x(y)_t = \boldsymbol{W}\boldsymbol{D}_t \tag{34.8}$$

where,

$$\boldsymbol{W} = \left[\frac{S_{squal_1}}{\sum_{i=1}^{N} S_{squal_i}} \ \frac{S_{squal_2}}{\sum_{i=1}^{N} S_{squal_i}} \ \cdots \ \frac{S_{squal_N}}{\sum_{i=1}^{N} S_{squal_i}} \right], \tag{34.9}$$

$$\boldsymbol{D}_t = [\, \Delta x(y)_{1,t} \ \Delta x(y)_{2,t} \ \cdots \ \Delta x(y)_{N,t} \,]^{\mathrm{T}}. \tag{34.10}$$

Combining Eqs. (34.7) and (34.8), the laser spot motion can be measured.

34.3.2 Measurement Noise Elimination

When try to tracking the laser spot motion by observing the speckle motion, measurement noise will induced when the dynamic speckles are boiling. We need a motion filtering method for noise elimination, and the Kalman filter seems a good solution. By using the prior state and the noisy observation of the current state, the filter estimates the current state of a moving target through recursive means [11]. Besides, an appropriate dynamic model of the laser spot must be selected for Kalman filter. Singer model [12] has been proved to be suitable for many maneuvering targets, including human motion [13]. Since the laser spot motion is manipulated by the user of laser pointer, we chose Singer model for modeling the laser spot motion.

34.4 Experimental Testing

Two experiments are given in this section to show the applicability of our laser spot tracking system for HCI input.

34.4.1 Moving and Tapping Test

To analyze the performance of the optical mouse sensor-based laser spot tracking system as a HCI input device, the ISO standard [14] tests are chosen for evaluating. One-direction and multi-direction tapping tests are used to test our system. Since the optical mouse is a most widely used HCI input device, an optical mouse is also been used in the tests for giving a comparative reference.

The main evaluating index for the ISO standard tests is *Throughput* (*TP*). *TP* is the rate of information transfer when a user is operating an HCI input device to control the pointer on a display [15]. The unit of *TP* is bits/s(bps), it implies the speed and accuracy of an input device.

$$TP = \frac{ID_e}{t_m}, \tag{34.11}$$

$$ID_e = \log_2(\frac{d + w_e}{w_e}), \tag{34.12}$$

$$w_e = 4.133 * s_x. \tag{34.13}$$

where, ID_e is the effective index of difficulty, t_m is the movement time, w_e is the effective target width, it can be deduced by the standard deviation of the tapping coordinates in the motion direction s_x. Figure 34.5 shows the graphical user interfaces we design for the tapping tests.

In the one-direction tapping test, the user must operate the device to control the cursor to make a make a click on the initial region then move the cursor to make another click on the target region as soon as possible to end a single operation. For the multi-direction tapping test, the 16 circular targets were displayed one at a time and the user must operate the input device to control the cursor for tapping in the as soon and accurately as possible. Hitting the current target the next target was displayed. For every test, we designed 4 kinds of difficulty level(ID_e) through changing d and w in the GUIs.

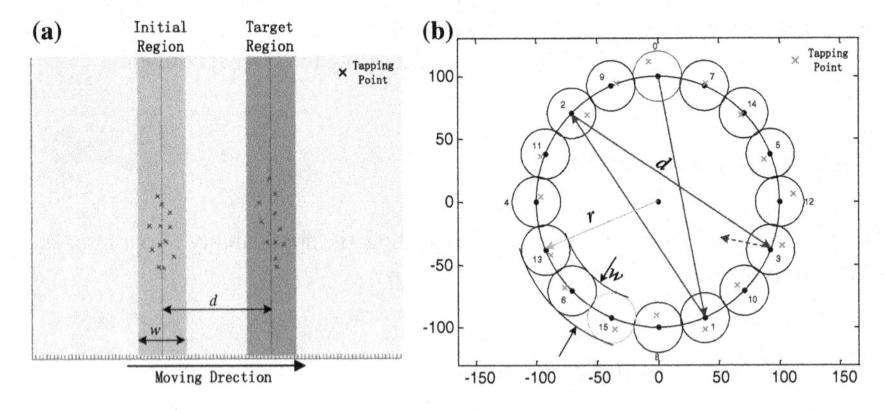

Fig. 34.5 The GUIs of ISO standard tests. **a** One-direction tapping test. **b** Multi-direction tapping test

Table 34.1 Experimental results of one-direction tapping test

d	w	ID (bits)	Optical mouse				Our system			
			w_e	$ID_e(bits)$	$\bar{t}_m(s)$	TP	w_e	$ID_e(bits)$	$\bar{t}_m(s)$	TP
90	4	4.56	1.95	5.56	1.443	3.85	4.67	4.34	3.809	1.14
70	7	3.46	2.50	4.86	1.305	3.72	7.46	3.38	3.052	1.11
50	10	2.58	8.28	2.82	1.268	2.22	9.89	2.60	3.010	0.86
30	13	1.73	11.72	1.83	1.147	1.60	11.73	1.83	2.605	0.70
Average			6.11	3.77	1.291	2.85	8.44	3.04	3.119	0.95

Table 34.2 Experimental results of multi-direction tapping test

d $2rcos(\frac{\pi}{16})$	w	ID (bits)	Optical mouse				Our system			
			w_e	$ID_e(bits)$	$\bar{t}_m(s)$	TP	w_e	$ID_e(bits)$	$\bar{t}_m(s)$	TP
196	20	3.43	11.14	4.22	0.915	4.61	15.93	3.73	3.405	1.10
196	40	2.56	21.47	3.34	0.747	4.47	23.10	3.25	3.336	0.97
98	20	2.56	12.05	3.19	0.787	4.06	16.64	2.78	2.091	1.33
98	40	1.79	16.25	2.81	0.717	3.93	29.72	2.10	1.524	1.38
Average			15.23	3.39	0.790	4.27	21.35	2.95	2.590	1.20

In the one-direction tapping test, for every kind of difficulty level we used the laser spot tracking system and optical mouse for 10 operations and recorded the time of each operation. The detailed experimental results are shown in Table 34.1. We can see that the *TP* of our device is about 33.3 % of the *TP* of optical mouse, and in [19] the *TP* of a camera-based laser spot tracking system is 34.6 % of the *TP* of mouse.

In the multi-direction tapping test, for every kind of difficulty level we used the laser spot tracking system and optical mouse for 5 sets of operations and recorded the time of each operation. The detailed experimental results are shown in Table 34.2. We can see that the *TP* of our device is about 28.1 % of the *TP* of optical mouse. Though the throughput of our laser spot tracking system is lower than the throughput of optical mouse, our system has advantages on completing mid-air HCI input task.

34.4.2 Dynamic Hand Gesture Recognition

The optical mouse sensor-based laser spot tracking system is used for giving a gesture-based interaction paradigm in this section. Figure 34.6 shows the experimental scenario, a projector was used for interactive display. A user operated a common laser pointer for writing hand gestures on the diffuse panel, and our system tracked the laser spot motion and send the date to computer. Meanwhile, the trajectory of

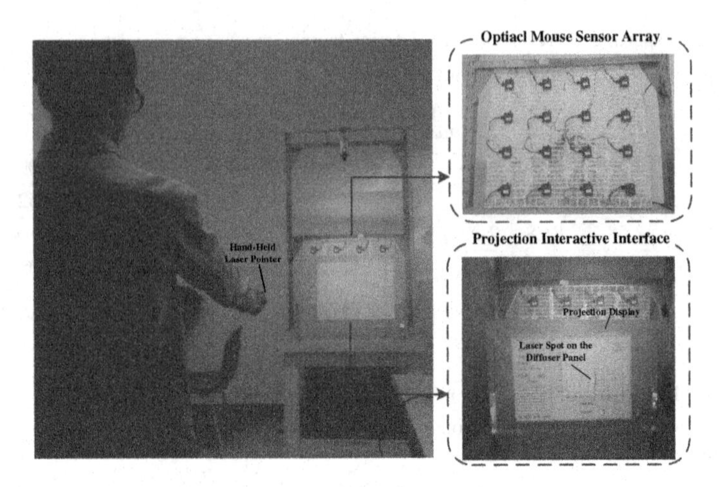

Fig. 34.6 Gesture-based interaction

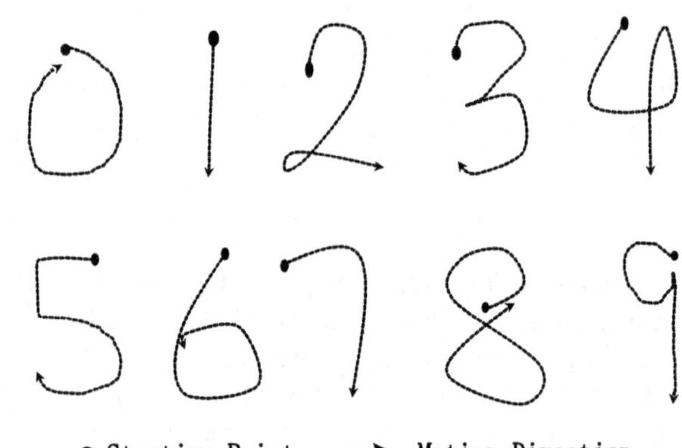

Fig. 34.7 Dynamic gestures

Table 34.3 The hidden states of every gesture

Gesture	0	1	2	3	4	5	6	7	8	9
N	6	3	5	7	7	5	6	4	5	6

laser spot motion and gesture recognition results were projected on the diffuse panel by the projector which was connected with the computer.

The famous HMM was employed for gesture recognition. 10 manuscript Arabic numerals('0'–'9') were used as sample gestures, shown as Fig. 34.7. The hidden states N of HMM is decided by the complexity of every kind of gesture, and with the increasement of N recognition rate is tending towards stability. But an excess of N may cause the overfitting and the computation will be large. The hidden states of every gesture in our work are shown in Table 34.3.

Table 34.4 Recognition results

Training Sample	Testing Sample	Gesture	Recognition Results										
			0	1	2	3	4	5	6	7	8	9	Recognition Rate
20	10	0	10	-	-	-	-	-	-	-	-	-	100%
20	10	1	-	9	-	-	-	-	1	-	-	-	90%
20	10	2	-	-	8	-	1	-	-	-	-	1	80%
20	10	3	-	-	-	10	-	-	-	-	-	-	100%
20	10	4	-	-	-	-	9	-	-	-	-	1	90%
20	10	5	-	-	-	3	-	7	-	-	-	-	70%
20	10	6	-	-	1	-	-	-	8	-	1	-	80%
20	10	7	-	-	-	-	-	-	-	10	-	-	100%
20	10	8	-	-	-	1	-	1	-	-	8	-	80%
20	10	9	-	-	-	-	-	-	-	1	-	9	90%
Averaging Recognition Rate													88%

For every kind of gesture, 30 samples were taken. 20 of them were used for training, and the rest were used as test data. The recognition results are shown in Table 34.4, the averaging recognition rate is 88 %. This gesture-based interaction input paradigm shows the potential of using our system for implementing gesture-based HCI input applications.

34.5 Conclusion

This thesis is inspired by the working principle of optical mouse, our research focuses on laser speckles, which is a special structured laser light. By combining laser speckles with the compact, low-cost and high-speed advantages of optical flow sensors, we propose a laser spot tracking panel system, which is based on optical mouse sensor array. We have used our system to complete some basic HCI tasks. Our research shows the potential of combing laser speckles with the commercial optical mouse sensors to make an ambient intelligent device for sensing human motion information and completing HCI input tasks.

Acknowledgments This work was supported by Science and Technology Program of Guangzhou under Grant No. 201510010017, and by the SYSU-CMU Shunde International Joint Research Institute Free Application Project under Grant No. 20130201.

References

1. Widodo RB, Chen W, Matsumaru T (2012) Laser spotlight detection and interpretation of its movement behavior in laser pointer interface. In: 2012 IEEE/SICE international symposium on system integration, pp 780–785. IEEE
2. Zhenying S, Yigang W, Lexiao Y (2008) Research on human-computer interaction with laser-pen in projection display. In: 11th IEEE international conference on communication technology, pp 620–622. IEEE
3. Ross R, Devlin J, Wang S (2012) Toward refocused optical mouse sensors for outdoor optical flow odometry. IEEE Sens J 12(6):1925–1932. IEEE Press
4. Mingxiao H, Xuemei G, Guoli W (2014) Enhanced positioning systems using optical mouse sensors. In: Xianmin Z, Honghai L, Zhong C, Nianfeng W (eds.) ICIRA 2014, Part II, vol 8918, pp 463–474, LNCS. Springer, Heidelberg
5. Ng TW, Carne M (2007) Optical mouse digital speckle correlation. Opt Commun 280(2): 435–437
6. Font D, Tresanchez M, Pallej T, Teixid M, Palacn J (2011) Characterization of a low-cost optical flow sensor when using an external laser as a direct illumination source. Sensors 11(12):11856–11870
7. Olwal A, Bardagjy A, Zizka J, Raskar R (2011) SpeckleEye: gestural interaction for embedded electronics in ubiquitous computing. In: CHI'12 extended abstracts on human factors in computing systems, pp 2237–2242. ACM
8. Asakura T, Takai N (1981) Dynamic laser speckles and their application to velocity measurements of the diffuse object. Appl. phys. 25(3):179–194
9. Wu Y (2013) Statistical properties of laser speckles and detection micro-motion for arbitrary with rough surface. Ph.D. thesis. Xi'an Electronic and Engineering University
10. Agilent ADNS-2610 Optical Mouse Sensor Data Sheet (2004) Agilent Technologies, Santa Clara
11. Welch G, Bishop G (2006) An Introduction to the Kalman Filter. University of North Carolina, Chapel Hill, North Carolina
12. Singer RA (1970) Estimating optimal tracking filter performance for manned maneuvering targets. IEEE Trans Aerosp Electron Syst (4):473–483. IEEE Press
13. Ricquebourg Y, Bouthemy P (2000) Real-time tracking of moving persons by exploiting spatio-temporal image slices. IEEE Trans Pattern Anal Mach Intell 22(8):797–808. IEEE Press
14. ISO: Reference Number ISO/TS 9241-411 (2012) Ergonomics of human-system interaction-part 411: evaluation methods for the design of physical input devices. International Organization for Standardization
15. Widodo RB, Matsumaru T (2013) Measuring the performance of laser spot clicking techniques. In: 2013 IEEE international conference on robotics and biomimetics (ROBIO), pp 1270–1275. IEEE Press

Chapter 35
Event-Triggered H_∞ Consensus Control for Multi-agent Systems with Disturbance

Yu Huan and Yang Liu

Abstract This paper is devoted to the event-triggered consensus control for a general linear multi-agent network system with external disturbances. First a controlled output is defined and a model transformation is conducted to transform the consensus problem into an H_∞ problem. Then a distributed event-based controller is proposed so that the system can reach the consensus results by only using the agent's own information and its neighbors'. The final conclusion is given in the form of a matrix inequality. Finally a simulation is introduced to verify the theoretical conclusions.

Keywords Event-triggered · Multi-agent system · H_∞ control

35.1 Introduction

Multi-agent system is a complex system consists of multiple intelligent agents which can operate itself independently, yet can still act in a consistent way due to the information connections among agents [1]. A decentralized control strategy is often used in such a system [2]. There has been respectable application success in the areas such as intelligent robot control and flight formation control. In a decentralized control strategy, each agent only use its own information and information from its neighbors through the communication network.

Y. Huan
School of Mathematics and Systems Science, Beihang University,
Beijing 100191, People's Republic of China

Y. Liu (✉)
The Seventh Research Division, Beihang University,
Beijing 100191, People's Republic of China
e-mail: ylbuaa@163.com

Y. Liu
School of Automation Science and Electrical Engineering, Beihang University,
Beijing 100191, People's Republic of China

Y. Jia et al. (eds.), *Proceedings of the 2015 Chinese Intelligent Systems Conference*, Lecture Notes in Electrical Engineering 360,
DOI 10.1007/978-3-662-48365-7_35

The consensus problem has been studied during the past decade in many ways because of its important position in the field of multi-agent systems [3–6]. Usually, in a multi-agent system, each agent has a digital micro-processor to measure the current information. And it brings us the problem like the waste of communication band and the high process ability demand for the microprocessor. The event-triggered control implies that the control task execution will be triggered by an event condition, which is usually related to the states of the system. It is obvious that the event-triggered strategy can reduce the use of resource in such a network system without the sacrifice of the system's behavior. Recently, many researchers have studied the distributed event-triggered consensus control for first-order, second-order and general linear systems such as in [7, 8]. However, the unavoidable disturbance effect is not considered in the present event-trigger consensus research.

Motivated by the above work and analysis, we take external disturbances into account and consider the event-triggered consensus problem of a disturbed general linear multi-agent system. To be specific, we first define a new controlled output to transform the state consensus problem into an H_∞ control problem. And then a distributed event-triggered controller is designed, in which the event-triggered times of agents are asynchronous. By model transformation technique and the H_∞ theory, a condition is established in the form of a matrix inequality to ensure that the closed-loop system can reach the consensus result with the desired disturbance attenuation ability. In the end, a simulation is given to show the efficiency of the theoretical result.

35.2 Problem Description and Reformulation

35.2.1 Problem Statement

Consider a multi-agent system which consists of n identical agents with the ith one described by the following linear system which has external disturbances

$$\dot{x}_i(t) = Ax_i(t) + B_1 w_i(t) + B_2 u_i(t) \tag{35.1}$$

where $x_i(t) \in \mathbb{R}^m$ is the state, $u_i(t) \in \mathbb{R}^{m_2}$ is control input and $w_i(t) \in \mathbb{R}^{m_1}$ is external disturbances, $w_i(t) \in \mathscr{L}_2[0,\infty)$. It is required that (A, B_2) is stabilized. The control input $u_i(t)$ is considered to asymptotically solve the consensus problem if all the agents' states satisfy the following equation

$$\lim_{t\to\infty}(x_i(t) - x_j(t)) = \mathbf{0}, \forall i,j \in \{1,\dots,n\} \stackrel{\text{def}}{=} \mathscr{N} \tag{35.2}$$

Undirected graphs are used to describe the agents' interaction network. Let $\mathscr{G} = (\mathscr{V}, \mathscr{E}, \mathscr{A})$ to be an undirected weighted graph with an order of n. $\mathscr{V} = \{v_1, v_2, \dots, v_n\}$ is defined as the set of nodes, $\mathscr{E} \subseteq \mathscr{V} \times \mathscr{V}$ are the set of undirected edges. $\mathscr{A} = [a_{ij}]$ is a symmetric adjacency matrix with $a_{ij} > 0$ to be the weighing factors. It is considered that (v_i, v_j) or $(v_j, v_i) \in \mathscr{E}$ if and only if $a_{ij} = a_{ji} > 0$. In particular, we set $a_{ii} = 0$ for all $i \in \mathscr{N}$. In graph $\mathscr{G}$, v_i represent the ith agent of the system, and there is information interaction between the ith and the jth agent if edge (v_i, v_j) exists. Define $\mathscr{N}_i = \{v_j \in \mathscr{V} : (v_i, v_j) \in \mathscr{E}\}$ as the set of v_i's neighbors. Diagonal matrix $\mathscr{D} = \text{diag}\{d_1, \dots, d_n\}$ is called the degree matrix of $\mathscr{G}$ with $d_i = \sum_{j=1}^{n} a_{ij}$. Then the Laplacian matrix of graph $\mathscr{G}$ is defined as $L = \mathscr{D} - \mathscr{A}$. An undirected path is a sequence of ordered edges of the form $(v_{i1}, v_{i2}), (v_{i2}, v_{i3}), \dots$ in an undirected graph, where $v_{ij} \in \mathscr{V}$. If there is an undirected path from each node to another, the graph is said to be connected.

Lemma 1 *[10] L is the Laplacian matrix of an undirected graph $\mathscr{G}$. Then L satisfies the following two statements.*

(1) There exists at least one zero eigenvalue of L and all the nonzero eigenvalues of L are positive.
(2) If $\mathscr{G}$ is connected, then L has exactly one zero eigenvalue and the eigenvector of 0 is **1**.

Lemma 2 *[11] $L_c = [L_{c_{ij}}] \in \mathbb{R}^{n \times n}$ is a symmetric matrix with each element defined as:*

$$L_{c_{ij}} = \begin{cases} \frac{n-1}{n}, & i = j \\ -\frac{1}{n}, & i \neq j \end{cases} \tag{35.3}$$

then we have the two following statements:

(1) L_c has only two eigenvalues, 0 and 1. The eigenvalue 0 has the multiplicity of 1 and the eigenvalue 1 has the multiplicity of $n-1$. Vectors $\mathbf{1}^T$ and $\mathbf{1}$ are the left and the right eigenvectors associated with the zero eigenvalue.
(2) There exists a specially defined orthogonal matrix $U \in \mathbb{R}^{n \times n}$, with its last column to be $\frac{1}{\sqrt{n}}$, such that $U^T L_c U = \begin{bmatrix} I_{n-1} & 0 \\ * & 0 \end{bmatrix}$, where $*$ denotes the symmetry part of a symmetry matrix. In particular, if $X \in \mathbb{R}^{n \times n}$ is the Laplacian matrix of an undirected graph, then $U^T X U = \begin{bmatrix} X_1 & 0 \\ * & 0 \end{bmatrix}$. $X_1 \in \mathbb{R}^{(n-1) \times (n-1)}$ is positive definite if and only if the graph is connected.

35.2.2 Problem Reformulation

To analyze the consensus performance, define a controlled output function as follows:

$$z_i(t) = x_i(t) - \frac{1}{n}\sum_{j=1}^{n} x_j(t) \tag{35.4}$$

Note that if $z_i(t) = \mathbf{0}$ is available for all $i \in \mathcal{N}$, then $x_i(t) = x_j(t), \forall i,j \in \mathcal{N}$, which means the system has reached the consensus result. Denote $x(t) = [x_1^T(t) \dots x_n^T(t)] \in \mathbb{R}^{mn}$, $w(t) = [w_1^T(t) \dots w_n^T(t)] \in \mathbb{R}^{m_1 n}$, $u(t) = [u_1^T(t) \dots u_n^T(t)] \in \mathbb{R}^{m_2 n}$ and $z(t) = [z_1^T(t) \dots z_n^T(t)] \in \mathbb{R}^{mn}$. Then the combining form of system (35.1) with controlled output (35.4) can be written as below in a compact form:

$$\begin{aligned} \dot{x}(t) &= (I_n \otimes A)x(t) + (I_n \otimes B_1)w(t) + (I_n \otimes B_2)u(t) \\ z(t) &= (L_c \otimes I_m)x(t) \end{aligned} \tag{35.5}$$

Because $z(t) = \mathbf{0}$ satisfies the consensus result, the object is to design controlled input $u(t)$ such that

$$\left\|T_{zw}(s)\right\|_\infty = \sup_{w\in\mathbb{R}} \bar{\sigma}(T_{zw}(jw)) = \sup_{w(t)\neq 0} \frac{\|z(t)\|_2}{\|w(t)\|_2} < \gamma$$

where $T_{zw}(s)$ is the transfer function matrix of the closed-loop system from $w(t)$ to $z(t)$. $\gamma > 0$ is a given H_∞ performance index. $\bar{\sigma}(\cdot)$ means the largest singular value. In the same way, the closed-loop system satisfies the inequality listed below

$$\int_0^\infty \|z(t)\|^2\, dt < \gamma^2 \int_0^\infty \|w(t)\|^2\, dt,\ \forall w(t) \in L_2[0,\infty).$$

Then the consensus problem of system (35.1) has been transformed into an H_∞ control problem.

35.3 Distributed Event-Triggered Protocol Design and Performance Analysis

35.3.1 Distributed Protocol Design

Set the event times for each agent i as $t_0^i, t_1^i, \dots$, and define the measurement error for agent i as

$$e_i(t) = x_i(t_k^i) - x_i(t),\ t \in [t_k^i, t_{k+1}^i) \tag{35.6}$$

Then using the neighbors' information, the distributed event-triggered control strategy of agent i is proposed as

$$u_i(t) = K\sum_{j\in\mathcal{N}_i} a_{ij}(x_i(t_k^i) - x_j(t_{k'(t)}^j)),\ t \in [t_k^i, t_{k+1}^i) \tag{35.7}$$

where $K \in \mathbb{R}^{m_2 \times m}$ is the feedback matrix, a_{ij} are elements of matrix A, and $k'(t) \triangleq \arg\min_{l \in N: t \geq t_l^j} \{t - t_l^j\}$. Thus for $t \in [t_k^i, t_{k+1}^i)$, $t_{k'(t)}^j$ is the last event time of agent j. itself Now we have

$$x_i(t_k^i) = x_i(t) + e_i(t),\ x_j(t_{k'(t)}^j) = x_j(t) + e_j(t) \tag{35.8}$$

Bringing (35.6) and (35.7) into system (35.5), then we get the closed-loop system in the following compact form:

$$\begin{aligned} \dot{x}(t) &= (I_n \otimes A + L \otimes B_2 K)x(t) + (I_n \otimes B_1)w(t) + (L \otimes B_2 K)e(t) \\ z(t) &= (L_c \otimes I_m)x(t) \end{aligned} \tag{35.9}$$

where L is the Laplacian matrix of $\mathcal{G}$, L_c is defined in Lemma 2, and $e(t) = [e_1^T(t) \ldots e_n^T(t)] \in \mathbb{R}^{mn}$.

35.3.2 Consensus Performance Analysis

From Lemma 1, it is obvious that the system (35.9) is unstable if A is an unstable matrix. In order to solve the problem, we conduct a model transformation in two steps before giving the consensus condition:

Step 1: Let

$$\bar{x}_i(t) = x_i(t) - \frac{1}{n}\sum_{j=1}^{n} x_j(t) \quad i = 1, \ldots, n$$

equivalently in a vector form:

$$\bar{x}(t) = x(t) - \mathbf{1} \otimes \frac{1}{n}\sum_{j=1}^{n} x_j(t) \tag{35.10}$$

where $\bar{x}(t) = [\bar{x}_1^T(t \ldots \bar{x}_n^T(t)]^T \in \mathbb{R}^{mn}$. Note that $\bar{x} = (L_c \otimes I_m)x$, $L_c\mathbf{1} = \mathbf{0}$, $L\mathbf{1} = \mathbf{0}$. Hence, we can get

$$\begin{aligned} \dot{\bar{x}}(t) &= (L_c \otimes I_m)\dot{x}(t) \\ &= (L_c \otimes A)\bar{x}(t) + (L_c L \otimes B_2 K)\bar{x}(t) + (L_c \otimes B_1)w(t) + (L_c L \otimes B_2 K)e(t) \\ z(t) &= (L_c \otimes I_m)x(t) \\ &= (L_c \otimes I_m)\bar{x}(t) \end{aligned} \tag{35.11}$$

Step 2: From Lemma 2, there exists an orthogonal matrix $U \in \mathbb{R}^{n\times n}$ satisfying the following equations:

$$U^T L_c U = \begin{bmatrix} I_{n-1} & 0 \\ * & 0 \end{bmatrix} \triangleq \bar{L}_c, \quad U^T L U = \begin{bmatrix} L_1 & 0 \\ * & 0 \end{bmatrix} \triangleq \bar{L}$$

Note that L_1 is positive definite if $\mathcal{G}$ is connected. Set $U = [U_1 \; U_2]$ with $U_2 = \frac{1}{\sqrt{n}}$ to be the last column. Performing orthogonal transformation to system (35.11):

$$\begin{aligned} \hat{x}(t) &= (U^T \otimes I_m)\bar{x}(t) = \begin{bmatrix} (U_1^T \otimes I_m)\bar{x}(t) \\ (U_2^T \otimes I_m)\bar{x}(t) \end{bmatrix} \triangleq \begin{bmatrix} \hat{x}^1(t) \\ \hat{x}^2(t) \end{bmatrix} \\ \hat{w}(t) &= (U^T \otimes I_{m_1})w(t) = \begin{bmatrix} (U_1^T \otimes I_{m_1})w(t) \\ (U_2^T \otimes I_{m_1})w(t) \end{bmatrix} \triangleq \begin{bmatrix} \hat{w}^1(t) \\ \hat{w}^2(t) \end{bmatrix} \\ \hat{z}(t) &= (U^T \otimes I_m)z(t) = \begin{bmatrix} (U_1^T \otimes I_m)z(t) \\ (U_2^T \otimes I_m)z(t) \end{bmatrix} \triangleq \begin{bmatrix} \hat{z}^1(t) \\ \hat{z}^2(t) \end{bmatrix} \\ \hat{e}(t) &= (U^T \otimes I_m)e(t) = \begin{bmatrix} (U_1^T \otimes I_m)e(t) \\ (U_2^T \otimes I_m)e(t) \end{bmatrix} \triangleq \begin{bmatrix} \hat{e}^1(t) \\ \hat{e}^2(t) \end{bmatrix} \end{aligned} \tag{35.12}$$

Applying transformation (35.12) into (35.11), we can get

$$\begin{aligned} \dot{\hat{x}}(t) &= (\bar{L}_c \otimes A + \bar{L}_c\bar{L} \otimes B_2K)\hat{x}(t) + (\bar{L}_c \otimes B_1)\hat{w}(t) + (\bar{L}_c\bar{L} \otimes B_2K)\hat{e}(t) \\ \hat{z}(t) &= (\bar{L}_c \otimes I_m)\hat{x}(t) \end{aligned} \tag{35.13}$$

and it can be easily verified that $\|T_{zw}(s)\|_\infty = \|T_{\hat{z}\hat{w}}(s)\|_\infty$. Note that system (35.13) consists of two subsystems:

$$\begin{aligned} \dot{\hat{x}}^1(t) =& (I_{n-1} \otimes A)\hat{x}^1(t) + (L_1 \otimes B_2K)\hat{x}^1(t) + (I_{n-1} \otimes B_1)\hat{w}^1(t) \\ &+ (L_1 \otimes B_2K)\hat{e}^1(t) \\ \hat{z}^1(t) =& \hat{x}^1(t) \end{aligned} \tag{35.14}$$

and $\dot{\hat{x}}^2(t) = \mathbf{0}$. Obviously, $\|T_{\hat{z}^1\hat{w}^1}(s)\|_\infty = \|T_{\hat{z}\hat{w}}(s)\|_\infty = \|T_{zw}(s)\|_\infty$.

Let $W = [I_{n-1} \; 0] \in \mathbb{R}^{(n-1)\times n}$, then

$$\begin{aligned} \hat{x}^1(t) &= (W \otimes I_m)\hat{x}(t) = (W \otimes I_m)(U^T \otimes I_m)\bar{x}(t) = (WU^T \otimes I_m)\bar{x}(t) \\ &= (WU^T \otimes I_m)(x(t) - \mathbf{1} \otimes \frac{1}{n}\sum_{j=1}^{n} x_j(t)) \\ &= (WU^T \otimes I_m)x(t) \\ \hat{e}^1(t) &= (W \otimes I_m)\hat{e}(t) = (W \otimes I_m)(U^T \otimes I_m)e(t) \\ &= (WU^T \otimes I_m)e(t) \end{aligned} \tag{35.15}$$

Equation (35.15) shows the relationship between $x(t)$, $e(t)$ and $\hat{x}^1, \hat{e}^1$ of the reduced system (35.14). Also, noticing that the second subsystem is not only independent from the external disturbance, but also the controlled output. And there is a fact that $\|T_{\hat{z}^1\hat{w}^1}(s)\|_\infty = \|T_{zw}(s)\|_\infty$, so we can study the H_∞ performance of the reduced system (35.14) instead of the original system (35.9).

Lemma 3 *[12] Consider a symmetric matrix $S \in \mathbb{R}^{n\times n}$*

$$S = \begin{bmatrix} S_{11} & S_{12} \\ S_{21} & S_{22} \end{bmatrix}$$

where $S_{11} \in \mathbb{R}^{r\times r}$, $S_{12} \in \mathbb{R}^{r\times(n-r)}$, $S_{21} \in \mathbb{R}^{(n-r)\times r}$, $S_{22} \in \mathbb{R}^{(n-r)\times(n-r)}$. Then $S < 0$ if and only if $S_{11} < 0$ and $S_{22} - S_{12}^T S_{11}^{-1} S_{12} < 0$, or equivalently, $S_{22} < 0$ and $S_{11} - S_{12} S_{22}^{-1} S_{12}^T < 0$

For the convenience of writing, in system (35.14), we set $\hat{A} \triangleq (I_{n-1} \otimes A)$, $\hat{B}_2 \triangleq (L_1 \otimes B_2)$, $\hat{K} \triangleq (I_{n-1} \otimes K)$, and $\hat{B}_1 \triangleq (I_{n-1} \otimes B_1)$. So system (35.14) is rewritten as below

$$\begin{aligned} \dot{\hat{x}}^1(t) &= (\hat{A} + \hat{B}_2\hat{K})\hat{x}^1(t) + \hat{B}_1\hat{w}^1(t) + \hat{B}_2\hat{K}\hat{e}^1(t) \\ \hat{z}^1(t) &= \hat{x}^1(t) \end{aligned} \tag{35.16}$$

Set the event trigger condition

$$\|e_i(t)\|^2 < \sigma\|x_i(t)\|^2, \ \sigma > 0 \tag{35.17}$$

Then the problem is to establish a condition to ensure that system (35.16) is asymptotically stable and satisfies the demanded disturbance attenuation property.

Theorem 1 *For a given performance index $\gamma > 0$, if there exist a positive definite matrix $X \in \mathbb{R}^{m(n-1)\times m(n-1)}$ and a positive scalar $\sigma > 0$ satisfying the following inequality*

$$\begin{bmatrix} X(\hat{A} + \hat{B}_2\hat{K}) + (\hat{A} + \hat{B}_2\hat{K})^T X + (\sigma + 1)I & X\hat{B}_2\hat{K} & X\hat{B}_1 \\ \hat{B}_2^T X & -I & 0 \\ \hat{B}_1^T X & 0 & -\gamma^2 I \end{bmatrix} < 0 \tag{35.18}$$

Then we have

(1) System (35.16) under the event trigger condition (35.17) is asymptotically stable.

(2) The transfer function matrix from $w(t)$ to $z(t)$ which is $T_{zw}(s)$, satisfies $\|T_{zw}(s)\|_\infty < \gamma$.

Proof From the event condition (35.17) we can get $\|e(t)\|^2 < \sigma\|x(t)\|^2$. Meanwhile, from Eq. (35.15) $\|\hat{e}^1(t)\|^2 = \|(WU^T \otimes I_m)e(t)\|^2 = e^T(t)(UW^T \otimes I_m)(WU^T \otimes I_m)$ $e(t) = e^T(t)(UW^T WU^T \otimes I_m)e(t) = e^T(t)e(t) = \|e(t)\|^2$, similarly, $\|\hat{x}^1(t)\|^2 = \|x(t)\|^2$. So $\|\hat{e}^1(t)\|^2 < \sigma\|\hat{x}^1(t)\|^2$ holds. Define $V(t) = \hat{x}^{1T}(t)X\hat{x}^1(t)$, then its derivative along the trajectory of system (35.16) with $\hat{w}^1(t) \equiv 0$ satisfies

$$\begin{aligned}\dot{V}(t) =& 2\hat{x}^{1T}(t)X[(\hat{A} + \hat{B}_2\hat{K})\hat{x}^1(t) + \hat{B}_2\hat{K}\hat{e}^1(t)]\\ =& \hat{x}^{1T}(t)[X(\hat{A} + \hat{B}_2\hat{K}) + (\hat{A} + \hat{B}_2\hat{K})^T X]\hat{x}^1(t) + 2\hat{x}^{1T}(t)X\hat{B}_2\hat{K}\hat{e}^1(t)\\ \le& \hat{x}^{1T}(t)[X(\hat{A} + \hat{B}_2\hat{K}) + (\hat{A} + \hat{B}_2\hat{K})^T X + X\hat{B}_2\hat{K}\hat{K}^T\hat{B}_2^T X]\hat{x}^1(t)\\ &+ \hat{e}^{1T}(t)\hat{e}^1(t)\end{aligned}$$

Since $\|\hat{e}^1(t)\|^2 < \sigma\|\hat{x}^1(t)\|^2$,

$$\dot{V}(t) \le \hat{x}^{1T}(t)[X(\hat{A} + \hat{B}_2\hat{K}) + (\hat{A} + \hat{B}_2\hat{K})^T X + X\hat{B}_2\hat{K}\hat{K}^T\hat{B}_2^T X + \sigma I]\hat{x}^1(t)$$

Using Schur complement formula (Lemma 3) into inequality (35.18) leads to

$$\begin{bmatrix}(\hat{A} + \hat{B}_2\hat{K})^T X + X(\hat{A} + \hat{B}_2\hat{K}) + \sigma I & X\hat{B}_2\hat{K}\\ \hat{K}^T\hat{B}_2{}^T X & -I\end{bmatrix} < 0 \quad (35.19)$$

which is equivalent to

$$X(\hat{A} + \hat{B}_2\hat{K}) + (\hat{A} + \hat{B}_2\hat{K})^T X + X\hat{B}_2\hat{K}\hat{K}^T\hat{B}_2{}^T X + \sigma I < 0$$

by Lemma 3. That implies $\dot{V}(t) < 0$. So system (35.16) is asymptotically stable.

Then let

$$J_T = \int_0^T \|\hat{z}^1(t)\|^2 dt - \gamma^2 \int_0^T \|\hat{w}^1(t)\|^2 dt$$

and we have the following result under the zero initial condition of system (35.16):

$$\begin{aligned}J_T =& \int_0^T [\hat{z}^{1T}(t)\hat{z}^1(t) - \gamma^2\hat{w}^{1T}(t)\hat{w}^1(t)]dt\\ =& \int_0^T [\hat{z}^{1T}(t)\hat{z}^1(t) - \gamma^2\hat{w}^{1T}(t)\hat{w}^1(t) + \dot{V}(t)]dt - V(\hat{x}^1(T))\\ =& \int_0^T [\hat{z}^{1T}(t)\hat{z}^1(t) - \gamma^2\hat{w}^{1T}(t)\hat{w}^1(t) + 2\hat{x}^{1T}(t)X((\hat{A} + \hat{B}_2\hat{K})\hat{x}^1(t) + \hat{B}_1\hat{w}^1(t)\\ &+ \hat{B}_2\hat{K}\hat{e}^1(t))]dt - V(\hat{x}^1(T))\end{aligned}$$

$$= \int_0^T \begin{bmatrix} \hat{x}^1(t) \\ \hat{e}^1(t) \\ \hat{w}^1(t) \end{bmatrix}^T \begin{bmatrix} X(\hat{A}+\hat{B}_2\hat{K}) + (\hat{A}+\hat{B}_2\hat{K})^T X + I & X\hat{B}_2\hat{K} & X\hat{B}_1 \\ \hat{B}_2^{\ T} X & 0 & 0 \\ \hat{B}_1^{\ T} X & 0 & -\gamma^2 I \end{bmatrix} \begin{bmatrix} \hat{x}^1(t) \\ \hat{e}^1(t) \\ \hat{w}^1(t) \end{bmatrix} dt$$

$$+ \hat{e}^{1T}(t)\hat{e}^1(t) - \hat{e}^{1T}(t)\hat{e}^1(t) - V(\hat{x}^1(T))$$

$$< \int_0^T \begin{bmatrix} \hat{x}^1(t) \\ \hat{e}^1(t) \\ \hat{w}^1(t) \end{bmatrix}^T \begin{bmatrix} X(\hat{A}+\hat{B}_2\hat{K}) + (\hat{A}+\hat{B}_2\hat{K})^T X + I & X\hat{B}_2\hat{K} & X\hat{B}_1 \\ \hat{B}_2^{\ T} X & 0 & 0 \\ \hat{B}_1^{\ T} X & 0 & -\gamma^2 I \end{bmatrix} \begin{bmatrix} \hat{x}^1(t) \\ \hat{e}^1(t) \\ \hat{w}^1(t) \end{bmatrix} dt$$

$$+ \sigma \hat{x}^{1T}\hat{x}^1 - \hat{e}^{1T}\hat{e}^1 - V(\hat{x}^1(T))$$

$$= \int_0^T \begin{bmatrix} \hat{x}^1(t) \\ \hat{e}^1(t) \\ \hat{w}^1(t) \end{bmatrix}^T \begin{bmatrix} X(\hat{A}+\hat{B}_2\hat{K}) + (\hat{A}+\hat{B}_2\hat{K})^T X + (\sigma+1)I & X\hat{B}_2\hat{K} & X\hat{B}_1 \\ \hat{B}_2^{\ T} X & -I & 0 \\ \hat{B}_1^{\ T} X & 0 & -\gamma^2 I \end{bmatrix}$$

$$\cdot \begin{bmatrix} \hat{x}^1(t) \\ \hat{e}^1(t) \\ \hat{w}^1(t) \end{bmatrix} dt - V(\hat{x}^1(T))$$

According to (35.18), we have

$$\int_0^T [\hat{z}^{1T}(t)\hat{z}^1(t) - \gamma^2 \hat{w}^{1T}(t)\hat{w}^1(t) + \dot{V}(t)]dt < 0$$

that is,

$$\hat{x}^{1T}(T)X\hat{x}^1(T) + \int_0^T \|\hat{z}^1(t)\|^2 dt < \gamma^2 \int_0^T \|\hat{w}^1(t)\|^2 dt$$

Because the system is asymptotically stable and let $T \to \infty$,

$$\|\hat{z}^1(t)\|_2^2 < \gamma^2 \|\hat{w}^1(t)\|_2^2$$

which, equivalently, is $\|T_{zw}(s)\|_\infty < \gamma$. Consequently, the stability and the desired disturbance attenuation ability of system (35.9) is realized due to the equivalence of systems (35.9) and (35.16) in terms of the H_∞ performance. This completes the proof. □

35.4 Simulation

Consider a multi-agent system consists of 4 agents with the ith one($i = 1, 2, 3, 4$) modeled by the following equations:

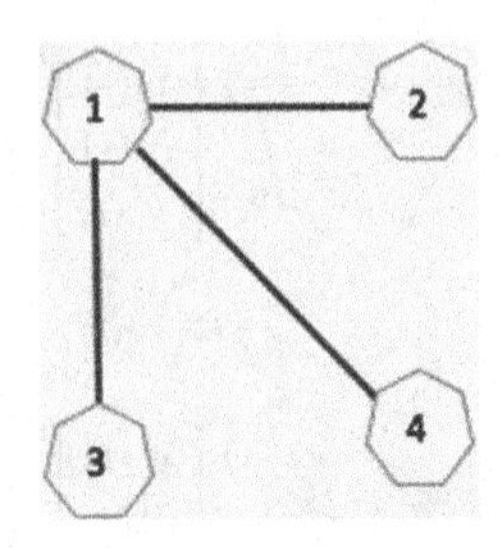

Fig. 35.1 Multi-agent network

$$\dot{x}_i(t) = \begin{bmatrix} 0.5 & 1 \\ 1 & 1 \end{bmatrix} x_i(t) + \begin{bmatrix} 0.1 \\ 0.1 \end{bmatrix} w_i(t) + \begin{bmatrix} 0 \\ 1 \end{bmatrix} u_i(t) \tag{35.20}$$

and the interaction network is shown in Fig. 35.1. Set all the nonzero weighing factors as 1, then we get the Laplacian matrix of Fig. 35.1 $L = \begin{bmatrix} 3 & -1 & -1 & -1 \\ -1 & 1 & 0 & 0 \\ -1 & 0 & 1 & 0 \\ -1 & 0 & 0 & 1 \end{bmatrix}$. The external disturbance $w(t)$ is given by $w = \text{randn}(4, 1)$ ($0 < t < 1/4T$, T is the whole time, otherwise $w = 0$) as white noise in the simulation. For the given H_∞ performance index $\gamma = 1$, we choose the feedback matrix $K = [-4.75 - 4.5]$, $\sigma = 0.0235$.

Figure 35.2 shows the trajectories of $z_i(t)$ for 4 agents, where $z(t) \to 0$ means that the states of all agents reach the consensus result. Figure 35.3 shows the triggering times for the 4 agents. It can be seen that each agent updates itself independently.

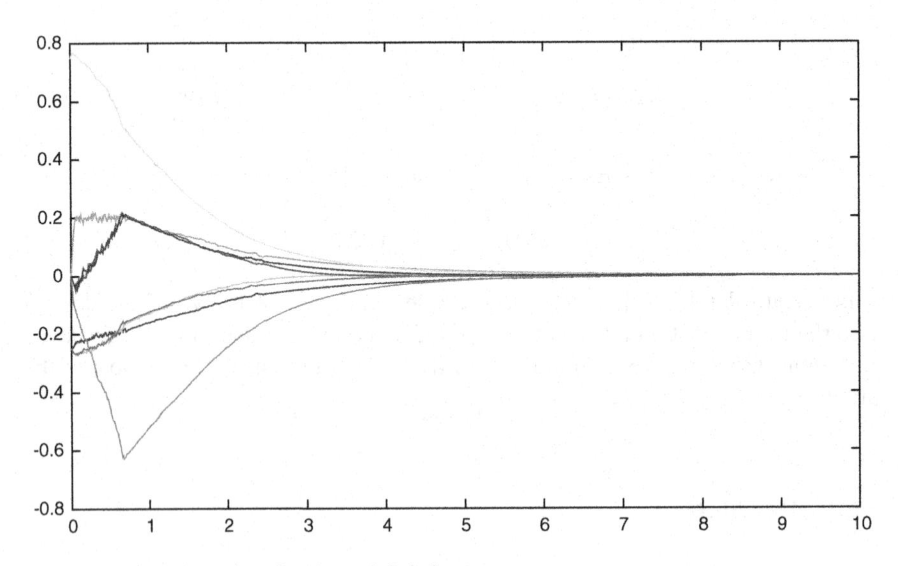

Fig. 35.2 The trajectories of $z_i(t)$, $i = 1, 2, 3, 4$

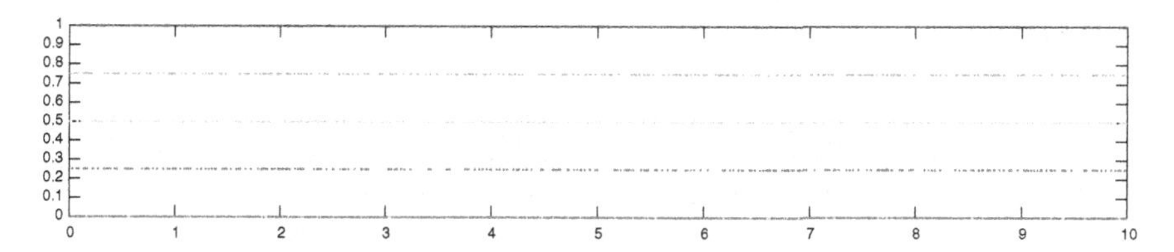

Fig. 35.3 Triggering times for each agent

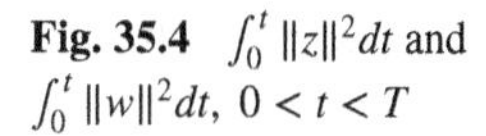

Fig. 35.4 $\int_0^t \|z\|^2 dt$ and $\int_0^t \|w\|^2 dt$, $0 < t < T$

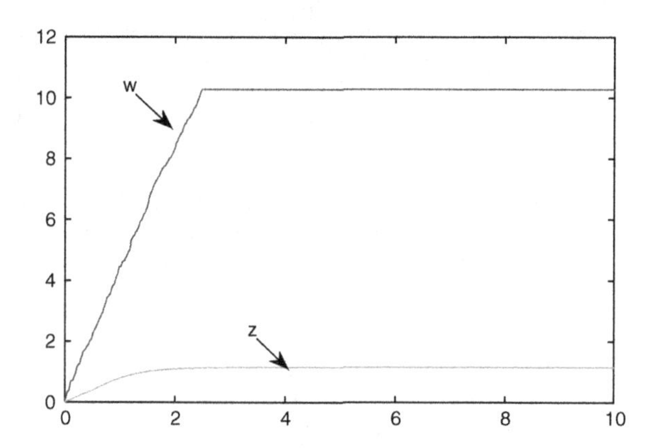

Figure 35.4 shows the trajectories of $\int_0^t \|z\|^2 dt$ and $\int_0^t \|w\|^2 dt$. It is clear that $\|T_{zw}(s)\|_\infty < \gamma = 1$ which means that the multi-agent system reaches consensus result with a required H_∞ performance.

35.5 Conclusions

This paper is devoted to the consensus problem of a general linear multi-agent system with external disturbances. By using the event-triggered control strategy, the frequency of the information update for each agent has been reduced. However, in this paper, time delay and the model uncertainty have not been taken into account, which would be the research topic of our future work on the event-triggered consensus control.

Acknowledgments This work was supported by the National Nature Science Foundation of China (61473015) and the Fundamental Research Funds for the Central Universities (YWF-14-SXXY-003).

References

1. Dimarogonas DV, Johansson KH (2000) Event-triggered control for multi-agent systems. In: [C] Proceedings of the 48th IEEE Conference on Decision and Control, 2009 Held Jointly with the 2009 28th Chinese Control Conference. CDC/CCC 2009, pp 7131–7136. IEEE
2. Dimarogonas DV, Frazzoli E, Johansson KH (2012) Distributed event-triggered control for multi-agent systems. [J] IEEE Trans Autom Control 57(5):1291–1297
3. Huang N, Duan Z, Zhao Y (2014) Event-triggered consensus for heterogeneous multi-agent systems. In: Control Conference (CCC), (2014) 33rd Chinese, pp 1259–1264. IEEE
4. Liu Y, Jia Y (2012) H_∞ consensus control for multi-agent systems with linear coupling dynamics and communication delays. Int J Syst Sci 43(1):50–62
5. Li L, Ho DWC, Huang C (2014) Event-triggered discrete-time multi-agent consensus with delayed quantized information. In: Control Conference (CCC), 2014 33rd Chinese, pp 1722–1727. IEEE
6. Hao J, Haochong Z, Yingmin J (2013) Event-triggered robust H control for linear systems with disturbance. In: Control Conference (CCC), 2013 32nd Chinese, pp 2102–2107. IEEE
7. Yan H, Shen Y, Zhang H et al (2014) Decentralized event-triggered consensus control for second-order multi-agent systems. Neurocomputing 133:18–24
8. Hu J (2014) Second-order event-triggered multi-agent consensus control. In: Control Conference (CCC), 2012 31st Chinese. pp 6339–6344. IEEE
9. Yan H, Shen Y, Zhang H et al (2014) Decentralized event-triggered consensus control for second-order multi-agent systems. Neurocomputing 133:18–24
10. Olfati-Saber R, Murray RM (2004) Consensus problems in networks of agents with switching topology and time-delays. IEEE Trans Autom Control 49(9):1520–1533
11. Lin P, Jia Y (2008) Average consensus in networks of multi-agents with both switching topology and coupling time-delay. Phys A: Stat Mech Appl 387(1):303–313
12. Jia Y (2007) Robust H_∞ Control, Chapter 6. Science Press, Beijing

Chapter 36
LQR-Based Optimal Leader-Follower Consensus of Second-Order Multi-agent Systems

Zonggang Li, Tongzhou Zhang and Guangming Xie

Abstract This paper considers an optimal consensus problem of second-order leader-follower multi-agent systems by using inverse optimality and LMI method. For a given control input, under the condition that the communication topology among followers is undirected connected, a positive definite matrix in a linear quadratic performance index function is obtained, which makes the linear quadratic performance index function to obtain the minimum value. Meanwhile, through theory analysis, we prove that the coefficient matrix of the given control input is the optimal feedback gain matrix. The simulation results show the effectiveness of our conclusions.

Keywords Multi-agent systems · Consensus · Inverse optimality · Linear quadratic regulator

36.1 Introduction

Recently, control of distributed cooperative multi-agent systems has become a hot topic of research. Its applications involve the multi-robot systems, military surveillance, intelligent transportation systems, smart grids, and so on. Consensus problem has been widely concerned and lots of scholars have proposed various types of control algorithms, but most of them have not considered the optimality [1–4]. Liu et al. [5] consider optimal formation control issue, by constructing the hamiltonian function, using the maximum principle to obtain finite time optimal control input. Semsar-Kazerooni [6] uses the methods of state space decomposition,

Z. Li (✉) · T. Zhang
School of Mechatronic Engineering, Lanzhou Jiaotong University,
Lanzhou 730070, China
e-mail: lizongg@126.com

Z. Li · G. Xie
Intelligent Control Laboratory, College of Engineering, Peking University,
Beijing 100871, China

Y. Jia et al. (eds.), *Proceedings of the 2015 Chinese Intelligent Systems Conference*, Lecture Notes in Electrical Engineering 360,
DOI 10.1007/978-3-662-48365-7_36

Riccati equation and LMI to investigate the optimal consensus. Cao et al. [7] use linear quadratic regulator method to investigate the optimal consensus and use algebraic Riccati equation to get the optimal control input and prove that the communication topology is a complete directed graph. Under the condition of an undirected connected graph, in order to obtain a distributed optimal consensus, the authors put forward an optimal global scale factor method. Consensus problem of a general linear system is investigated by Kristian Hengster Movric [8], using inverse optimality and partial stability. Under the condition of continuous-time and discrete-time, Wang et al. [9] research the optimal consensus of first-order and second-order multi-agent systems, respectively, by solving Riccati equation and find out the optimal topology is a star topology.

We research on optimal leader-follower consensus based on LQR theory, LMI and inverse optimality. By choosing the positive definite matrix Q in the quadratic performance index, the given control input makes all agents achieve consensus, while the coefficient matrix of the control input is the optimal feedback gain matrix.

36.2 Preliminaries

Consider an undirected communication topology $G = (V, E, A)$ with N nodes, $V = \{v_1, v_2, \ldots, v_N\}$ represents N agents, a set of edges $E = \{(v_i, v_j)|i, j = 1, \ldots, N\}$, $(v_i, v_j) \in E$ indicates that node v_j can get information from node v_i, and the associated adjacency matrix $A = [a_{ij}] \in R^{N\times N}$. If $(v_j, v_i) \in E$ then $a_{ij} > 0$, else $a_{ij} = 0$. If $(v_j, v_i) \in E$ we call node j is a neighbor of node i. The collection of neighbors of node i is expressed as $N_i = \{j|a_{ij} > 0, j = 1, \ldots, N\}$. The in-degree matrix is defined as $D = diag\{d_1, \ldots, d_N\}$, with $d_i = \sum_{j\in N_i} a_{ij}$ named the in-degree of node i and the Laplacian matrix as $L = D - A$. A path from node i to node j is a series of edges in the form $\{(v_i, v_l), (v_i, v_l), \ldots, (v_k, v_j)\}$. In this paper, suppose leader has no neighbors and is able to transfer its information to at least one follower, meanwhile, also suppose there are N followers and one leader represents by node $N + 1$. For second-order multi-agent systems, not only the position information but also velocity information are exchanged, so we use Laplacian matrix L^A and L^W to describe the exchange of them, respectively. We can see L^A and L^W can be decomposed into the following form

$$L^A = \begin{pmatrix} L^A_{ff} & -\mathbf{b} \\ \mathbf{0}^T & 0 \end{pmatrix}; L^W = \begin{pmatrix} L^W_{ff} & -\mathbf{d} \\ \mathbf{0}^T & 0 \end{pmatrix},$$

where $\mathbf{b} = [b_1, b_2, \ldots, b_N]^T$, $\mathbf{d} = [d_1, d_2, \ldots, d_N]^T$. $b_i > 0$ indicates that the ith agent can receive the leader's position information and $d_i > 0$ indicates that the ith agent can receive the leader's velocity information, $i = 1, \ldots, N$. So $L^A_{ff}\mathbf{1}_N - \mathbf{b} = \mathbf{0}; L^W_{ff}\mathbf{1}_N - \mathbf{d} = \mathbf{0}$ are obtained.

36.3 Problem Formulation and Some Useful Lemmas

In this section, we consider the following system and the dynamics of the followers are formulated as

$$\begin{cases} \dot{x}_i = v_i \\ \dot{v}_i = u_i \end{cases}, i = 1, \ldots, N, \tag{36.1}$$

and the dynamics of the leader is described by

$$\begin{cases} \dot{x}_{N+1} = v_{N+1} \\ \dot{v}_{N+1} = 0 \end{cases}, \tag{36.2}$$

where $x_i(t) \in R$, $v_i(t) \in R$ and $u_i(t) \in R$ are the position, the velocity and the control input of the follower i, respectively. In this paper, we propose the following control protocol

$$u_i = \gamma \sum_{j \in N_i} \left[a_{ij}(x_j - x_i) + w_{ij}(v_j - v_i) - b_i(x_i - x_{N+1}) - d_i(v_i - v_{N+1}) \right]. \tag{36.3}$$

Define $\varepsilon_i = x_i - x_{N+1}$, $\bar{\varepsilon}_i = v_i - v_{N+1}$ then $\varepsilon = [x_1 - x_{N+1}, \ldots, x_N - x_{N+1}]^T$ and $\bar{\varepsilon} = [v_1 - v_{N+1}, \ldots, v_N - v_{N+1}]^T$. Let $\varepsilon_0 = [\varepsilon^T, \bar{\varepsilon}^T]^T$, the cost function is proposed as follows

$$J = \int_0^\infty (\varepsilon_0^T Q \varepsilon_0 + U^T R U) dt, \tag{36.4}$$

where $U(t) = \left[u_1, u_2, \ldots, u_N\right]^T$, Q is a positive definite matrix to be designed and R is an identity matrix with compatible dimension.

The problem we consider is how to choose a suitable positive definite matrix Q and parameter γ to make the given control protocol not only can realize leader-follower consensus but also make the cost function (36.4) achieve a minimum.

Lemma 1 ([10]) *For a completely controllable or completely stabilizable linear system $\dot{x}(t) = Ax(t) + Bu(t), x(t_0) = x_0$, its cost function is described as $J = \int_0^\infty (x^T Qx + u^T Ru)dt$, where the control input is unconstrained, namely, $u(t) \in R^r$, $x(t) \in R^n$, $A \in R^{n \times n}$ and $B \in R^{n \times r}$. Let $Q \in R^{n \times n}$ and $R \in^{n \times n}$ be symmetric nonnegative and positive definite, respectively. Then the unique optimal control $u^*(t) = -R^{-1}B^T Px^*(t)$, where P is the solution of Riccati equation $A^T P + PA - PBR^{-1}B^T P + Q = \mathbf{0}$, in addition, the minimum of cost function J is $J^* = x_0^T Px_0$.*

Lemma 2 ([11]) *Given a block matrix $M = \begin{bmatrix} \bar{A} & \bar{B} \\ \bar{C} & \bar{D} \end{bmatrix}$, it is known that $det(M) = det(\bar{A}\bar{D} - \bar{C}\bar{B})$ if $\bar{A}$ and $\bar{C}$ can commute, where $det(\cdot)$ denotes the determinant of a matrix.*

36.4 Main Results

Theorem 1 *Consider the systems (36.1), (36.2) and cost function (36.4), under the condition of control protocol (36.3) if and only if*

$$Q = \begin{bmatrix} \gamma^2 L_{ff}^2 & \gamma^2 L_{ff}^2 - P_{11} \\ \gamma^2 L_{ff}^2 - P_{11} & \gamma^2 L_{ff}^2 - 2\gamma L_{ff} \end{bmatrix},$$

consensus is achieved and the cost function (36.4) is minimal, where γ, P_{11} are the solutions of the following LMIs

$$\begin{cases} P_{11} > \gamma L_{ff} \\ \begin{bmatrix} 2P_{11} - 2\gamma L_{ff} & P_{11} \\ P_{11} & \gamma^2 L_{ff}^2 \end{bmatrix} > 0 \end{cases}$$

Proof Denote $X(t) = [x_1, \ldots, x_N, v_1, \ldots, v_N]^T$, the matrix form of (36.1), (36.2) and (36.3) are

$$\dot{X}(t) = AX(t) + BU(t), \tag{36.5}$$

$$U(t) = -\gamma([L_{ff}^A \ L_{ff}^W]X(t) + \mathbf{b}x_{N+1} + \mathbf{d}v_{N+1}), \tag{36.6}$$

where

$$A = \begin{bmatrix} \mathbf{0}_{N\times N} & I_{N\times N} \\ \mathbf{0}_{N\times N} & \mathbf{0}_{N\times N} \end{bmatrix}, B = \begin{bmatrix} \mathbf{0}_{N\times N} \\ I_{N\times N} \end{bmatrix}. \tag{36.7}$$

Also note that

$$\varepsilon_0 = \left[\varepsilon^T \ \bar{\varepsilon}^T\right]^T = X(t) - \begin{bmatrix} 1_N \otimes x_{N+1} \\ 1_N \otimes v_{N+1} \end{bmatrix}. \tag{36.8}$$

Take the derivative both sides of (36.8) and simplify it

$$\begin{aligned} \dot{\varepsilon}_0 &= \dot{X}(t) - \begin{bmatrix} \mathbf{1}_N \otimes v_{N+1} \\ \mathbf{1}_N \otimes 0 \end{bmatrix} \\ &= A\varepsilon_0 + \begin{bmatrix} \mathbf{1}_N \otimes v_{N+1} \\ \mathbf{1}_N \otimes 0 \end{bmatrix} + BU(t) - \begin{bmatrix} \mathbf{1}_N \otimes v_{N+1} \\ \mathbf{1}_N \otimes 0 \end{bmatrix} \\ &= A\varepsilon_0 + BU(t). \end{aligned} \tag{36.9}$$

From (36.6), we have

$$U(t) = -\gamma \left[L_{ff}^A \ L_{ff}^W\right] \varepsilon_0, \tag{36.10}$$

substituting (36.10) into (36.9), the following formula is obtained

$$\dot{\varepsilon}_0 = A\varepsilon_0 - \gamma B \left[L_{ff} \; L_{ff} \right] \varepsilon_0. \tag{36.11}$$

Remark 1 For the convenience of demonstration, we assume position graph and velocity graph are the same, while we'll consider the difference between them in the future.

It follows from (36.7) that error system (36.11) can be simplified as

$$\dot{\varepsilon}_0 = A_1 \varepsilon_0(t), \tag{36.12}$$

where

$$A_1 = \begin{bmatrix} \mathbf{0}_{N\times N} & I_{N\times N} \\ -\gamma L_{ff} & -\gamma L_{ff} \end{bmatrix}.$$

It is known that error system (36.12) is asymptotically stable when and only when matrix A_1 is Hurwitz. Now, we prove A_1 is Hurwitz. From Lemma 2, we have

$$det(\lambda I - A_1) = det(\lambda^2 I + \gamma \lambda L_{ff} + \gamma L_{ff}) = 0,$$

that is $\prod_{i=1}^{n}(\lambda_i^2 + \gamma\lambda_i\mu_i + \gamma\mu_i) = 0$, where μ_i is the ith eigenvalue of the matrix L_{ff}. Note that $\lambda_i^{\pm} = \frac{-\gamma\mu_i \pm \sqrt{\gamma^2\mu_i^2 - 4\gamma\mu_i}}{2}$ and according to the Gerschgorin disk theorem, for all $i = 1, \ldots, N$, we can get $\mu_i > 0$. Also note that if $0 < \mu_i \leq \frac{4}{\gamma}$, $Re(\lambda_i^{\pm}) = -\frac{\gamma\mu_i}{2} < 0$ and if $\mu_i > \frac{4}{\gamma}$, $\lambda_i^{-} = \frac{-\gamma\mu_i - \sqrt{\gamma^2\mu_i^2 - 4\gamma\mu_i}}{2} < 0$, $\lambda_i^{+} = \frac{-\gamma\mu_i + \sqrt{\gamma^2\mu_i^2 - 4\gamma\mu_i}}{2} < \gamma\frac{-\mu_i + \mu_i}{2} = 0$. Therefore, A_1 is a Hurwitz matrix and error system (36.12) is asymptotically stable.

Following we use inverse optimality method to find a positive definite matrix Q such that the corresponding cost function (36.4) is minimized. Note that

$$K^* = -\gamma \left[L_{ff} \; L_{ff} \right] = -R^{-1}B^T P, \tag{36.13}$$

and denote

$$P = \begin{bmatrix} P_{11} & P_{12} \\ P_{12}^T & P_{22} \end{bmatrix}. \tag{36.14}$$

Thus,

$$\gamma \left[L_{ff} \; L_{ff} \right] = \left[\mathbf{0}_{N\times N} \; I_{N\times N} \right] \begin{bmatrix} P_{11} & P_{12} \\ P_{12}^T & P_{22} \end{bmatrix}. \tag{36.15}$$

Then, the following formula is obtained

$$P_{12} = \gamma L_{ff}, P_{22} = \gamma L_{ff}. \tag{36.16}$$

Substituting matrix A, B, P into the following Riccati equation

$$PA + A^T P - PBR^{-1}B^T P + Q = \mathbf{0}, \tag{36.17}$$

then, we have

$$\begin{bmatrix} \mathbf{0}_{N\times N} & \mathbf{0}_{N\times N} \\ P_{11} & P_{12} \end{bmatrix} + \begin{bmatrix} \mathbf{0}_{N\times N} & P_{11} \\ \mathbf{0}_{N\times N} & P_{12}^T \end{bmatrix} - \begin{bmatrix} \mathbf{0}_{N\times N} & P_{12} \\ \mathbf{0}_{N\times N} & P_{22} \end{bmatrix} \begin{bmatrix} P_{11} & P_{12} \\ P_{12}^T & P_{22} \end{bmatrix} + Q = \mathbf{0}.$$

So we can get

$$Q = \begin{bmatrix} P_{12}P_{12}^T & P_{12}P_{22} - P_{11} \\ P_{22}P_{12}^T - P_{11} & P_{22}^2 - P_{12} - P_{12}^T \end{bmatrix}. \tag{36.18}$$

Substituting (36.16) into (36.18), we have

$$Q = \begin{bmatrix} \gamma^2 L_{ff}^2 & \gamma^2 L_{ff}^2 - P_{11} \\ \gamma^2 L_{ff}^2 - P_{11} & \gamma^2 L_{ff}^2 - 2\gamma L_{ff} \end{bmatrix}. \tag{36.19}$$

If $Q > \mathbf{0}$ and $P > \mathbf{0}$, then from (36.14) and (36.19) we have

$$\begin{cases} \gamma L_{ff} > \mathbf{0} \\ P_{11} - \gamma L_{ff} > \mathbf{0} \end{cases} \tag{36.20}$$

and

$$\begin{cases} L_{ff}^2 > \mathbf{0} \\ L_{ff}^2 - (L_{ff}^2 - P_{11})(L_{ff})^{-1}(L_{ff}^2 - P_{11}) > \mathbf{0} \end{cases} \tag{36.21}$$

by Schur complement Lemma. Conditions (36.20) and (36.21) should be satisfied together, we can combined them as a one condition as follows.

$$\begin{cases} \gamma L_{ff} > \mathbf{0} \\ \gamma^2 L_{ff}^2 > \mathbf{0} \\ P_{11} - \gamma L_{ff} > \mathbf{0} \\ 2P_{11} - 2\gamma L_{ff} - P_{11}(\gamma^2 L_{ff}^2)^{-1} P_{11} > \mathbf{0} \end{cases} \tag{36.22}$$

Since the communication topology of followers is an undirected connected graph, in which at least one follower can receive the leader's information, we have $L_{ff} > \mathbf{0}$, $L_{ff}^2 > \mathbf{0}$. Clearly, if

$$\begin{cases} P_{11} - \gamma L_{ff} > \mathbf{0} \\ 2P_{11} - 2\gamma L_{ff} - P_{11}(\gamma^2 L_{ff}^2)^{-1} P_{11} > \mathbf{0} \end{cases} \tag{36.23}$$

holds, then (36.22) is satisfied. By Schur complement Lemma, rewrite (36.22) as

$$\begin{cases} P_{11} > \gamma L_{ff} \\ \begin{bmatrix} 2P_{11} - 2\gamma L_{ff} & P_{11} \\ P_{11} & \gamma^2 L_{ff}^2 \end{bmatrix} > \mathbf{0} \end{cases} \tag{36.24}$$

That is, $P > \mathbf{0}$, $Q > \mathbf{0}$ is equivalent to the LMIs (36.24) holds. □

Remark 2 In order to transform (36.24) into a standard LMIs, we use the method of traverse parameter γ, by repeated search parameter γ and solve LMIs, we can get solutions of γ and matrix P_{11}.

36.5 Simulations

In this section, we consider the following system which consists of four followers, named from 1 to 4, and a leader, named by 5, the interaction topology is depicted in Fig. 36.1.

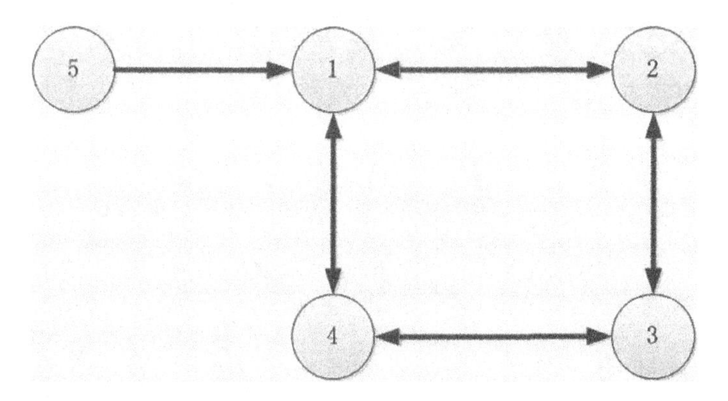

Fig. 36.1 The interaction topology among agents

Consider the dynamics of leader as follows

$$\begin{cases} \dot{x}_{N+1} = 2 \\ \dot{v}_{N+1} = 0 \end{cases}, x_{N+1}(0) = 1, v_{N+1}(0) = 1,$$

the intial state of the followers position and velocity are $x(0) = \begin{bmatrix} 2 & 1 & -2 & -1 \end{bmatrix}^T$ and $v(0) = \begin{bmatrix} -2 & -3 & 2 & 3 \end{bmatrix}^T$, respectively. The simulation results are shown in Figs. 36.2 and 36.3.

By repeated search parameter γ, if $\gamma > 10.75$, the matrix Q which satisfies Theorem 1 is obtained. Choose $\gamma = 11$ then we get

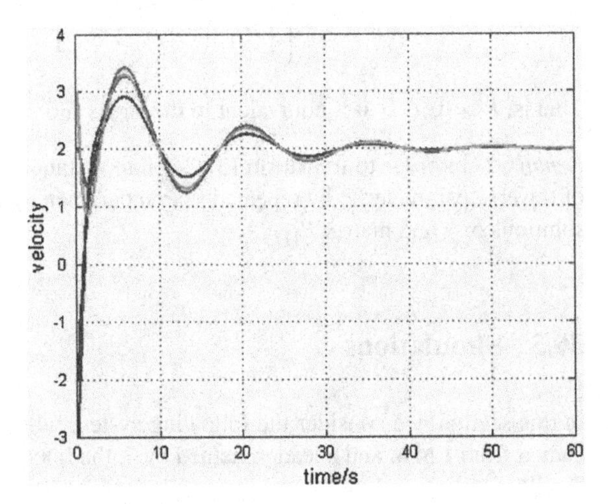

Fig. 36.2 The velocities of followers

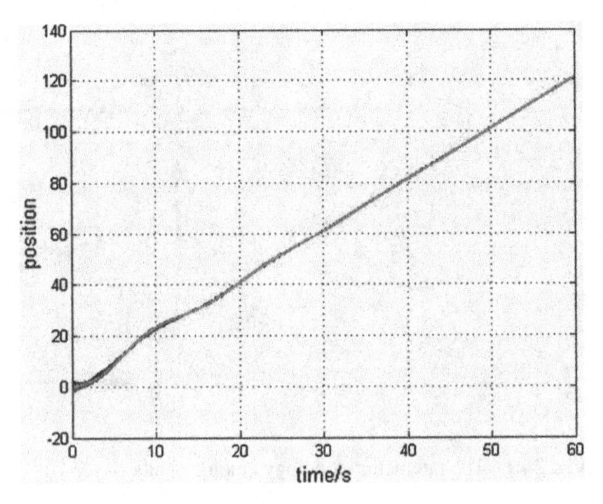

Fig. 36.3 The positions of followers

$$Q = \begin{bmatrix} 1.3310 & -0.6050 & 0.2420 & -0.6050 & -0.4171 & 0.1934 & -0.0819 & 0.1934 \\ -0.6050 & 0.7260 & -0.4840 & 0.2420 & 0.1934 & -0.2238 & 0.1524 & -0.0818 \\ 0.2420 & -0.4840 & 0.7260 & -0.4840 & -0.0819 & 0.1524 & -0.2237 & 0.1524 \\ -0.6050 & 0.2420 & -0.4840 & 0.7260 & 0.1934 & -0.0818 & 0.1524 & -0.2238 \\ -0.4171 & 0.1394 & -0.0819 & 0.1934 & 1.2650 & -0.5830 & 0.2420 & -0.5830 \\ 0.1934 & -0.2238 & 0.1524 & -0.0818 & -0.5830 & 0.6820 & -0.4620 & 0.2420 \\ -0.0819 & 0.1524 & -0.2237 & 0.1524 & 0.2420 & -0.4620 & 0.6820 & -0.4620 \\ 0.1934 & -0.0818 & 0.1524 & -0.2238 & -0.5830 & 0.2420 & -0.4620 & 0.6820 \end{bmatrix} \times 10^3,$$

we know that matrix Q is positive definite and the simulation results show the effectiveness of our results.

36.6 Conclusions

The optimal leader-follower consensus is considered in this paper. Under the condition that communication topology among followers is undirected connected and at least one follower can receive the information of leader, for a given control protocol, by choosing a positive definite matrix Q we get a linear quadratic performance index function to obtain the minimum value, while the coefficient matrix of the control input is the optimal feedback gain matrix based on the theory of linear quadratic regulator. In the future, we will consider the optimal leader-follower consensus with different velocity graph and position graph.

References

1. Vicsek T, Czirok A, Ben-Jacob E (1995) Novel type of phase transition in a system of self-driven particles. Phys Rev Lett 75(6):1226–1229
2. Jadbabaie A, Lin J, Morse AS (2003) Coordination of groups of mobile autonomous agents using nearest neighbor rules. IEEE Trans Autom Control 48(6):988–1001
3. Savkin AV (2004) Coordinated collective motion of groups of autonomous mobile robots: analysis of Vicsek model. IEEE Trans Autom Control 39(6):981–983
4. Olfati-Saber R, Murray RM (2004) Consensus problems in networks of agents with switching topology and time-delays. IEEE Trans Autom Control 49(9):1520–1533
5. Liu YF, Geng ZY (2014) Finite-time optimal formation control for second-order multiagent systems. Asian J Control 16(1):138–148
6. Semsar-Kazerooni E, Khorasani K (2010) Optimal consensus seeking in a network of multiagent systems: an LMI approach. IEEE Trans Syst Man Cybern 40(2):540–547
7. Cao YC, Ren W (2010) Optimal linear-consensus algorithms: an LQR perspective. IEEE Trans Syst Man Cybern 40(3):819–830
8. Movric KH, Lewis FL (2014) Cooperative optimal control formulti-agent systems on directed graph topologies. IEEE Trans Autom Control 59(3):769–774
9. Ma JY, Zheng YS, Wang L, LQR-based optimal topology of leader-following consensus. Int J Robust Nonlinear Control. doi:10.1002/rnc.3271
10. Xie XS (1984) Optimal control theory and application, pp 385–389. Tsinghua University Press, Beijing
11. Ren W, Beard RW (2008) Distributed consensus in multi-vehicle cooperative control theory and applications. Springer, Berlin

Chapter 37
An Activity Recognition Algorithm Based on Multi-feature Fuzzy Cluster

Huile Xu, Yi Chai, Wangli Lin, Feng Jiang and Shuaihui Qi

Abstract In this paper an activity recognition algorithm based on multi-feature fuzzy cluster is designed to find out more details of the activities so as to achieve an accurate classification among them. Firstly, it is proved that distribution of feature vectors vary from activity to activity. And then, a multi-feature extraction algorithm is designed to extract the feature vectors of each activity which makes up a standard activity class. Finally, an activity recognition algorithm based on similarity measurement is brought up and the misjudgment rate turns out to be acceptable, which proves that this algorithm is accurate and highly feasible.

Keywords Activity recognition · Multi-feature extraction · Similarity measurement

37.1 Introduction

Population aging is one of the world's population development tendencies and many countries are facing this problem. This phenomenon is serious, especially in Europe. According to the report [1, 2], in Europe, the percentage of aging people is projected to increase greatly in the next 50 years. Therefore, a wearable activity recognition system for elderly care is urgently needed. Undoubtedly, monitoring and assessment are the most basic functions in these systems [3], which can give us more comprehensive knowledge of the physical condition. Even lots of efforts have been devoted in the last years to the activity recognition field, there are still several open issues respectively referring to systems reliability, robustness, pervasiveness

H. Xu · Y. Chai (✉) · W. Lin · F. Jiang · S. Qi
College of Automation, Chongqing University, Chongqing 400044, China
e-mail: chaiyi@cqu.edu.cn

Y. Chai
State Key Laboratory of Power Transmission Equipment and System Security and New Technology, Chongqing University, Chongqing 400044, China

Y. Jia et al. (eds.), *Proceedings of the 2015 Chinese Intelligent Systems Conference*, Lecture Notes in Electrical Engineering 360,
DOI 10.1007/978-3-662-48365-7_37

and seamless of usage [4]. Among them, the reliability is of great importance. Many of these users criticize the accuracy of the original activity recognition system. To that end, many methods are put up with to promote the accuracy of the activity recognition.

There are lots of different lab-based methods refer to activity recognition. Energy expenditure assessment is the way which was largely used by calculating direct or indirect calorimeters. Then, the intensity of activities can derive. Besides, due to the accurate classification and great stability, support vector machines(SVMs) are being used to solve problems in real worlds including activity recognition and hand-written characters recognition. Such as Shaopeng Liu et al., who assessing physical activity of human subjects by a sensor fusion method based on SVMs. However, the principle of this fusion method is to analyze the data from multiple sensors and extract one feature of them for the classifiers, which will lose some feature about the activity and decrease the accuracy [5]. To increase recognition accuracy, the autoregressive(AR) analysis also applied by Adil Mehmood Khan et al. They not only proposed the model for the first time, but also proved the feasibility of it. The accuracy can arrive 99 % when the accelerometer to be strongly attached to one's chest. But without the above condition, the accuracy decreased sharply 47 % [6]. Beyond these methods, another common method is back propagation algorithms, which train artificial neural networks and try to minimize the gradient of a loss function. It has applied to chronic disease management and disease prevention [8, 9]. But these ways only can perform well in laboratories due to requiring sophisticated lab equipment and home environment [7]. At the same time, wearable activity recognition systems are more popular and can be seen everywhere. But they require to place sensors in the location specified and the displacement of sensors will influence the accuracy of judgment greatly [10].

In recent years, multi-sensors data fusion technology has become a hot issue. Researchers have done many relevant experiments. Owing to its fast response and high accuracy, the accelerometer was considered as the most powerful sensor in this field. So, in the experiment, there are many accelerometers are placed in different parts of the body [12]. Meanwhile, to increase the accuracy, the gyroscope and some other sensors often used with accelerometers [11]. Then by using data fusion to synthesize all kinds of data sources come from accelerometers or other types of sensors. Compared with single sensors, this technology will promote the reliability of data and improve prediction accuracy. But most of the paper only extracts one feature of all processed data to represent the activity and go on activity recognition. However, the activity is a process which will change with time and in each period of time the feature may have some difference.

This paper, firstly, exploring the difference between each activity's feature and prove the feature can be used in classification. Then the misjudgment percent are computed after extracting one feature from each activity. However, the experimental result is not satisfied when compare it with situations extracting multiple features from each activity in the later research. So, in order to gain more information and to extract more features of an activity, this paper developed methods by dividing the activity process into several periods of time. Then, using a feature value to represent the state over each period of time. Thus, for each activity, getting multiple feature values to represent it. To describe the similarity between two feature values, method based on the Euclidean distance method and close degree is applied. The back-substitute has been used to check the accuracy of our method. Compared with the original method, these results demonstrate that the method is more effective for promoting the accuracy of activity recognition system.

37.2 The Proposed Approach

37.2.1 Data Collecting

It's common that there is much difference between the elderly and the young men. Usually, the old will not do strenuous exercises like young men especially the one who is in bad physical condition. Therefore, it's much easier to recognize the activity of the elderly. So, this paper just needs to distinguish the unforeseen circumstances, such as fall down, from the normal activities such as walking and jogging. In this way, whether the old are in unforeseen circumstances or not can be known.

Banos and his partners do a series of experiments for movement recognition [13–15]. They let a dozen volunteers each wear 9 transducer in the different parts of their body. The transducer outputs 13 values at each measurement, which leads to an overall set of 117 recorded signals. The recordings were sampled at 50 Hz. The recorded data including acceleration in x-axis, y-axis and z-axis, and angular acceleration in x-axis, y-axis and z-axis and quaternion etc. Then, they let these volunteers to do 32 activities which include walking, jogging, cycling, swimming randomly. They repeatedly do this experiment and gain lots of dataset for activity recognition. They are uploading these data onto a database in the internet. So this paper uses these data to design an algorithm to recognize activity accurately.

In this paper, an algorithm which can recognize simple activity was put up with. If readers want to recognize some other activity, you can do the same experiment and put your data onto a standard dataset. The algorithm also can compare and recognize it.

37.2.2 Signal Processing and Feature Extraction

According to analysis of these data, the different physical quantity has different sensitivity for the activity. The acceleration exists a big difference between various activities. However, quaternion changes small when recognize different activities. By analyzing those data, the acceleration and angular acceleration is more sensible than the quaternion. On the basis of above discussion, in order to simplify the analysis, this paper use the acceleration and angular acceleration to describe the feature of the activity.

There are several methods of standardization and maximum difference normalization method be applied in this paper. The formula for maximum difference normalization method is:

$$x'_{ij} = \frac{x_{ij} - \min\limits_{1 \le i \le n} \{x_{ij}\}}{\max\limits_{1 \le i \le n} \{x_{ij}\} - \min\limits_{1 \le i \le n} \{x_{ij}\}} \quad (37.1)$$

Among them, x'_{ij} is the value after normalization. $\min\limits_{1 \le i \le n} \{x_{ij}\}$ represent finding the minimum among x_{ij} and $\max\limits_{1 \le i \le n} \{x_{ij}\}$ represent finding the maximum among x_{ij}.

Then, according to the above introduction, 117 data gained per 20 ms. Actually, the physical process is slow, there is no need uses so much data to describe a state. Thus, using a feature value to represent the whole physical process.

Meanwhile, the goal is to design an activity recognition algorithm which is suitable for wearable devices, such as a wrist watch. So, this paper use only the sensors in the right and left lower arms, instead of all 9 sensors.

According to the physical knowledge, the resultant acceleration is the combination of three directions of the acceleration. Furthermore, the resultant angular velocity can be obtained in the same way. When recognizing an unknown activity, the acceleration in three directions is used, which shows us more information and be used to promote the accuracy. Firstly, using the resultant acceleration and resultant angular velocity to demonstrate these two physical quantity can show the difference between different activities.

In conclusion, four variables will be used to describe an activity as follows:

$$Activity\_j = (ACC_R \; GYR_R \; ACC_L \; GYR_L) \quad (37.2)$$

Which represents the feature vector of activity. ACC represents the resultant acceleration in the right lower arm. GYR means the resultant angular velocity in the right lower arm. Similarly, ACC is the resultant acceleration in the left lower arm, GYR represents the resultant angular velocity in the left lower arm.

In order to describe the process of standardizing, we define as follows:

$$Activity\_matrix = \begin{pmatrix} Activity\_0 \\ Activity\_1 \\ \vdots \\ Activity\_X \end{pmatrix} \quad (37.3)$$

Among them, aggregate feature vectors of all activities. Then using above formula of normalization to standardize every column.

37.2.3 *Similarity Measurement*

For the evaluation of the proposed model, an activity recognition benchmark dataset is used. These records are translated into raw and unprocessed signals that numerically represent the magnitude measured. And when someone does fitness exercise, a dataset of this process can get. Using the above method to process the dataset and a row vector contains the feature of activity can get. Then, standardizing the row vector to eliminate the influence of the magnitude and comparing its feature parameters with 33 feature vectors.

Compared with other vectors, the similarity can be described in several ways. In this paper, Euclidean distance method, absolute distance method and close degree be applied.

37.2.3.1 Euclidean Distance Method

The principle of Euclidean distance method is to calculate distance between row vector and some features vector. Among them, there will be a shortest distance which means the row vector is the most similar with it and it's no doubt the row vector belongs to that type of activity. The Euclidean distance is:

$$d(x_i, x_j) = \sqrt{\sum_{k=1}^{m} (x_{ik} - x_{jk})^2} \tag{37.4}$$

Among them, $d(x_i, x_j)$ represents the Euclidean distance between vector x_i and vector x_j. x_{ik} represents the kth element of vector x_i.

37.2.3.2 Absolute Distance Method

The principle of absolute distance method is similar as Euclidean distance method. The only difference between them is the formula which was used. The absolute distance is:

$$d(x_i, x_j) = \sum_{k=1}^{m} |x_{ik} - x_{jk}| \tag{37.5}$$

Among them, $d(x_i, x_j)$ represents the absolute distance between vector x_i and vector x_j. x_{ik} also represents the kth element of vector x_i.

37.2.3.3 Close Degree Method

Close degree is an index which was used to describe the closeness between fuzzy set. There are several practical close degree formulas for the actual work. Using two of them to describe the closeness. The bigger the numerical value is, the more the similarity is. The close degree formulas that are applied to this paper are:

$$\sigma_1(\tilde{A},\tilde{B}) \triangleq \sum_{k=1}^{m} (\tilde{A}(x_k) \wedge \tilde{B}(x_k)) / \sum_{k=1}^{m} (\tilde{A}(x_k) \vee \tilde{B}(x_k)) \tag{37.6}$$

$$\sigma_2(\tilde{A},\tilde{B}) \triangleq \frac{1}{m} \sum_{k=1}^{m} \left|(\tilde{A}(x_k) - \tilde{B}(x_k))\right| \tag{37.7}$$

Among them, $\vee_{i=1}^{n}$ represents become the maximum of two numbers. On the contrary, $\wedge_{i=1}^{n}$ means become the minimum of two numbers.

At last, considering the results of these methods overall and giving the final result. Because one way has limitations, this paper decreases the judge error by synthesizing each kind of situation.

37.2.4 *Misjudgment Percentages*

When recognizing an unknown activity, this paper just needs to derive its feature vector in the same way and compare it with our standard feature vectors. By using distance measurement to define the similarity between the unknown feature vector and standard feature vectors. And the smaller the distance is, the bigger the similarity. Based on the similarity, the unknown activity is derived to belong to which standard activity.

Back substitution method is commonly used way for testing the accuracy of the fuzzy recognition. Specific idea is putting the original data onto the model and calculate the misjudgment percentage. The misjudgment percentage was defined as:

$$\alpha = \frac{m}{n} \times 100\,\% \tag{37.8}$$

Among them, α means misjudgment percentages, m means the number of misjudgments, n means total number of judges.

37.3 Experiment and Results

37.3.1 *Feature Analysis*

The curves that describe the change of acceleration and angular velocity of each activity with time plotted. For example, this paper chooses the processed data come from the first volunteer to show this change.

As Fig. 37.1 shows, there is huge difference between acceleration and angular velocity about various activities. In Fig. 37.1, the vertical axis only represents the processed data come from sensors. That is saying, it just to express the degree of size does not represent a specific number. And the horizontal axis represents the increase in the time, also do not represent specific time value. In the first part of Fig. 37.1a, acceleration of walking changes tiny and the size of that are also small. But, the acceleration of jogging will have larger oscillation, at the same time, numerical also become bigger. Jumping up will have more violent changes in acceleration. These are also consistent with our intuitive sense. Meanwhile, even the angular velocity does not have so big difference like the acceleration in the Fig. 37.1b, the angular velocity of each activity exists different features, which also can be used in activity recognition and promote the accuracy.

Then, getting all feature vectors of 17 volunteers for every activity and to demonstrate the difference between every activity, taking the feature value which derived from sensors in the left lower arm as an example. Firstly, processing these data to derive the resultant acceleration and the resultant angular velocity as feature

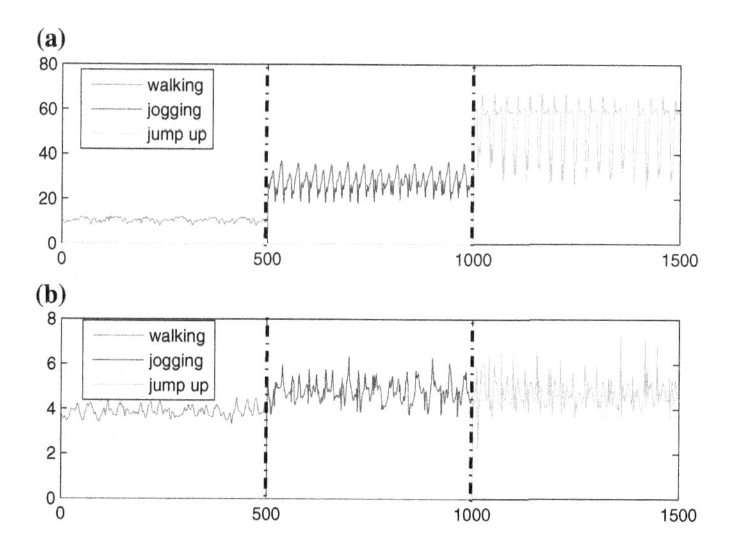

Fig. 37.1 The changing process of acceleration and angular velocity for walking, jogging and jump up. **a** Acceleration. **b** Angular velocity

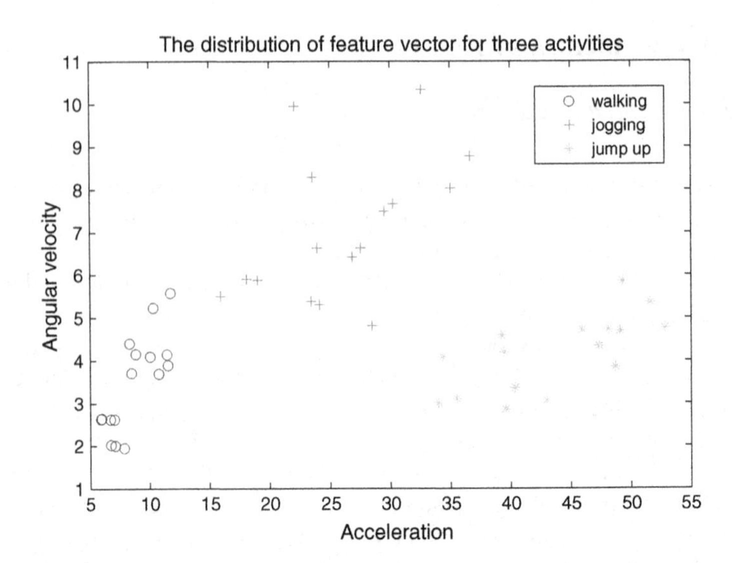

Fig. 37.2 The distribution of feature vectors for three activities ("O", "+", "*" represent a feature vector for walking, jogging and jump up of one volunteer respectively.)

vectors. To present it intuitively, these feature vectors are put into coordinating graphs. Using two physical quantity acceleration and angular velocity to present the feature of activity. Then plot the coordinate graphs by acceleration on the horizontal axis and angular velocity on the vertical. For 17 volunteers, each one can give three feature vectors respectively represent walking, jogging and jump up. All these feature vectors are put into coordinating graphs as follows.

In Fig. 37.2, the feature vectors of walking all concentrate on the lower left corner, while the feature vectors of jumping up all concentrate on the lower right corner. And the feature vectors of jogging are relatively dispersive and mainly locate at the center of graphs. So, Fig. 37.2 intuitively demonstrates the feature difference between each activity. To recognize an unknown activity, just need to derive its feature vector by the same way and put it into Fig. 37.2. The type of activity can be confirmed by comparing the similarity with existing point.

According to our further research, it is clear that one feature value may lose some information about one activity. Moreover, with time going by, the physical action is changing. The activity always has more than one feature and the activity cannot be finished in transient, so it's not reasonable to use a feature value to represent the whole process. So, dividing the process into several periods of time. Then, using a feature value to represent the state over a period of time. Thus, for each activity, a feature dataset is got to represent it. Then, also using the average as a feature value to represent the state over a period of time.

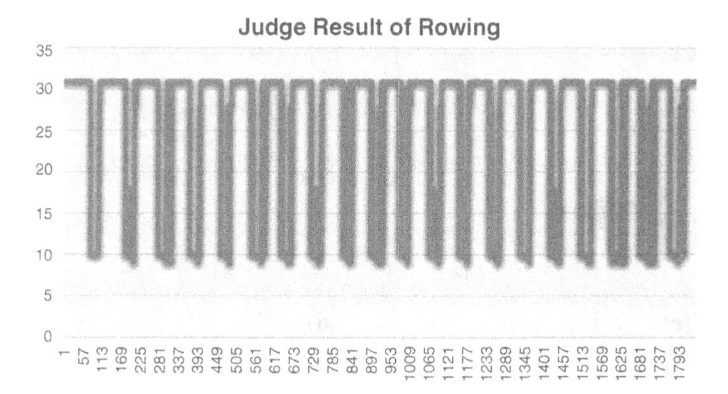

Fig. 37.3 Judge results of rowing (The vertical axis represents the judgment result. Every number is a kind of activity. For example, number 31 represents the activity rowing and the number 10 represents the activity trunk twist.)

In the rest part of the paper, the rationality of the above discussion could be demonstrated intuitively. For an activity, it can be described by using a feature dataset. Then putting all feature vectors of one activity come from various periods of time together into one coordinating graphs as follows.

Taking the activity rowing as an example, the judge result from above method was presented in the following figure.

As shown in Fig. 37.3, the horizontal is time series, representing the increase in time. Meanwhile, the vertical axis represents the judgment result. It is obvious that most of the results are rowing and some of the results are trunk twists. These two cases constitute nearly all judgment results. Therefore, considering whether an activity has more than one feature vector. Maybe there are more feature vectors that are similar to the feature vector of trunk twists. So when classify these datasets, the judgment result of some dataset is trunk twists. If this feature vector found out, the error can be fixed greatly.

The feature value based on the time series is obtained and some classical distribution of feature value is taken as examples to analyze the feature of activity. The first one is walking, most of the feature vectors are located together, only few vectors are deviating from them. Actually, the walking is a relatively steady process, the acceleration and angular velocity do not change much. For the jogging process, the feature vectors also distribute over a small range. It is obvious that the range is bigger than that of walking. Then, the activity lateral bend arm up is a very interesting and typical example which can intuitively tell why some situations have misjudgment. It is clear that the activity has two or more feature vectors. If one feature vector is used to represent the whole process, it is no doubt that some features of this activity were ignored. And this feature vector may close some other activity. The misjudgment situation will happen. For the last activity repetitive forwards stretching, which is a drastic action that moves the body frequently.

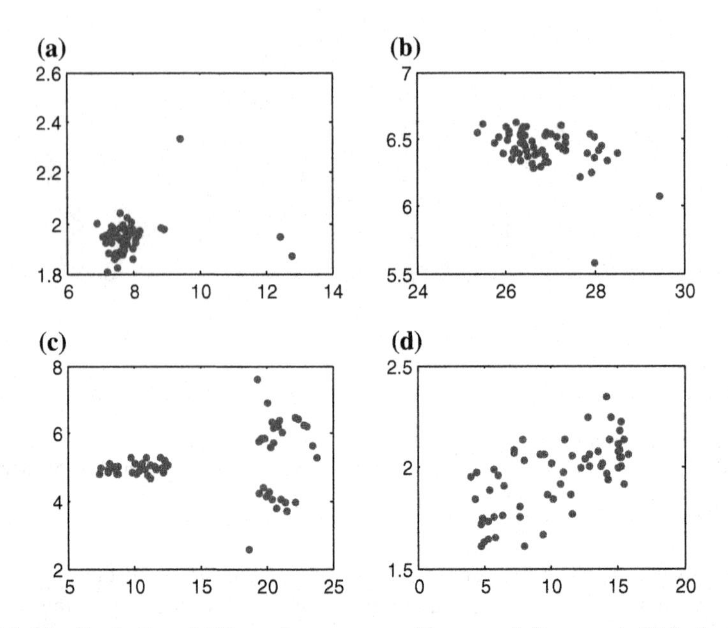

Fig. 37.4 The distribution of different feature vectors (For one activity, we gain 60 feature vectors based on the time series and plot them all in same coordinate graphs. We plot four activities respectively are walking, jogging, lateral bend arm up and repetitive forward stretching). **a** Walking. **b** Jogging. **c** Lateral bend arm up. **d** Repetitive forward stretching

From Fig. 37.4, acceleration and regular velocity changes greatly and the distribution is scattered.

According to the discussion, one activity in different periods of time may have different features. These features may differ from each other greatly. If one activity just represented by one feature vector will lose much relevant information, which results in inaccurate description of activity and misjudgment. So, in the rest of this paper, we use 60 feature vectors to present one activity and compare the result with the situation when only use one feature.

37.3.2 Results Based on a Feature Vector

Taking all feature vectors into account, the misjudgment percentage of some activities was obtained and the results were summarized as follows.

As indicated in Table 37.1, there are significant differences depending on the type of activity. Rowing has the minimum misjudgment percentage only be 6.7 %. Whereas, the maximum misjudgment percentage of jumping up is 75 %. What's more, the average misjudgment percentage of all activity is about 43.1 % which is too big to judge the type of the activity accurately.

Table 37.1 The misjudgment percentage of each activity

The misjudgment percentage of each activity				
Activity type	Walking	Jogging	Running	Upper trunk and body opposite twist
Percentage	0.313	0.375	0.375	0.357
Activity type	Arms frontal crossing	Arms inner rotation	Rotation on the knees	Jump up
Percentage	0.467	0.375	0.236	0.750

Table 37.2 The misjudgment percentage by advanced method

The misjudgment percentage by advanced method				
Activity type	Walking	Jogging	Running	Upper trunk and body opposite twist
Percentage	0.0035	0.2974	0.2251	0.0729
Activity type	Arms frontal crossing	Arms inner rotation	Rotation on the knees	Rowing
Percentage	0.1119	0.0906	0.1361	0.067

37.3.3 Result Based on Multiple Feature Vectors

The improved method of the last experiment has been putting forward, which is to increase the number of feature vectors. Then in this paper, 60 feature vectors were used to present an activity. Then the same judge method was used to recognize unknown activities. The results are summarized in Table 37.2.

37.3.4 Result Comparisons

Taking some activities to intuitively compare the difference between two methods in Fig. 37.5.

As the Fig. 37.5 shows, the misjudgment percent by using multiple feature vectors all less than the former one, especially walking. The conclusion was drawing that this method improved the accuracy of judgment. What's more, the result in this way is much acceptable than the last one, and can be used in reality. However, this algorithm is more complex than the original method and increases the calculating time.

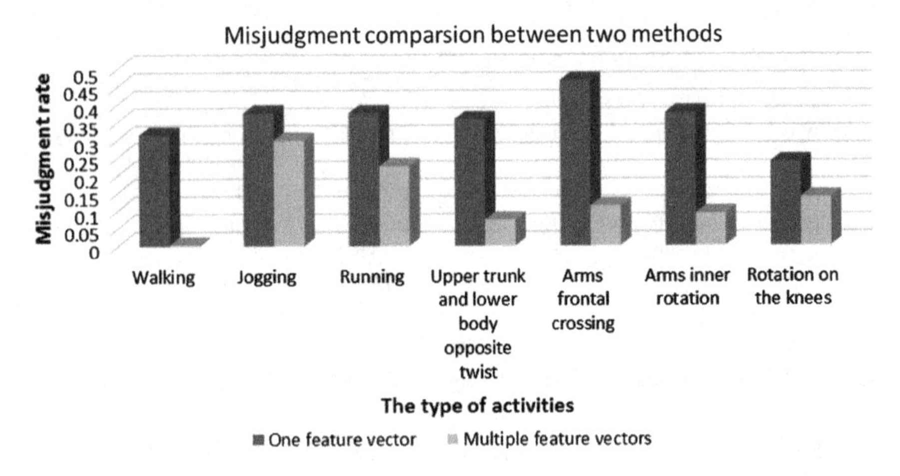

Fig. 37.5 Misjudgment comparison between two methods

37.4 Conclusion and Future Work

With the development of wearable devices in our society, the portable monitoring devices for the elderly are more and more popular in our society. Meantime, these devices can prevent the elderly from some illness and benefit for them. But, the existing related devices cannot recognize activities accurately. This paper design a multi-feature fuzzy cluster algorithm based on multi-sensors to find out more details of activities to achieve accurate classification. Firstly, the distribution of feature value was analyzed to prove that can be used in activity recognition. As Fig. 37.4 shows, there are several classical distributions of activities and it's related to the frequency and amplitude of activities. Secondly, this paper designed an activity recognition algorithm based on similarity measurement and several methods for similarity measurement, such as Euclidean distance method and close degree method.

Finally, this paper did some experiment to show the effect of the proposed method. Compared with Tables 37.1 and 37.2, the advanced method that uses multi-feature can improve the accuracy of judgment. What's more, the result in some way is more acceptable than the first one and can be used in reality. At the same time, our paper focus on activity recognition algorithms for the elderly and the elderly just will do some simple actions, which are much easier to be recognized. As our result shows, the accuracy for simple actions is great. Actually, the proposed method in this paper has been applied to a monitoring device for the elderly. By the repetitive tests, the devices can recognize the activities of the elderly accurately.

By analyzing the distribution of feature vectors, some more features can be found. Some activities have obvious periodicity and linearity. It is consistent with the reality that some activities are performed periodically and some other activities are changing with time linearly. If we can find out a good way to present these features, it can increase the accuracy greatly. It is of great importance for activity recognition to dig out more useful feature information.

In another aspect, it is necessary for us to find out a more effective method to present the similarity between two activities. Although the distance method always was used in this field, there is no doubt it will ignore some details of an activity process. It is more suitable to present the characteristic of the whole process. So, it's necessary to find out a method to present the feature of each period of time. If above improved methods are finished, the activity recognition can be more accurate.

Acknowledgments We would like to thank the supports by the National College Students' Innovative Experiment Project (201410611054).

References

1. DeSA UN (2013) World population prospects: the 2012 revision. Population Division of the Department of Economic and Social Affairs of the United Nations Secretariat, New York
2. Corchado JM, Bajo J, De Paz Y et al (2008) Intelligent environment for monitoring Alzheimer patients, agent technology for health care. Decis Support Syst 44(2):382–396
3. Chernbumroong S, Cang S, Yu H (2014) A practical multi-sensor activity recognition system for home-based care. Decis Support Syst 66:61–70
4. Baños O, Damas M, Pomares H et al (2013) Activity recognition based on a multi-sensor meta-classifier. In: Rojas I, Joya G, Cabestany J (eds) Advances in Computational Intelligence. Springer, Heidelberg, pp 208–215
5. Mo L, Liu S, Gao RX et al (2013) Multi-sensor ensemble classifier for activity recognition. J Softw Eng Appl 5(12):113
6. Khan AM, Lee YK, Lee S et al (2010) Accelerometer's position independent physical activity recognition system for long-term activity monitoring in the elderly. Med Biol Eng Comput 48 (12):1271–1279
7. Fang H, He L, Si H et al (2014) Human activity recognition based on feature selection in smart home using back-propagation algorithm. ISA Trans 53(5):1629–1638
8. Amft O, Tröster G (2008) Recognition of dietary activity events using on-body sensors. Artif Intell Med 42(2):121–136
9. Sazonov ES, Fulk G, Hill J et al (2011) Monitoring of posture allocations and activities by a shoe-based wearable sensor. IEEE Trans Biomed Eng 58(4):983–990
10. Baños O, Damas M, Pomares H et al (2012) A benchmark dataset to evaluate sensor displacement in activity recognition. In: Proceedings of the 2012 ACM Conference on Ubiquitous Computing. ACM, pp 1026–1035
11. Preece SJ, Goulermas JY, Kenney LPJ et al (2009) A comparison of feature extraction methods for the classification of dynamic activities from accelerometer data. IEEE Trans Biomed Eng 56(3):871–879
12. Parkka J, Ermes M, Korpipaa P et al (2006) Activity classification using realistic data from wearable sensors. IEEE Trans Inf Technol Biomed 10(1):119–128
13. Banos O, Damas M, Pomares H et al (2013) Human activity recognition based on a sensor weighting hierarchical classifier. Soft Comput 17(2):333–343
14. Banos O, Damas M, Pomares H et al (2012) On the use of sensor fusion to reduce the impact of rotational and additive noise in human activity recognition. Sensors 12(6):8039–8054
15. Chavarriaga R, Sagha H, Calatroni A et al (2013) The Opportunity challenge: A benchmark database for on-body sensor-based activity recognition. Pattern Recogn Lett 34 (15):2033–2042

Chapter 38
Classifying the Epilepsy EEG Signal by Hybrid Model of CSHMM on the Basis of Clinical Features of Interictal Epileptiform Discharges

Shanbi Wei, Jian Tang, Yi Chai and Weifeng Zhao

Abstract Many methods of processing epileptic EEG signals are concentrated in the classification, and most of them use the wavelet transform and SVM classification algorithm. Although these algorithms acquire the high accuracy, it is still unable to provide a good explanation of quantitative difference and physical meaning between epileptic EEG and normal EEG. This paper presents a new hybrid algorithm (CWT-SVM-HMM) to classify epileptic EEG signal. By the results of classification of HMM, we can track back abnormal signal frequency sources, through the analysis of the sources of seizures during different frequency band, we can get a seizure of accurate quantitative analysis according to clinical feature of interictal epileptiform discharges.

Keywords Wavelet transform · SVM · HMM · Epileptic · EEG · IED

38.1 Introduction

A seizure is a transient occurrence of signs and/or symptoms due to abnormal excessive or synchronous neuronal activity in the brain. An EEG signal is classified as having a delta component if its dominant frequency component f is < 4 Hz, a theta component if $4 < f < 8$ Hz, an alpha component when $8 < f < 12$ Hz, a beta component when $12 < f < 30$ Hz, or a gamma component when $f \geq 30$ Hz [1, 2]. Based on the different behaviors, the EEG signal presents composite wave with variable frequencies. Most of analyses about EEG signal focus on the classification

S. Wei · J. Tang (✉) · Y. Chai · W. Zhao
College of Automation, Chongqing University, Chongqing 400044, China
e-mail: weishanbi@cqu.edu.cn

S. Wei
Key Laboratory of Power Transmission Equipment and System Security, Chongqing 400044, China

Y. Jia et al. (eds.), *Proceedings of the 2015 Chinese Intelligent Systems Conference*, Lecture Notes in Electrical Engineering 360,
DOI 10.1007/978-3-662-48365-7_38

and prediction [3–7], through three stages: De-noising process, feature extraction and classification; there are lots of approaches to obtain quite remarkable classification accuracy.

This paper describes the application of CWT-SVM-HMM (CHMM) model. Through the multi-channel signal of EEG signal for classification extraction, to obtain the binary form {0, 1} as the basic physical meaning (e.g. Open Eyes State and Close Eyes State). Recoding these binary data as hidden state matrix of HMM, the final physical performance is seen as observation matrix, and then the last work is to estimate these two matrices. We use the obtained matrices which calculated by SVM to estimate the HMM model, because this method not only can avoid the local convergence, but also has better explanation for physical meaning. The algorithm proposed by this paper is based on enough accuracy rates, more importantly.

38.2 Background

38.2.1 Epileptic-Form Abnormalities

The patterns which are associated with epileptic-form abnormalities and with special significance are discussed in this chapter [2].

Spike waves: Shaped likes a thorn, a spike wave time of <70 ms.

Sharp waves: Sharp wave likes a peak, time between 70 and 200 ms.

Spike and sharp wave complex: Composed by a spike wave or sharp wave.

Poly spikes: Two or more spike wave appears continuously.

Poly spike wave complex: By two or more than 1 spike wave and slow wave

Epileptiform discharges appear in different morphologies. IEDs are commonly identified as spikes and sharp waves with or without after-going slow waves.

The discharge frequency band of slow-spike (or sharp)-and-wave (SSW) is 0–3 Hz and it is presented in patients with the Lennox-Gastaut syndrome (LGS). It often contains a biphasic or triphasic surface-negative sharp wave which followed by a slow wave in a symmetrical, bilateral, synchronous, front and central complex. Patients with seizures are mostly with symptoms of running frequencies ranging from 1.5 to 2.5 Hz and often appear as repetitive bursts, in addition, they are usually prolonged, shifting and asymmetrical in sleep [2].

38.2.2 Wavelet Transform

Wavelet transform is a method used commonly in the aspect of electrical treatment [8–10]. Localization of wavelet transform inherits and develops the advantages of Short Time Fourier Transform (STFT), at the same time; it overcomes the shortcoming that the window size doesn't change over with the frequency of STFT.

So wavelet can focus on any details of signal. For such characteristics of wavelet transform, we can use it to extract the suitable frequency range of EEG. According to the last section, we know the time and frequency domain characteristics of the epilepsy. In this paper, CWT is chosen to gain the detail from each scale of the EEG signal, rather than DWT(Discrete wavelet transform), which will lose some details caused by reducing half of coefficients. The CWT we use is different from [8–12], which lies that we choose the scale of the filter for different reasons, we pay more attention to the epilepsy feature of waveform and frequency, the more detail will be described in Sect. 38.3.

38.2.3 Support Vector Machine

Support vector machine (SVM) is a very mature theory which has been applied in many fields. SVM uses nuclear techniques, the input implicit mapping in high dimensional feature space to solve nonlinear classification problem. Based on this reason, SVM is a good way to deal with the EEG signal.

SVM theory is developed and improved continuously with using in different fields. Such as least square support vector machine (LSSVM), which uses the LSSVM to classify EEG signal, and achieves a good classification accuracy. Even so, SVM is used as classifier to deal with EEG signal. In this paper, we use it as a part of filter and will show how to transform the result of SVM to a binary code, and we will recode these binary codes, which can be processed by time series algorithm. More detail will be discussed in Sect. 38.3.

38.2.4 Hidden Markov Model

Hidden Markov Model (HMM) is one of the Markov chain, and its status cannot be directly observed, but can be observed through the observation vector sequence, each observation vector is distributed of various states, though probability density.

Reference [13] describes a method that jointly employ PCA and HMM for EEG pattern classification. Although it uses BCI dataset, it improves and develops HMM.

We use HMM to generate the probability of hidden state. These states are defined by binary code which generated by SVM. The probabilities of these states describe the different frequency band scales corresponding to the probability of epileptic seizures. More details are described by Sect. 38.3.

38.3 Methodology

In order to reverse the result of classification, this paper proposes the hybrid model of WSHMM (Wavelet-SVM-HMM). This algorithm contains the following steps:

Step 1: Choose the appropriate wavelet function
Step 2: Compute the coefficients of wavelet transformation
Step 3: Train the SVM model with these coefficients
Step 4: Recode the results of SVM classifications
Step 5: Train the HMM model with outcome of step 4
Step 6: Test the hybrid algorithm
Step 7: Reverse the result of Step 6. De-noising process

38.3.1 Noise-Reduction

In step 1, these wavelet functions are measured by following method:

Use different wavelet functions to reconstruct the original signal, and then calculate the reconstruction error and relative signal-to-noise ratio (RSNR).

The formula of the RSNR is (3.1):

$$SNR = \log 10(sig^2/N^2) \tag{3.1}$$

where

$$N = \sum (sig - resig)$$

sig is original signal, N is noise, resig is reconstructed signal by wavelet function.

In step 2, we use CWT to de-noising the origin signal, In order to make the algorithm reversibly, the signal will reconstruct in different frequency scale. These sub-signal retain the information of each frequency scale, by rebuilt the signal respectively. This method allows us to backtrack. The principles of selection scale are described in Sect. 38.4.

38.3.2 Feature Extraction and Recoding

In step 3, we obtained the model of SVM by train it with reconstructed signals.

In step 4, we get a binary data set $\{0, 1\}$, this group of data set operated by using the following formula (3.2):

$$X = \sum_{i=1}^{N} 2^{D_i} sgn\{\sum_{i=1}^{l} \alpha_i^* y_i K(w_f, w_i) + b^*\} \quad (3.2)$$

where D_i the layers of decomposition of wavelet. N is the total number of layers decomposition. α_i denote Lagrange multipliers. K is RBF kernel function. According to Eq. (3.2), we can change the original signal into the number within $0 - 2^N$. These numbers are obtained by SVM classification of each scale of wavelet transform, and form a set of comprehensive discrete states.

This data set encoded by SVM represents the feature of EEG in every frequency domain. According to time series algorithm, the data set can be classified by these features.

38.3.3 Classification and Backtracking

In step 5, we train the HMM model with outcome of Step 4; we calculate the maximum likelihood estimate of transition and emission matrices.

In step 6, we use the testing dataset to evaluate the algorithm. The results of this step contain the probability distribution of hidden state which can make our backtracking process feasible. The largest probability of hidden state during epileptic seizures indicates that the frequency scale corresponding to it is abnormally, in other words, such abnormally frequency scales can be quantified during the time of seizure by probability of hidden state. The more details of reversing process are discussed in the next section.

38.4 Experiments and Results

The EEG dataset come from Bonn University open source database. The dataset has five sets which denoted as A–E. Each of the sets has 23.6 s duration and includes 100 single-channel EEG signals, and with the sampling frequency of 173.61 Hz. Set A and B are taken from surface EEG recordings of five healthy volunteers whose eyes open and close respectively. The signals in sets C and D are measured intracranial from five patients in the epileptogenic zone (D) in seizure-free intervals and from the hippocampal formation of the opposite hemisphere of the brain (C). Set E only includes seizure activity. For a more detailed description of the data please refer to the manuscript [14]. Typical EEG signals (one from set D and E) are shown in Fig. 38.1. In this work, the dataset A and E are used to form the non-seizure and seizure classes respectively.

Daubechies wavelet function was chosen to transform both training set and testing set. The central frequencies of each scale are shown in Table 38.1.

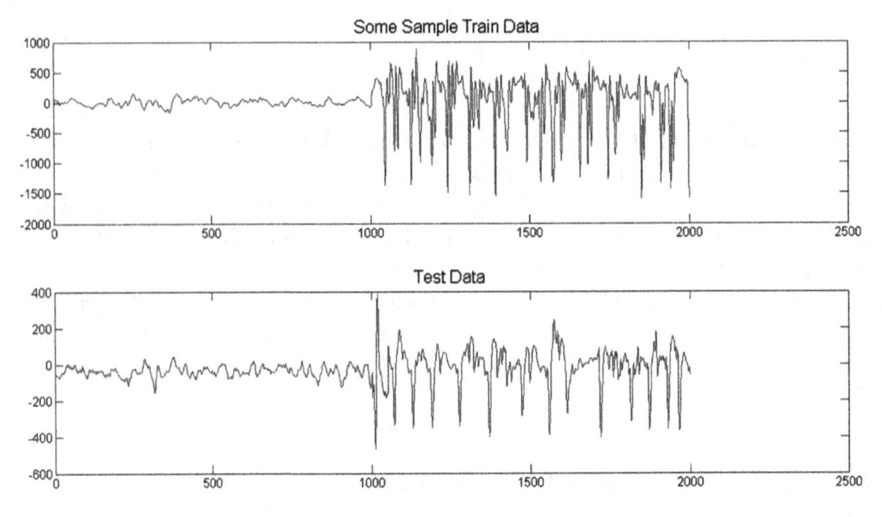

Fig. 38.1 Sample of EEG signals (part of set D and E)

Table 38.1 Central frequencies of each scale

Layers	Scales	Frequencies
D1	2	59.3929
D2	4	29.6964
D3	8	14.8482
D4	16	7.4241
D5	32	3.7121
D6	64	1.8560
D7	128	0.9280
D8	256	0.4640

Figure 38.2 shows the coefficient of training set transformed by db10 from scales D3 to D6.

Computing these coefficient statistical features as the training set of SVM; statistical features contain mean and max of coefficient. Recoding the result of SVM model as Eq. (3.2) and training the HMM model, obtaining the transition matrix and emission matrix.

These matrices contain the key information of reversing process, we will describe it later. Then, SVM model and HMM model are evaluated by testing set.

Table 38.2 lists the correct classification rates of existing methods and proposed method in this paper.

It can be seen that our method achieves 99.11 correct classification rate, which is better than other methods. However, our algorithm can analyze through clinical experience though it reverses these results, the others ignore it. The reversing procedure is mainly focus on the transition matrix. Figure 38.3 shows the emission matrix.

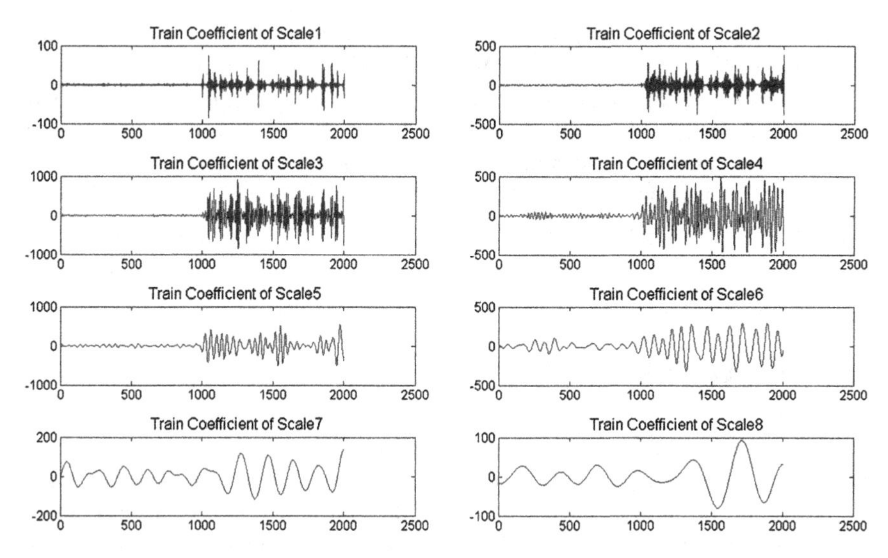

Fig. 38.2 Coefficient of training set

Table 38.2 The accuracy of different methods

Method	Classification problems	Classification accuracy (%)
Neural networks [17]	A, B, C, D–E	97.73
Decision tree [18]	A, E	98.72
Expert system [19]	A, E	95
This work	A, B, C, D, E	99.11

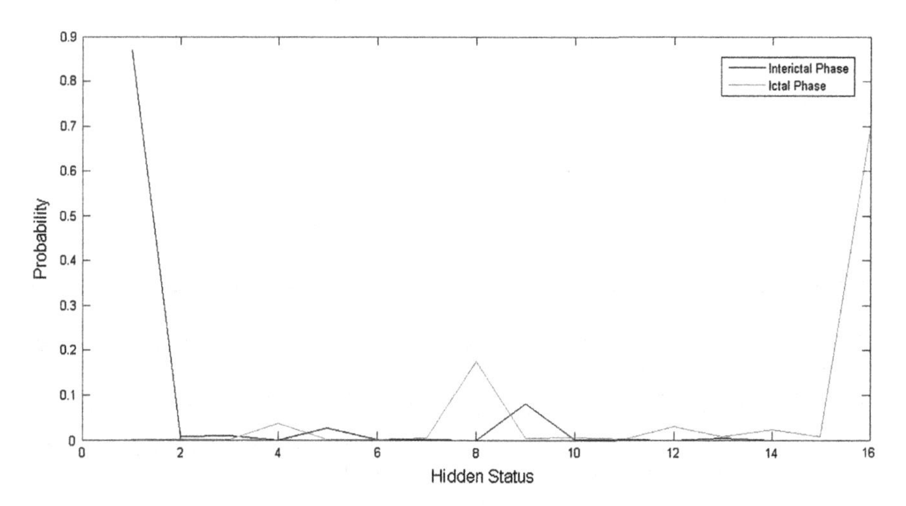

Fig. 38.3 Probability of hidden status

According to Eq. (3.2) and the counts of valid scales, the number of hidden status and observation status of HMM are 16 and 2. In interictal stage, the probability of hidden status is mainly concentrated in state 1 and this state corresponds to wavelet decomposition scale in 14.8482 Hz. In ictal phase, the probability of hidden status is mainly concentrated in state 8, 16 and these states corresponding wavelet decomposition scale in 7.4241, 3.7121 and 1.8560 Hz. The distribution of these hidden states conforms to the clinical characteristics of epilepsy seizures sample discharge.

38.5 Conclusion

This paper proposes a hybrid CSHMM. It provides the reversing analysis and maintains a high accuracy of classification. Clinical experience can be brought into the algorithm for these advantages. We test the above algorithms on EEG dataset and observe that they favorably reverse from the end to beginning of the schedule. Our method not only achieves a high accuracy of classification, but also can trace the result back. This is good for clinical experience.

In the future work, we will commit to use parallel processing to accelerate the CSHMM. Another important issue is about the development of clinical experience basic our results. As mentioned above, the main frequency of ictal is mainly distributed in 1–7 Hz, we still need to find more details about this kind of clinic experience. So extending the proposed algorithm by reverse analysis, following the ideas in [11, 12, 15, 16], and using it to solve to choose EEG features is another possible research direction. Finally, we are going to use the CSHMM along with techniques and criteria to estimate the optimal number of classification.

Acknowledgments This research is supported by the National Natural Science Foundation of China (Grant No.61203084 and 61374135) and the National Natural Science Foundation of Chongqing China (cstc2011jjA40013).

References

1. Shoeb AH (2009) Application of machine learning to epileptic seizure onset detection and treatment. Massachusetts Institute of Technology
2. Tatum WO, Tatum WO, Benbadis SR, Benbadis SR (2008) Handle book of EEG interpertation, Handle Book of EEG Interpertation
3. Schuyler R, White A, Staley K et al (2007) Epileptic seizure detection. IEEE Eng Med Biol Mag 26(2):74–81
4. Joshi V, Pachori RB, Vijesh A (2014) Classification ofictal and seizure-free EEG signals using fractional linear prediction. Biomed Sig Process Contr 9:1–5
5. Pachori RB, Sircar P (2007) A new technique to reduce cross terms in the Wigner distribution. Digit Sig Process 17(2):466–474

6. Shoeb A, Edwards H, Connolly J, Bourgeois B, Treves T, Guttagf J (2004) Patient-specific seizure onset detection. Epilepsy Behav 5:483–498
7. Cristianini N, Shawe-Taylor J (2000) Support vector machines and other kernel-based learning methods. Cambridge University Press, Cambridge
8. Liu S (2007) Gearbox fault diagnosis based on wavelet-AR model. In: International conference on machine learning and cybernetics, 08/2007
9. Lin J, Feature extraction based on MORLET wavelet and its application for mechanical fault diagnosis. J Sound Vibr 20000629
10. Su W, Rolling element bearing faults diagnosis based on optimal Morlet wavelet filter and autocorrelation enhancement. Mech Syst Sig Proces 201007
11. Chen Guangyi (2014) Automatic EEG seizure detection using dual-tree complex wavelet-Fourier features. Expert Syst Appl 41:2391–2394
12. Jahankhani P, Kodogiannis V, Revett K, EEG signal classification using wavelet feature extraction and neural networks. Mod Comput (JVA'06) 0-7695-2643-8/06
13. Lee H, Choi S (2003) PCA + HMM + SVM for EEG pattern classification, signal processing and its applications. Proceedings, vol 1, 1–4 July 2003, pp 541–544
14. Andrzejak RG, Lehnertz K, Mormann F et al (2001) Indications of nonlinear deterministic and finite-dimensional structures in time series of brain electrical activity: dependence on recording region and brain state. Phys Rev E 64(6):061907
15. Güler NF et al (2005) Recurrent neural networks employing Lyapunov exponents for EEG signals classification. Expert Syst Appl 29(3):506–514
16. Shen M, Lin L, Chen J, Chang CQ (2010) A prediction approach for multichannel EEG signals modeling using local wavelet SVM. In: IEEE transactions on instrumentation and measurement, vol 59, no 5, May 2010
17. Tzallas AT, Tsipouras MG, Tsalikakis DG, Karvounis EC, Astrakas L, Konitsiotis S, Tzaphlidou M (2012) Automated epileptic seizure detection methods: a review study. In: Stevanovic D (ed) Epilepsy—histological, electroencephalographic and psychological aspects. ISBN:978-953-51-0082-9
18. Polat K, Gunes S (2007) Classification of epileptiform EEG using a hybrid system based on decision tree classifier and fast Fourier transform. Appl Math Computati 187(2):1017–1026
19. Subasi A (2007) EEG signal classification using wavelet feature extraction and a mixture of expert model. Expert Syst Appl 32(4):1084–1093

Chapter 39
Coordinated Control for Multi-WTGs Wind-Diesel Hybrid Power System Based on Disturbance Observer

Yang Mi and Vanninh Hoang

Abstract Wind diesel hybrid system is an important new energy supply mode, but the output power of wind turbine generator (WTG) is fluctuated depending on weather conditions. Especially, the wind-diesel hybrid system with multiple WTGs for remote areas and islands operation, it will inevitably lead to the fluctuated output power to produce the large frequency deviation. So it's crucial to coordinate the WTGs for providing high quality of electricity. Based on the designed disturbance observer, the coordination control strategy for the multi-WTGs wind-diesel hybrid system is proposed. The load variation is allocated to the WTGs output power reference by using the observer. Here, the constructed controller is compared with the traditional method for every WTG system with only PID control. The simulation results show that of frequency deviations are reduced and output power of every WTGs are controlled effectively.

Keywords Coordination control · Wind-diesel hybrid power system · Observer

39.1 Introduction

Energy supply for remote area is very difficult for long distance transport. Renewable energy such as wind power is often near the load side and getting more attention for its non-pollution and inexhaustible [1–3].

Because wind energy is affected by weather conditions, so the output power has stochastic characteristics. Therefore, it affects the stability of the hybrid power

Y. Mi
Electrical Engineering and Automation, Tianjin University, Tianjin 300072, China
e-mail: miyangmi@163.com

Y. Mi · V. Hoang (✉)
School of Electrical Engineering, Shanghai University of Electric Power, Shanghai 20090, China
e-mail: huangnynk@163.com

Y. Jia et al. (eds.), *Proceedings of the 2015 Chinese Intelligent Systems Conference*, Lecture Notes in Electrical Engineering 360,
DOI 10.1007/978-3-662-48365-7_39

system. There are many articles offer multiple methods to smooth WTG output power [4–9].

Based on the above analysis, a new coordination control method is proposed based on disturbance observer. The change of the load is calculated according to the estimated load value, and the load variation is allocated to the WTGs output power reference. The designed control strategy is tested by matlab simulation.

39.2 Small Hybrid Power System Configuration

The wind-diesel hybrid system with multiple WTGs is shown as Fig. 39.1. It includes diesel generator, multiple WTGs systems and load. Where $\boldsymbol{P}_{\mathrm{d}}$ is diesel generator output, $\boldsymbol{R}$ is speed adjustment coefficient, $\boldsymbol{T}_{\mathrm{g}}$ is controller constant, $\boldsymbol{T}_{\mathrm{p}}$ is diesel generator time constant, $\boldsymbol{K}_{\mathrm{p}}$ is related to gain, $\boldsymbol{P}_{\mathrm{L}}$ is the load, $\Delta \boldsymbol{f}$ is the frequency variation. The constructer of WTG control system is shown as Fig. 39.2.

The WTG output power $\boldsymbol{P}_{\mathrm{w}}$ [12, 13] can be calculated as,

$$P_{\mathrm{w}} = \frac{1}{2}\rho\pi \boldsymbol{R}_{\mathrm{r}}^{2} C_{\mathrm{p}}(\lambda, \beta)\boldsymbol{V}_{\mathrm{w}}^{3} \tag{39.1}$$

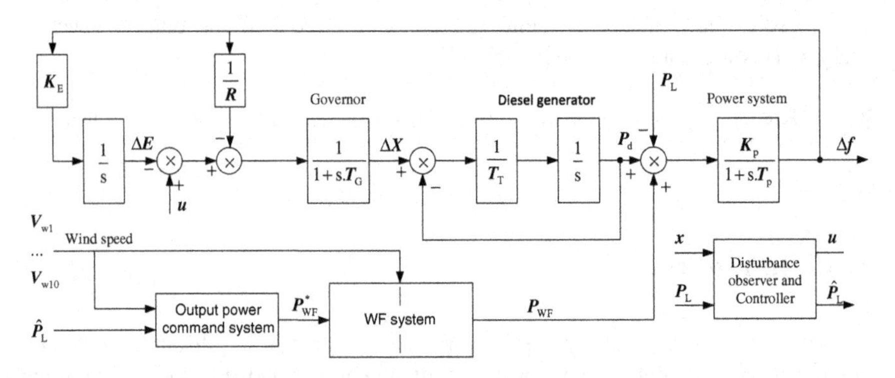

Fig. 39.1 Small power system configuration model

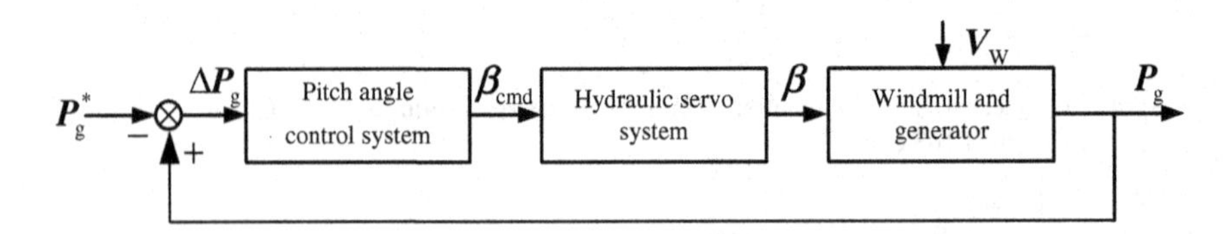

Fig. 39.2 Wind generator systems block diagram

In this paper the squirrel cage induction generator is used, and its output power $\boldsymbol{P}_{\mathrm{g}}$ is computed as [5],

$$\boldsymbol{P}_{\mathrm{g}}=\frac{-3V^{2}s(1+s)\boldsymbol{R}_{2}}{(\boldsymbol{R}_{2}-s\boldsymbol{R}_{1})^{2}+s^{2}(\boldsymbol{X}_{1}+\boldsymbol{X}_{2})^{2}} \tag{39.2}$$

39.3 Output Power Control System

39.3.1 Design Disturbance Observer

The load changing value is estimated by the proposed disturbance observer [14–17]. From Fig. 39.1, the state space model of the power system is satisfied as,

$$\begin{cases} \dot{\boldsymbol{x}}(\mathrm{t})=\boldsymbol{A}\boldsymbol{x}(\mathrm{t})+\boldsymbol{B}\boldsymbol{u}(\mathrm{t})+\boldsymbol{H}\boldsymbol{P}_{\mathrm{L}} \\ \boldsymbol{y}(\mathrm{t})=\boldsymbol{C}\boldsymbol{x}(\mathrm{t}) \end{cases} \tag{39.3}$$

where $\boldsymbol{x}(\mathrm{t})=[\Delta f,\ \Delta \boldsymbol{P}_{\mathrm{d}},\ \Delta \boldsymbol{X},\ \Delta \boldsymbol{E}]^{\mathrm{T}}$ is the state vector, Δf is the power system frequency deviation, $\Delta \boldsymbol{P}_{\mathrm{d}}$ is the output power of the diesel generator, $\Delta \boldsymbol{X}$ is the governor output, $\Delta \boldsymbol{E}$ is the integral control. The system parameter matrixes are as following,

$$\boldsymbol{A}=\begin{bmatrix} -\frac{1}{T_{\mathrm{p}}} & \frac{K_{\mathrm{p}}}{T_{\mathrm{p}}} & 0 & 0 \\ 0 & -\frac{1}{T_{\mathrm{T}}} & \frac{1}{T_{\mathrm{T}}} & 0 \\ -\frac{1}{RT_{\mathrm{G}}} & 0 & -\frac{1}{T_{\mathrm{G}}} & -\frac{1}{T_{\mathrm{G}}} \\ K_{\mathrm{E}} & 0 & 0 & 0 \end{bmatrix},\ \boldsymbol{B}=\begin{bmatrix} 0 \\ 0 \\ \frac{1}{T_{\mathrm{G}}} \\ 0 \end{bmatrix},\ \boldsymbol{H}=\begin{bmatrix} -\frac{K_{\mathrm{p}}}{T_{\mathrm{p}}} \\ 0 \\ 0 \\ 0 \end{bmatrix},\ \boldsymbol{C}=[1\quad 0\quad T_{\mathrm{G}}\quad 0]$$

Define $\boldsymbol{g}(\mathrm{t})=\boldsymbol{H}\boldsymbol{P}_{\mathrm{L}}$ in the system model. In order to design the disturbance observer, the original system is reconstructed as,

$$\begin{bmatrix} \dot{\boldsymbol{x}}(\mathrm{t}) \\ \dot{\boldsymbol{y}}(\mathrm{t}) \\ \dot{\boldsymbol{g}}(\mathrm{t}) \end{bmatrix}=\begin{bmatrix} \boldsymbol{A} & 0 & \boldsymbol{I} \\ 0 & 0 & 0 \\ 0 & 0 & 0 \end{bmatrix}\begin{bmatrix} \boldsymbol{x}(\mathrm{t}) \\ \boldsymbol{y}(\mathrm{t}) \\ \boldsymbol{g}(\mathrm{t}) \end{bmatrix}+\begin{bmatrix} \boldsymbol{B} \\ 0 \\ 0 \end{bmatrix}\boldsymbol{u}(\mathrm{t})+\begin{bmatrix} 0 \\ \dot{\boldsymbol{y}}(\mathrm{t}) \\ 0 \end{bmatrix}+\begin{bmatrix} 0 \\ 0 \\ \dot{\boldsymbol{g}}(\mathrm{t}) \end{bmatrix} \tag{39.4}$$

The corresponding disturbance observer is designed as:

$$\begin{bmatrix} \dot{\hat{\boldsymbol{x}}}(\mathrm{t}) \\ \dot{\hat{\boldsymbol{y}}}(\mathrm{t}) \\ \dot{\hat{\boldsymbol{g}}}(\mathrm{t}) \end{bmatrix}=\begin{bmatrix} \boldsymbol{A} & 0 & \boldsymbol{I} \\ 0 & 0 & 0 \\ 0 & 0 & 0 \end{bmatrix}\begin{bmatrix} \hat{\boldsymbol{x}}(\mathrm{t}) \\ \hat{\boldsymbol{y}}(\mathrm{t}) \\ \hat{\boldsymbol{g}}(\mathrm{t}) \end{bmatrix}+\begin{bmatrix} \boldsymbol{B} \\ 0 \\ 0 \end{bmatrix}\boldsymbol{u}(\mathrm{t})-\begin{bmatrix} \boldsymbol{L}_{1} \\ \boldsymbol{L}_{2} \\ \boldsymbol{L}_{3} \end{bmatrix}\tilde{\boldsymbol{x}}(\mathrm{t})-\begin{bmatrix} \boldsymbol{K}_{1} \\ \boldsymbol{K}_{2} \\ \boldsymbol{K}_{3} \end{bmatrix}\tilde{\boldsymbol{y}}(\mathrm{t}) \tag{39.5}$$

where $\hat{\boldsymbol{x}}(\mathrm{t})$ is the state estimated value, $\tilde{\boldsymbol{x}}(\mathrm{t}) = \hat{\boldsymbol{x}}(\mathrm{t}) - \boldsymbol{x}(\mathrm{t})$ is the estimated error value, $\hat{\boldsymbol{y}}(\mathrm{t})$ is the system output estimated value, $\tilde{\boldsymbol{y}}(\mathrm{t}) = \hat{\boldsymbol{y}}(\mathrm{t}) - \boldsymbol{y}(\mathrm{t})$ is the system output estimated, $\hat{\boldsymbol{g}}(\mathrm{t})$ is the disturbance estimated value, $\tilde{\boldsymbol{g}}(\mathrm{t}) = \hat{\boldsymbol{g}}(\mathrm{t}) - \boldsymbol{g}(\mathrm{t})$ is the estimated value, $\boldsymbol{L}_1$, $\boldsymbol{L}_2$, $\boldsymbol{L}_3$ $\boldsymbol{K}_1$, $\boldsymbol{K}_2$, $\boldsymbol{K}_3$ are the designed observer gain matrix respectively. The error equation is derived through Eq. (39.5) minus (39.4) as following,

$$\begin{bmatrix} \dot{\tilde{\boldsymbol{x}}}(\mathrm{t}) \\ \dot{\tilde{\boldsymbol{y}}}(\mathrm{t}) \\ \dot{\tilde{\boldsymbol{g}}}(\mathrm{t}) \end{bmatrix} = \begin{bmatrix} \boldsymbol{A} - \boldsymbol{L}_1 & -\boldsymbol{K}_1 & \boldsymbol{I} \\ -\boldsymbol{L}_2 & -\boldsymbol{K}_2 & 0 \\ -\boldsymbol{L}_3 & -\boldsymbol{K}_3 & 0 \end{bmatrix} \begin{bmatrix} \tilde{\boldsymbol{x}}(\mathrm{t}) \\ \tilde{\boldsymbol{y}}(\mathrm{t}) \\ \tilde{\boldsymbol{g}}(\mathrm{t}) \end{bmatrix} \tag{39.6}$$

39.3.2 Design WTG Output Power Control System [16]

The wind farm (WF) system configuration is shown in Fig. 39.3. The estimated load variation term $\Delta \boldsymbol{P}_{\mathrm{L}}$ [10, 11, 14] is calculated by:

$$\Delta \boldsymbol{P}_{\mathrm{L}} = \boldsymbol{P}_{\mathrm{L}} - \frac{1}{\boldsymbol{T}} \int_{\mathrm{t}-\boldsymbol{T}}^{\mathrm{t}} \boldsymbol{P}_{\mathrm{L}} \mathrm{dt} \tag{39.7}$$

where $\boldsymbol{T}$ is integral time constant. The WF system output power control is given by the following equation [18, 19],

$$\boldsymbol{P}^{*}_{\mathrm{WF}} = \boldsymbol{P}_{\mathrm{base}} + \Delta \boldsymbol{P}_{\mathrm{L}} \tag{39.8}$$

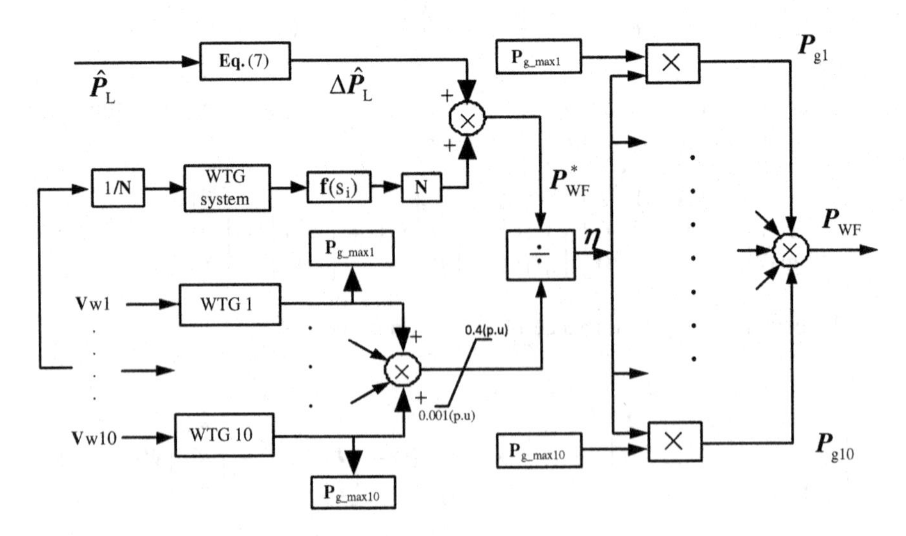

Fig. 39.3 WF system

where P_{base} is calculated as,

$$P_{\text{base}} = \sum_{i=1}^{I=10} P_{g\_maxi} \tag{39.9}$$

where P_{g_maxi} is the WTG_i system maximum output power [18, 19].

39.3.3 Design Sliding Mode Controller

To improve the system dynamic performance and robustness, the sliding mode control is designed for the power system with matched and unmatched perturbations.

The PI switching surface [20] is selected as,

$$s(t) = Cx(t) - \int_0^t (CA - CBF)x(\tau)d\tau \tag{39.10}$$

The sliding mode reaching condition is selected as,

$$\dot{s}(t) = -as(t) - b\text{sgn}(s(t)) \tag{39.11}$$

where a, b are non-negative constant, sgn(*) is the sign function.

So the controller can be derived from (39.11) as,

$$u(t) = -Fx(t) - (CB)^{-1}C\hat{g}(t) - a.s(t) - b.\text{sgn}(s(t)) \tag{39.12}$$

Based on the proposed observer, the designed sliding mode controller can improve the robustness for the system.

39.4 Simulation Results

In this paper, in order to prove the control effect, the system model parameters are selected as [5, 16]. The WTG rated power is 275 kW. The proposed controller and the PID controller are used for the WTG system [7, 12, 14, 15, 17] respectively. From Fig. 39.4, the estimated value can track the actual values accurately. The wind speed models are shown in Fig. 39.5. The pitch angle of WTG1 is proposed in Fig. 39.6. The output power of the WF is presented Fig. 39.7. From Fig. 39.8, the frequency deviation of the power system with the proposed controller is smaller than the PID controller.

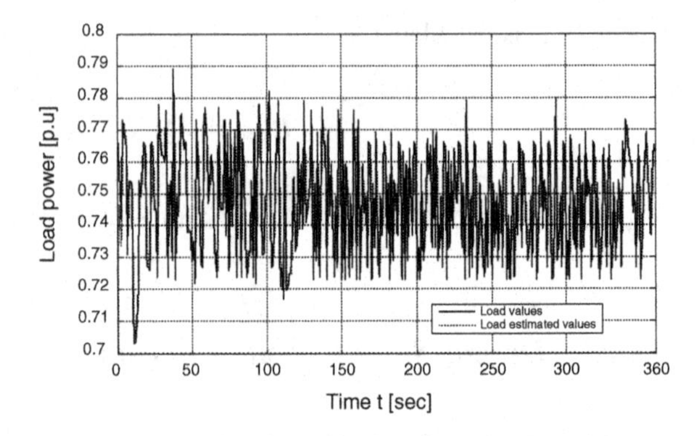

Fig. 39.4 Load and load estimated values (p.u)

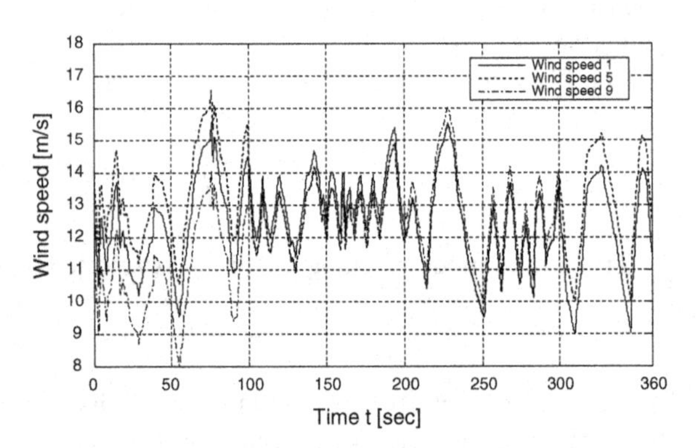

Fig. 39.5 Wind speed (m/s)

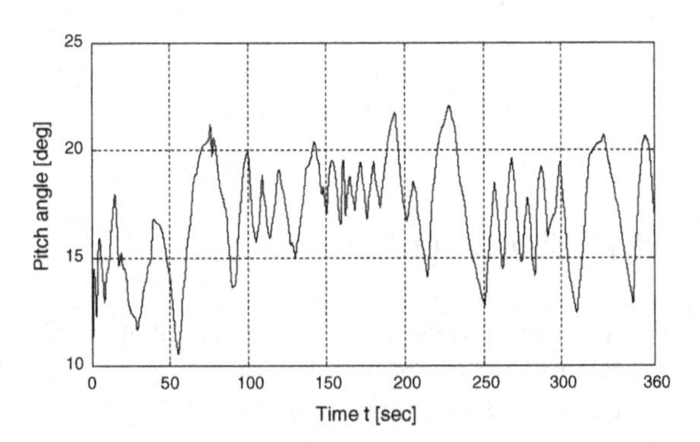

Fig. 39.6 Pitch angle (deg)

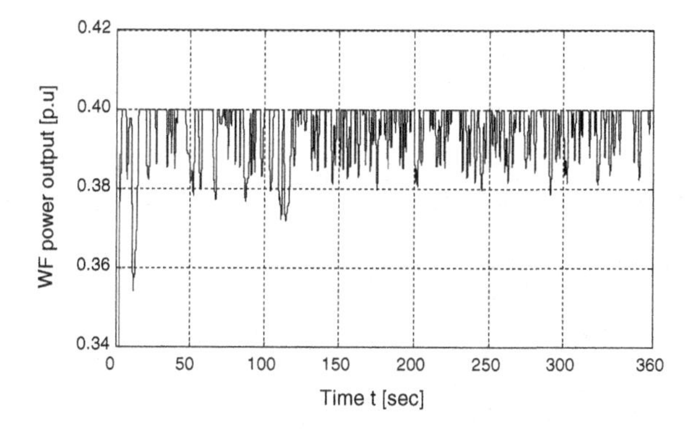

Fig. 39.7 WF power output (p.u)

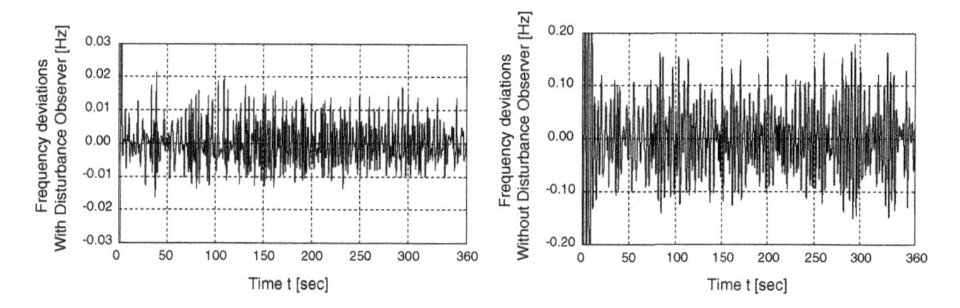

Fig. 39.8 Frequency deviations (Hz)

39.5 Conclusion

The coordinating control method for the wind-diesel hybrid system with multiple WTGs is proposed to reduce frequency deviations. The disturbance observer is used to estimate the load changes in the hybrid power system and then the load estimated values is used to allocate to the WTGs output power reference. The control strategy is used to regulate the WTGs output power. The simulation results show that the frequency deviation is reduced effectively.

Acknowledgment This work was supported in part by National Natural Science Foundation of China (No.61403246), China Postdoctoral Science Foundation funded project (No.2014M560187), the Innovation Program of Shanghai Municipal Education Commission (No.15ZZ085) and Shanghai Engineering Research Center of Green Energy Grid- Connected Technology (No.13DZ2251900).

References

1. Slootweg JG, Kling WL (2003) Is the answer blowing in the wind. IEEE Power Energ Mag. 26–33
2. Gubina AF, Xiangyang X, Zhengmin S (2006) Analysis and support policy recommendation of renewable energy sources in Western China. In: International conference on power system technology, pp 1–8
3. Nagaraj R (2012) Renewable energy based small hybrid power system desalination applications in remote locations. In: 2012 IEEE 5th India international conference on power electronics (IICPE), p 1–5
4. Muljadi E, Butterfield CP (2001) Pitch-controlled variable speed wind turbine generation. IEEE Trans Ind Appl 37(1):240–246
5. Senjyu Tomonobu, Sakamoto Ryosei, Urasaki Naomitsu et al (2006) Output power leveling of wind turbine generator for all operating regions by pitch angle control. IEEE Trans Energy Convers 21(2):467–475
6. Sakamoto Ryosei, Senjyu Tomonobu, Kaneko Toshiaki et al (2008) Output power leveling of wind turbine generator by pitch angle control using H∞ control. Electr Eng Jpn 126 (4):2044–2049
7. Sakamoto R, Senjyu T, Kinjo T et al (2004) Output power leveling of wind turbine generator by pitch angle control using adaptive control method. In: 2004 International conference on power system technology, pp 834–839
8. Le-Ren Chang-Chien, Yin Yao-Ching (2009) Strategies for operating wind power in a similar manner of conventional power plant. IEEE Trans Energy Convers 24(4):926–934
9. Conroy James F, Wastson Rick (2008) Frequency response capability of full converter wind turbine generators in comparison to conventional generation. IEEE Trans Power Syst 23 (2):649–656
10. Kaneko Toshiaki, Uehara Akie, Yona Atsushi et al (2011) A new control methodology of wind turbine generators for frequency control of power system in isolated island. Wind Energy 14:407–423
11. Uehara A, Senjyu T, Yona A et al (2009) Frequency control by coordination control of WTG and battery using load estimation. In: International conference on power electronics and drive systems, PEDS 2009, pp 216–221
12. Yang M, Xiaowei B, Yang Y et al (2014) The sliding mode pitch angle controller design for squirrel-cage induction generator wind power generation system. In: Proceedings of the 33rd Chinese control conference, pp 8113–8117
13. Senjyu T, Sakamoto R, Urasaki N et al (2006) Output power leveling of wind farm using pitch angle control with fuzzy neural network. In: Power engineering society general meeting
14. Senjyu T, Tokudome M, Uehara A et al (2008) A new control methodology of wind farm using short-term ahead wind speed prediction for load frequency control of power system. In: 2nd IEEE international conference on power and energy, pp 425–430
15. Kaneko T, Uehara A, Senjyu T et al (2011) An integrated control method for a wind farm to reduce frequency deviations in a small power system. Energy 88:1049–1058
16. Uehara A, Senjyu T, Kaneko T et al (2010) Output power dispatch control for a wind farm in a small power system. Wind Energy in press, doi:10.1002/we.388
17. Uehara A, Senjyu T, Yona A et al (2010) A Frequency control method by wind farm and battery using load estimation in isolated power system. Int J Emerg Electr Power Syst 11:1–20
18. Senjyu T, Ochi Y, Kikunaga Y et al (2009) Sensor-less maximum power point tracking control for wind generation system with squirrel cage induction generator. Renew Energy 34:994–999
19. Hongwen L, Xiangyan R, Yubin X et al (2014) 5 MW wind turbine research on control strategy and simulation. Wind Energy 8(75):70–76
20. Shyu Kuo-Kai, Shieh Hsin-Jang (1996) A new switching surface sliding-mode speed control for induction motor drive systems. IEEE Trans Power Electron 11(4):660–667

Chapter 40
Sketch-Based 3D Model Shape Retrieval Using Multi-feature Fusion

Dianhui Mao, Huanpu Yin, Haisheng Li and Qiang Cai

Abstract At present, the application of 3D models is becoming more widely, which makes it very important and crucial to retrieve 3D model effectively. With the method based on content to search for those 3D models, rather than textual annotations, it is very important. For this purpose, this paper presented an effective sketch-based 3D model shape retrieval approach. The algorithm compares multi-view rendering of 3D models with the 2D sketch, and the feature vector is defined as a combination of global feature and local feature. We perform experiments, the results of which show a significant increase in precision for 3D model retrieval.

Keywords Sketch-based retrieval · Shape retrieval · Multi-view · 3D model

40.1 Introduction

With the development of 3D graphics modeling technology and 3D data acquisition technology, it has produced an ever increasing number of 3D model databases of objects, such as virtual reality, 3D game, and industrial entity CAD model library and so on. In order to fully utilize the existing 3D model, it is necessary to develop convenient, efficient, and reliable 3D shape (model) search engine. The 3D shape retrieval is through specific interactions from the database to find out a 3D shape with user's intention. Most 3D model retrieval system using existing 3D models as queries, but such models may not be around. A feasible alternative in 3D model

D. Mao (✉) · H. Yin · H. Li · Q. Cai
School of Computer and Information Engineering, Beijing Technology and Business University, Beijing 100048, China
e-mail: amaode@gmail.com

D. Mao · H. Yin · H. Li · Q. Cai
Beijing Key Laboratory of Big Data Technology for Food Safety, Beijing Technology and Business University, Beijing 100048, China

Y. Jia et al. (eds.), *Proceedings of the 2015 Chinese Intelligent Systems Conference*, Lecture Notes in Electrical Engineering 360,
DOI 10.1007/978-3-662-48365-7_40

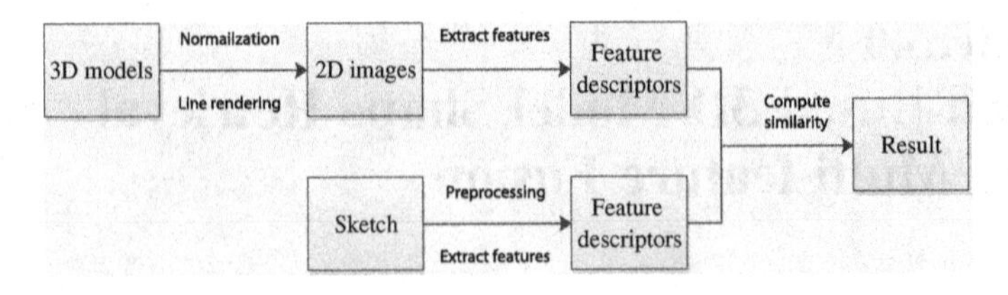

Fig. 40.1 The process of retrieving 3D model based on 2D sketch

retrieval is to use the two dimensional, user-drawn sketches as the query. In particular tablets and smart phones are everywhere now, it will be a very potential choice.

As we knows, most algorithms of retrieving 3D model based on 2D sketch as show in Fig. 40.1: First, rendering models in the database from different viewpoints, the line rendering types generally as follows: (1) silhouette, describe the 2D closed boundary of the rendering; (2) occluding contours, describe all points on the mesh in the direction normal orthogonal to the view; (3) suggestive contours [1], convey an object's shape consistently and precisely. Second, extract features from rendering images and sketches. The feature descriptors categorized into two groups: (1) global feature, like [2, 3]; (2) local feature, includes Shape Contexts [4], HOG (Histogram of Oriented Gradients) [5], SIFT (Scale Invariant Feature Transform) [6] and so on. Finally, compute distance or similarity between the sketch and the model.

In 2003, Funkhouser et al. [2] developed the Princeton Shape Search Engine which support 2D and 3D queries based on 3D spherical harmonics. The feature presentation has been proposed to represent both queries and model sketches globally. Such global sketch representations are able to encode high-level shape information, but sensitive to intra-class variation and shape deformation. Other approaches use statistics about local descriptors for sketch representation. For example, Yoon et al. [7] project 3D models to suggestive contour images from 14 different viewpoints, then extract HOG descriptors and compute similarity. Saavedra et al. [8] developed their sketch-based 3D model retrieval algorithms based on suggestive contours [1] feature views sampling and diffusion tensor fields feature representation or structure-based local approach (STELA). Aono and Iwabuchi [9] proposed an image-based 3D model retrieval algorithm based on the Zernike moments and PCA-HOG features (apply Principal Components Analysis for each block of HOG features). Eitz et al. [10] proposed BoF-GALIF which dense sampling Gabor local line-based feature and use Bag of Feature framework that is proven more robust against the variations in the sketch and model rendering images. Thousands of Gabor features of each image are integrated into a feature vector, so that position of each local feature is ignored. However, approaches based on this representation may easily return locally similar but globally very different models.

In this paper, we present a new approach combined with the global feature and local feature for sketch-based 3D model retrieval. Our proposal, extract three effective and popular features, and use different feature coding approach for different feature. Lastly, compute the final retrieval result according to fusion of three distance matrixs. We experimentally evaluated the proposed algorithm by using one retrieval scenarios. The set of experiments used SHape REtrieval Content 2012 (SHREC) Sketch-Based 3D Shape Retrieval [11] benchmark. The result shows that the approach proposed in this paper outperforms the state-of-the-art methods in both speed and accuracy on the SHREC12 benchmark.

The remaining part of this document is organized as follows. Section 40.2 describes our approach in detail. Section 40.3 discusses the experiments and analyses the results of the experiment. Finally, Sect. 40.4 presents some conclusions.

40.2 Our Work

The basic idea of this approach is learn features based on three shape descriptors. This acquires several steps:

1. Pre-processing and extract three features i.e. PHOG [12] (Pyramid Histogram of Oriented Gradients), PHOW (a variant of dense SIFT descriptors, extracted at multiple scales), and GALIF (Gabor local line-based feature) [10].
2. Clustering the features and encoding features.
3. Fuse the distance matrixes and compute final results.

40.2.1 Pre-processing and Feature Extraction

In the pre-processing of the sketch image, we extract contour and use Gauss filter for image denoising. Then, for each model in the 3D model dataset, we used 102 viewpoints on earth to project line drawings, and line rendering type is suggestive contours.

We extract GALIF, which that is proven more robust against the variations in the sketch and model rendering images.

Besides, we extract PHOG to represent an image by the spatial layout of the shape and local shape. This descriptor has three advantages: (i) it can keep a good invariance to the geometric and optical deformation of the image. More significantly; (ii) PHOG is a compact descriptor appropriate for standard learning algorithms with kernels; (iii) PHOG can detect the characteristics of different scales, and the expression ability is stronger.

At the same time, we extract a large number of PHOW descriptors. Even with clutter and under partial occlusion, PHOW has strong robustness to identify objects. Because the POHW feature descriptor has Scaling invariance, directional invariance and partially affine invariance.

40.2.2 Feature Encoding

Due to the different characteristic of different features, we use different feature coding approach for different feature:

(1) The PHOG descriptor is a weighted concatenation of all vectors for different scales, we directly feature matching for PHOG.

(2) We extract the number of PHOW for every image may not be same, so we use GMM (Gaussian Mixture Model) [13] to construct a visual word dictionary, and encoding feature using the Fisher Coding [14].

Compared with BoF, Fisher vector is not limited to the number of happening of visual word and also includes other information, like the distribution of the descriptors. And to generate a similar size histogram vectors, it requires a smaller codebook than BoF that smaller computational cost.

Let $X = \{x_1, x_2, \ldots, x_T\}$ be the set of T PHOW descriptors for an image, we can describe X by the gradient vector of the likelihood function:

$$G_\lambda^{\mathrm{X}} = \nabla_\lambda \log \mu_\lambda(\mathrm{X}) \tag{40.1}$$

The gradient vector of the likelihood function describes the contribution of parameters to better fit the current parameters, i.e. We transforms a sample X which has an indefinite length to a gradient vector has a fixed length, and the dimensionality of this vector equals to the number of parameters in λ.

To normalize the gradient vector, we use Fisher information matrix F_λ:

$$F_\lambda = E_{x \sim \mu_\lambda}\left[G_\lambda^{\mathrm{X}} G_\lambda^{\mathrm{X}'}\right] \tag{40.2}$$

And the normalization gradient vector g_λ^{X} as the Fisher vector of X:

$$g_\lambda^{\mathrm{X}} = F_\lambda^{-1/2} G_\lambda^{\mathrm{X}} \tag{40.3}$$

A Fisher kernel K(X, Y) denotes the similarity between X and Y:

$$K(X, Y) = G_\lambda^{\mathrm{X}'} F_\lambda^{-1} G_\lambda^{\mathrm{Y}} = g_\lambda^{\mathrm{X}'} g_\lambda^{\mathrm{Y}} \tag{40.4}$$

We assume that the above distribution obeys the Gaussian mixture model (GMM), the parameters $\lambda = \{\omega_k, \mu_k, \sum k | k = 1, \ldots, K\}$, there is:

$$\mu_\lambda(x) = \sum_{k=1}^{K} \omega_k \mu_k(x) \tag{40.5}$$

Where ω_k,μ_k and $\sum k$ are respectively the mixture weight, mean vector and covariance matrix of Gaussian μ_k. The probability of descriptor x_t belongs to Gaussian model i is:

$$\gamma_t(i) = \frac{\omega_i \mu_i(x_t)}{\sum_{k=1}^{K} \omega_k \mu_k(x_t)} \tag{40.6}$$

(3) for GALIF, we use Bag of Feature.

To our knowledge, BoF-GALIF [10] is one of the best performing methods among existing sketch-based 3D retrieval algorithms.

40.2.3 Feature Distance

We use a common Euclidean distance metric to compute the distances of PHOW descriptors (d_W), PHOG descriptors (d_G), GALIF descriptors (d_F).

$$d = \min \sqrt{\sum_{k=1}^{n} (X_k - Y_{ijk})^2} \tag{40.7}$$

Where X is a sketch vector, and Y_{ij} is the jth image of ith model, n is the dimension of vector. The final distance D between sketch and model is:

$$D = \omega_1 d_W + \omega_2 d_G + \omega_3 d_F \tag{40.8}$$

Where ω_1, ω_2, ω_3 is the weight of distances.

40.3 Experiments

40.3.1 Dataset

For our experiments, 3D benchmark dataset is the Watertight Model Benchmark (WMB) dataset [11], it divided into 20 classes, and includes 400 watertight models, as show Fig. 40.2 (left).

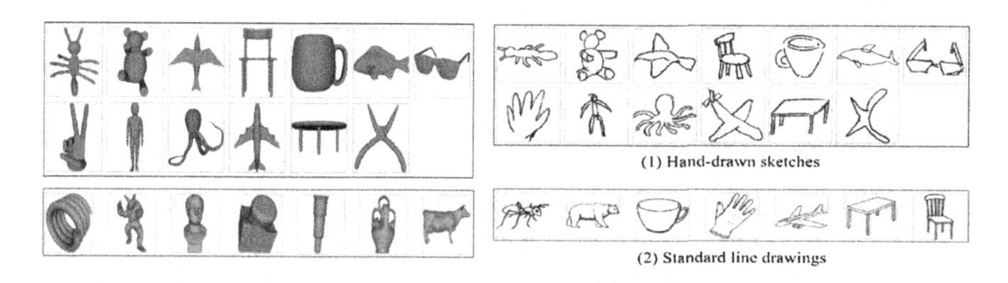

Fig. 40.2 3D watertight models classes (*left*) and 2D query images (*right*) [11]

Additionally, the 2D sketch set includes two different types of subsets, which are used as input for the retrieval task evaluation: (1) Hand-drawn sketches: It divided into 13 classes (human, cup, glasses, plane, ant, chair, octopus, table, teddy, hand, pliers, fish, bird) and contains 250 hand-drawn sketches; (2) Standard line drawings: It contains 12 standard line drawings(human, cup, glasses, plane, ant, chair, table, teddy, hand, pliers, fish, bird), as show Fig. 40.2 (right).

40.3.2 Evaluation Metrics

We mainly compared our framework with SHERC12, and six performance metrics are adopted: Precision-Recall plot (PR), First Tier (FT), E-Measures (E), Nearest Neighbor (NN), Second Tier (ST), and Discounted Cumulated Gain (DCG).

40.3.3 Result

(1) Parameters of Experiment:

PHOG: we set pyramid level L = 3, the angle to 360, and the number of histogram bins to 40.

PHOW: we set the dimensions of SIFT descriptor to 128, and GMM codebook K = 12.

GALIF: we set the number of filter orientation to 4, sample 25*25 = 625 key-points, and set cluster center count K = 1000.

Fuse the distance matrixes: we set $\omega_1 = 1$, $\omega_2 = 0.1$, $\omega_3 = 0.05$ for a good performance (Standard line drawings queries); $\omega_1 = 1$, $\omega_2 = 5$, $\omega_3 = 0.08$ (Hand-drawn sketch queries).

(2) Result:

The PR evaluation results on the WMB dataset are as shown in Fig. 40.3. The PR curve of the our method is about half lower than that of SBR-2D-3D [15] on SHREC12 which the query images are standard line drawings, while the query images are hand-drawn sketch queries, the PR curve of our method is closed to SBR-2D-3D. The different performances due to the different styles of the query sketches in SHREC12.

And for other performance metrics listed in Tables 40.1 and 40.2, we can draw the same conclusion, for hand-drawn sketch queries, our method is better than BOF-SBR and closed to SBR-2D-3D.

However, it should be noted that the computational-complexity of the SBR-2D-3D is much more than our method (SBR-2D-3D takes more than 19 s on SHREC12 for each query, while our method finish a query within 1 s).

To conclude, (1) for Standard line drawings queries, our method is best; (2) for Hand-drawn sketch queries, SBR-2D-3D is much better than our method in retrieval accuracy but our method takes less time than SBR-2D-3D.

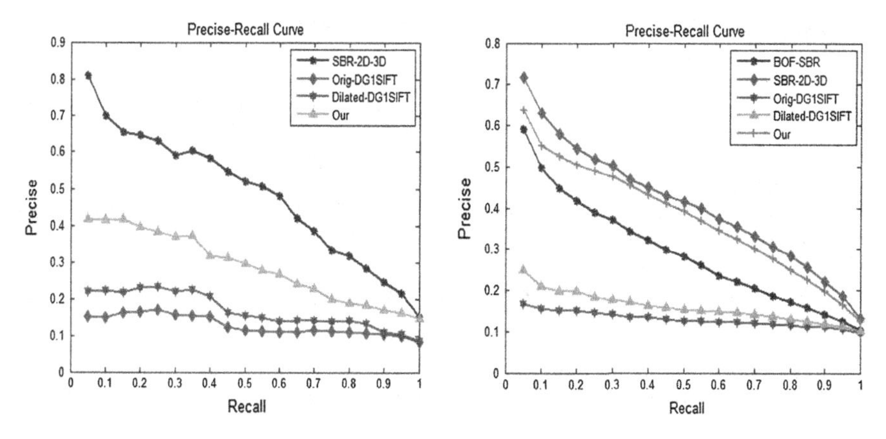

Fig. 40.3 Precision-recall plot performance comparison on the standard line drawings queries (*left*) and hand-drawn sketch queries (*right*)

Table 40.1 Other performance metrics comparison on the standard line drawings queries

	NN	FT	ST	E	DCG
SBR-2D-3D	0.75	0.454	0.625	0.442	0.75
Orig_DG1SIFT	0.083	0.1	0.163	0.106	0.426
Dilated_DG1SIFT	0.167	0.133	0.229	0.144	0.465
Our	0.333	0.258	0.379	0.263	0.582

Table 40.2 Other performance metrics comparison on the hand-drawn sketch queries

	NN	FT	ST	E	DCG
BOF-SBR	0.46	0.278	0.412	0.281	0.614
SBR-2D-3D	0.628	0.371	0.52	0.364	0.692
Orig_DG1SIFT	0.1	0.092	0.158	0.1	0.426
Dilated_DG1SIFT	0.168	0.12	0.212	0.137	0.462
Our	0.548	0.348	0.492	0.34	0.662

40.4 Conclusion

We have presented an effective sketch-based 3D model retrieval approach, which the feature vector is defined as a combination of global feature and local feature from line rendering images of models. And use different feature coding for different feature. Through our comparative experiments, it indicated that our method have a significant increase in precision for 3D model retrieval. In the future, we plan to investigate more about the optimal method for 3D shape retrieval from a collection of 2D sketches, especially Standard line drawings.

References

1. DeCarlo D, Finkelstein A, Rusinkiewicz S et al (2003) Suggestive contours for conveying shape. ACM Trans Graph 22(3):848–855
2. Funkhouser T, Min P, Kazhdan M (2003) A search engine for 3D models. ACM Trans Graph 22:83–105
3. Li B, Lu Y, Johan H (2013) Sketch-based 3D model retrieval by viewpoint entropy-based adaptive view clustering. In: Sixth Eurographics Workshop on 3D Object Retrieval. Eurographics Association, pp 49–56
4. Puzicha J, Belongie S, Malik J (2000) Shape context: a new descriptor for shape matching and object recognition. NIPS 13:831–837
5. Yoon SM, Yoon G, Schreck T (2014) User-drawn sketch-based 3D object retrievalusing sparse coding. Multimedia Tools Appl 74:1–16
6. Dg L (2004) Distinctive image features from scale-invariant keypoints. Int J Comput Vis 60 (2):91–110
7. Yoon SM, Scherer M, Schreck T et al (2010) Sketch-based 3D model retrieval using diffusion tensor fields ofsuggestive contours. ACM Multimedia Conference
8. Saavedra, JM et al (2011) STELA: sketch-based 3D model retrieval using a structure-based local approach. In: Proceedings of the 1st ACM International Conference on Multimedia Retrieval
9. Aono M, Iwabuchi H (2012) 3D shape retrieval from a 2D image as query. In: Asia-Pacific Signal & Information Processing Association Summit & Conference. IEEE, pp 1–10
10. Eitz M, Richter R, Boubekeur T et al (2012) Sketch-based shape retrieval. ACM Trans Graph Proc SIGGRAPH 31(4):13–15
11. Li B, Schreck T, Godil A et al (2012) SHREC 2012 Track: Sketch-Based 3D Shape Retrieval, 13 May 2012. doi:10.2312/3DOR/3DOR12/109-118
12. Bosch A, Zisserman A, Munoz X. Representing shape with a spatial pyramid kernel. In: CIVR Proceedings of ACM International Conference on Image & Video Retrieval
13. He X, Cai D, Shao Y et al (2010) Laplacian regularized gaussian mixture model for data clustering. Knowl Data Eng IEEE Trans 23(9):1406–1418
14. Perronnin F, Dance C (2007) Fisher kernels on visual vocabularies for image categorization. In: IEEE Conference on Computer Vision & Pattern Recognition, pp 1–8
15. Li B, Johan H (2013) Sketch-based 3D model retrieval by incorporating 2D–3D alignment. Multimedia Tools Appl 65(3):363–385

Chapter 41
Autonomous Navigation Based on Sequential Images for Planetary Landing

Chao Xu, Dayi Wang and Xiangyu Huang

Abstract A new autonomous navigation scheme for planetary landing is presented. The navigation system contains an inertial measurement unit (IMU) and a stereo camera which can measure unit directional vectors and range information from the camera to detected landmarks. The lander's motion is estimated by a algorithm known as vision-aided inertial navigation (VAIN). The algorithm uses the unit directional vectors and range measurements of features tracked in two sequential images and the lander's corresponding poses derived from the IMU and it does not require any a priori terrain information. An augmented implicit extended Kalman filter (IEKF) tightly integrates measurements from the stereo camera and the IMU to produce an accurate estimation of the lander's pose and velocity and to correct the IMU constant biases. The results of a numerical simulation show that the proposed VAIN method can vastly improve the navigation accuracy of the INS and satisfy the requirements of future planetary exploration missions.

Keywords Autonomous navigation · Sequential images · Planetary landing · Implicit extended Kalman filter

41.1 Introduction

Autonomous planetary landing navigation is one of the most important technologies for future planetary exploration. In the past few decades, Many navigation methods have been developed to improve landing precision. Typically, IMU is the main navigation sensor used for planetary landing. Due to the accumulation of navigation errors, the navigation solution derived from IMU are often needed to be corrected by external measurements to improve navigation accuracy. In previous planetary explorations with landing missions, the landers have determined their poses by fusing

C. Xu · D. Wang (✉) · X. Huang
National Laboratory of Space Intelligent Control, Beijing Institute of Control Engineering, Beijing 100190, China
e-mail: dayiwang@163.com

Y. Jia et al. (eds.), *Proceedings of the 2015 Chinese Intelligent Systems Conference*, Lecture Notes in Electrical Engineering 360,
DOI 10.1007/978-3-662-48365-7_41

measurements from IMUs with velocity and altitude information from Doppler radar [1, 2]. However, this method does not allow for the accurate estimation of a lander's horizontal position.

An alternative method for improving the basic navigation accuracy of an inertial navigation system (INS) is VAIN, which has received substantial attention in recent years. Numerous VAIN approaches have been proposed based on the assumption of a priori landmark information [3–5]. These methods rely on obtaining 3D positioning information of detected landmarks by matching descent images with orbiter images. Nevertheless, in certain cases, it is impossible to obtain accurate 3D positioning information for such landmarks tracked between images. The general solution in such situations is to estimate the 3D coordinates of landmarks simultaneously, which is referred to as simultaneous localization and mapping (SLAM) [6]. However, the computational requirements continuously increase with increasing number of landmarks. In [7], a visual odometer based on stereo camera is presented for autonomous navigation.

In this paper, a VAIN scheme based on two sequential images captured by a stereo camera for autonomous planetary landing navigation is proposed. The algorithm does not require any a priori terrain information and is implemented by an augmented IEKF.

41.2 Method Overview

In our work, we consider that a lander is equipped with an IMU and a stereo camera. The IMU provides acceleration and angular rate measurements that are processed into a navigation solution. The stereo camera can sense unit directional vectors and range information from the camera to detected landmarks.

During landing, the INS integrates the differential equations describing the lander's dynamics from the outputs of the IMU. Meanwhile, the camera-captured images associated with the INS navigation solutions are stored. Once a set of landmarks observed in two sequential images has been identified, the measurements of detected landmarks and the INS pose estimations at the moments of observation are used to calculate the implicit measurement equations developed in next Section. The IEKF and the implicit measurement equations estimate the navigation errors and the IMU drift biases, which are used to correct the navigation solution and the subsequent IMU readings, respectively. In the following sections, we present the method in detail, especially the derivation of the implicit measurements.

41.3 VAIN System Description

In this section, we describe the proposed VAIN system in detail. To formulate problem clearly, we introduce the landing frame $\{L\}$, the body frame $\{B\}$ and the camera frame $\{C\}$, which are defined as in [8]. For simplicity, we assume the frame $\{C\}$ is concentric with the frame $\{B\}$.

41.3.1 The State Equation

The vehicle landing dynamics used for dead reckoning in INS are given by [9]

$$
\begin{aligned}
{}^{L}\dot{\boldsymbol{r}} &= {}^{L}\boldsymbol{v} \\
{}^{L}\dot{\boldsymbol{v}} &= \boldsymbol{C}^{T}_{{}^{B}_{L}\boldsymbol{q}}(\boldsymbol{a}_{imu} - \boldsymbol{b}_{a}) - 2\lfloor\boldsymbol{\omega}_{L}\times\rfloor{}^{L}\boldsymbol{v} - \lfloor\boldsymbol{\omega}_{L}\times\rfloor^{2}\,{}^{L}\boldsymbol{r} + \boldsymbol{g} \\
{}^{B}_{L}\dot{\boldsymbol{q}} &= \frac{1}{2}\Omega(\boldsymbol{\omega}_{imu} - \boldsymbol{b}_{\omega} - \boldsymbol{C}_{{}^{B}_{L}\boldsymbol{q}}\boldsymbol{\omega}_{L}){}^{B}_{L}\boldsymbol{q}
\end{aligned} \tag{41.1}
$$

where ${}^{L}\boldsymbol{r}$ and ${}^{L}\boldsymbol{v}$ are the lander's position and velocity vectors; ${}^{B}_{L}\boldsymbol{q}$ is the attitude quaternion that describes the rotation from the $\{L\}$ to the $\{B\}$; $\boldsymbol{b}_{a}$ and $\boldsymbol{b}_{w}$ are the IMU bias vectors. $\boldsymbol{\omega}_{L}$ is the planet's rotational velocity in the $\{L\}$, $\boldsymbol{g}$ is the local gravity vector fluctuated by Gauss white noise $\boldsymbol{n}_{g}$, and $\lfloor\cdot\times\rfloor$ and $\Omega(\cdot)$ are defined as in [9].

For brevity, the superscript and subscript of ${}^{B}_{L}\boldsymbol{q}$ can be omitted, that is, ${}^{B}_{L}\boldsymbol{q} = \boldsymbol{q}$. Then, the accelerometer and gyroscope measurements are given by

$$
\boldsymbol{a}_{imu} = \boldsymbol{C}_{q}({}^{L}\boldsymbol{a} - \boldsymbol{g} + 2\lfloor\boldsymbol{\omega}_{L}\times\rfloor{}^{L}\boldsymbol{v} + \lfloor\boldsymbol{\omega}_{L}\times\rfloor^{2}\,{}^{L}\boldsymbol{r}) + \boldsymbol{b}_{a} + \boldsymbol{n}_{a} \tag{41.2}
$$

$$
\boldsymbol{\omega}_{imu} = \boldsymbol{\omega} + \boldsymbol{C}_{q}\boldsymbol{\omega}_{L} + \boldsymbol{b}_{\omega} + \boldsymbol{n}_{\omega} \tag{41.3}
$$

respectively, ${}^{L}\boldsymbol{a}$ is the lander's acceleration expressed in the {L}; $\boldsymbol{\omega}$ is the lander's angular velocity expressed in the {B}; $\boldsymbol{C}_{q}$ denotes the direction-cosine matrix; $\boldsymbol{n}_{a}$ and $\boldsymbol{n}_{\omega}$ are the drift noises modeled as Gaussian white noise. The drift biases are disturbed by Gauss white noise, given by

$$
\dot{\boldsymbol{b}}_{a} = \boldsymbol{n}_{wa}, \quad \dot{\boldsymbol{b}}_{\omega} = \boldsymbol{n}_{w\omega} \tag{41.4}
$$

where $\boldsymbol{n}_{wa}$ and $\boldsymbol{n}_{w\omega}$ are drift bias noises.

The state vector is chosen as

$$
\boldsymbol{x} = \left[{}^{L}\boldsymbol{r}^{T}\ {}^{L}\boldsymbol{v}^{T}\ {}^{B}_{L}\boldsymbol{q}^{T}\ \boldsymbol{b}^{T}_{a}\ \boldsymbol{b}^{T}_{w}\right]^{T} \in \Re^{16} \tag{41.5}
$$

According to the chosen system state in Eq. (41.5), the error-state vector $\tilde{\boldsymbol{x}}$ is given by

$$
\tilde{\boldsymbol{x}} = \left[{}^{L}\tilde{\boldsymbol{r}}^{T}\ {}^{L}\tilde{\boldsymbol{v}}^{T}\ \delta\boldsymbol{\theta}^{T}\ \tilde{\boldsymbol{b}}^{T}_{a}\ \tilde{\boldsymbol{b}}^{T}_{w}\right]^{T} \in \Re^{15} \tag{41.6}
$$

It is worth noting that the quaternion error is described by $\delta\boldsymbol{\theta}$, as $\delta\boldsymbol{q} \simeq \left[\frac{1}{2}\delta\boldsymbol{\theta}^{T}\ 1\right]^{T}$. Then, the error-state equation can be obtained as follows:

$$
\dot{\tilde{\boldsymbol{x}}} = \boldsymbol{F}\tilde{\boldsymbol{x}} + \boldsymbol{G}\boldsymbol{n} \tag{41.7}
$$

where $\boldsymbol{n} = \left[\boldsymbol{n}_{w\omega}^T\ \boldsymbol{n}_{\omega}^T\ \boldsymbol{n}_{wa}^T\ \boldsymbol{n}_a^T\ \boldsymbol{n}_g^T\right]^T$ is the system noise with a covariance matrix $\boldsymbol{Q}$, while the matrices $\boldsymbol{F}$ and $\boldsymbol{G}$ are given by

$$\boldsymbol{F} = \begin{bmatrix} \boldsymbol{0}_{3\times3} & \boldsymbol{I}_{3\times3} & \boldsymbol{0}_{3\times3} & \boldsymbol{0}_{3\times3} & \boldsymbol{0}_{3\times3} \\ -\lfloor\boldsymbol{\omega}_L\times\rfloor^2 & -2\lfloor\boldsymbol{\omega}_L\times\rfloor & -\boldsymbol{C}_{\hat{q}}^T\lfloor(\boldsymbol{a}_{imu}-\hat{\boldsymbol{b}}_a)\times\rfloor & -\boldsymbol{C}_{\hat{q}}^T & \boldsymbol{0}_{3\times3} \\ \boldsymbol{0}_{3\times3} & \boldsymbol{0}_{3\times3} & -\lfloor(\boldsymbol{\omega}_{imu}-\hat{\boldsymbol{b}}_\omega-\boldsymbol{C}_{\hat{q}}\boldsymbol{\omega}_L)\times\rfloor & \boldsymbol{0}_{3\times3} & -\boldsymbol{I}_{3\times3} \\ \boldsymbol{0}_{3\times3} & \boldsymbol{0}_{3\times3} & \boldsymbol{0}_{3\times3} & \boldsymbol{0}_{3\times3} & \boldsymbol{0}_{3\times3} \\ \boldsymbol{0}_{3\times3} & \boldsymbol{0}_{3\times3} & \boldsymbol{0}_{3\times3} & \boldsymbol{0}_{3\times3} & \boldsymbol{0}_{3\times3} \end{bmatrix}_{15\times15}$$

$$\boldsymbol{G} = \begin{bmatrix} \boldsymbol{0}_{3\times3} & \boldsymbol{0}_{3\times3} & \boldsymbol{0}_{3\times3} & \boldsymbol{0}_{3\times3} & \boldsymbol{0}_{3\times3} \\ \boldsymbol{0}_{3\times3} & \boldsymbol{0}_{3\times3} & -\boldsymbol{C}_{\hat{q}}^T & \boldsymbol{0}_{3\times3} & \boldsymbol{I}_{3\times3} \\ -\boldsymbol{I}_{3\times3} & \boldsymbol{0}_{3\times3} & \boldsymbol{0}_{3\times3} & \boldsymbol{0}_{3\times3} & \boldsymbol{0}_{3\times3} \\ \boldsymbol{0}_{3\times3} & \boldsymbol{0}_{3\times3} & \boldsymbol{0}_{3\times3} & \boldsymbol{I}_{3\times3} & \boldsymbol{0}_{3\times3} \\ \boldsymbol{0}_{3\times3} & \boldsymbol{I}_{3\times3} & \boldsymbol{0}_{3\times3} & \boldsymbol{0}_{3\times3} & \boldsymbol{0}_{3\times3} \end{bmatrix}_{15\times15}$$

41.3.2 Implicit Measurement Equation

Implicit measurement equations can be obtained by combining the observations from sequential images with the navigation data associated with each image. Suppose that a single landmark p_j is observed in two sequential images at time instances t_k and t_{k+1}. $\boldsymbol{y}_i^j$ and l_i^j denote the unit directional vector and the distance from the camera to the landmark p_j at t_{k+i-1}, which can be obtained by the stereo camera. Assume that the lander's position at time t_{k+i-1} in the $\{L\}$ reference frame is $\boldsymbol{r}_i$, with $i = 1, 2$. Then, the position of the landmark p_j, expressed in the $\{L\}$ reference frame, can be written as

$$^{L}\boldsymbol{p}_j = {}^{L}\boldsymbol{r}_1 + l_1^j\boldsymbol{C}_{q_1}^T\boldsymbol{y}_1^j \tag{41.8}$$

Let $\boldsymbol{\hbar}_j = \boldsymbol{C}_{q_2}\left({}^{L}\boldsymbol{p}_j - {}^{L}\boldsymbol{r}_2\right)$, then the unit directional vector $\boldsymbol{y}_2^j$ and the range l_2^j can be computed as follows:

$$\boldsymbol{y}_2^j = \frac{\boldsymbol{\hbar}_j}{\sqrt{\boldsymbol{\hbar}_j^T\boldsymbol{\hbar}_j}}, \quad l_2^j = \sqrt{\boldsymbol{\hbar}_j^T\boldsymbol{\hbar}_j} \tag{41.9}$$

Combining Eqs. (41.8) and (41.9) yield the following constraint:

$$z_j = l_2^j\boldsymbol{y}_2^j - \boldsymbol{C}_{q_2}\left({}^{L}\boldsymbol{r}_1 - {}^{L}\boldsymbol{r}_2 + l_1^j\boldsymbol{C}_{q_1}^T\boldsymbol{y}_1^j\right) = \boldsymbol{0}_{3\times1} \tag{41.10}$$

When measurement noise and navigation error are considered, the implicit measurement constraints are inaccurate. Suppose that the measurement $\boldsymbol{\eta}_i^j$ is contaminated by Gaussian white noise $\boldsymbol{v}_i^j = \begin{bmatrix} \boldsymbol{v}_{y_i^j}^T & v_{l_i^j} \end{bmatrix}^T$ with a covariance matrix $\boldsymbol{R}_\eta \in \Re^{4\times4}$.

Linearizing Eq. (41.10) and retaining the first-order terms yield

$$\begin{aligned}\hat{z}_j &\simeq -\boldsymbol{H}_{r_1}^j\,{}^L\tilde{\boldsymbol{r}}_1 - \boldsymbol{H}_{r_2}^j\,{}^L\tilde{\boldsymbol{r}}_2 - \boldsymbol{H}_{\delta\theta_1}^j\delta\boldsymbol{\theta}_1 - \boldsymbol{H}_{\delta\theta_2}^j\delta\boldsymbol{\theta}_2 - \boldsymbol{D}_1^j\boldsymbol{v}_1^j - \boldsymbol{D}_2^j\boldsymbol{v}_2^j \\ &= -\boldsymbol{H}_{x_{t_k}}^j\tilde{\boldsymbol{x}}(t_k) - \boldsymbol{H}_{x_{t_{k+1}}}^j\tilde{\boldsymbol{x}}(t_{k+1}) - \boldsymbol{v}_j \end{aligned} \tag{41.11}$$

where

$$\begin{gathered}\boldsymbol{H}_{r_1}^j = -\boldsymbol{C}_{q_2}, \quad \boldsymbol{H}_{r_2}^j = \boldsymbol{C}_{q_2} \\ \boldsymbol{H}_{\delta\theta_1}^j = l_1^j\boldsymbol{C}_{q_2}\boldsymbol{C}_{q_1}^T\lfloor\boldsymbol{y}_1^j\times\rfloor, \quad \boldsymbol{H}_{\delta\theta_2}^j = -\lfloor\boldsymbol{C}_{q_2}({}^L\boldsymbol{r}_1 - {}^L\boldsymbol{r}_2 + l_1^j\boldsymbol{C}_{q_1}^T\boldsymbol{y}_1^j)\times\rfloor \\ \boldsymbol{D}_1^j = -\boldsymbol{C}_{q_2}\boldsymbol{C}_{q_1}^T\left[l_1^j\boldsymbol{I}_{3\times3}\;\boldsymbol{y}_1^j\right], \quad \boldsymbol{D}_2^j = \left[l_2^j\boldsymbol{I}_{3\times3}\;\boldsymbol{y}_2^j\right], \; \boldsymbol{v}_j = \boldsymbol{D}_1^j\boldsymbol{v}_1^j + \boldsymbol{D}_2^j\boldsymbol{v}_2^j \\ \boldsymbol{H}_{x_{t_k}}^j = \left[\boldsymbol{H}_{r_1}^j\;\boldsymbol{0}_{3\times3}\;\boldsymbol{H}_{\delta\theta_1}^j\;\boldsymbol{0}_{3\times6}\right], \quad \boldsymbol{H}_{x_{t_{k+1}}}^j = \left[\boldsymbol{H}_{r_2}^j\;\boldsymbol{0}_{3\times3}\;\boldsymbol{H}_{\delta\theta_2}^j\;\boldsymbol{0}_{3\times6}\right]\end{gathered}$$

The constraint equation described in Eq. (41.11) is used as implicit measurement equation to update the IEKF state [10], which is discussed in the next subsection.

41.3.3 Navigation Filter

The estimation algorithm of choice is an IEKF, which is an EKF that has been adapted for application to implicit measurements.

Based on Eq. (41.11), the current implicit measurement equation not only depends on the current sate and measurements but also depends on the previous state and measurements. Hence, an augmented method known as Stochastic Cloning is used to estimate the current state [11]. The augmented error state and its propagation are given by

$$\bar{\tilde{\boldsymbol{x}}}(t_k) = \left[\tilde{\boldsymbol{x}}(t_k)^T \quad \tilde{\boldsymbol{x}}(t_k)^T\right]^T \tag{41.12}$$

$$\bar{\tilde{\boldsymbol{x}}}(t_{k+1}) = \bar{\boldsymbol{\Phi}}_{k+1,k}\bar{\tilde{\boldsymbol{x}}}(t_k) + \bar{\boldsymbol{\Gamma}}_{k+1,k}\boldsymbol{n}(t_k) \tag{41.13}$$

where

$$\bar{\boldsymbol{\Phi}}_{k+1,k} = \begin{bmatrix}\boldsymbol{I}_{15\times15} & \boldsymbol{0}_{15\times15} \\ \boldsymbol{0}_{15\times15} & \boldsymbol{\Phi}_{k+1,k}\end{bmatrix}, \bar{\boldsymbol{\Gamma}}_{k+1,k} = \begin{bmatrix}\boldsymbol{0}_{15\times15} \\ \boldsymbol{\Gamma}_{k+1,k}\end{bmatrix}$$

$\boldsymbol{\Phi}_{k+1,k}$ and $\boldsymbol{\Gamma}_{k+1,k}$ can be calculated from Eq. (41.7).

Correspondingly, the covariance matrix and its propagation are

$$\bar{\boldsymbol{P}}_{k|k} = \begin{bmatrix} \boldsymbol{P}_{k|k} & \boldsymbol{P}_{k|k} \\ \boldsymbol{P}_{k|k} & \boldsymbol{P}_{k|k} \end{bmatrix} \tag{41.14}$$

$$\bar{\boldsymbol{P}}_{k+1|k} = \bar{\boldsymbol{\Phi}}_{k+1,k}\bar{\boldsymbol{P}}_{k|k}\bar{\boldsymbol{\Phi}}^{T}_{k+1,k} + \bar{\boldsymbol{\Gamma}}_{k+1,k}\boldsymbol{Q}_k\bar{\boldsymbol{\Gamma}}^{T}_{k+1,k} \tag{41.15}$$

where $\bar{\boldsymbol{P}}_{k|k} = E(\bar{\tilde{\boldsymbol{x}}}(t_k)\bar{\tilde{\boldsymbol{x}}}(t_k)^T)$ and $\boldsymbol{P}_{k|k} = E(\tilde{\boldsymbol{x}}(t_k)\tilde{\boldsymbol{x}}(t_k)^T)$.

The implicit measurement equation in Eq. (41.11) can be rewritten as

$$\hat{z}_j = -\bar{\boldsymbol{H}}_j\bar{\tilde{\boldsymbol{x}}}(t_{k+1}) - \boldsymbol{v}_j \tag{41.16}$$

where

$$\bar{\boldsymbol{H}}_j = \begin{bmatrix} \boldsymbol{H}^j_{x_{t_k}} & \boldsymbol{H}^j_{x_{t_{k+1}}} \end{bmatrix}, \boldsymbol{R}_j = \boldsymbol{D}^j_1\boldsymbol{R}_\eta(\boldsymbol{D}^j_1)^T + \boldsymbol{D}^j_2\boldsymbol{R}_\eta(\boldsymbol{D}^j_2)^T$$

Generally, there is some set of landmarks that are observed in two sequential images. Under the assumption that N landmarks are included in the set, N constraints, each of the form given in Eq. (41.16), can be combined to obtain

$$\hat{z} = -\bar{\boldsymbol{H}}\bar{\tilde{\boldsymbol{x}}}(t_{k+1}) - \boldsymbol{v} \tag{41.17}$$

where

$$\hat{z} = \begin{bmatrix} \hat{z}^T_1 \cdots \hat{z}^T_N \end{bmatrix}^T, \bar{\boldsymbol{H}} = \begin{bmatrix} \bar{\boldsymbol{H}}^T_1 \cdots \bar{\boldsymbol{H}}^T_N \end{bmatrix}^T, \boldsymbol{v} = \begin{bmatrix} \boldsymbol{v}^T_1 \cdots \boldsymbol{v}^T_N \end{bmatrix}^T$$

The Kalman gain matrix is given by

$$\bar{\boldsymbol{K}} = \bar{\boldsymbol{P}}_{k+1|k}\bar{\boldsymbol{H}}^T(\bar{\boldsymbol{H}}\bar{\boldsymbol{P}}_{k+1|k}\bar{\boldsymbol{H}}^T + \boldsymbol{R})^{-1} \tag{41.18}$$

Then, the augmented error state and the corresponding covariance matrix can be updated as follows:

$$\bar{\tilde{\boldsymbol{x}}}_{k+1|k+1} = \bar{\tilde{\boldsymbol{x}}}_{k+1|k} - \bar{\boldsymbol{K}}\hat{z} \tag{41.19}$$

$$\bar{\boldsymbol{P}}_{k+1|k+1} = \bar{\boldsymbol{P}}_{k+1|k} - \bar{\boldsymbol{K}}\bar{\boldsymbol{H}}\bar{\boldsymbol{P}}_{k+1|k} \tag{41.20}$$

while the lander's state error and the corresponding covariance can be computed by

$$\tilde{\boldsymbol{x}}_{k+1|k+1} = \bar{\tilde{\boldsymbol{x}}}_{k+1|k+1}(16:30) \tag{41.21}$$

$$\boldsymbol{P}_{k+1|k+1} = \bar{\boldsymbol{P}}_{k+1|k+1}(16:30, 16:30) \tag{41.22}$$

41.4 Simulation and Results

A simulation demonstrating the validity of the suggested scheme was performed in the MATLAB environment. A low quality IMU performance parameters are listed in Table 41.1. The initial state parameters of the lander are listed in Table 41.2. Suppose that the Moon is the target planetary object, with a planned landing span of $\tau = 200$ s.

Using the simulation parameters mentioned above, the proposed VAIN scheme was simulated and verified. The estimated trajectory by the VAIN based on two sequential images follows the true trajectory very well, while the estimated trajectory only by INS gradually steps away from the true trajectory, as shown in Fig. 41.1. Figures 41.2, 41.3 and 41.4 compare the position errors, velocity errors, and attitude angle errors of the INS and the VAIN scheme based on two sequential images. These figures demonstrate that the INS estimation errors are much larger than those of the proposed navigation scheme. Furthermore, Figs. 41.5 and 41.6 show the estimation errors of the IMU biases; the presented results indicate that the IMU biases have been significantly corrected through the use of observations of landmarks in two sequential images from stereo camera.

Table 41.1 The IMU performance parameters used in simulation

	Drift bias	Drift noise (Std.)	Bias noise (Std.)
Accelerometer bias (m/s^2)			
x	1.2×10^{-2}	10^{-6}	10^{-8}
y	1.2×10^{-2}	10^{-6}	10^{-8}
y	1×10^{-2}	10^{-6}	10^{-8}
Gyroscope bias (deg/s)			
x	1.67×10^{-2}	10^{-6}	10^{-8}
y	1.67×10^{-2}	10^{-6}	10^{-8}
y	2×10^{-2}	10^{-6}	10^{-8}

Table 41.2 Initial state parameters

Initial state parameters	
$\boldsymbol{r}_0$	$[300\ 500\ 3000]^T$(m)
$\boldsymbol{v}_0$	$[-3\ -2\ -20]^T$(m/s)
$\boldsymbol{\omega}_0$	$[0\ 0\ 0.2]^T$(deg/s)
$\boldsymbol{q}_0$	$[0.99387\ 0.060855\ 0.069392\ 0.060855]^T$
$\boldsymbol{b}_a$	$[0\ 0\ 0]^T$(m/s^2)
$\boldsymbol{b}_\omega$	$[0\ 0\ 0]^T$(deg/s)

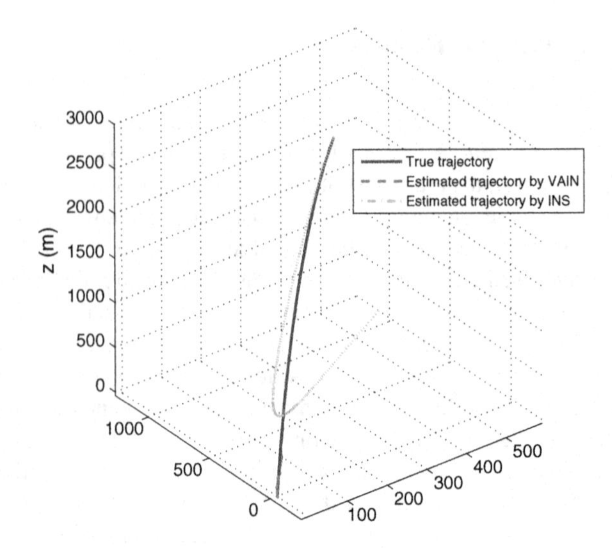

Fig. 41.1 Landing trajectory used in the simulation

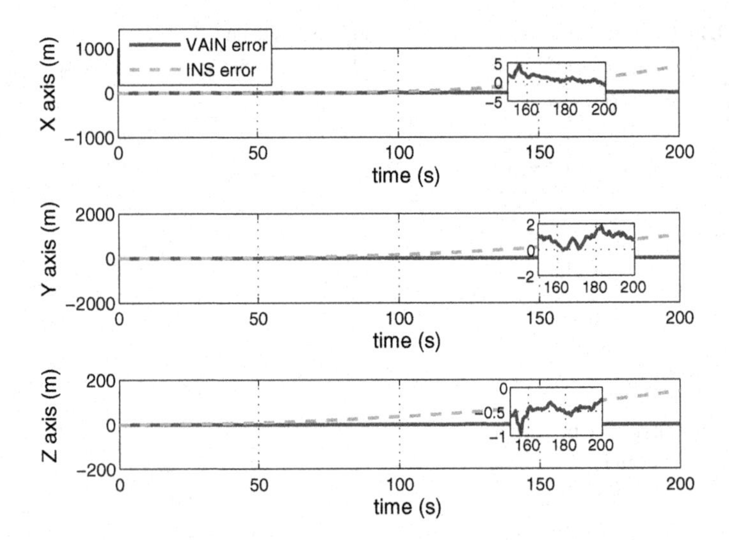

Fig. 41.2 Position estimation errors

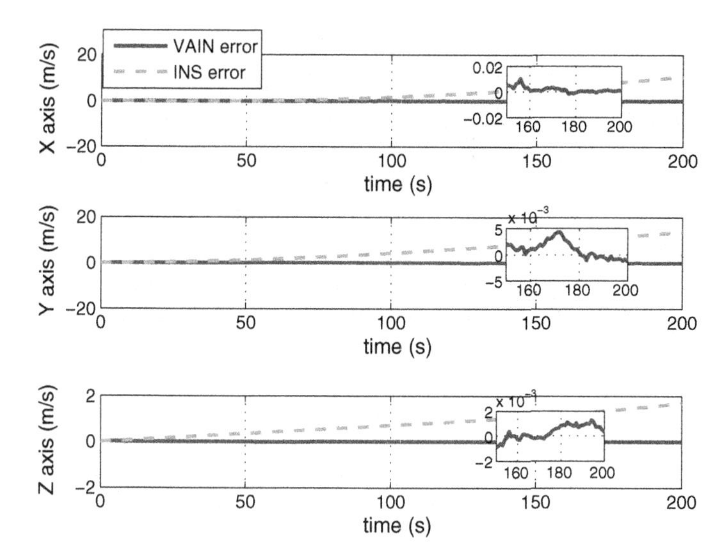

Fig. 41.3 Velocity estimation errors

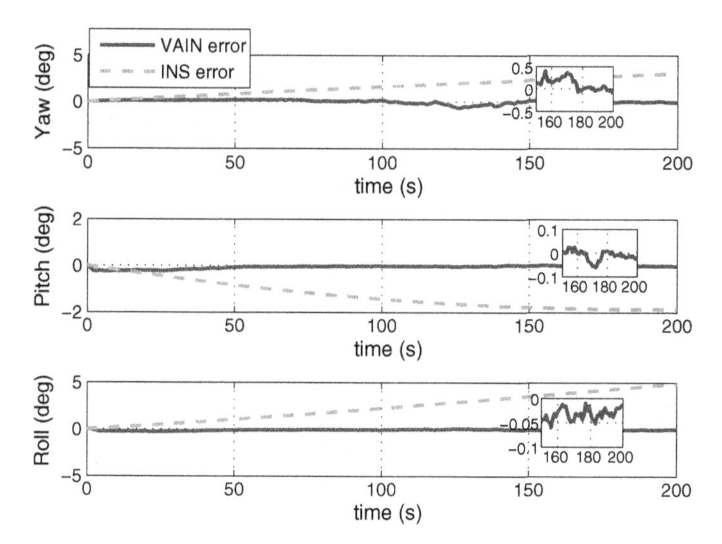

Fig. 41.4 Attitude angle estimation errors

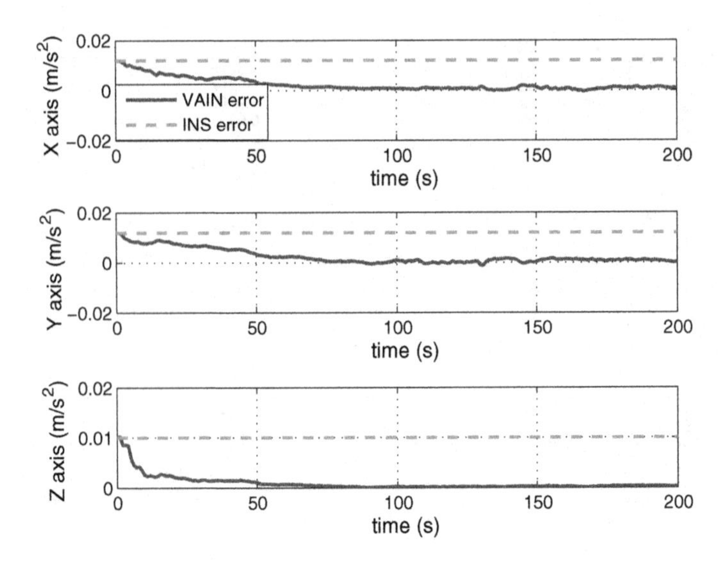

Fig. 41.5 Accelerometer bias estimation errors

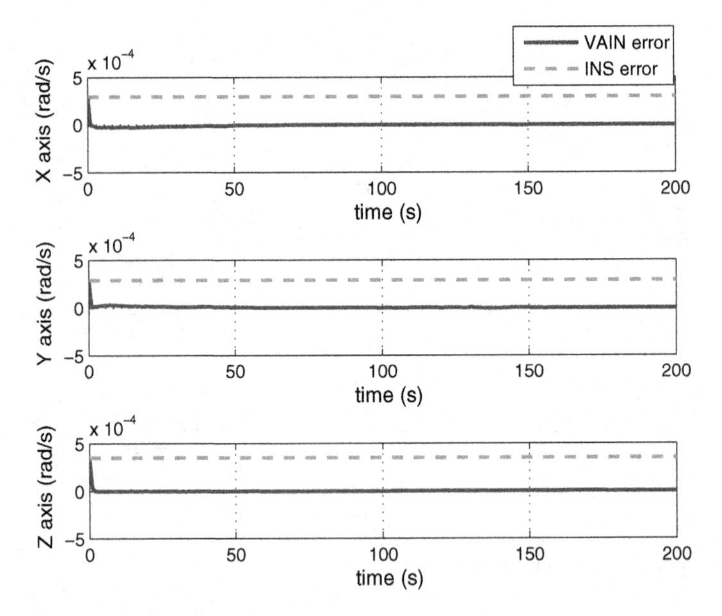

Fig. 41.6 Gyroscope bias estimation errors

41.5 Conclusions

A new autonomous navigation scheme based on two sequential images captured by a stereo camera with an IMU for planetary landing in unknown terrain environment is proposed. In this scheme, the IMU outputs are exploited for dead reckoning in an INS. Observations of landmarks observed in two sequential images by the stereo camera are utilized in combination with the corresponding INS pose estimations to formulate implicit measurement equations. Then, the implicit measurement equations and an augmented IEKF are used to estimate the lander's pose and velocity and to correct the drift biases of the IMU . The simulation results demonstrated that the proposed navigation scheme vastly improves the planetary landing navigation accuracy compared to the INS only used.

References

1. Wang D, Huang X, Guan Y (2008) GNC system scheme for lunar soft landing spacecraft. Adv Space Res 42(2):379–385
2. Busnardo DM, Aitken ML, Tolson RH et al (2011) LIDAR-aided inertial navigation with extended Kalman filtering for pinpoint landing. In: 49th AIAA aerospace science meeting including the new horizons forum and aerospace exposition, Orlando, Florida
3. Trawny N, Mourikis AI, Roumeliotis SI (2007) Vision-aided inertial navigation for pin-point landing using observations of mapped landmarks. J Field Robot 24(5):357–378
4. Pham VB, Lacroix S, Devy M (2012) Vision-based absolute navigation for descent and landing. J Field Robot 29(4):627–647
5. Yu M, Cui HT, Tian Y (2014) A new approach based on crater detection and matching for visual navigation in planetary landing. Adv Space Res 53(12):1810–1821
6. Kim J, Sukkarieh S (2004) Autonomous airborne navigation in unknown terrain environments. IEEE Trans Aerosp Electron Syst 40(3):1031–1045
7. Amidi O, Kanade T, Fujita K (1999) A visual odometer for autonomous helicopter flight. Robot Auton Syst 28:185–193
8. Indelman V, Gurfil P, Rivlin E et al (2012) Real-time vision-aided localization and navigation based on three-view geometry. IEEE Trans Aerosp Electron Syst 48(3):2239–2259
9. Mourikis AI, Trawny N, Roumeliotis SI, Johnson AE et al (2009) Vision-aided inertial navigation for spacecraft entry, descent, and landing. IEEE Trans Robot 25(2):264–280
10. Pini G, Hector R (2001) Partial aircraft state estimation from visual motion using the subspace constraints approach. J Guid Control Dyn 24(5):1016–1028
11. Roumeliotis SI, Burdick JW (2002) Stochastic cloning: a generalized framework for processing relative state measurements, In: IEEE international conference on robotics and automation, Washington, DC, pp. 1788–1795

Chapter 42
Strong Structural Controllability and Leader Selection for Multi-agent Systems with Unidirectional Topology

Peng Liu, Yuping Tian and Ya Zhang

Abstract For unidirectional communication topology of multi-agent systems, this paper studies its strong structural controllability. When the topology of agents is a pabud graph, we prove that the multi-agent systems can be strongly structurally controllable by selecting only one agent as leader. When the topology is partitioned to disjoint basic controllable graphs, the system can be strongly structurally controllable via selecting corresponding number of agents as leaders. A method to select leaders is presented to ensure the strong structural controllability of multi-agent systems. Finally, the effectiveness of the proposed method is verified with two examples.

Keywords Strong structural controllability · Multi-agent systems · Leader selection

42.1 Introduction

Recently, the dynamical graph-based method has become popular because of its potential application to various problems such as complex networks [1, 2], multi-agent systems [3–7], etc. For example, Liu, Slotine and Barabasi [1] investigated the structural controllability of complex directed networks, and exploited the graph-theoretic technique of maximum matching to determine the minimum number of inputs for maintaining full control over the network. And the network controllability is decided by the nodes with low density [2].

Multi-agent system is a special kind of networked system. In order to study its controllability, we choose one or more agents as leaders and other individuals as followers. The leaders are controlled by external input while the followers are controlled by other followers and leaders. Previous literature indicated that connectivity had some effect on controllability. Tanner [3] has shown that a path topology is controllable. At the same time, he also pointed out that a complete graph was uncontrollable.

P. Liu · Y. Tian (✉) · Y. Zhang
School of Automation, Southeast University, Nanjing 210096, China
e-mail: yptian@seu.edu.cn

Y. Jia et al. (eds.), *Proceedings of the 2015 Chinese Intelligent Systems Conference*, Lecture Notes in Electrical Engineering 360,
DOI 10.1007/978-3-662-48365-7_42

Rahmani and Mesbahi extended Tanner's results and pointed out that the symmetrical structure of the network influenced the network controllability [4, 5]. Moreover, the controllability was investigated by equitable partitions [6] and relaxed equitable partitions [7]. These two methods provided the interpretation of controllability with graph theory.

How to control the multi-agent systems was investigated by Laplacian matrix decomposition [3–7]. Sometimes, we only concerned with the existence of links and didn't care about the concrete weight values. So the structured system was a tool to analyze this problem [8, 9]. On the structural controllability, Lin [8] gave the graphic characteristics of single-input systems, and Shields and Pearson [9] generalized Lin's result to multi-input systems, for details see [10]. Zamani and Lin [11] showed that the controllability completely depended on the topology of the communication scheme. And they provided a necessary and sufficient condition based on graph. For systems with agents loss or links fail, some results were stated in [12, 13]. Structural controllability could be studied by graph theory, and if a topology is structurally controllable, it is controllable for almost all choice of the weights of the edges [10, 13]. But they may fail to hold if certain symmetries or constraints are presented [6, 7]. These disadvantages can be avoided by strong structural controllability which was firstly put forward in [14], and a graphic characteristic was given for one dimension input systems. This result was extended to multiple dimensions input systems and applied to inverted pendulum [15]. Another method was to use knowledge of cycle families. And the graphic conditions and algebraic conditions were presented for the strong structural controllability of more general systems [16, 17]. Chapman and Mesbahi [18] used constraint matching method to investigate the strong structural controllability, and an algorithm was presented to verify the controllability. Recently, Monshizadeh et al. [19] established a one-to-one correspondence between the set of leaders rendering the network controllable and zero forcing sets.

There are only a few results on strong structural controllability of multi-agent systems. Here we mainly focus on the 'pabud' and a class of topology which can be partitioned to disjoint paths and pabuds, and solve two problems. One is to investigate the minimum number of agents being leaders, which are controlled by external inputs, for maintaining full control. The other is how to select leaders. Our works are inspired by [20], which gave three well-defined clusters to study the control properties of complex network. Monshizadeh et al. [19] have applied the zero forcing sets algorithm to control all the initial nodes of disjoint paths. From their approach, the system with pabud topology requires two leaders. This paper applies graph theory and algebraic criterion of controllability to discuss the strong structural controllability. We prove that for systems with 'pabud' graph only one leader is needed for network controllability. A method how to place the needed leaders is presented, and the number is minimum for some special topologies.

42.2 Preliminaries and Problem Formulation

In this section, preliminary knowledge of graph theory and controllability criterion of linear systems are presented for convenience, and the structural knowledge is also stated for the next sections.

42.2.1 Graph Theory and System Graph

In multi-agent systems, we use a directed graph to represent the relationships among agents. A directed edge (i,j) implies that node i transmit information to node j, the nodes i and j are called the parent and child of (i,j) respectively. We denote node i's in-neighbors as $\mathcal{N}_i = \{j|(j,i) \in \mathcal{E}\}$. If $i = j$, this edge is called loop or self-loop. A path $x \rightarrow y$ is a sequence of distinct vertices where consecutive vertices are adjacent. And its length r means that $r + 1$ distinct vertices starting with x and ending with y. Two paths from $v_1 \in \mathcal{V}$ to $v_2 \in \mathcal{V}$ are called disjoint if they consist none common node except v_1 and v_2. When exists l paths between $\mathcal{V}_1$ and $\mathcal{V}_2$, they are disjoint means any two of them disjoint. A path is called a cycle if it have the same starting vertex and ending vertex. And a acyclic graph is the graph which contains no any cycle. If the child of a edge is the node of a cycle or the child have a loop, then the graph of this form will be called a bud. A node is called root if it have no parent. The nodes of no child in a graph are called leaves. For a directed tree, the paths that start with the root node are called the first-level branch, the paths that start with the first-level branches are called the second-level branch, and so on.

Given system matrix A, there exists a directed edge (j, i) if the component $a_{ij} \neq 0$. And there is a link from input u to state i, if the ith row $b_{i,\cdot}$ of input matrix B is nonzero. Without loss of generality, we use A to represent the numerical matrix, $\bar{A}$ to denote the structural matrix (i.e. the entries are zero or nonzero). In order to describe the structured system, we usually denote by $\mathcal{X} = \{x_1, \ldots, x_n\}$ and $\mathcal{U} = \{u_1, \ldots, u_m\}$ the sets of state and input vertices, respectively. And by $\mathcal{E}_{\mathcal{X},\mathcal{X}} = \{(x_i, x_j) : \bar{a}_{ji} \neq 0\}$, $\mathcal{E}_{\mathcal{U},\mathcal{X}} = \{(u_j, x_i) : \bar{b}_{ij} \neq 0\}$, the sets of edges between the vertex sets in subscript. We called the directed graph $\mathcal{D}(\bar{A}) = (\mathcal{X}, \mathcal{E}_{\mathcal{X},\mathcal{X}})$ as state digraph and $\mathcal{D}(\bar{A}, \bar{B}) = (\mathcal{X} \cup \mathcal{U}, \mathcal{E}_{\mathcal{X},\mathcal{X}} \cup \mathcal{E}_{\mathcal{U},\mathcal{X}})$ system digraph. Note that the isolated vertices of the digraph $\mathcal{D}(\bar{A}, \bar{B})$ representing the input vertices associated with zero columns of $\bar{B}$. A state vertex is called input-accessible if there exists a path from some input vertex to this vertex in the graph $\mathcal{D}$. Two systems $(\bar{A}, \bar{B})$ and $(\tilde{A}, \tilde{B})$ have the same structure, it means that for every fixed (zero) entry of matrix $[\bar{A}, \bar{B}]$ the corresponding entry of the matrix $[\tilde{A}, \tilde{B}]$ is fixed (zero) and reverse is also right [8].

42.2.2 Problem Formulation

In this paper, we will study leader selection problem when the agents's topology is known. Consider a set of agents which are modeled with single integrator by

$$\dot{x}_i(t) = u_i(t), i \in \mathbb{N} = \{1, 2, \ldots, n\} \tag{42.1}$$

where $x_i(t)$ is called the agent's state and $u_i(t)$ is called the control input of agent i, respectively. Assume there are m agents directly controlled by external control inputs $u^s_{ext}(t), s \in \mathbb{N}^* = \{1, \ldots, m\}$. These m agents are called leaders, the remaining agents are called followers. The value of m will be determined by the topology of agents, see [19] for details.

The followers' input $u_i(t)$ has the following form

$$u_i(t) = \sum\nolimits_{j \in \{\mathcal{N}_i \cup \{i\}\}} a_{ij}x_j(t) + \sum\nolimits_{j=1}^{m} b_{ij}u^j_{ext}(t),$$

where the coefficients $a_{ij}, b_{ij} \in \mathbb{R}$ are zero or indeterminate component. $a_{ij} = 0$ denotes agent j be not the neighbor of agent i. If the external input j doesn't act on agent i directly, then $b_{ij} = 0$, otherwise $b_{ij} \neq 0$.

Given the control law and stacking all the variables, the dynamic equation of the whole system can be described as

$$\dot{x}(t) = \bar{A}x(t) + \bar{B}u(t) \tag{42.2}$$

where $x(t) = [x_1(t), \ldots, x_n(t)]^T \in \mathbb{R}^n$ denotes the state vector stacked with all followers and leaders; $u(t) = [u^1_{ext}(t), \ldots, u^m_{ext}(t)]^T \in \mathbb{R}^m$ denotes the input vector composed with all external control. The system matrix $\bar{A} = [a_{ij}] \in \mathbb{R}^{n\times n}$, the input matrix $\bar{B} = [b_{ij}] \in \mathbb{R}^{n\times m}$. We assume that there's at most one nonzero element in $b_{i1}, \ldots, b_{im}$.

In this paper, we consider the strong structural controllability of $(\bar{A}, \bar{B})$. We firstly give the definition and criterion of structural controllability for linear systems.

Definition 1 ([8, 9]) The system $(\bar{A}, \bar{B})$ is said to be structurally controllable if there exists a controllable system (A_0, B_0) which has the same structure with the system $(\bar{A}, \bar{B})$.

Definition 2 ([14]) The system $(\bar{A}, \bar{B})$ is said to be strongly structurally controllable if any system (A_0, B_0) which has the same structure with the system $(\bar{A}, \bar{B})$ is controllable.

Obviously, when the systems have the same structure, the strong structural controllability implies the structural controllability, but the reverse is not true.

In the following, we will study how many leaders are needed and how to select agents as leaders. In other words, we consider how to get matrix $\bar{B}$ such that $(\bar{A}, \bar{B})$ be strongly structurally controllable. Obviously, the Eq. (42.2) is a structured system [8, 14]. All agents will be controllable when the system (42.2) is strongly structurally controllable.

42.3 Main Results

In this section, a basic graph of agents is introduced. And we show that it just needs one leader such that system (42.2) is strongly structurally controllable. There have been some results known to control the 'path' graph ([19, 20]).

Here we focus on the 'pabud' graph and the topology which can be partitioned to paths and pabuds.

42.3.1 Basic Graphs and Graph Partition

We introduce two basic graphs and the definition of graph partition. In this paper, the basic graphs are path and pabud.

Definition 3 A directed graph is called pabud, if a bud is linked to some node of a path.

Remark 1 The graph 'pabud' is composed of a bud and a path, and the definition of bud see [8]. The pabud is different from the 'cactus' in [8]. The bud can link to any vertex in the path to form the pabud (see Fig. 42.1) while the bud can't link to the last node in 'cactus' ([8]). Adding one edge to some branch of the Y-dilation we get the graph of pabud which contains a self-loop or cycle.

Definition 4 Disjoint subsets $C_1, \ldots, C_k$ is a partition of C if $C = \cup_{i=1}^{k} C_i$ and $C_i \neq \emptyset$ for every i. We say subgraphs $\mathcal{G}_1, \ldots, \mathcal{G}_k$ are partitions of graph $\mathcal{G}$, if $\mathcal{V}(\mathcal{G}_i) \cap \mathcal{V}(\mathcal{G}_j) = \emptyset$ and $\cup_{i=1}^{k} \mathcal{V}(\mathcal{G}_i) = \mathcal{V}(\mathcal{G})$.

For simplicity of proof, this paper only investigates the case with unidirectional links between C_i and C_j, $i \neq j$. Here some examples are presented to clarify the definition of partition. First, the topology can be partitioned to the union of disjoint paths (see Fig. 42.2). The dashed blue rectangle and dashed red ellipse are two different partitions. The zero forcing number in [19] is the minimum number of paths by partition. Second, the topology can be partitioned to the union of disjoint paths and cycles (see Fig. 42.3). The dashed blue rectangle is a partition of the graph. And for any cycle, it contains no chords. Here, the chord of a cycle is the edge between

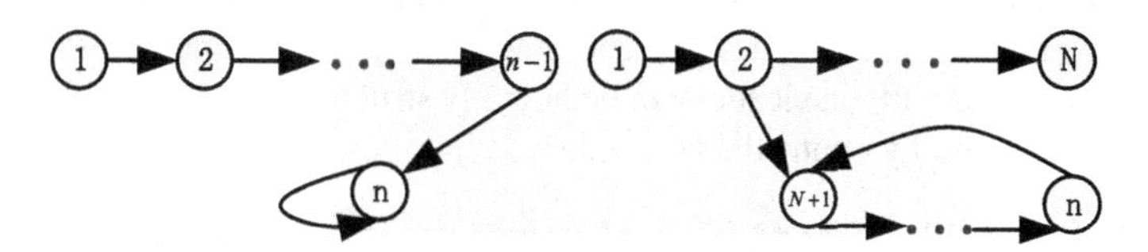

Fig. 42.1 The pabud communication topology

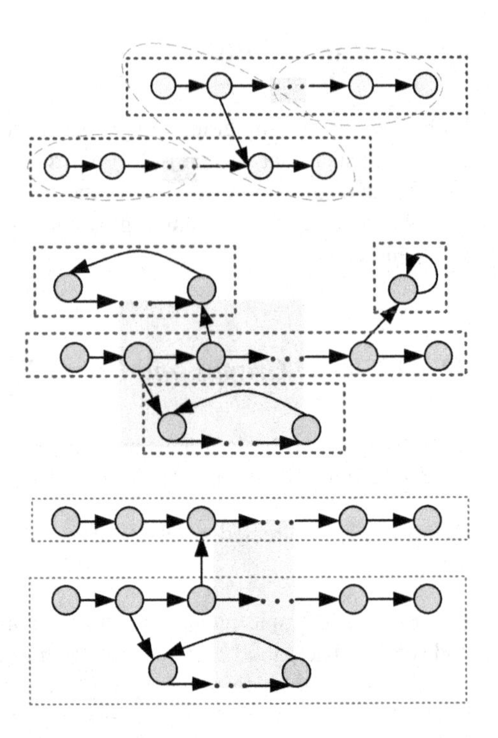

Fig. 42.2 A topology is partitioned to paths

Fig. 42.3 A topology is partitioned to paths and cycles

Fig. 42.4 A topology is partitioned to paths and pabuds

different adjacency nodes. Third, the topology can be partitioned to the union of disjoint paths and pabuds (see Fig. 42.4). The dashed red rectangle is a partition of the graph. Here 'partition' means that there are directed edges from a basic topology configuration to another one, and the parent and child nodes of these directed edges could't be composed of a basic topology. So it can be avoided of being a new basic topology. We known that there maybe one or more partitions for a topology. In order to select minimal leaders in multiple agents, we assume the graph is partitioned with minimum different basic controllable graphs. Note that any graph can be partitioned to one or more types of the above partitions.

42.3.2 *Leader Selection for Strong Structural Controllability*

This subsection focuses on two issues. One is to prove the strong structural controllability of pabud with one leader. The other is to provide a method to place the leaders. The pabud with one leader can be not only structurally controllable, but also be strongly structurally controllable.

Proposition 1 *1. When the topology of agents is a path or pabud, we only need one leader such that the system (42.2) is strongly structurally controllable.*

2. *When the digraph of agents is partitioned to the union of disjoint paths and cycles, then we need at most N_D leaders such that the system (42.2) is strongly structurally controllable, where N_D is the number of the disjoint paths and cycles.*

Proof (1) We use the algebraic criterion of controllability to derive the results. When the graph is a path, we set the starting node as leader and get the matrix $[\bar{A}, \bar{B}]$ as the following form:

$$\bar{A} = \begin{bmatrix} 0 & 0 & 0 & \cdots & 0 & 0 \\ a_{21} & 0 & 0 & \cdots & 0 & 0 \\ 0 & a_{32} & 0 & \cdots & 0 & 0 \\ \vdots & \vdots & \vdots & \cdots & \vdots & \vdots \\ 0 & 0 & 0 & \ldots & a_{nn-1} & 0 \end{bmatrix}, \quad \bar{B} = \begin{bmatrix} b_1 \\ 0 \\ 0 \\ \vdots \\ 0 \end{bmatrix}.$$

and the associated controllability matrix of structured system $(\bar{A}, \bar{B})$ is

$$C_M = [\bar{B}\ \bar{A}\bar{B}\ \bar{A}^2\bar{B}\ \cdots \bar{A}^{n-1}\bar{B}] = \begin{bmatrix} b_1 & 0 & 0 & \cdots 0 & 0 \\ 0 & a_{21}b_1 & 0 & \cdots 0 & 0 \\ 0 & 0 & a_{32}a_{21}b_1 & \cdots 0 & 0 \\ \vdots & \vdots & \vdots & \cdots \vdots & \vdots \\ 0 & 0 & 0 & \ldots 0 & a_{nn-1} \cdots a_{21}b_1 \end{bmatrix}.$$

No matter what the value of indeterminate entries take, the diagonal entries are nonzero as long as all the indeterminate entries are nonzero. So the system (42.2) is strongly structurally controllable.

Next we discuss the case when the graph of followers is a pabud. If we set the initial agent as leader, we can get the matrix $\bar{B}$. The system graph associated with the matrix pair $(\bar{A}, \bar{B})$ has four different forms as follows: a self-loop links to the last node or any other node on a path with exactly one edge; a directed cycle links to the last node or any other node on a path with exactly one edge. Without loss of generality, we prove the second and the fourth cases. For simplicity, we firstly consider a papud composed of a path with four nodes and a bud with a self-loop linking to the second node. So

$$\bar{A}_1 = \begin{bmatrix} 0 & 0 & 0 & 0 & 0 \\ a_{21} & 0 & 0 & 0 & 0 \\ 0 & a_{32} & 0 & 0 & 0 \\ 0 & 0 & a_{34} & 0 & 0 \\ 0 & a_{52} & 0 & 0 & a_{55} \end{bmatrix}, \bar{B}_1 = \begin{bmatrix} b_1 \\ 0 \\ 0 \\ 0 \\ 0 \end{bmatrix}$$

and the controllability matrix

$$C_M = \left[\bar{B}_1\ \bar{A}_1\bar{B}_1\ \bar{A}_1^2\bar{B}_1\ \bar{A}_1^3\bar{B}_1\ \bar{A}_1^4\bar{B}_1\right] =$$
$$\begin{bmatrix} b_1 & 0 & 0 & 0 & 0 \\ 0 & a_{21}b_1 & 0 & 0 & 0 \\ 0 & 0 & a_{32}a_{21}b_1 & 0 & 0 \\ 0 & 0 & 0 & a_{43}a_{32}a_{21}b_1 & 0 \\ 0 & 0 & a_{52}a_{21}b_1 & a_{55}a_{52}a_{21}b_1 & a_{55}^2a_{52}a_{21}b_1 \end{bmatrix}.$$

Next, we consider a papud composed of a path with three nodes and a bud with a cycle linking to the second node. For simplicity, we assume the cycle contains two nodes. The matrices are as following:

$$\bar{A}_2 = \begin{bmatrix} 0 & 0 & 0 & 0 & 0 \\ a_{21} & 0 & 0 & 0 & 0 \\ 0 & a_{32} & 0 & 0 & 0 \\ 0 & a_{42} & 0 & 0 & a_{45} \\ 0 & 0 & 0 & a_{54} & 0 \end{bmatrix}, \bar{B}_2 = \begin{bmatrix} b_1 \\ 0 \\ 0 \\ 0 \\ 0 \end{bmatrix}$$

and the controllability matrix

$$C_M = \left[\bar{B}_2\ \bar{A}_2\bar{B}_2\ \bar{A}_2^2\bar{B}_2\ \bar{A}_2^3\bar{B}_2\ \bar{A}_2^4\bar{B}_2\right] =$$
$$\begin{bmatrix} b_1 & 0 & 0 & 0 & 0 \\ 0 & a_{21}b_1 & 0 & 0 & 0 \\ 0 & 0 & a_{32}a_{21}b_1 & 0 & 0 \\ 0 & 0 & a_{42}a_{21}b_1 & 0 & a_{45}a_{54}a_{42}a_{21}b_1 \\ 0 & 0 & 0 & a_{54}a_{42}a_{21}b_1 & 0 \end{bmatrix}.$$

Obviously, for any nonzero value of indeterminate entries, the diagonal entries of the above two controllability matrices are nonzero. And then the above two systems are controllable.

Note that here we consider a pabud with five nodes. In fact this result can be extended to a pabud composed of a path with N nodes and a cycle with m nodes. The bud links to the ith node which has three cases: $N - i > m, N - i = m$ and $N - i < m$. No matter which case, we can find appropriate permutation matrices P_1 and P_2 such that $P_1 C_M P_2$ be full rank. So the system (42.2) is strongly structurally controllable. For brevity, the proof of another two cases is omitted.

(2) From the properties of path and pabud and the results of reference [22] (i.e. an uncontrollable system can be transformed to controllable via adding appropriate external input), we only need to control the initial node of every path and any node of the cycle such that the system (42.2) is strongly structurally controllable. Thus, we need at most N_D leaders. □

Remark 2 When the topology of agents is a pabub, by applying the zero forcing sets method in [19], two leaders are needed. While Proposition 1 shows that in our approach just one leader can be satisfactory.

For notational convenience, we denote $\mathcal{V}_L$ as the set of needed leaders, and $\mathcal{V}(\mathcal{G}_{\bar{A}})$ denotes the set of all agents.

Proposition 2 *If the topology of agents $\mathcal{G}_{\bar{A}}$ can be partitioned to the union of disjoint paths and pabuds, and let $\mathcal{V}_L(\subseteq \mathcal{V}(\mathcal{G}_{\bar{A}}))$ be the set of all initial nodes of the paths and pabuds, then the system $(\bar{A}, \bar{B})$ is strongly structurally controllable.*

Proof For the graph which is partitioned to the union of disjoint paths and pabuds, rearranging the nodes, matrix $\bar{A}$ can be transformed to block lower triangular matrix. Then the proof follows from the simple inspection and it is omitted here. □

Remark 3 If the topology of agents $\mathcal{G}_{\bar{A}}$ is tree, then the minimum number of leaders equals to the leaves' number. And if the topology of agents $\mathcal{G}_{\bar{A}}$ is the union of disjoint paths and pabuds, then the minimum number of leaders is the number of paths and pabuds.

In the following subsection, we give a method to get the number of the leaders and their placement. Here, we review the notion of zero forcing sets together with the involved notations and terminology which will be used in the next method. For more details see e.g. [19, 21].

Let $\mathcal{G}$ be a given digraph, where each vertex is colored either black or white. Consider the following coloring rule:

$\star$: Assume u is a black vertex and it has exactly one white out-neighbor v, we change the color of v to black.

Now we present a method to determine the number of leaders by finding the spanning forest with most leaves. The number of spanning trees of this spanning forest is equivalent to the number of nodes that have no in-degree. Assume the topology of agents is known, which has the sparse structure (i.e. the basic controllable graph or their combination). The following method is to choose an appropriate topology, and derives the number and positions of leaders.

Algorithm 1.

Step 1: Finding a spanning forest with most leaves on the topology of agents.

Step 2: Let the leaves of the spanning forest as the starting number of leaders. Denoting all the roots and the first node in each branch as black nodes.

Step 3: For each subgraph, which contains some spanning tree and all the edges between these nodes on the original topology, run the following iteration: First, choose one path from all the first-level branches. Next, the black node forcing the next node in the topology with the coloring rule. If there exists one cycle or self-loop containing the black nodes, or one forcing way passes through other black nodes, then the number of leaders minus one corresponding number of black nodes. Continue the process on the second-level branches, until check the final-level branches.

Remark 4 For the pabud topology, the nonzero elements of $\bar{B}$ is one with this method, while the nonzero elements of $\bar{B}$ is two with zero forcing sets in [19]. For the topologies which is the union of disjoint minimum paths and pabuds, the method can give the number of leaders and is also minimum. And for general directed topology, the method could provide enough number of leaders for strong structural controllability. But this number is not the minimum, as some cycles maybe have chords when decomposing the graph to a directed acyclic graph. And what effect of the chords to the strong structural controllability will be investigated in future.

42.4 Illustrative Examples

This section aims to show the correctness of our method, and two examples are given to illustrate the effectiveness of the algorithm. Assuming there are seven agents, and the topology structure is shown in Fig. 42.5.

Obviously, we can partition the graph to two disjoint basic controllable subgraphs. From Remark 3, if the system (42.2) is strongly structurally controllable, the minimum number of leaders is not larger than two. While the minimum number of leaders is three and the leaders are agent 1, 4 and 7 from the algorithm in [19].

By the placement method in this paper, we find a spanning forest in Fig. 42.6. It contains two spanning trees: $\mathcal{T}_1 = (\mathcal{V}_1, \mathcal{E}_1)$ with $\mathcal{V}_1 = \{1, 2, 3\}$ and $\mathcal{E}_1 = \{(1, 2),$ $(2, 3)\}$, $\mathcal{T}_2 = (\mathcal{V}_2, \mathcal{E}_2)$ with $\mathcal{V}_2 = \{4, 5, 6\}$ and $\mathcal{E}_2 = \{(4, 5), (5, 6), (5, 7)\}$. Firstly, from the step 1 and 2 of the method, agent 1, 4 and 7 are the initial set of leaders. Then running step 3 in the method on the subgraph composed by agents 4, 5, 6, 7 and edges set $\{\mathcal{E}_2 \cup (7, 7)\}$, it knows that agent 7 can be deleted in the leaders set $\{1, 4, 7\}$ when the system (42.2) is strongly structurally controllable.

It is shown that the number of leaders and their placement can be determined with the method. This number can be smaller than that obtained by the zero forcing sets algorithm in [19].

Next, we consider five different graphs of agents to compare the number of leaders with the method in this paper and the algorithm in reference [19]. $\mathcal{G}_p$ denotes a path with unidirectional, $\mathcal{G}_c$ denotes a cycle with unidirectional, $\mathcal{G}_t$ denotes a directed tree with n leaves, $\mathcal{G}_{ca}$ denotes a like-cactus with n branches, here the like-cactus is a directed graph that different pabud linking to the path of another pabud, and $\mathcal{G}_{tc}$

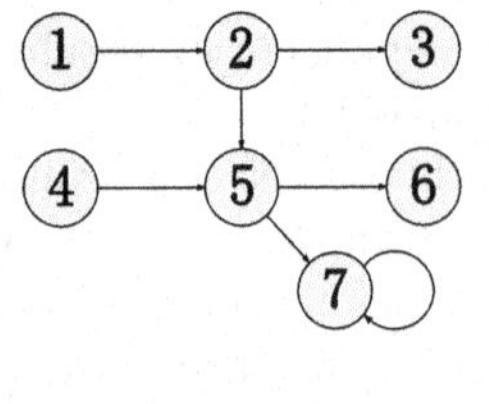

Fig. 42.5 The combination of a path and a pabud

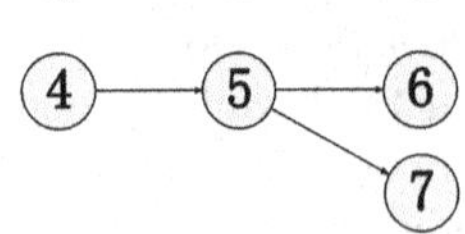

Fig. 42.6 The spanning forest

Table 42.1 The number of leaders

$\mathcal{G}$	[19]	Algorithm 1
$\mathcal{G}_p/\mathcal{G}_c$	1	1
$\mathcal{G}_t$	n	n
$\mathcal{G}_{ca}$	$2n$	n
$\mathcal{G}_{tc}$	n	$n-N$

denotes a directed binary tree with n leaves and N ($N \leq n/2$) cycles or self-loops in the n leaves associated with different parents.

From Table 42.1, we know that when the digraph has no cycles and self-loops the numbers of leaders needed in [19] and this paper are same. When the digraph of followers has cycles or self-loops, the leaders in our approach are less.

42.5 Conclusion

This article mainly discusses the strong structural controllability of multi-agent systems. And how to get the number of leaders and their placement is given under the assumption that the topology of agents is known. The number of leaders for achieving the strong structural controllability is the number of paths and pabuds by graph partition, and the leaders can be the initial agents of corresponding paths and pabuds. Furthermore, the positions of leaders are given via combining the first search and coloring rule. At last, we use two examples to demonstrate the feasibility of the proposed methods.

References

1. Liu YY, Slotine JJ, Barabasi AL (2011) Controllability of complex networks. Nature 473(7346):167–173
2. Menichetti G, DallAsta L, Bianconi G (2014) Network controllability is determined by the density of low in-degree and out-degree nodes. Phys Rev Lett 113:078701
3. Tanner H (2004) On the controllability of nearest neighbor interconnection. In: Proceedings of the 43rd IEEE conference on decision and control. pp. 2467–2472
4. Rahmani A, Mesbahi M (2006) On the controlled agreement problem. In: Proceedings of the american control conference. pp. 1376–1381
5. Rahmani A, Mesbahi M (2007) Pulling the strings on agreement: anchoring, controllability and graph automorphisms. In: Proceedings of the american control conference. pp. 2738–2743
6. Rahmani A, Ji M, Mesbahi M, Egerstedt M (2009) Controllability of multi-agent systems from a graph-theoretic perspective. SIAM J Control Optim 48(1):162–186
7. Martini S, Mesbahi M, Rahmani A (2010) Controllability analysis of multi-agent systems using relaxed equitable partitions. Int J Syst Control Commun 2(1–3):100–121
8. Lin C-T (1974) Structural controllability. IEEE Trans Autom Control 19(3):201–208
9. Shields RW, Pearson JB (1976) Structural controllability of multiinput linear systems. IEEE Trans Autom Control 21(2):203–212

10. Dion JM, Commault C, Van Der Woude J (2003) Generic properties and control of linear structured systems: a survey. Automatica 39(7):1125–1144
11. Zamani M, Lin H (2009) Structural controllability of multiagent systems. In: Proceedings of the american control conference. pp. 5743–5748
12. Jafari S, Ajorlou A, Aghdam AG (2011) Leader localization in multi-agent systems subject to failure: a graph-theoretic approach. Automatica 47(8):1744–1750
13. Rahimian MA, Aghdam AG (2013) Structural controllability of multi-agent networks: Robustness against simultaneous failures. Automatica 49(11):3149–3157
14. Mayeda H, Yamada T (1979) Strong structural controllability. SIAM J Control Optim 17(1):123–138
15. Bowden C, Holderbaum W, Becerra V (2012) Strong structural controllability and the multilink inverted pendulum. IEEE Trans Autom Control 57(11):2891–2896
16. Reinschke KJ, Svaricek F, Wend HD (1992) On strong structural controllability of linear systems. In: Proceedings of the 31st IEEE conference on decision and control. pp. 203–208
17. Jarczyk JC, Svaricek F, Alt B (2011) Strong structural controllability of linear systems revisited. In: Proceedings of the 50th IEEE conference on decision and control. pp. 1213–1218
18. Chapman A, Mesbahi M (2013) On strong structural controllability of networked systems: a constrained matching approach. In: Proceedings of the american control conference. pp. 6126–6131
19. Monshizadeh N, Zhang S, Camlibel MK (2014) Zero forcing sets and controllability of dynamical systems defined on graphs. IEEE Trans Autom Control 59(9):2562–2567
20. Ruths J, Ruths D (2014) Control profiles of complex networks. Science 343:1373
21. Barioli F, Barrett W, Fallat SM, Hall HT, Hogben L, Shader B, Van Den Driessche P, Van Der Holst H (2010) Zero forcing parameters and minimum rank problems. Linear Algebra Appl 433(2):401–411
22. Commault C, Dion JM (2013) Input addition and leader selection for the controllability of graph-based systems. Automatica 49(11):3322–3328

Chapter 43
Group Consensus Control in Uncertain Networked Euler-Lagrange Systems Based on Neural Network Strategy

Jinwei Yu, Jun Liu, Lan Xiang and Jin Zhou

Abstract This paper investigates the group consensus problem for a network consisting of Euler-Lagrange systems under directed topology with acyclic partition via neural network strategy. The neural network based controller achieves group consensus for uncertain networked Euler-Lagrange systems. By exploiting thoroughly the specific structure of the network topology, the stable analysis of the group consensus problem for such uncertain networked systems is also provided. Furthermore, a necessary and sufficient condition for ensuring that the systems reach group consensus is established. Finally, examples and simulations are given to show the effectiveness of the presented theoretical results.

Keywords Group consensus · Networked Euler-Lagrange systems · Neural network · Stable analysis

43.1 Introduction

As a specific consensus problem, group consensus in networked multi-agent systems has successfully given rise to tremendous attentions from diverse field backgrounds [1–3]. So-called group consensus demands that all the agents in the same group converge to the same trajectory, but the final states of different groups may not coincide. This mechanism can well formulate many phenomena in the real world. For example, in a large-scale industrial manufacturing, a large swarm of robots is required to

J. Yu · J. Liu · J. Zhou (✉)
Shanghai Institute of Applied Mathematics and Mechanics,
Shanghai University, Shanghai 200072, China
e-mail: jzhou@shu.edu.cn

J. Liu
Department of Mathematics, Jining University, Qufu 273155, Shandong, China

L. Xiang
Department of Physics, School of Science, Shanghai University,
Shanghai 200444, China

Y. Jia et al. (eds.), *Proceedings of the 2015 Chinese Intelligent Systems Conference*, Lecture Notes in Electrical Engineering 360,
DOI 10.1007/978-3-662-48365-7_43

achieve a cooperative task of multiple sophisticated subtasks, where the designed group consensus control scheme is effectively implemented to divide a large swarm of robots into multiple groups such that all agents in different groups can perform the corresponding sophisticated subtasks. Consequently, many valuable results on this subject have been reported in the last decade [1–3].

As is well known that, many physical or mechanical systems can be well characterized by the Euler-Lagrange dynamics. The networked Euler-Lagrange systems possess extensive engineering applications, especially involving complex and integrated production processes, such as coordination of multiple manipulators, formation of flying spacecrafts, mobile sensor networks and so on. This has resulted in tremendous amount of interest for the consensus problem in networked Euler-Lagrange systems [4–6, 8]. On the other hand, it is quite common in practice for a group of robots to accomplish a cooperative work, the dynamics of robotic systems change in time due to uncertain parameters caused by a large variety of physical environments and external disturbance, etc. In addition, it is accepted that the neural network (NN) control has been extensively used to deal with nonlinear dynamical systems with uncertain dynamics for its multifunction distinguishing feature, for instance, strong learning and concurrent computation ability in NN technology [7, 9, 11–13].

With the aforementioned background, in this brief a distributed group consensus protocol for uncertain networked Euler-Lagrange systems based on neural network control will be designed. A feedforward neural network will be employed to guarantee that all agents with acyclic segmentation reach group consensus, or even cluster consensus. The notable features of the proposed control scheme are described as follows: (i) both the exact knowledge of system dynamics and the tedious preliminarily analysis to determine a regression are not required; (ii) the feedforward NN used in control strategy can learn the uncertain nonlinear mechanical systems adaptively; (iii) the global stable analysis can be established for such NN-like controller under some reasonable assumptions.

The structure of this brief is as follows. Section 43.2 gives the mathematical preliminary. Section 43.3 focuses on the group consensus of networked Euler-Lagrange systems. Examples and simulations are given to validate the corresponding NN control algorithm in Sect. 43.4. Section 43.5 is finally left for conclusions.

43.2 Problem Formulation

Some basis notations and necessary graph theory have been employed in [10], now they are just introduced into this brief.

Suppose a network consist of d Euler-Lagrange systems, and the motion equation of ith Euler-Lagrange system is compactly given by

$$M_i(q_i)\ddot{q}_i + C_i(q_i, \dot{q}_i)\dot{q}_i + g_i(q_i) = \tau_i, i = 1, 2, \ldots, d, \tag{43.1}$$

where q_i represents the generalized coordinate vector, $M_i(q_i)$ represents the symmetric positive definite inertia matrix, $C_i(q_i, \dot{q}_i)$ represents the Coriolis and centrifugal matrix, $g_i(q_i)$ represents the gravitational torque, and τ_i represents the generalized control force for ith agent. The dynamics of each agent enjoys the assumptions below [6, 8].

Assumption 1 There are constants $k_i > 0$, $K_i > 0$, $k_{gi} > 0$ and $c_i > 0$, such that $0 < k_i I_p \leq M_i(q_i) \leq K_i I_p$, $\|g_i(q_i)\| \leq k_{gi}$ and $\|C_i(x, y)z\| \leq c_i \|y\| \|z\|$, for any x, y and $z \in \mathbb{R}^p$.

Assumption 2 The matrix $\dot{M}_i(q_i) - 2C_i(q_i, \dot{q}_i)$ is skew symmetric, i.e., for an any vector $v \in R^N$, there is $v^T \left(\dot{M}_i(q_i) - 2C_i(q_i, \dot{q}_i)\right) v = 0$.

Assumption 3 The dynamics (43.1) can be written as

$$M_i(q_i)\ddot{q}_i + C_i(q_i, \dot{q}_i)\dot{q}_i + g_i(q_i) = W_i^{\mathrm{T}} h_i, \tag{43.2}$$

where W_i is the NN weights for the vehicle i, h_i is the gauss active function.

Let d agents constitute the graph $\mathcal{G}$ with the network topology. The node set $\mathcal{V} = \{1, \ldots, d\}$ has a partition which takes the form as $\{\mathcal{V}_1, \ldots, \mathcal{V}_k\}$, i.e., $\mathcal{V}_i \neq \emptyset$, $\cup_{l=1}^{k} \mathcal{V}_l = \mathcal{V}$, $\mathcal{V}_i \cap \mathcal{V}_j = \emptyset$, $i \neq j$. For $i \in \mathcal{V}_s$ and $j \in \mathcal{V}_t$, if $s = t$, we state that agents i and j belong to the same group. Without loss of generality, we can index $\mathcal{V}_l = \{\sum_{j=0}^{l-1} n_j + 1, \ldots, \sum_{j=0}^{l} n_j\}$ as the node set of each group, where $1 \leq l \leq k$, $n_0 = 0$ and $\sum_{l=1}^{k} n_l = d$. We also denote $o_j = \sum_{i=1}^{j} n_j$, $j = 1, 2, \ldots, k$, that is $o_1 = n_1$, $o_2 = n_1 + n_2, \ldots, o_k = n_1 + n_2 + \cdots + n_k$.

Definition 1 The control protocols τ_i, $i = 1, \ldots, d$, are said to be solve group consensus problem of (43.1) with d agents under partition $\{\mathcal{V}_1, \ldots, \mathcal{V}_k\}$ if

$$\lim_{t\to\infty} \|q_i - q_j\| = \lim_{t\to\infty} \|\dot{q}_i - \dot{q}_j\| = 0, \quad i, j \in \mathcal{V}_l, \quad l = 1, 2, \ldots, k.$$

43.3 Group Consensus Based on NN Control

In this section, a necessary and sufficient condition is established to guarantee group consensus when the directed topology of $\mathcal{G}$ is acyclic. To do so, we need the following assumptions [10].

Assumption 4 $\{\mathcal{V}_1, \ldots, \mathcal{V}_k\}$ is an acyclic partition of node set $\mathcal{V}$.

In this case, the corresponding Laplician matrix of $\mathcal{G}$ has the form as [3]

$$\mathcal{L} = [l_{ij}] = \begin{bmatrix} \mathcal{L}_{11} & \cdots & \mathbf{0}_{n_1 \times n_k} \\ \vdots & \ddots & \vdots \\ \mathcal{L}_{k1} & \cdots & \mathcal{L}_{kk} \end{bmatrix}, \tag{43.3}$$

where $\mathcal{L}_{ii}$ is associated with the graph $\mathcal{G}_i$, and $\mathcal{L}_{ij}$ denotes the information flow from group $\mathcal{G}_j$ to group $\mathcal{G}_i$, where $i,j = 1, 2, \ldots, k$.

Just like the most of existing literature, e.g., [1, 3], we assume that the effect between $\mathcal{G}_i$ and $\mathcal{G}_j$ is balanced.

Assumption 5 Each row sum of $\mathcal{L}_{ij}$ is equal to zero.

Assumption 6 Each group $\mathcal{G}_i$ has a directed spanning tree.

Construct the following decomposition transform matrix $C \in \mathbb{R}^{d \times d}$ as

$$C = \begin{bmatrix} & \pi_1^{\mathrm{T}} & \\ & \vdots & \\ & \pi_k^{\mathrm{T}} & \\ \begin{matrix} -1 & 1 & \cdots & 0 \\ \vdots & & \ddots & \vdots \\ -1 & 0 & \cdots & 1 \end{matrix} & & \\ & \ddots & \\ & & \begin{matrix} -1 & 1 & \cdots & 0 \\ \vdots & & \ddots & \vdots \\ -1 & 0 & \cdots & 1 \end{matrix} \end{bmatrix}, \tag{43.4}$$
$$\underbrace{\qquad\qquad}_{n_1} \cdots \underbrace{\qquad\qquad}_{n_k}$$

where $\pi_i, i = 1, 2, \ldots, k$ are the left eigenvectors of $\mathcal{L}$ associated with zero eigenvalue defined in [10].

Lemma 1 ([10]) *Under Assumption 4, then* $C\mathcal{L}C^{-1} = \mathrm{diag}\{\mathbf{0}_{2\times 2}, \mathcal{L}_r\}$, *where* $-\mathcal{L}_r$ *is Hurwitz stable.*

We define a sliding variable reference velocity $\dot{q}_{ri} \in \mathbb{R}^p, i \in \mathcal{V}$, for ith Euler-Lagrange system,

$$\dot{q}_{ri} = -\sum_{j \in \mathcal{V}} a_{ij}(q_i - q_j).$$

Then

$$\ddot{q}_{ri} = -\sum_{j \in \mathcal{V}} a_{ij}(\dot{q}_i - \dot{q}_j).$$

Define slide vector $s_i \in \mathbb{R}^p$,

$$s_i = \dot{q}_i - \dot{q}_{ri}. \tag{43.5}$$

Thus, the distributed algorithm of (43.1) is designed as

$$\tau_i = \hat{W}_i h_i - K_i s_i, \tag{43.6}$$

and

$$\dot{\hat{W}}_i = -\gamma(h_i s_i^{\mathrm{T}} - v\hat{W}_i^{\mathrm{T}}), \tag{43.7}$$

where K_i represents a symmetric positive definite matrix and $\hat{W}_i$ represents the estimate of the desired weight W_i for ith agent, , γ and v are positive values.

Therefore, from (43.6) and (43.7), the closed-loop system (43.1) can be expressed as

$$M_i(q_i)\dot{s}_i = -C_i(q_i, \dot{q}_i)s_i - h_i\tilde{W}_i^{\mathrm{T}} - K_i s_i, \tag{43.8}$$

where $\tilde{W}_i = W_i - \hat{W}_i$ is the weight estimation error.

Next, we can rewrite system (43.5) in a vector form,

$$\dot{\mathbf{q}} = \mathbf{s} + \dot{\mathbf{q}}_r = \mathbf{s} - (\mathcal{L} \otimes I_p)\mathbf{q}, \tag{43.9}$$

where $\otimes$ denotes Kronecker product, $\mathbf{q}$, $\mathbf{s}$ and $\mathbf{q}_r$ are the column stack vectors of q_i, s_i and q_{ri}, $i \in \mathcal{V}$, respectively.

Consider the coordinate transformation

$$\omega = (C \otimes I_p)\mathbf{q}, \tag{43.10}$$

then the vector ω can be expressed as $\omega = (\omega_1^{\mathrm{T}}, \ldots, \omega_k^{\mathrm{T}}, \omega_R^{\mathrm{T}})^{\mathrm{T}}$, in which $\omega_i = (\pi_i^{\mathrm{T}} \otimes I_p)\mathbf{q}$, $i = 1, 2, \ldots, k$, and $\omega_R = [\omega_{R_1}^{\mathrm{T}}, \omega_{R_2}^{\mathrm{T}}, \ldots, \omega_{R_k}^{\mathrm{T}}]^{\mathrm{T}}$, where $\omega_{R_1} = [(q_2 - q_1)^{\mathrm{T}}, (q_3 - q_1)^{\mathrm{T}}, \ldots, (q_{o_1} - q_1)^{\mathrm{T}}]^{\mathrm{T}}, \omega_{R_2} = q_{o_1+2} - q_{o_1+1})^{\mathrm{T}}, \ldots, (q_{o_2} - q_{o_1+1})^{\mathrm{T}}]^{\mathrm{T}}, \ldots, \omega_{R_k} = [(q_d - q_{o_{k-1}+1})^{\mathrm{T}}, \ldots, (q_d - q_{o_{k-1}+1})^{\mathrm{T}}]^{\mathrm{T}}$ (o_i ($i = 1, 2, \ldots, k$) are defined in Sect. 43.2).

By (43.10), system (43.9) can be written as

$$\dot{\omega} = -(C\mathcal{L}C^{-1} \otimes I_p)\omega + (C \otimes I_p)\mathbf{s}. \tag{43.11}$$

Thus, by (43.11) and Lemma 1, we have

$$\dot{\omega}_i = (\pi_i \otimes I_p)\mathbf{s}, \ i = 1, 2, \ldots, k \tag{43.12}$$

and

$$\dot{\omega}_R = -(\mathcal{L}_r \otimes I_p)\omega_R + s_R, \tag{43.13}$$

where $s_R = \left((s_2 - s_1)^{\mathrm{T}}, \ldots, (s_{o_1} - s_1)^{\mathrm{T}}, (s_{o_1+2} - s_{o_1+1})^{\mathrm{T}}, \ldots, (s_{o_2} - s_{o_1+1})^{\mathrm{T}}, \ldots, (s_d - s_{o_{k-1}+1})^{\mathrm{T}}\right)^{\mathrm{T}}$.

Theorem 1 *Under Assumption 5, by using the NN adaptive control law* (43.6)*, the networked Euler-Lagrange systems* (43.1) *can reach group consensus if and only if Assumption 6 is satisfied.*

Proof (*Sufficient*) Now we can rewrite the closed loop system (43.10) as

$$M(q)\dot{s} = -C(q, \dot{q})s - \tilde{W}^{\mathrm{T}}h - Ks.$$

Considering the Lyapunov-like function candidate for system (43.8) as

$$V(t) = \frac{1}{2}s^{\mathrm{T}}M(q)s + \frac{1}{2\gamma}\sum_{i\in\mathcal{V}} tr(\tilde{W}^{\mathrm{T}}\tilde{W}).$$

From Assumption 2 and (43.7), and by taking derivative of V, we obtain

$$\dot{V} = -s^{\mathrm{T}}Ks + s^{\mathrm{T}}\tilde{W}^{\mathrm{T}}h - \sum_{i\in\mathcal{V}} tr(\tilde{W}_i^{\mathrm{T}}h_i s_i^{\mathrm{T}}) + v\sum_{i\in\mathcal{V}} tr(\tilde{W}_i^{\mathrm{T}}\hat{W}_i).$$

By employing the same analysis procedure in [8], we have $s_i(t) \to \mathbf{0}_p$, as $t \to \infty$. From the structure of $s_{_R}$, we can get $s_{_R} \to \mathbf{0}_{p(d-2)}$. By using the input-to-state stable property of system (43.13), we can get $\omega_{_R} \to 0$, which implies that, $q_i - q_j \to 0$, as $t \to \infty$, $\forall i, j \in \mathcal{V}_i$, $k = 1, 2, \ldots, k$. Following (43.12) and (43.13), we can get $\dot{\omega} \to 0$, as $t \to \infty$. Consequently, by differentiating (43.10), we can get $\dot{q}_i \to 0$, $i \in \mathcal{V}$, so the sufficiency part is proved.

(Necessity) If not, then there is at least one agent which cannot get any information from another group. Therefore, group consensus cannot be achieved for arbitrary initial conditions. The proof is proved completely.

43.4 Simulations

We here consider a network consisting of five two-link revolute joint manipulators (43.1) with an acyclic partition $\{\mathcal{V}_1, \mathcal{V}_2\}$, in which $\mathcal{V}_1 = \{1, 2\}$, $\mathcal{V}_2 = \{3, 4, 5\}$. The corresponding network graph is shown in Fig. 43.1a. It is easy to check that Theorem 1 holds, and it follows that the group consensus can be achieved by using protocol (43.6). In this simulation, the initial values are randomly selected in $[-2.1, 4]$, the initial weight matrix is chosen as $\hat{w}_i(0) = \mathbf{0}_{5\times 2}$, and the NN controller parameters with five neuron are selected as $v = 10$, $\gamma = 1000$, $K_i = [25 \quad 0; 0 \quad 25]$, $b_i = 9$, $c_i = 0.8 * [-2 \quad -1 \quad 0 \quad 1 \quad 2; \quad -2 \quad -1 \quad 0 \quad 1 \quad 2]$. Figure 43.1b, c clearly visualize the group consensus process of generalized coordinates q_i and their velocities, respectively. It turns out that the positions of the agents in the same group will asymptotically converge to a specific consensus state, meanwhile the velocities of all the agents will asymptotically approach zero at last.

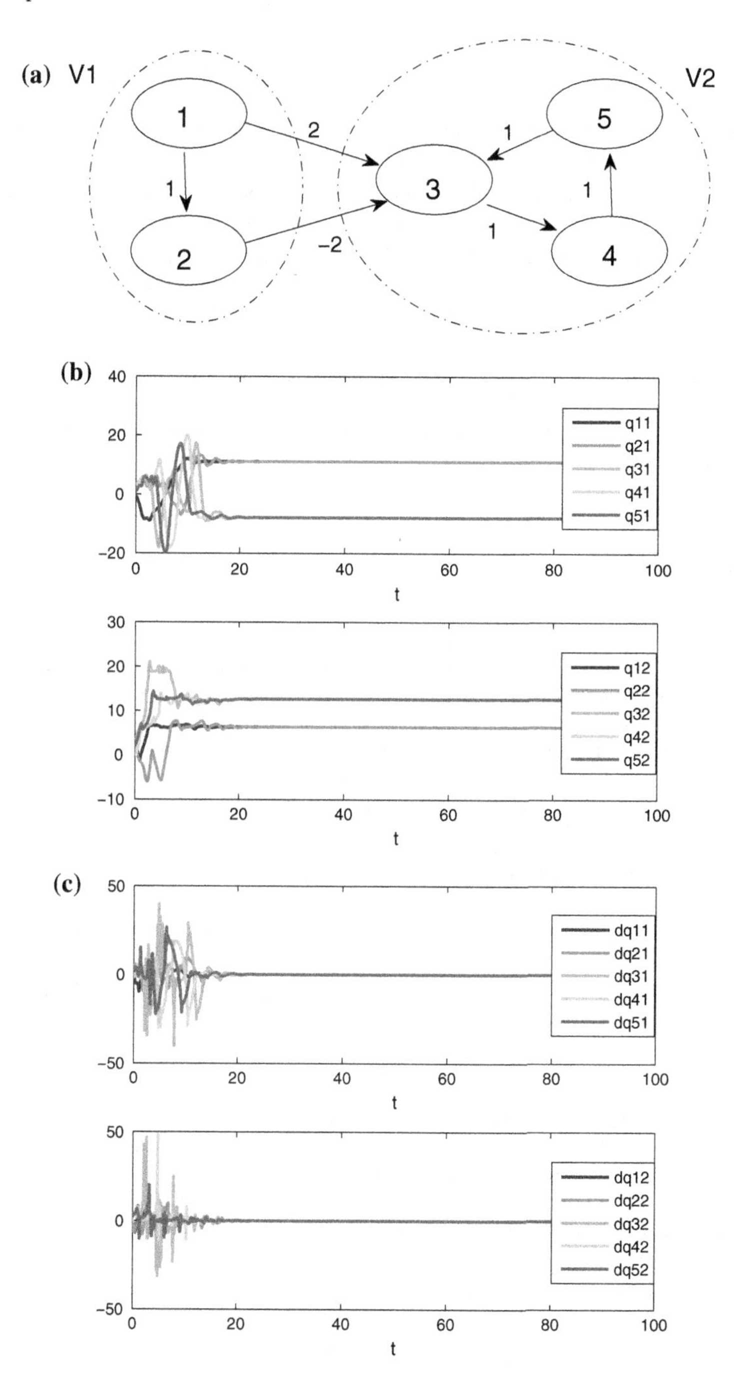

Fig. 43.1 **a** The topology of network. **b** Position group consensus of the revolute arms agents (first and second coordinates). **c** Velocities of the revolute arms agents (first and second coordinates)

43.5 Conclusions

In this brief, the group consensus issue of networked Euler-Lagrange systems under directed topology graph with acyclic partition has been studied An NN based group consensus law has deduced and the corresponding stability analysis is conducted based on the designed decomposition method. A necessary and sufficient condition for ensuring that the systems achieve group consensus is then presented. To this end, numerical examples and simulations are given to verify the theoretical analysis results.

Acknowledgments This work was supported by the National Natural Science Foundation of China under Grant 10972129, Science and Technology Project of High Schools of Shandong Province under Grant J15LJ07 and the Natural Science Foundation of Shandong Province Grant ZR2015FL026.

References

1. Yu J, Wang L (2009) Group consensus of multi-agent systems with undirected communication graphs. In: 7th Asian Control Conference Hong Kong, China, pp. 105–110
2. Feng Y, Xu S, Zhang B (2014) Group consensus control for double-integrator dynamic multiagent systems with fixed communication topology. Int J Robust Nolinear Control 24(7):532–547
3. Qin J, Yu C (2013) Cluster consensus control of generic linear multi-agent systems under directed topology with acyclic partition. Automatica 49(9):2898–2905
4. Chen G, Lewis FL (2011) Distributed adaptive tracking control for synchronization of unknown networked lagrangian systems. IEEE 41(3):805–816
5. Chung S, Slotine JJE (2009) Cooperative robot control and concurrent synchronization of Lagrangian systems. IEEE Trans Robot 25(3):686–700
6. Nuño E, Ortega R, Basañez L, Hill D (2011) Synchronization of networks of nonidentical euler-lagrange systems with uncertain parameters and communication delays. Automatica 56(4):935–941
7. Kwan C, Dawson DM, Lewis FL (2001) Robost adaptive control of robots using neural network: global atability. Asian J control 3(2):111–121
8. Mei J, Ren W, Ma G (2012) Distributed containment control for lagrangian networks with parametric uncertainties under a directed graph. Automatica 48:653–659
9. Kumar N, Panwar V (2011) Neural network-based nonlinear tracking control of kinematcally redundant robot manipulators. Math Comput Model 53:1889–1901
10. Liu J, Zhou J (2015) Distributed impulsive group consensus in second-order multi-agent systems under directed topology. Int J Control 88(5):910–919
11. Slotine JJE, Li W (1991) Applied nonlinear control. Prentice-Hall, Englewood Cliffs
12. Kriesel D (1991) A brief introduction neural networks
13. Haykin S (2011) Neural networks and learning machines

Chapter 44
Evaluation Strategy of Mine Gas Outburst Risk Level Based on Fuzzy Neural Network

Haibo Liu and Fuzhong Wang

Abstract Because the complicated non-linear relation between the coal gas outburst and its affecting factors, it is difficult to establish model with traditional mathematical method. The fuzzy system and the neural network were organically combined to establish evaluation strategy of coalmine gas outburst risk level based on fuzzy neural network. This paper made use of fuzzy mathematics to express and deal with the imprecise data and fuzzy information, and utilized self-adaptive neural networks system to solve the problems. Simulation results show that the model is reliable and precise and outburst risk level can be accurately predicted with proposed method and the mean error is small.

Keywords Fuzzy neural network • Gas outburst • Evaluation strategy • Information fusion

44.1 Introduction

Currently, coal accounted for about 70 % for national energy production and consumption structure in China's energy industry, which remains China's major energy for a long period of time. The usage of Coal will still account for 50 % as the primary energy until around 2050, but the gas explosion occurred in China with yet another per year, it caused incalculable loss of National interest and People's lives and property. So prevention and control of mine gas outburst occurred has become a major problem in the country need to be resolved, and it is the core task of work

H. Liu (✉) · F. Wang
School of Electrical Engineering and Automation, Henan Polytechnic University, Jiaozuo 454000, China
e-mail: liuhaibo09@hpu.edu.cn

Y. Jia et al. (eds.), *Proceedings of the 2015 Chinese Intelligent Systems Conference*, Lecture Notes in Electrical Engineering 360,
DOI 10.1007/978-3-662-48365-7_44

safety. How to predict mine gas outburst accurately, and the risk level of coal mining face for scientific evaluation is an important research topic in coal mine safety technology.

44.2 Evaluation Method of Mine Gas Outburst Risk Level

Coal and gas outburst is a very complex process, which exist many affecting factors. According to the combined effects of the hypothesis, it occurred under the action of crustal stress, Gas Pressure, Physical and mechanical properties of coal and external factors mainly. Our country adopts damage type of seam, initial velocity of gas emission, consistent coefficient of seam, gas pressure and other parameters to evaluate the risk of level. From now on, we have many methods that include professional system, nervous network, and fuzzy theory to evaluate. Among these, professional system is robotics diagnoses system based on knowledge, but it needs much of knowledge and proper reasoning system to creative flexibility and applicability. According to the fuzzy relationship matrix judged the characteristics of the object on the fuzzy vector and the state set and the mapping relation of fuzzy vector, it is based on ambiguous theory, and judging it based on judgment principle. Because of low efficiency for the prominent status of emerging judgment, neural network methods have strong learning ability. We can deal with complicated nonlinear problems. And it has been wildly applied in evaluation strategy of coal mine gas outburst risk level.

The framework of mine gas outburst evaluation strategy based on fuzzy neural network is shown in Fig. 44.1. It is made of three parts; the part 1 is to make use of many sensors to collect the key index of the mining working face. The part 2 is to preprocess indexical information that is measured by sensor. The part 3 is to keep the processed data as input, input the fuzzy neural network to make judgment, the order of evaluation has 3 kinds, which is the status of security, dangerous, very dangerous.

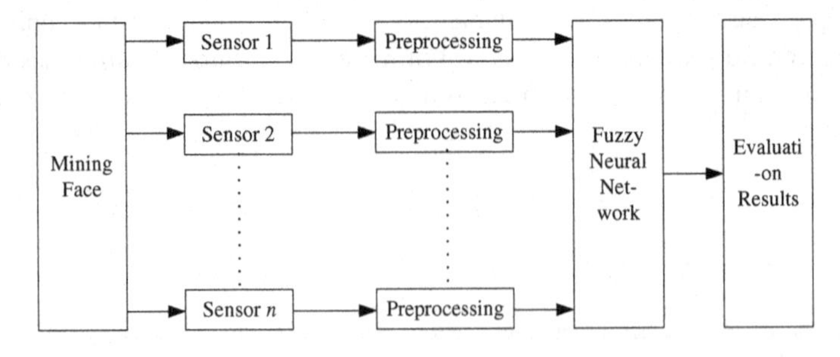

Fig. 44.1 Evaluation strategy based on fuzzy neural network

44.3 Fuzzy Neural Network Structure

Combined with the present prediction situation of coal mine gas, there is selected eight key indicators that have the influxes on gas outburst rank, that is respectively the pressure of the coal seam gas, gas content of coal, the thickness of coal seam, the radiation initial velocity Δp of coal gas and the consistence coefficient of coal, that is 'f', the geological structure type of coal, the destruction of the coal types, coal seam permeability coefficient. Using many sensors to collect the parameter values of key indicators, and there is a preprocessing for the collected data, use the fuzzy neural network to judge the collected data, and make the judgment for the state of mining working face.

There are different types of fuzzy neural network. The reason why this paper chooses the forward fuzzy neural network is that to predict the gas safety was designed to find the relevant laws between all kinds of influence factors and the gas safety state, in order to realize the prediction. Usually a forward fuzzy neural network consists of input layer, fuzz fiction layer, fuzzy inference layer, to blur layer and output layer 5 layers, and its topology structure is shown in Fig. 44.2.

44.4 BP Learning Algorithm

Fuzzy neural network adopts the BP learning algorithm. The mathematical indicators we need to learn are mainly weight. $w_{\mathrm{ij}} = (i = 1, 2, \ldots, z; j = 1, 2, \ldots, u)$ membership *f* unction c_{ij}'s center and width $(i = 1, 2, \ldots, z; j = 1, 2, \ldots, u)$. Firstly initialize the input/output the center and width of membership function, c_{ij}, σ_{ij}'s starting value is chosen randomly in (0,1), So is the w_{ij}, the system error is

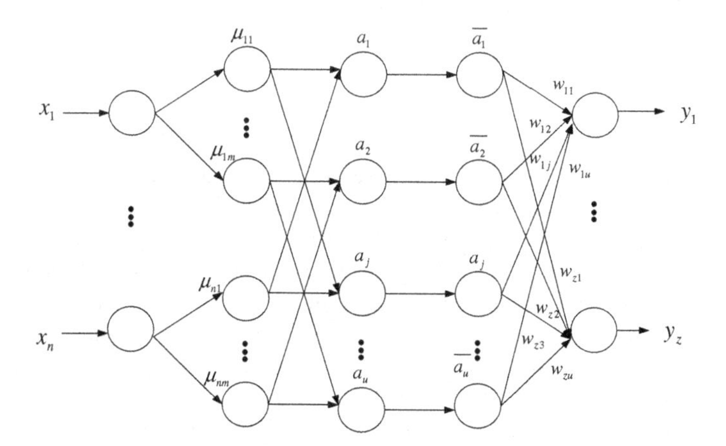

Fig. 44.2 Architecture of fuzzy neural network

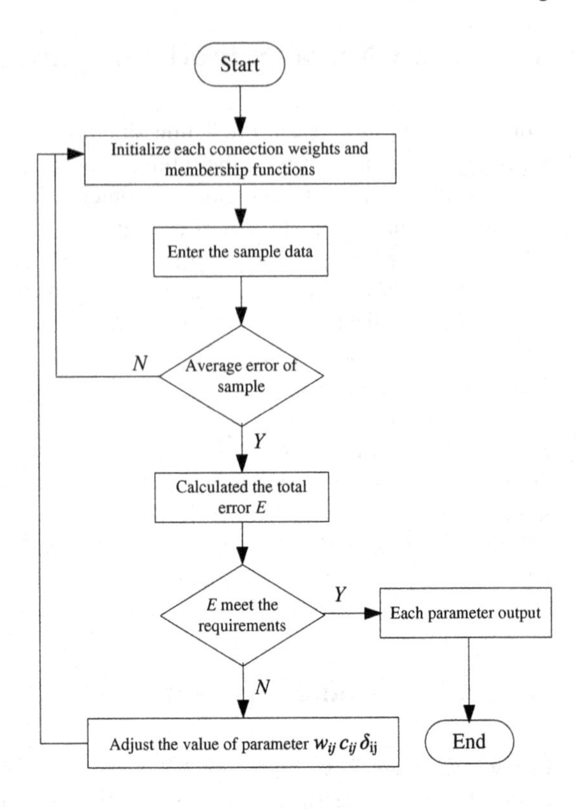

Fig. 44.3 Flow diagram of fuzzy neural network

0.0001. After the training, 0.5 is the most appropriate disturbance quantity η. Fuzzy neural network algorithm flow chart is shown in Fig. 44.3.

44.5 Initial Data Preprocessing

At the time of the excavation face to judge the level of danger outburst, parameter values must first select a typical outburst of impact parameters, including the value of the normal state and a variety of prominent state. Based on the measurement parameters and site conditions of each sensor, the judge selected eight key indicators, namely, coal seam gas pressure, gas content of coal, coal seam thickness, coal gas irradon initial velocity Δp, consistent coefficient of coal f, the geological structure of, the type of coal, coal type of damage, coal seam permeability coefficient. Safe, more dangerous, dangerous example for analysis of three evaluation state, read the data stream parameters, and these data streams readings normalized, taking to account the fluctuation detection parameter detection point, a single change parameter is set to 0.25 and 0.62 normal, the normal two way change parameter is set to 0.5, sample data is shown in Table 44.1.

Table 44.1 Initial data of key indicators

No	p_1	p_2	p_3	p_4	p_5	p_6	p_7	p_8	Evaluation
1	12.54	0.5	0.70	0.9	0.55	5.062	9.059	0.1	Security
2	11.17	0.5	1.60	0.5	0.34	5.081	16.326	0.1	Security
3	10.17	0.5	1.65	0.6	0.38	5.093	16.54	0.2	Security
4	10.05	0.4	0.80	0.7	0.37	5.054	12.34	0.1	Security
5	18.24	0.5	0.65	0.5	0.33	2.650	9.814	0.2	More dangerous
6	18.14	0.6	1.75	0.6	0.32	2.653	17.23	0.2	More dangerous
7	17.08	0.6	0.86	0.8	0.30	2.243	12.76	0.3	More dangerous
8	17.23	0.6	1.74	0.6	0.28	2.321	13.54	0.3	More dangerous
9	20.34	0.7	1.20	0.9	0.24	0.502	13.305	0.5	Very dangerous
10	21.34	0.7	1.55	0.5	0.25	0.502	14.982	0.4	Very dangerous

44.6 Primary Diagnosis Based on Fuzzy Neural Network

Fuzzy neural network adopts the BP learning algorithm, and its network structure includes five parts. Network architecture model in MATLAB is shown in Fig. 44.4. Where in the input nodes are affected eight key indicators gas outburst. The hidden nodes are selected by experience; they were safe, more dangerous and very dangerous, but with (001) represent a very dangerous, with (010) on behalf of the more dangerous, with (100) on behalf of security, network error set to 10^{-4}.

In fuzzy neural network training, choose 10 groups of samples, the samples are the first 5 groups of training samples, after three sets of data as the test sample simulation. The actual output result of the fuzzy neural network is shown in Table 44.2, you can see the result is basically correct, but there is a certain expectation of the output error. After training the network training, error curve obtained is shown in Fig. 44.5. The figure shows that after 1200 after learning steps to meet the accuracy requirements, and the curve is very smooth.

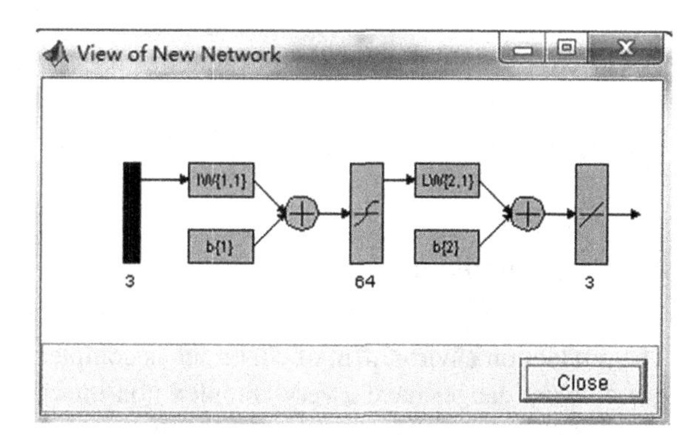

Fig. 44.4 Simulation model of fuzzy neural network

Table 44.2 The result of fuzzy neural network output

Evaluation samples		Actual output		
Sample1	1	0.987301	0.00430	0.00814
	2	0.998905	0.00783	0.00679
	3	0.981257	0.00874	0.00423
	4	0.973218	0.00627	0.00329
Sample2	1	0.007691	0.98375	0.00341
	2	0.006372	0.99873	0.00463
	3	0.003789	0.98986	0.00631
	4	0.004125	0.97549	0.00542
Sample3	1	0.003817	0.00480	0.99382
	2	0.005698	0.00631	0.98123
	3	0.003287	0.00518	0.97783
	4	0.002437	0.00438	0.98238

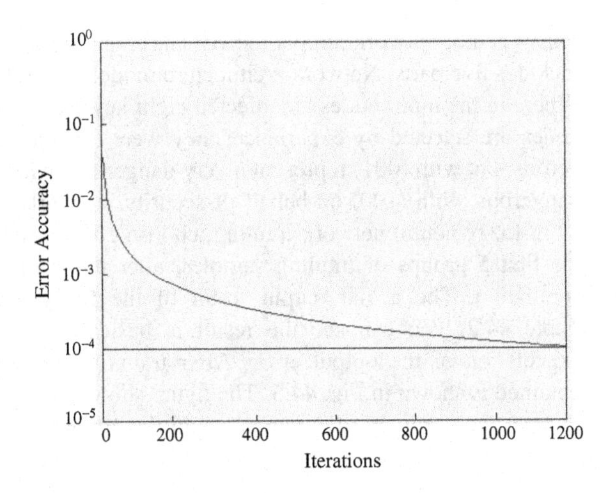

Fig. 44.5 Training error curve of fuzzy neural network

The error test sample is 0.0016. From this result and the training error curve can be seen, the use of the network to judge the surface mining of coal mine gas prominent is feasible and has a very high accuracy.

44.7 Conclusion

The production environment of coal mine is complex. Outburst indexes of coal and gas outburst danger have a very complex non-linear relationship. We combine the fuzzy mathematics and adaptive neural network closely and establish the coal mine gas outburst dangerous grade evaluation strategy based on fuzzy neural network.

The simulation experiment indicate that this method can predict the coal mine gas outburst risk levels precisely, which provide a new idea and method of the area of coal and gas outburst hazard prediction.

References

1. Xueqiu H, Yuan E, Tong LZ et al (2003) coal and gas outburst prediction technology research status and development trend [J] China Saf Sci 13(6):40–43
2. Jinzhi F, Yong G, Chi MS (2004) Work face fuzzy comprehensive evaluation of coal mine gas outburst risk [J] Mine Saf 35(3):17–19
3. Jian W, Honghua W (2004) Fuzzy logic self-learning rate adjustment BP neural network [J]. Jilin University 7:153–156
4. Sheng Z, fan YB, Hui MD (2000) Coal mine gas outburst hazard classification method [J] Min Saf Environ Prot 4, 27(2):4–8
5. Zhengrong P, Qun W (2011) COD soft measurement technique based on fuzzy neural networks [J]. Comput Measure Control 19(7):1572–1574
6. Xun YF, Yue HX, Ke LJ (2009) Based on fuzzy neural network information fusion technology [J] Autom Instrum 14(2):5–7
7. Fu H, Sun S, Xu Zet al (2006) Information fusion method based on fuzzy neural network in mine supply air system [J] coal 31(2):264–267
8. Hai J (2012) Fuzzy multi sensor information fusion algorithm in coal mine gas monitoring [J] Coal Technol 31(8):82–84

Chapter 45
Adaptive Control with Prescribed Performance for Nonlinear Two-Inertia System

Shubo Wang and Xuemei Ren

Abstract In this paper, a new adaptive neural dynamic surface control technique with an improved prescribed performance function is proposed for the nonlinear two-inertia system. An improved error transformation function is used to ensure the prescribed output tracking performance, and the neural network (NN) is utilized to estimate the unknown disturbance. The dynamic surface technique simplifies the controller by introducing first-order filters to eliminate "explosion of complexity" inherent in backstepping approach. Simulation results demonstrate the control scheme is effective.

Keywords Dynamic surface control (DSC) · Neural network · Prescribed performance · Two-inertia system

45.1 Introduction

The two-inertia system is composed of a motor connected to a load via a flexible shaft, which is widely used in industrial applications. This configuration may cause mechanical vibration. Mechanical vibration suppression has been studied from many different point of views, including state space control, load disturbance observer, model predictive control, immersion and invariance method [2, 8]. Proportional-integral-derivative (PID) speed controller is used for two-inertia system [12]. This algorithm adopts speed feedback from motor to suppress the torsional vibration, however, this method is not effective. Then, an improved PI speed controller is proposed by changing the controller parameters [11]. To improve the performance of the PI speed controller, additional feedback is presented for two-inertia system to damp the torsional vibration [3]. Recently, with the development of intelligent technology, neural network and fuzzy logic system have been utilized to damp torsional vibration for two-inertia system [6]. In [7], neural network observer is employed to

S. Wang · X. Ren (✉)
School of Automation, Beijing Institute of Technology, Beijing 100081, China
e-mail: xmren@bit.edu.cn

Y. Jia et al. (eds.), *Proceedings of the 2015 Chinese Intelligent Systems Conference*, Lecture Notes in Electrical Engineering 360,
DOI 10.1007/978-3-662-48365-7_45

estimate the unmeasured state variables, such as load speed and torsional torque, to damp torsional vibration. With the development of the nonlinear technique, the sliding-mode technique combined with neural-fuzzy method is utilized to suppress the mechanical oscillation of the two-inertia system [5]. The backstepping control technique has been utilized to design a stabilizing controller for nonlinear system using recursive procedure. However, the virtual controller designed at each step needs repeated differentiation that exists "explosion of complexity". To avoid the disadvantage, the dynamic surface control technique is proposed to eliminate disadvantage by introducing a first-order filter in controller design [10]. Sun [9] proposed a modified neural DSC by introducing a high-order differentiator instead of the first-order filter in MIMO nonlinear system.

Recently,a new method named prescribed performance function is proposed to guarantee the tracking error within a prescribed region in [1]. Na [4] proposed an adaptive control with improved prescribed performance function to compensate nonlinear friction for turntable servo system. Therefore, inspired by [1], an improved prescribing performance function (PPF) for transient and asymptotic tracking is proposed for two-inertia system. We briefly summarize the contributions of this paper as follows. (1) Aforementioned the two-inertia system, most researchers consider the two-inertia system as a linear system. As far as we know, it is the first time that the two-inertia system is considered as a nonlinear system. (2) An improved prescribed performance function based on [4] is proposed, the prescribed performance function is simplified without being divided into two conditions, and the predefined performance bounded on the tracking error is integrated into controller design.

45.2 Problem Formulation

45.2.1 *Mathematical Model of Two-Inertia System*

A typical two-inertia system can be depicted as Fig. 45.1. The dynamic equations of the two-inertia model can be described as

$$J_m \omega_m = T_e - d_1 - T_s, J_l \omega_l = T_s - d_2 - T_l, T_s = K_f(\theta_m - \theta_l) \tag{45.1}$$

where θ_m and θ_l are positions of the motor and the load, ω_m and ω_l are the velocity, respectively; J_m and J_l are the inertias. T_e, T_s, and T_l are the electromagnetic torque, shaft torque and load torque, respectively; K_f is the stiffness coefficient. The term d_1 is nonlinear function, which contains the Coulomb friction, torque ripple, back-

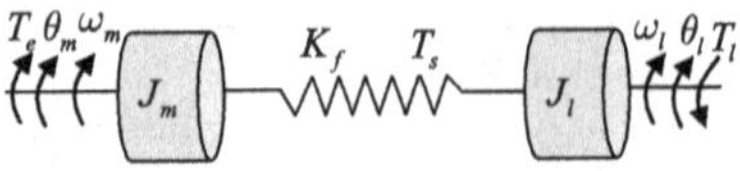

Fig. 45.1 Two-inertia system model

lash, damping, and disturbance, and the term d_2 contains the coriolis and centrifugal terms, gravity, backlash, and disturbance.

Defining, $x_1 = \theta_l, x_2 = \omega_l, x_3 = \theta_m, x_4 = \omega_m, u = T_e$, the system equation (45.1) indicates can be simplified as

$$\dot{x}_1 = x_2, \dot{x}_2 = f_1(x_1, x_3) + x_3$$
$$\dot{x}_3 = x_4, \dot{x}_4 = f_2(x_1, x_3) + \frac{1}{J_m}u \tag{45.2}$$

where the nonlinear function $f_1(x_1, x_3) = \frac{1}{J_l}(K_f(x_3 - x_1) - d_2 - T_l - x_3), f_2(x_1, x_3) = \frac{1}{J_m}(-K_f(x_3 - x_1) - d_1)$.

Assumption 1 The reference $y_d, \dot{y}_d$ and $\ddot{y}_d$, are continuous and bounded, that is, there exists a known compact set $\Omega_0 = \{(y_d, \dot{y}_d, \ddot{y}_d) : y_d^2 + \dot{y}_d^2 + \ddot{y}_d^2 \leq \delta\}$, where δ is a positive constant.

45.2.2 Prescribed Performance Function

In order to illustrate the prescribed performance of tracking error $e_i(t) = [e_1(t), e_2(t), \ldots e_i(t)]$, a smooth decreasing function $\lambda_i(t) : R^+ \longrightarrow R^+$ with $\lim_{t \longrightarrow \infty} \lambda_i(t) = \lambda_{i\infty}$ will be used. In this paper, the $\lambda_i(t)$ is given as

$$\lambda_i(t) = (\lambda_{i0} - \lambda_{i\infty})e^{-c_i t} \tag{45.3}$$

where $\lambda_{i0} > \lambda_{i\infty}$, and c_i are design parameters. According to [4], the prescribed performance is given as

$$-\underline{\delta}_i \lambda_i(t) < e_i(t) < \overline{\delta}_i \lambda_i(t), \forall t > 0 \tag{45.4}$$

where $-\underline{\delta}_i$ and $\overline{\delta}_i$ are constant design parameters. We define a smooth strictly increasing function $T_i(z_i)$, which possess the following properties:

1. $-\underline{\delta}_i < T_i(z_i) < \bar{\delta}_i$, 2. $\lim_{z_i \longrightarrow +\infty} T_i(z_i) = \overline{\delta}_i$, and $\lim_{z_i \longrightarrow -\infty} T_i(z_i) = -\underline{\delta}_i$. From the propoerties of $T_i(z_i)$, the (45.4) is equal to

$$e_i(t) = \lambda_i(t) T_i(z_i). \tag{45.5}$$

Then, the z_i can be written as

$$z_i = T_i^{-1}(\frac{e_i(t)}{\lambda_i(t)}). \tag{45.6}$$

In this paper, an improved prescribed performance function is chosen as

$$T_i(z_i) = \frac{\overline{\delta}_i e^{z_i} - \underline{\delta}_i e^{-z_i}}{e^{z_i} + e^{-z_i}}. \tag{45.7}$$

Then, from the (45.7), the transformed error z_i is derived as

$$z_i = T_i^{-1}(\frac{e_i(t)}{\lambda_i(t)}) = \frac{1}{2}\ln\frac{\frac{e_i(t)}{\lambda_i(t)} + \underline{\delta}_i e^{-z_i}}{\overline{\delta}_i - \frac{e_i(t)}{\lambda_i(t)}}. \tag{45.8}$$

45.3 Controller Design

In this section, we will design the output feedback adaptive controller for system (45.1) based on DSC, the design steps are given as follows.

Step 1: To start, let us define the tracking error as

$$e_1 = y - y_d. \tag{45.9}$$

Define a new variable α_{2d} and input a virtual controller χ_1 to a first-order filter, then

$$\tau_2\dot{\alpha}_{2d} + \alpha_{2d} = \chi_1, \quad \alpha_{2d}(0) = \chi_1(0) \tag{45.10}$$

where α_{2d} is output variable. Define a filter error

$$\xi_1 = \alpha_{2d} - \chi_1. \tag{45.11}$$

Choose a Lyapunov function as follows:

$$V_1 = \frac{1}{2}e_1^2 + \frac{1}{2}\xi_1^2. \tag{45.12}$$

The time derivative of (45.12) is given as

$$\dot{V}_1 = e_1\dot{e}_1 + \xi_1\dot{\xi}_1 = re_1[x_2 - \dot{y}_d - e_1\frac{\dot{\lambda}_1}{\lambda_1}] \leq re_1[e_2 + \chi_1 + \xi_1 - \dot{y}_d - e_1\frac{\dot{\lambda}_1}{\lambda_1}] \tag{45.13}$$

where $r = (1/2\lambda)[1/(\rho + \underline{\delta}) - 1/(\rho - \bar{\delta})]$, and $\rho = e(t)/\lambda(t)$.

Choose the virtual variable as

$$\chi_1 = -k_1e_1 + y_d + e_1\frac{\dot{\lambda}_1}{\lambda_1}. \tag{45.14}$$

Step 2: The second error surface is defined as

$$e_2 = x_2 - \alpha_{2d}. \tag{45.15}$$

Define a new variable α_{3d} and input a virtual controller χ_2 to a first-order filter, then

$$\tau_3 \dot{\alpha}_{3d} + \alpha_{3d} = \chi_2, \quad \alpha_{3d}(0) = \chi_2(0). \tag{45.16}$$

Define the output error of this filter as

$$\xi_2 = \alpha_{3d} - \chi_2. \tag{45.17}$$

Choose a Lyapunov function as

$$V_2 = \frac{1}{2}e_2^2 + \frac{1}{2}\xi_2^2 + \frac{1}{2}\tilde{W}_1{}^T \Gamma^{-1} \tilde{W}_1 \tag{45.18}$$

where $\tilde{W}_1^T = W_1^* - W_1$. The derivative of (45.18) is

$$\begin{aligned} \dot{V}_2 &= e_2 \dot{e}_2 + \tilde{W}_1{}^T \Gamma^{-1} \dot{\hat{W}}_1 + \xi_2 \dot{\xi}_2 \\ &\le e_2 [W_1^{*T} \varphi_1(x_1, x_3) + \chi_2 - \dot{\alpha}_{2d}] + e_2(e_3 + \xi_2) \\ &\quad + \tilde{W}_1 \varphi_1(x_1, x_3)(e_2 - \Gamma_1^{-1} \dot{\hat{W}}_1) + \xi_2(-\frac{\xi_2}{\tau_3} - \dot{\chi}_2). \end{aligned} \tag{45.19}$$

Choose the virtual control χ_2 and adaptation law for $\hat{W}_1$ as

$$\begin{aligned} \chi_2 &= -re_1 - k_2 e_2 - \hat{W}_1^T \varphi_1(x_1, x_3) + \dot{\alpha}_{2d} \\ \dot{\hat{W}}_1 &= \Gamma_1(e_2 \varphi_1(x_1, x_3) - \Gamma_1^{-1} \hat{W}_1) \end{aligned} \tag{45.20}$$

Step 3: The third error surface is defined as

$$e_3 = x_3 - \alpha_{3d}. \tag{45.21}$$

Define a new variable α_{4d} and input a virtual controller χ_3 to a first-order filter, then

$$\tau_4 \dot{\alpha}_{4d} + \alpha_{4d} = \chi_3, \quad \alpha_{4d}(0) = \chi_3(0). \tag{45.22}$$

Define the output error of this filter as

$$\xi_3 = \alpha_{4d} - \chi_3. \tag{45.23}$$

Choose a Lyapunov function candidate as

$$V_3 = \frac{1}{2}e_3^2 + \frac{1}{2}\xi_3^2. \tag{45.24}$$

The derivative of V_3 is

$$\begin{aligned}\dot{V}_3 &= e_3\dot{e}_3 + \xi_3\dot{\xi}_3 = e_3(\chi_3 - \dot{\alpha}_{3d}) + e_3(e_4 + \xi_3) + \xi_3(-\frac{\xi_3}{\tau_4} - \chi_3)\\ &\leq e_3(\chi_3 - \dot{\alpha}_{3d}) + e_3(e_4 + \xi_3) - \frac{\xi_3^2}{\tau_3} + |\ \xi_3 \ || \ \dot{\chi}_3 \ |\end{aligned} \tag{45.25}$$

Choose the virtual control χ_3 as

$$\chi_3 = e_2 - k_3e_3 + \dot{\alpha}_{3d}. \tag{45.26}$$

Step 4: The last error surface is defined as

$$e_4 = x_4 - \alpha_{4d}. \tag{45.27}$$

The time derivative of e_4 is

$$\dot{e}_4 = \dot{x}_4 - \dot{\alpha}_{4d} = W_2^{*T}\varphi_2(x_1, x_3) + \varepsilon_2(x_1, x_3) + (1/J_m)u - \dot{\alpha}_{4d}. \tag{45.28}$$

Consider the Lyapunov function candidate as

$$V_4 = \frac{1}{2}e_4^2 + \frac{1}{2}\tilde{W}_2^T\Gamma_2^{-1}\tilde{W}_2 \tag{45.29}$$

where $\tilde{W}_2 = W_2^* - W_2$. The derivative of V_4 is

$$\begin{aligned}\dot{V}_4 &= e_4\dot{e}_4 + \tilde{W}_2^T\Gamma_2^{-1}\dot{\tilde{W}}_2\\ &= e_4[W_2^{*T}\varphi_2(x_1, x_3) + \varepsilon_2(x_1, x_3) + (1/J_m)u - \dot{\alpha}_{4d}] + \tilde{W}_2^T\varphi_2(x_1, x3)(e_4 - \Gamma_2^{-1}\dot{\tilde{W}}_2)\\ &\leq e_4[\hat{W}_2^T\varphi_2(x_1, x_3) + (1/J_m)u - \dot{\alpha}_{4d}] + \tilde{W}_2^T(e_4\varphi_2(x_1, x_3) - \Gamma_2^{-1}\dot{\hat{W}}_2).\end{aligned} \tag{45.30}$$

Choose the controller u and adaptation law for $\hat{W}_2$ as

$$\begin{aligned}u &= -e_3 + J_m(-k_4e_4 - \hat{W}_2^T\varphi_2(x_1, x_3) + \dot{\alpha}_{4d})\\ \dot{\hat{W}}_2 &= \Gamma_2(e_4\varphi_2(x_1, x_3) - \Gamma_2^{-1}\hat{W}_2)\end{aligned} \tag{45.31}$$

45.4 Stability Analysis

Consider system (45.1) composed of DSC controller and adaptive laws (45.14), (45.20), (45.26), (45.31), guarantees the closed-loop system is semi-global stability.

Proof We consider Lyapunov function candidate as

$$\begin{aligned} V &= \sum_{i=1}^{4}(V_i) = V_1 + V_2 + V_3 + V_4 \\ \dot{V} &\leq -rk_1e_1^2 - k_2e_2^2 - k_3e_3^3 - k_4e_4^4 + re_1\xi_1 + e_2\xi_2 \\ &\quad + e_3\xi_3 + e_4\xi_4 + \tilde{W}_1^T\hat{W}_1 + \tilde{W}_2^T\hat{W}_2 + \sum_{i=1}^{3}(-\frac{\xi_i^2}{\tau_{i+1}} + |\xi_i||\chi_i|). \end{aligned} \tag{45.32}$$

Using the Young's inequality, one has

$$\begin{aligned} \tilde{W}_j^T\hat{W}_j &\leq -\frac{\tilde{W}_j^T\tilde{W}_j}{2} + \frac{\tilde{W}_j^{*T}\tilde{W}_j^*}{2} \\ e_i\xi_i &\leq \frac{1}{2}e_i^2 + \frac{1}{2}\xi_i^2 \\ |\xi_i||\chi_i| &\leq \frac{\xi_i^2\dot{\chi}_i^2}{2\gamma} + \frac{\gamma}{2}. \end{aligned} \tag{45.33}$$

Substituting (45.33) into (45.32) results in

$$\begin{aligned} \dot{V} &\leq -(rk_1 - \frac{1}{2})e_1^2 - (k_2 - \frac{1}{2})e_2^2 - (k_3 - \frac{1}{2})e_3^2 - k_4e_4^2 \\ &\quad - \sum_{j=1}^{2}\frac{\tilde{W}_j^T\tilde{W}_j}{2} - \sum_{i=1}^{3}(\frac{1}{\tau_{i+1}} - \frac{1}{2} - \frac{\dot{\chi}_i^2}{2\gamma})\xi_i^2 + \varrho \end{aligned} \tag{45.34}$$

where $\varrho = \Sigma_{j=1}^2\frac{\tilde{W}_j^{*T}\tilde{W}_j^*}{2} + \frac{3\gamma}{2}$. Let Ξ_n be feasible set such that

$$\Xi_n = \{\frac{1}{2}\sum_{i=1}^{n}(e_i^2 + \tilde{W}_j^T\tilde{W}_j + \frac{1}{2}\sum_{i=1}^{n}\xi_i^2) \leq p\}. \tag{45.35}$$

Since Ξ_n is a compact set and $\phi_i(\cdot)$ ($\phi_i(\cdot) = \dot{\chi}_i$) is a continuous function, there is a constant $M_i > 0$ and $|\phi(\cdot)| \leq M_i$, we can obtain

$$|\xi_i\phi(i)| \leq \frac{1}{2}\xi_i^2\phi_i^2 + \frac{1}{2} \leq \frac{1}{2}\xi_i^2M_i^2 + \frac{1}{2} \tag{45.36}$$

Substituting (45.36) into (45.34)

$$\dot{V} \leq -(rk_1 - \frac{1}{2})e_1^2 - (k_2 - \frac{1}{2})e_2^2 - (k_3 - \frac{1}{2})e_3^2 - k_4 e_4^2$$
$$- \sum_{j=1}^{2} \frac{\tilde{W}_j^T \tilde{W}_j}{2} - \sum_{i=1}^{3} (\frac{1}{\tau_{i+1}} - \frac{1}{2} - \frac{M_i^2}{2\gamma})\xi_i^2 + \varrho. \tag{45.37}$$

Choose the design parameters $rk_1 - \frac{1}{2} > 0$, $k_2 - \frac{1}{2} > 0$, $k_3 - \frac{1}{2} > 0$, $\frac{1}{\tau_{i+1}} - \frac{1}{2} - \frac{M_i^2}{2\gamma} > 0$, then define $\pi = \min\{2(rk_1 - \frac{1}{2}),\ 2(k_2 - \frac{1}{2}),\ 2(k_3 - \frac{1}{2}),\ 2(\frac{1}{\tau_{i+1}} - \frac{1}{2} - \frac{M_i^2}{2\gamma}),\ k_4\}$. Then, (45.37) can be written as

$$\dot{V} \leq -\pi V + \varrho. \tag{45.38}$$

45.5 Simulation Results

In this paper, we utilize the proposed control scheme to control system (45.1)and two control methods are compared. The system parameter are given as $J_m = 0.005\,\text{kg}\cdot\text{m}$, $J_l = 0.15\,\text{kg}\cdot\text{m}$, $K_f = 5$, the prescribed performance function parameters are $\lambda_0 = 0.3$, $\lambda_\infty = 0.013$, $c_i = 0.4$, and $\underline{\delta} = \overline{\delta} = 1$. The desired trajectory is defined as $y_d = 0.8sin(\pi t)$. The controller parameters are $\tau_2 = \tau_3 = 0.01$, $\tau_3 = 0.02$, $k_1 = 5, k_2 = 40, k_3 = 25, k_4 = 5$. The PID parameters are given as $k_p = 100, k_i = 1$ and $k_d = 50$. The simulation results are depicted in Fig. 45.2. Figure 45.2a shows the tracking performance of PID control scheme. From the Fig. 45.2, one can see that the proposed ANDSC method achieves faster tracking than PID approach, and the tracking error of the ANDSC satisfies the prescribed output error performance constraints.

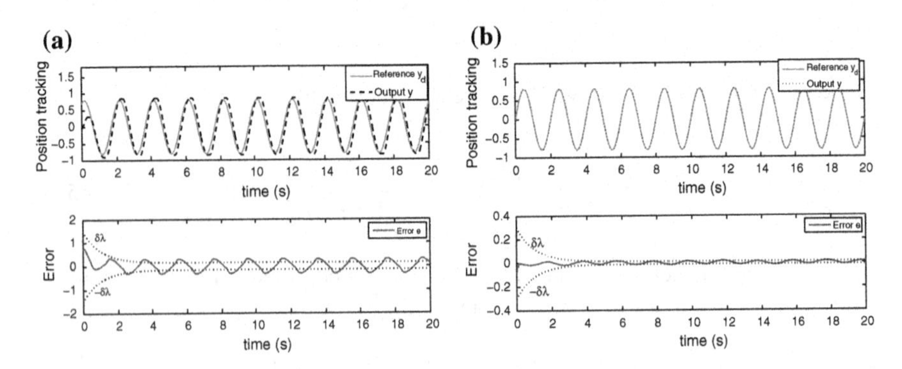

Fig. 45.2 Output tracking and Tracking error: **a** PID controller, **b** ANDSC controller

45.6 Conclusion

Adaptive neural DSC combined with the prescribed performance function for the nonlinear two-inertia system with uncertainty and unknown disturbance has been investigated in this paper. An improved error transformation function is used to ensure the prescribed output tracking performance, and the adaptive NN is used to estimate the unknown disturbance. In comparison with the simulation results using PID control scheme, the proposed control approach shows better results.

Acknowledgments This work is supported by National Nature Science Foundation of China (No.61433003, No.612173150, No.61321002).

References

1. Bechlioulis C, Rovithakis G (2008) Robust adaptive control of feedback linearizable mimo nonlinear systems with prescribed performance. IEEE Trans Autom Control 53(9):2090–2099
2. Cychowski M, Szabat K, Orlowska-Kowalska T (2009) Constrained model predictive control of the drive system with mechanical elasticity. IEEE Trans Ind Electron 56(6):1963–1973
3. Ji JK, Sul SK (1995) Kalman filter and LQ based speed controller for torsional vibration suppression in a 2-mass motor drive system. IEEE Trans Ind Electron 42(6):564–571
4. Na J, Chen Q, Ren X, Guo Y (2014) Adaptive prescribed performance motion control of servo mechanisms with friction compensation. IEEE Trans Ind Electron 61(1):486–494
5. Orlowska-Kowalska T, Dybkowski M, Szabat K (2010) Adaptive sliding-mode neurofuzzy control of the two-mass induction motor drive without mechanical sensors. IEEE Trans Ind Electron 57(2):553–564
6. Orlowska-Kowalska T, Szabat K (2007) Control of the drive system with stiff and elastic couplings using adaptive neuro-fuzzy approach. IEEE Trans Ind Electron 54(1):228–240
7. Orlowska-Kowalska T, Szabat K (2007) Neural-network application for mechanical variables estimation of a two-mass drive system. IEEE Trans Ind Electron 54(3):1352–1364
8. Saarakkala S, Hinkkanen M (2014) State-space speed control of two-mass mechanical systems: analytical tuning and experimental evaluation. IEEE Trans Ind Appl 50(5):3428–3437
9. Sun G, Li D, Ren X (2014) Modified neural dynamic surface approach to output feedback of MIMO nonlinear systems. IEEE Trans Neural Netw Learn Syst PP(99), 1–1 (2014)
10. Swaroop D, Hedrick J, Yip P, Gerdes J (2000) Dynamic surface control for a class of nonlinear systems. IEEE Trans Autom Control 45(10):1893–1899
11. Szabat K, Orlowska-Kowalska T (2007) Vibration suppression in a two-mass drive system using PI speed controller and additional feedbacks mdash; comparative study. IEEE Trans Ind Electron 54(2):1193–1206
12. Zhang G (2000) Speed control of two-inertia system by PI/PID control. IEEE Trans Ind Electron 47(3):603–609

Chapter 46
Neural Network Observer Based Optimal Tracking Control for Multi-motor Servomechanism with Backlash

Minlin Wang and Xuemei Ren

Abstract In this paper, a new neural network observer based optimal tracking control is presented to attenuate the effect of backlash and other uncertainty for the position tracking of multi-motor servomechanism (MMS). By adopting a continuously differentiable function instead of the non-differential dead-zone model of the backlash, the state space representation of MMS is set up by using a linear part of the differentiable function. Based on the state space representation, the optimal neural network (NN) observer is used to estimate the uncertainties and unmeasured states, which combines with the optimal state feedback to synthesis the actual control law. Finally, Lyapunov theory is utilized to certify that the tracking error, the observed error and neural network weights are all semi-globally uniformly ultimately bounded (SGUUB). Simulation results validate the effectiveness of this method.

Keywords Multi-motor servomechanism · Backlash compensation · Uncertainty estimation · Neural network · Optimal state feedback

46.1 Introduction

In many industrial control systems, as demands of large inertia and power are gradually increasing, it is necessary to employ multi-motor to drive servomechanisms [1]. Precise tracking control is important to guarantee the normal operation of the servomechanisms. However, there are many nonlinear factors influencing the dynamic and static performance of control systems such as the friction nonlinearity, backlash nonlinearity and other nonlinear disturbances. Adaptive compensation control design has been significantly advanced for the nonlinear system with friction [2, 3], while many control methodologies also have been devoted to eliminate the effects of backlash nonlinearity, e.g., Tao [4] designed an adaptive backlash inverse controller for unknown plants with backlash and proved the global boundedness of

M. Wang · X. Ren (✉)
School of Automation, Beijing Institute of Technology, Beijing 100081, China
e-mail: xmren@bit.edu.cn

Y. Jia et al. (eds.), *Proceedings of the 2015 Chinese Intelligent Systems Conference*, Lecture Notes in Electrical Engineering 360,
DOI 10.1007/978-3-662-48365-7_46

the closed-loop signals. Furthermore, zhao [5] presented a switching control scheme with PTO (proximate time optimal) compensation method for the systems containing inner backlash. All the above-mentioned methods for anti-backlash were based on a single-motor, but researches on the multi-motor were rare.

To handle the unknown nonlinearities and disturbances, neural networks (NN) are effective tools due to their universal approximation abilities, learning and adaptation abilities. Approximating the unknown nonlinearities by a high-order neural network, Na [6] proposed an adaptive tracking controller with a priori prescribed performance. Moreover, the problems of stability and tracking control for the multi-motor servomechanism (MMS) with unmodeled dynamics were addressed by the active disturbance rejection control (ADRC) based on a neural network observer in [7].

In this paper, we prove that the tracking performance of the MMS can be guaranteed by using the optimal NN observer and the optimal state feedback. Since the MMS is equivalently transformed as n subsystems, the optimal NN observer is designed to estimate the uncertainties and the unmeasured states of each subsystem. Then, the optimal feedback controller is proposed to make each subsystem track the desired signal in an optimal way. Simulation results are conducted to validate the desired system tracking performance and estimated ability.

46.2 Problem Statement

Without losing generality, the mechanical dynamics of multi-motor servomechanism can be described by

$$\begin{cases} J_i\dot{\theta}_i + T_{fi} + T_{di} + T_i = u_i \\ J_l\dot{\theta}_l + T_{fl} + T_{dl} = \sum\limits_{i=1}^{n} T_i \end{cases} \tag{46.1}$$

where θ_i ($i = 1, 2, \ldots n$, n is the motor total number) is the angular position of each motor, θ_m is the load angular position; J_i is the inertia of each motor, J_m is the load inertia; u_i is the control input of each motor; T_{fi} and T_{di} represent the friction and unknown external disturbance of each motor, respectively; T_{fl} and T_{dl} represent the friction and unknown disturbance of the load, respectively. Affected by the backlash, the motor-load interaction torques T_i can be expressed as

$$T_i = \begin{cases} k(\Delta\theta_i(t) - \alpha) + c(\Delta\theta_i(t)) & \Delta\theta_i \geq \alpha \\ 0 & |\Delta\theta_i| < \alpha \\ k(\Delta\theta_i(t) + \alpha) + c(\Delta\theta_i(t)) & \Delta\theta_i \leq -\alpha \end{cases} \tag{46.2}$$

where $\Delta\theta_i(t) = \theta_i(t) - \theta_m(t)$, $k > 0$ is the stiffness coefficient, $c > 0$ is the damping coefficient of the contact force between the load and motor, and $2\alpha > 0$ is the backlash width parameter.

The control objective is to make the load position track the desired trajectory y_d. According with [5], the dead-zone model for backlash is applicable in (46.1) and can be expressed as a smooth, continuous and differentiable function

$$T_i = k\Delta\theta_i + d_\alpha(\Delta\theta_i) = k\Delta\theta_i - \alpha k(\frac{2}{1-e^{-r\Delta\theta_i}} - 1) + c\Delta\theta_i(t)(1 - 2\alpha\frac{e^{-r\Delta\theta_i}}{\left(1-e^{-r\Delta\theta_i}\right)^2}). \tag{46.3}$$

where r is a positive constant.

Defining the state variable $x(t)^T = [x_1(t) \quad x_2(t) \quad \ldots \quad x_{2n+1}(t) \quad x_{2n+2}(t)] = [\theta_1(t) \quad \theta_1(t) \quad \ldots \quad \theta_l(t) \quad \theta_l(t)]$, the input $u(t)^T = [u_1(t) \quad u_2(t) \quad \ldots \quad u_n(t)]$, the nonlinearity $d_\alpha(x)^T = [d_\alpha(\Delta\theta_1) \quad d_\alpha(\Delta\theta_2) \quad \ldots \quad d_\alpha(\Delta\theta_n)]$, the unknown disturbance $T_d^T = [T_{d1} \quad T_{d2} \quad \ldots \quad T_{dn} \quad T_{dl}]$ and the friction $T_f^T = [T_{f1} \quad T_{f2} \quad \ldots \quad T_{fn} \quad T_{fl}]$, the state-space representation of system (46.1)–(46.3) yields to

$$\dot{x}(t) = Ax(t) + Bu(t) + F(T_f + T_d - d_\alpha(x)) \tag{46.4}$$

where

$$A = \underbrace{\begin{bmatrix} 0 & 1 & 0 & 0 & 0 & 0 \\ -\frac{k}{J_1} & 0 & 0 & 0 & \frac{k}{J_1} & 0 \\ 0 & 0 & 0 & 1 & 0 & 0 \\ 0 & 0 & -\frac{k}{J_n} & 0 & \frac{k}{J_n} & 0 \\ 0 & 0 & 0 & 0 & 0 & 1 \\ \frac{k}{J_l} & 0 & \frac{k}{J_l} & 0 & -\frac{nk}{J_l} & 0 \end{bmatrix}}_{2n+2}, B = \underbrace{\begin{bmatrix} 0 & 0 \\ \frac{1}{J_1} & 0 \\ 0 & 0 \\ 0 & \frac{1}{J_n} \\ 0 & 0 \\ 0 & 0 \end{bmatrix}}_{n}, F = \underbrace{\begin{bmatrix} 0 & 0 & 0 \\ -\frac{1}{J_1} & 0 & 0 \\ 0 & 0 & 0 \\ 0 & -\frac{1}{J_n} & 0 \\ 0 & 0 & 0 \\ 0 & 0 & -\frac{1}{J_l} \end{bmatrix}}_{n+1},$$

Through a simple calculation, the rank of the system controllability matrix can be obtained as: $rank(M) = rank\left(\left[B \quad AB \quad \ldots \quad A^{2n+2}B\right]\right) = 2n+1 < 2n+2$. Hence, the system is an uncontrollable system.

46.3 The Equivalent System Transformation

As the multi-motor servomechanism is an uncontrollable system, we cannot simply design a controller to make the system stably. Combing the load tracking performance with the multi motors synchronization, the control target is transformed to track the desired signal of each motor with an external transformed term, respectively. Therefore, the state-space representation of the whole system can be simplified as a state-space representation of each motor.

Theorem 1 *If each of multiple motors can track the desired trajectory*

$$x_{id} = \frac{J_m}{nk}\ddot{y}_d + y_d + (-1)^{\frac{i-1}{2}}\alpha - \frac{1}{n}\sum_{i=1}^{n} d_\alpha(\Delta\theta_i) + \frac{1}{n}(T_{fl} + T_{dl}),\ (i = 1, 3, 5, \ldots 2n-1) \tag{46.5}$$

the load can track its desired trajectory y_d *and the following multi-motor synchronization are satisfied.*

$$x_2 \to x_4 \to \cdots \to x_{2n} \tag{46.6}$$

Proof The state equation of the load is

$$\begin{aligned} \dot{x}_{2n+1} &= x_{2n+2} \\ \dot{x}_{2n+2} &= \frac{1}{J_l}\left[\sum_{i=1}^{n} k(x_{2i-1} - x_{2n+1}) + \sum_{i=1}^{n} d_\alpha(\Delta\theta_i) - (T_{fl} + T_{dl})\right] \end{aligned} \tag{46.7}$$

Substituting $x_i = x_{id}$ into (46.7) yields

$$\begin{aligned} \ddot{x}_{2n+1} = &\frac{1}{J_l}\left[\sum_{i=1}^{n} k(\frac{J_l}{nk}\ddot{y}_d + y_d - \frac{1}{n}\sum_{i=1}^{n} d_\alpha(\Delta\theta_i) + \frac{1}{n}(T_{fl} + T_{dl}) - x_{2n+1})\right] \\ &+ \frac{1}{J_l}\left[\sum_{i=1}^{n} d_\alpha(\Delta\theta_i) - (T_{fl} + T_{dl})\right] = \ddot{y}_d + \frac{nk}{J_l}(y_d - x_{2n+1}) \end{aligned} \tag{46.8}$$

Define tracking error as $e = x_{2n+1} - y_d$, Eq. (46.8) can be rewritten as $\ddot{e} + \kappa e = 0$ where $\kappa = \frac{nk}{J_l} > 0$. Obviously, we have $e_{vi} \to 0$. Define synchronization error as

$$e_{vi} = x_{2i} - x_{2i+2}\ (i = 1, 2, \ldots n-1). \tag{46.9}$$

Derivative (46.5) and substitute in (46.9), we have $e_{vi} \to 0$. That completes the proof.

As the external disturbance term $-\frac{1}{n}\sum_{i=1}^{n} d_\alpha(\Delta\theta_i) + \frac{1}{n}(T_{fl} + T_{dl})$ in (46.5) is unknown, we transform it from the load side to the motor side and design a NN observer to approximate it. Therefore, the MMS (46.4) can be simplified as a state-space equation of each motor

$$\dot{x}_i(t) = A_i x_i(t) + B_i u_i(t) + F_i[(T_{fi} + T_{di}) + o((T_{fl} + T_{dl})) - d_\alpha(\Delta x_i) - o(d_\alpha(\Delta x_i))] \quad (46.10)$$

where $A_i = \begin{bmatrix} 0 & 1 \\ -\frac{k}{J_i} & -\frac{b_i}{J_i} \end{bmatrix}, B_i = \begin{bmatrix} 0 \\ \frac{1}{J_i} \end{bmatrix}, F_i = \begin{bmatrix} 0 \\ -\frac{1}{J_i} \end{bmatrix}$ and $o(d_\alpha(x_i))$, $o((T_f + T_d))$ are the higher order terms of $d_\alpha(x_i)$, $(T_f + T_d)$, which are produced from (46.5). Then, the desired trajectory (46.5) of each motor can be rewritten as

$$x_{id} = \frac{J_m}{nk}\ddot{y}_d + y_d + (-1)^{\frac{i-1}{2}}\alpha. \quad (46.11)$$

Therefore, the system (46.4) with the load control objective (46.6) is equivalently transformed to the system (46.10) with each motor control objective (46.11). □

46.4 Optimal NN Observer Design

In this section, a linear optimal observer with neural network is designed to estimate the unmeasured states and uncertainties for the MMS.

Defining the uncertainties $D(x_i) = (T_{fi} + T_{di}) + o(d_\alpha(T_{fl} + T_{dl})) - d_\alpha(\Delta x_i) - o(d_\alpha(\Delta x_i))$, system (46.10) is expressed as

$$\dot{x}_i = A_i x_i + B_i u_i + F_i D(x_i). \quad (46.12)$$

The uncertainty $D(x_i)$ can be approximated by an ideal neural network

$$D(x_i) = W^T \varphi(x_i) + \varepsilon \quad (46.13)$$

where W is a constant ideal weight, $\varphi(x_i)$ is the basis vector which is chosen as

$$\varphi(x_i) = e^{-\frac{(x_i - x_{ic})^2}{4\sigma^2}} \quad (46.14)$$

where x_{ic} and σ denote the center and the width of the basis vector and ε is the approximated error satisfied $\varepsilon > 0$.

As the uncertainty $D(x_i)$ is unknown but related to the state, a neural network is proposed to estimate $D(x_i)$ which is given by

$$\hat{D}(x_i) = \hat{W}^T \varphi(x_i) \quad (46.15)$$

where $\hat{W} \in \Re^{q \times m}$ is a the estimated weight.

Then, the optimal neural network observer is designed as

$$\dot{\hat{x}}_i = A_i\hat{x}_i + B_iu_i + F_i\hat{D}(\hat{x}_i) + K(x_i - \hat{x}_i) \tag{46.16}$$

where K is the observer gain which is chosen as

$$K = -P_0R_0^{-1} \tag{46.17}$$

and P_0 is the solution of the following Riccati equation:

$$P_0A_i + A_i^TP_0 - P_0R_0^{-1}P_0 + Q_0 = 0 \tag{46.18}$$

where Q_0 and R_0 are the positive definite weighting matrices.

Chosen the estimation error as $e_{ai} = x_i - \hat{x}_i$, its derivative can be written as

$$\dot{e}_{ai} = (A_i - K)e_{ai} + F_i(\tilde{W}^T\varphi(x_i) + \varepsilon) \tag{46.19}$$

where $\tilde{W} = W - \hat{W}$ is the error between the ideal and estimated weights.

46.5 Optimal Feedback Controller Design

For convenience, we take subsystem i as a control object to design optimal feedback controller. Consider the system (46.10) without backlash nonlinearity and other uncertainties as a nominal system

$$\dot{x}_i = A_ix_i + B_iu_i. \tag{46.20}$$

The control objective of each subsystem is to track the desired signal $x_{id} \in \Re$ which is satisfied with the following dynamics.

$$\dot{x}_{id} = A_ix_{id} + B_iu_{id}. \tag{46.21}$$

where $u_{id} \in \Re$ is the desired control signal. If we define the tracking error as $e_i = x_i - x_{id}$, the tracking error dynamics of system (46.20) can be written as

$$\dot{e}_i = A_ie_i + B_iv_i. \tag{46.22}$$

where $v_i = u_i - u_{id}$.

According to the optimal control theory, the optimal feedback controller is designed to minimize the following performance index

$$J(t_0) = \frac{1}{2}\int_{t_0}^{\infty} (e_i Q_i e_i + v_i R_i v_i) dt \tag{46.23}$$

where $Q_i \geq 0, R_i \geq 0$ are the positive definite weighting matrix. Thus, the optimal feedback controller is chosen as the following form

$$v_i = -R_i^{-1} B_i^T P_i e_i \tag{46.24}$$

where P_i is a positive definite weighting matrix calculated from the Riccati equation

$$P_i A_i + A_i^T P_i - P_i B_i R_i^{-1} B_i^T P_i + Q_i = 0 \tag{46.25}$$

Theorem 2 *Consider the estimated error system* (46.19) *and the tracking error system* (46.22). *If the optimal feedback controller is designed as* (46.24) *and the weight updated rule of W is selected as*

$$\dot{\hat{W}} = \varphi(x_i) e_{ai}^T P_0^{-1} F_i + \varphi(x_{id}) e_i^T P_i F_i - \sigma \hat{W}, \tag{46.26}$$

the estimation error e_{ai}, the tracking error e_i and the adaptive weight $\hat{W}$ are all SGUUB.

Proof Consider the following Lyapunov function candidate

$$V_i = e_{ai}^T P_0^{-1} e_{ai} + e_i^T P_i e_i + \tilde{W}^T \tilde{W}. \tag{46.27}$$

The time derivative of the Lyapunov function (46.27) is derived by

$$\dot{V}_i = e_{ai}^T P_0^{-1} \dot{e}_{ai} + \dot{e}_{ai}^T P_0^{-1} e_{ai} + e_i^T P_0 \dot{e}_i + \dot{e}_i^T P_0 e_i + \tilde{W}^T \dot{\tilde{W}} + \dot{\tilde{W}}^T \tilde{W}. \tag{46.28}$$

The tracking error of each subsystem is

$$\dot{e}_i = (\dot{x}_i - \dot{x}_{id}) = (A_i - B_i R^{-1} B_i^T P) e_i + F_i(D(x_i) - \hat{D}(x_{id})). \tag{46.29}$$

Substituting (46.19) and (46.29) into (46.28), we have

$$\begin{aligned}\dot{V} = {} & e_{ai}^T P_0^{-1} (A_i P_0 + P_0^T A_i - 2 P_0 R_0^{-1} P_0) P_0^{-1} e_{ai} + 2 e_{ai}^T P_0^{-1} F_i (\tilde{W} \varphi(x_i) + \varepsilon) \\ & + e_i^T (A_i P_i + P_i^T A_i - 2 P_i B_i R_i^{-1} B_i^T P_i) e_i + e_i^T P_i F_i (\tilde{W} \varphi(x_{id}) + \varepsilon) \\ & + e_i^T P_i F_i D(x_i) - e_i^T P_i F_i D(x_{id}) + \tilde{W}^T \dot{\tilde{W}} + \dot{\tilde{W}}^T \tilde{W}\end{aligned} \tag{46.30}$$

Using the updated law (46.26) and Young' inequality yields

$$\begin{aligned}\dot{V} \leq & -\lambda_{\min}(P_0^{-1}Q_0P_0^{-1})\|e_{ai}\|^2-\lambda_{\min}(Q_i)\|e_i\|^2+2\varepsilon^2 \\ & +2\|P_0^{-1}F_i\|^2\|e_{ai}\|^2+\|P_iF_i\|^2\|e_i\|^2+D_{\max}^2+W_{\max}^2\varphi_{\max}^2-2\sigma\hat{W}^2 \\ \leq & -\eta V+\mu\end{aligned} \tag{46.31}$$

where

$$\begin{aligned}&\eta=\min\{\lambda_{\min}(P_0^{-1}Q_0P_0^{-1})-2\|P_0^{-1}F_i\|^2,\lambda_{\min}(Q_i)-\|P_iF_i\|^2,2\sigma\} \\ &\mu=2\varepsilon^2+D_{\max}^2+W_{\max}^2\varphi_{\max}^2\end{aligned} \tag{46.32}$$

Integrating both side of (46.31), one have

$$0\leq V(t)\leq\frac{\mu}{\eta}+(V(0)-\frac{\mu}{\eta})e^{-\eta t} \tag{46.33}$$

which implies that all of the signals in the closed-loop system are SGUUB. That completes the proof □

46.6 Simulation Result

In this section, simulated result is performed to demonstrate the tracking performance of the proposed method for the dual-motor servomechanism with backlash. According to the analysis in Sect. 2, the dual-motor servomechanism can be equivalently transformed as two subsystems of each motor and their physical parameters are selected as $J_1=J_2=0.0026, b_1=b_2=0.015, J_m=0.0113, b_m=0.02, k=1, c=0.15, \alpha=0.2$.

The optimal feedback controllers for all subsystems are designed by the linear quadratic regulator (LQR) with index matrices $Q_i=diag([2000,0.05])$ and $R_i=1$, adaption rates $P_i=130I$ and $\sigma=10^{-4}$, and an observer gain $K=diag([20,220])$. A sinusoidal signal $y_d=2\sin(0.4\pi t)$ is employed as the desired trajectory of the load.

It can be seen in Fig. 46.1 that the optimal feedback controller can provide an excellent tracking performance with the unknown disturbances estimated by the optimal neural network observer accurately. Moreover, Fig. 46.2 demonstrates that the dual motors can achieve speed synchronization when driving the load to track the desired trajectory.

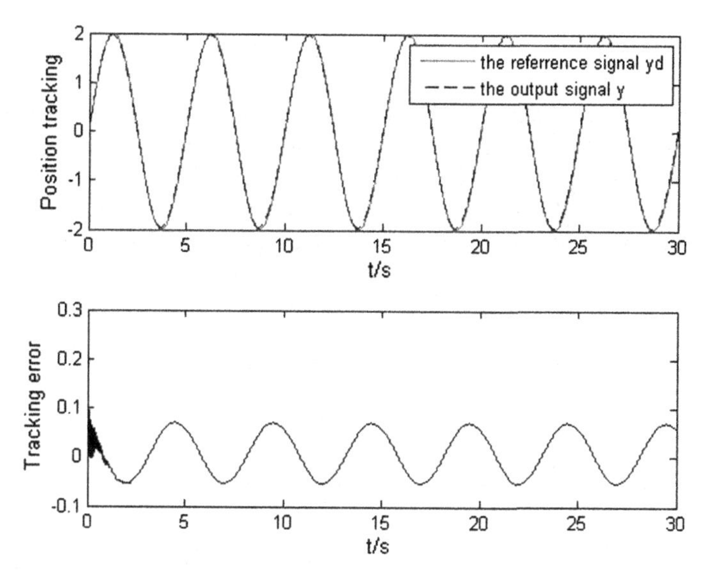

Fig. 46.1 The tracking performance of the load

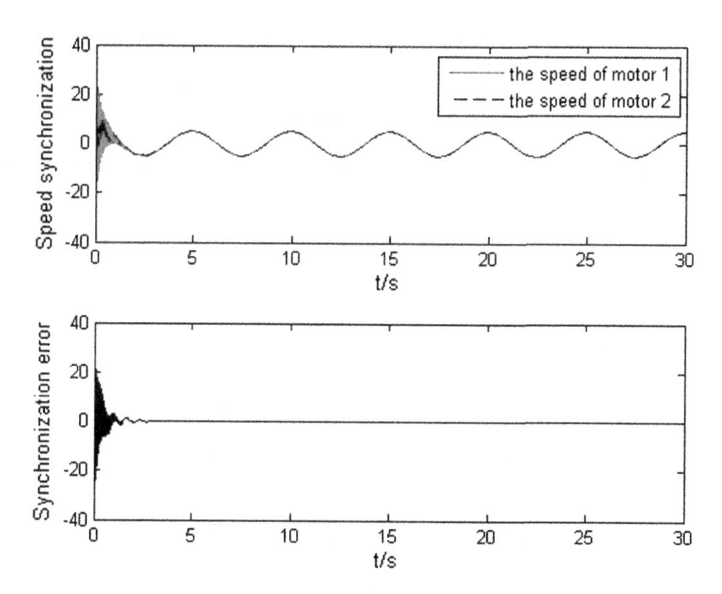

Fig. 46.2 The synchronization performance of dual motors

46.7 Conclusion

This paper presented a novel optimal tracking control for the multi-motor servomechanism containing backlash nonlinearity, friction nonlinearity and other uncertainties. Simulation results indicated that with the unknown disturbances handled by the neural network, the linear optimal observer was able to accurately estimate the unmeasured states. Furthermore, the optimal feedback controller guaranteed the better transient and steady tracking performance for the MMS.

References

1. Su CY, Tan YH, Stepanenko Y (2000) Adaptive control of a class of nonlinear systems preceded by an unknown backlash-like hysteresis. [J] In: IEEE, conference on decision and control 2000, pp 1459–1464
2. Gawronski W, Beech-Brandt JJ, Ahlstrom HG, Maneri E (2000) Torque-Bias profile for improved tracking of the deep space network antennas. [J] Antennas and Propag Mag IEEE 42 (6):35–45
3. Ren L, Qiang F (2010) Dual-drive anti-backlash system based on torque compensation. [J] J Mech Electr Eng 27(4):17–27
4. Tao G, Kokotovic PV (1993) Adaptive control of system with backlash [J]. Automatica 29 (2):323–325
5. Zhao G, Song Y, Guo J, Hu W (2006) Backlash nonlinearity switching control with proximate time optimal compensation.[J] Introducing J China Ordnance 27(2):325–329
6. Jing Na, Qiang Chen, XueMei Ren, Guo Yu (2014) Adaptive prescribed performance motion control of servo mechanisms with friction compensation. [J]. IEEE Trans Industr Electron 61 (1):486–494
7. Sun G, Ren X, Li D (2015) Neural active disturbance rejection output control of multi-motor servomechanism. [J] IEEE Trans Control Syst Technol 23(2):746–753

Chapter 47
Application of Fractal on the Fluids

Bo Qu and Paul S. Addison

Abstract The applications of fractal have been involved in a wide range of areas. As a fractal brunch—fractional Brownian motion (fBm) has also been used and applied in numerous physical sciences. The fBm technique overcome the traditional non-memory random walk-Brownian motion, it could be used on modelling particle tracking dispersion in ocean surface. The technique gives more accurate simulation on tracking particle diffusion and dispersion. Here we use a developed fBm model (FBMINC) as a diffusion process, simulate an idealized coastal bay surface trajectories and particle cloud dispersions using fBm particle tracking techniques. Compared to the traditional Brownian motion particle tracking model, the newly developed fBm particle tracking model produces patterns more close to the reality. Due to its flexibility, the fBm particle tracking model can be widely used in pollutant dispersion on difference size of water bodies.

Keywords Fractal · Fractional Brownian motion · Particle tracking model · Hurst exponent · Ocean dispersion

47.1 Introduction

Hydraulic researches have drawn more attention on modeling of the contaminants dispersion in fluids. More accurate numerical techniques are required on the research. Accurate prediction of the spread of contaminants requires methods which can mimic the dynamics of the dispersion in the moving fluids. One of traditional diffusion research methods is random walk, where each step is independent and

B. Qu (✉)
School of Science, Nantong University, Nantong 226017, China
e-mail: qubo62@gmail.com

P.S. Addison
Elvingston Science Centre, Cardio Digital Ltd., East Lothian, Edinburgh EH33 1EH, UK

Y. Jia et al. (eds.), *Proceedings of the 2015 Chinese Intelligent Systems Conference*, Lecture Notes in Electrical Engineering 360,
DOI 10.1007/978-3-662-48365-7_47

there is no memory associate with it. In Brownian motion, a cloud of particles (it's standard deviation is $\sigma_c(t)$) follows Eq. (47.1):

$$\sigma_c^2(t) = 2Dt \tag{47.1}$$

where D is diffusion coefficient, t is time.

However, unlike traditional Brownian motion, the turbulent diffusions are often non-Fickian in the fluids, especially in coastal water and open ocean [5–7]. It is found that the mass trajectories on the ocean surface have non-Fickian properties [9].

The fBm has already been used in a wide variety of studies in the physical sciences. The most common application of fBm is to model surfaces or landscapes. There are many applications about pollutant dispersion in fluids using the traditional techniques (Advection-Diffusion equation models) on application of pollutant dispersion in fluids. The more flexible and quicker method is used here with fractional Brownian motion as diffusion process. The previous work by authors can be found in Qu and Addison [8] followed by previous works Addison and Qu [1, 2]. A novel fBm particle tracking model has been developed by the authors that can be used in a large range of pollutant dispersion applications in varies size of water bodies (river, coastal regions and open ocean surface). Here we will introduce the process of development of the fBm particle tracking model and emphasis on the application of the model into coastal bay particle dispersions. The techniques introduced here is novel and practical and useful for hydraulic engineer and environmental researchers.

47.2 Fractional Brownian Motion

Fractional Brownian motion is generated from Brownian motion [4]. Mandelbrot and Van Ness [3] defined the random function $B_H(t)$ using increment $dB_H(t)$, and weighted by the kernel $(t-s)^{H-\frac{1}{2}}$, as

$$B_H(t) = \frac{1}{\Gamma\left(H+\frac{1}{2}\right)} \int_{-\infty}^{t} (t-s)^{H-\frac{1}{2}} dB(s) \tag{47.2}$$

Where gamma function denoted as $\Gamma(x)$ and Hurst exponent of the trace is H. Equation (47.2) shows that value of $B_H(t)$ dependent on the all previous increments $dB(s)$ (here $s < t$). Here B(s) is a random walk.

Mandelbrot and Van Ness [3] replaced the Eq. (47.2) with another function in which forces the function through the origin. Given the value $B_t(t=0)$, hence

$$B_H(t) - B_H(0) = \frac{1}{\Gamma\left(H+\frac{1}{2}\right)}\left(\int_{-\infty}^{t}(t-s)^{H-\frac{1}{2}}dB(s) - \int_{-\infty}^{0}(0-s)^{H-\frac{1}{2}}dB(s)\right)$$
$$= \frac{1}{\Gamma\left(H+\frac{1}{2}\right)}\left[\left[\int_{-\infty}^{0}\left((t-s)^{H-\frac{1}{2}} - (-s)^{H-\frac{1}{2}}\right)dB(s) + \int_{0}^{t}(t-s)^{H-\frac{1}{2}}dB(s)\right]\right. \tag{47.3}$$

$B_H(t)$ in Eq. (47.3) is the improved definition of fBm, There is an additional parameter H called 'Hurst exponent' H (where $0 < H < 1$), Here the standard deviation of a group of particles at time t scales with t^H

$$\sigma_c(t) \propto t^H \tag{47.4}$$

Actually, the Hurst exponent H was first introduced by Edwin Hurst [10] when he tried to find a solution of the problem of determining the reservoir storage required on a given stream. H is a parameter which can control the 'smoothness' of the trace and is related to the fractal dimension of the trace as D_f:

$$H = 2 - D_f \tag{47.5}$$

The cases where $H \neq \frac{1}{2}$ are proper fractional Brownian motions; where $H = \frac{1}{2}$ is the special case of independent increments valid for Brownian motion.

However, the fBm model (47.3) has two faults: one is that there are quite large errors when memory M less than number of steps; Another is the each step of the standard deviation increment $\sigma(B_H(t_i) - B_H(t_{i-1}))$ always increases with the time step i. An improved FBMINC model was developed by the Authors which overcame these two faults and gave a better approximation of fBm [8]:

$$B_H(i) - B_H(i-1) = \frac{1}{\Gamma\left(H+\frac{1}{2}\right)}\left[\sum_{j=i-M}^{i-2}[(i-j)^{H-1/2} - (i-j-1)^{H-1/2}]R(j) + R(i-1)\right] \tag{47.6}$$

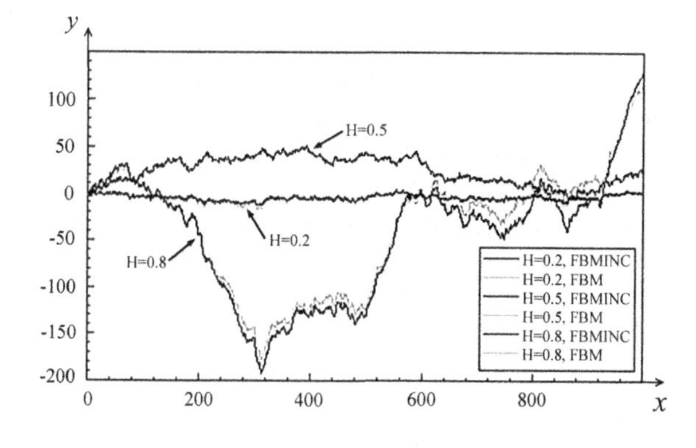

Fig. 47.1 FBM and FBMINC model's comparison, where $M = 5000$, $D = 1$, $\Delta t = 1$

$$B_H(t) - B_H(0) = \sum_{i=1}^{t} [B_H(i) - B_H(i-1)] \tag{47.7}$$

Although the FBM and FBMINC models synthesis fBm using different methods, they produce similar fBm traces. Figure 47.2 is the comparisons between FBM model and FBMINC model. The figure shows the two models are similar with little differences. However, FBMINC model proved a better model by increased accuracy when memory is small and standard deviation of each step jump $\sigma(B_H(t_i) - B_H(t_{i-1}))$ does not increase with the time step i [8].

47.3 The fBm Particle Tracking Model

The pollutant dispersion in river or ocean is determined by combination of two kind of movements: advection and diffusion. Equations (47.8) and (47.9) show the combination of the two movements:

$$\Delta x_i = U(i)\Delta t + \Delta B_{H_x}(i) \tag{47.8}$$

$$\Delta y_i = V(i)\Delta t + \Delta B_{H_y}(i) \tag{47.9}$$

Here $\Delta B_{Hx}(i)$ and $\Delta B_{Hy}(i)$ are the diffusion displacement of each time step in both x and y directions, they are two independent fBms (defined by Eqs. 47.6 and 47.7). Combination of (47.8) and (47.9) are the fBm particle tracking model. We will use this model to simulate the particle movement on the water surface.

47.4 Particle Trajectories in an Idealized Coastal Bay

An open coastal bay with its width of 2000 m and length of 4775 m was set up by Glassgow University, UK. The main flow area has grid sizes of 50 m × 100 m. The grid sizes are finer and not evenly spaced in the bay area of the model. The bay area is on the right hand side. The maximum speed is 0.67 m/s in the main flow and the average speed is between 0.4 and 0.5 m/s. The bay area contains a recirculating flow field driven by the main flow.

In the numerical model, a set of massless particles representing the pollutant cloud is injected at time $t = 0$ from a point source on the surface. The particle cloud is then advected through the grid domain in the following manner. Suppose that for time step i (time $= i \times \Delta t$) the particle P(x_i, y_i) has gone into grid (k, l), where the velocities at the four grid points A, B, C, D are:

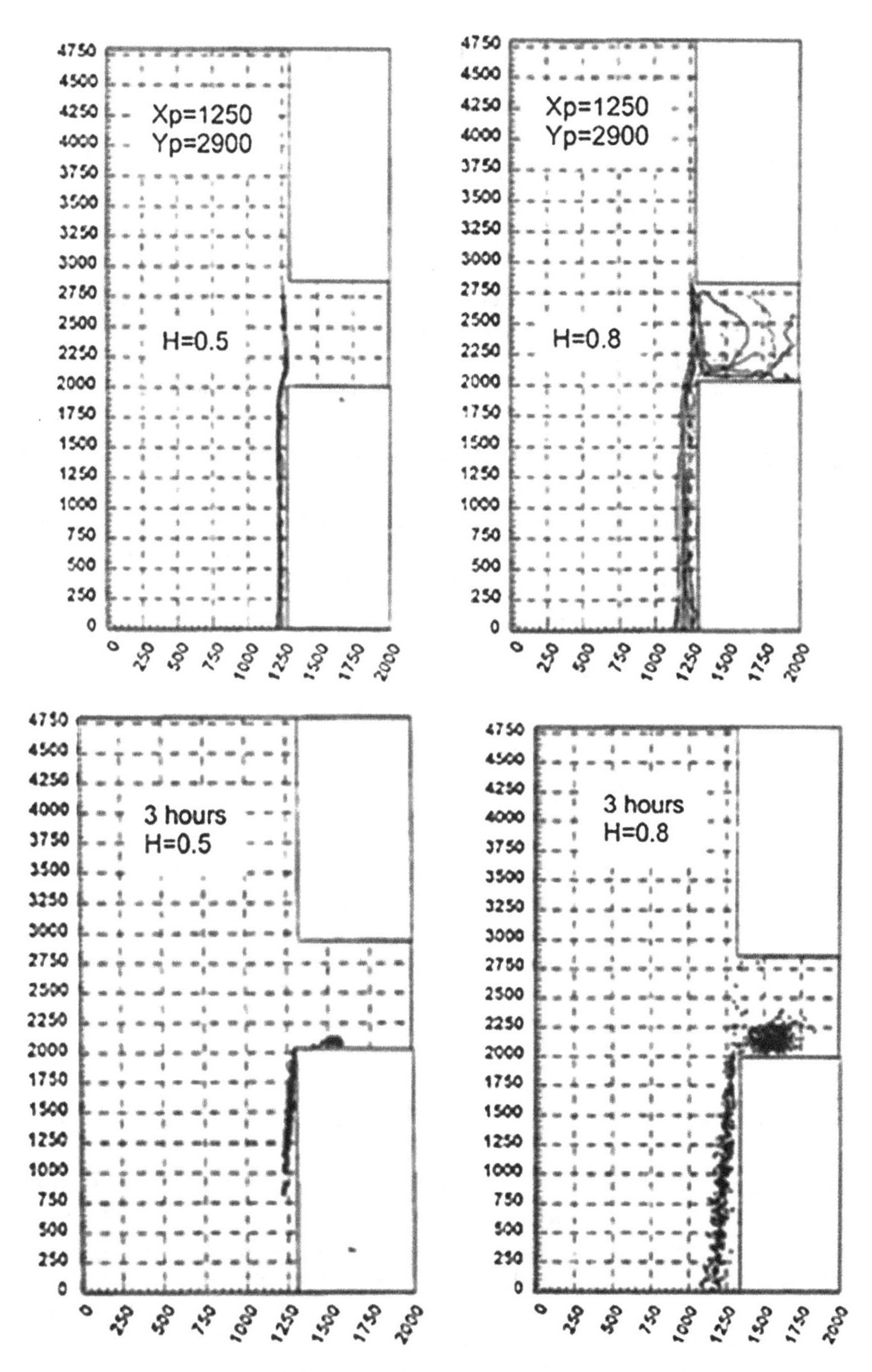

Fig. 47.2 *Upper plots* 20 particles trajectories released from the same location (1250, 2900); *Lower plots* 400 particles released from the same point and recorded the spreading after 3 h; for $H = 0.5$ (*left*) and $H = 0.8$ (*right*) respectively

$$\mathrm{A}(U(k,l),V(k,l)),\quad \mathrm{B}(U(k,l+1),V(k,l+1)),$$

$$\mathrm{C}(U(k+1,l+1),V(k+1,l+1)),\quad \mathrm{D}(U(k+1,l),V(k+1,l)).$$

When a particle $P(x_i, y_i)$ is in grid (k, l) with corner ABCD, the velocity in x direction is $U(x_i, y_i)$ and in the y direction is $V(x_i, y_i)$. The advective component used an interpolation scheme for an off-grid points of velocity. For time, a simple linear interpolation was used. The bilinear interpolation in space was selected.

The fBm particle tracking model (47.8) and (47.9) are used to simulate the particles trajectories on the flow surface. Figure 47.2 shows a 20 particles released from the same location in the coastal bay. The fBm particle tracking model is used with two different H values ($H = 0.5$: left plots and $H = 0.8$: right plots).

The Fickian ($H = 0.5$) and non-Fickian ($H = 0.8$) cases are shown in Fig. 47.3. We use mass particles (400 particles) represent a pollutant cloud. Considering the same release point, less particles enter the recirculation zone in the case of $H = 0.5$. The particles are much more spread out in the case of $H = 0.8$. More particles were observed in the recirculation zone. For more particles, more realistic patterns appeared in the figure. This shows that we could simulate different spreading of pollutant by changing the Hurst exponent value. The natural flow patterns are more likely related with H values greater than 0.5. The flexibility of fBm is far more useful than the traditional Brownian motion.

47.5 Conclusions

Ocean surface particle diffusion can be simulated by fractal process using fractional Brownian motion. An fBm particle tracking model is derived by combining both advection and diffusion processes. The diffusion process is a random process and could be simulated by memory related fractional Brownian motion, rather than traditional Brownian motion. The new fBm model FBMINC improved from the Mandelbrot model which is more accurate to simulate the diffusion processes. Here, we introduced the more accurate fBm particle tracking method. The new fBm particle tracking model is used for simulating particle trajectories on both coastal bay and ocean surface. The particle clouds would be released in a coastal bay and the fBm particle tracking model is used for simulating the particle cloud spreading in the bay. The different results were compared for Hurst exponent $H = 0.5$ and $H = 0.8$.

More theoretical and numerical research concerning pollutant dispersions in ocean surface using fBm particle tracking model are also studied [9]. An accelerated fBm model was developed for an large water body. In that case, Hurst exponent value could exceed 1. Apart from an idealized bay, a more natural coastal bay is a challenging task with combination of wind, tides and wave currents factors. Future works related to the fBm particle tracking model could be interesting and more practical for a wide range of users.

References

1. Addison SP, Qu B (1998) A non-Fickian, particle tracking diffusion model based on fractional Brownian motion. Int J Numer Meth Fluids 25(12):1373–1384
2. Addison SP, Qu B (2000) A fast non-Fickian particle-tracking diffusion simulator and the effect of shear on the pollutant diffusion process. Int J Numer Meth Fluids 34(2):145–166
3. Mandelbrot BB, Van Ness JW (1968) Fractional Brownian motions, fractal noises and applications. SIAM Rev 10(4):422–437
4. Mandelbrot BB, Wallis JR (1969) Computer experiments with fractional gaussian noises. Water Resour Res 5(1):228–267
5. Okubo A (1971) Oceanic diffusion diagrams. Deep-Sea Res 18:789–802
6. Osborne AR, Kirwan AD Jr, Provenzale A, Bergamasco L (2010) Fractal drifter trajectories in the Kuroshio extension. Tellus 41A(5):416–435
7. Qu B (1999) The use of fractional Brownian motion in the modelling of the dispersion of contaminants in fluids. Ph.D. thesis, Napier University, UK
8. Qu B, Addison PS (2009) Development of FBMINC model for particle diffusion in fluids. Int J Sedim Res 24(4):439–454
9. Qu B, Addison PS, Mead CT (2009) Application of fractional Brownian motion particle tracking model to coastal contaminant dispersion. J Hydraul Eng 40: 1517–1523
10. Hurst HE (1950) Long-Term Storage Capacity of reservoirs. Am Soc Civil Eng 116:770–808
11. Sanderson BG, Booth DA (1991) The fractal dimension of drifter trajectories and estimates of horizontal Eddy-diffusivity. Tellus 43A(5):334–349

Printed in the United States
By Bookmasters